Heide Schatten
Editor

Cell and Molecular Biology of Breast Cancer

Humana Press

Editor
Heide Schatten
Department of Veterinary Pathobiology
University of Missouri-Columbia
Columbia, MO, USA

ISBN 978-1-4939-6277-8 ISBN 978-1-62703-634-4 (eBook)
DOI 10.1007/978-1-62703-634-4
Springer New York Heidelberg Dordrecht London

Printed on acid-free paper

Humana Press is a brand of Springer
Springer is part of Springer Science+Business Media (www.springer.com)

Cell and Molecular Biology of Breast Cancer

Preface

New research on finding effective treatment for breast cancer patients has led to a wealth of new data on a number of different levels that allowed a new understanding of the disease. It further allowed development of new strategies to treat the heterogeneous disease with different and patient-specific approaches. A variety of different efforts on genetic, cell, and molecular levels have been focused on understanding causes that lead to cellular abnormalities, cell migration, epithelial–mesenchymal transition points, and metastasis that have become possible with new research methods. The advent of molecular technologies has significantly improved our understanding of the biological processes underlying breast cancer; targeted therapies are now available to inhibit specific signaling pathways that are aberrant in breast cancer cell populations, and we are now able to image signaling molecules with specific markers in live cells. Progress has also been made in designing nanoparticles that can be utilized for imaging and for targeted breast cancer treatment. The joint initiatives and efforts of advocate patients, breast cancer survivors, basic researchers, statisticians, epidemiologists, and clinicians with specific and combined expertise have allowed close communication for more effective and targeted treatment. Furthermore, reliable animal models are available for specific experimentation, and biopsies from hundreds of patients are now available through a number of different resources including the large Translational Breast Cancer Research Consortium (TBCRC) from fourteen research centers with extensive tissue-banking components.

This book highlights recent advances in our understanding of breast cancer, and it includes review articles of genetics, epigenetics, various aspects of cell and molecular biology, and several other areas of breast cancer that are aimed at determining new intervention sides for treatments and cures of breast cancer. The chapters are written by internationally recognized experts in their specific fields of expertise and include reviews of key topics in the field. Cutting-edge new information is balanced with background information that will be readily understandable for the newcomer, for breast cancer patients, and for the experienced breast cancer researcher alike. All articles will highlight new aspects of specific research topics and impacts on designing new strategies or identify new targets for therapeutic intervention. The topics addressed are selected to be of interest to patients, scientists, students, and teachers and to all who are interested in expanding their knowledge related to breast cancer imaging, diagnostics, therapeutics, or basic biomedical research on breast cancer.

The book is intended for a large audience as a reference book on the subject and includes the following chapters: Histopathology and Grading of Breast Cancer; Multicentric/Multifocal Breast Cancer: Overview, Biology, and Therapy; The Immune System in Breast Cancer Initiation and Progression: Role of Epithelial to Mesenchymal Transition; Remodeling of the Extracellular Matrix: Implications for Cancer; Biology and Treatment of Basal-Like Breast Cancer; Re-excision After Lumpectomy for Breast Cancer; Novel Antiangiogenic Therapies Using Naturally Occurring and Synthetic Drugs to Combat Progestin-Dependent Breast Cancer; New Insights on Estrogen Receptor Actions in Hormone-Responsive Breast Cancer Cells by Interaction Proteomics; Reprogramming Breast Cancer Cells with Embryonic Microenvironments: Insights from Nodal Signaling; Metastatic Determinants: Breast Tumor Cells in Circulation; Breast Cancer Epigenetics: Biomarkers and Therapeutic Potential; The Impact of Centrosome Abnormalities on Breast Cancer Development and Progression with a Focus on Targeting Centrosomes for Breast Cancer Therapy; A New Perspective on Cyclin D1: Beyond Cell Cycle Regulation; Gene Signatures of Inflammatory Breast Cancer: Epithelial Plasticity and a Cancer Stem Cell Phenotype; An Integrated Human Mammary Epithelial Cell Culture System for Studying Carcinogenesis and Aging; and New Breast Cancer Treatment Considerations: A Brief Review of the Use of Genetically Modified (Attenuated) Bacteria as Therapy for Advanced and Metastatic Breast Cancer.

It has been a special privilege to edit this book on breast cancer, and I would like to sincerely thank all contributors for their outstanding chapters and for sharing their unique expertise with the breast cancer community. I hope that this book will stimulate further advances in breast cancer research leading to new treatment strategies to effectively treat the disease in early as well as advanced stages.

Columbia, MO, USA Heide Schatten

Contents

Chapter 1
Histopathology and Grading of Breast Cancer

Magda Esebua

Abstract Carcinoma of the breast is the most common non-skin malignancy in women. It is now understood that breast cancer is not a single disease but rather many different diseases, each with its own clinical, morphologic, and molecular characteristics. Greater than 95 % of breast malignancies are adenocarcinomas, which are divided into in situ carcinomas and invasive carcinomas. Carcinoma in situ (CIS) refers to a neoplastic proliferation that is limited to ducts and lobules by the basement membrane. Invasive carcinoma (synonymous with "infiltrating" carcinoma) has penetrated through the basement membrane into stroma. Here, the cells have the potential to invade into the vasculature and thereby reach regional lymph nodes and distant sites.

Ductal carcinoma in situ (DCIS) is characterized by a proliferation of abnormal cells confined within the mammary ductal system.

DCIS is commonly classified according to architectural and cytologic features and cell necrosis as low and intermediate grade (papillary, cribriform, and solid) and high grade (comedo).

- DCIS represents a precursor to invasive breast cancer.

- The invasive breast carcinomas consist of several histologic subtypes.
  - Infiltrating ductal carcinoma is the most common type of invasive breast cancer, accounting for 70–80 % of invasive cancers. (See Sect. 1.3.).
  - Infiltrating lobular carcinoma is the second most common invasive breast cancer, accounting for 5–10 % of invasive cancers. (See Sect. 1.4.).
  - As compared with infiltrating ductal carcinomas, infiltrating lobular carcinomas tend to be multicentric and/or bilateral, more differentiated, and hormone receptor positive, arise in older women, metastasize later, and spread to unusual locations, such as meninges, peritoneum, or gastrointestinal tract.

M. Esebua (✉)
Department of Pathology and Anatomical Sciences, University of Missouri,
Columbia, MO 65211, USA
e-mail: esebuam@health.missouri.edu

H. Schatten (ed.), *Cell and Molecular Biology of Breast Cancer*,
DOI 10.1007/978-1-62703-634-4_1,

  - Other less common invasive breast carcinoma histologies include tubular, mucinous, and medullary carcinomas.

- The only uniformly accepted prognostic markers that provide critical information necessary for treatment decisions are TNM stage, axillary lymph node status, tumor size and grade, hormone receptor status, and HER2 receptor status.
  - Guidelines for measurement and use of prognostic markers are available from a 2007 ASCO expert panel on tumor markers in breast cancer (Harris L et al. J Clin Oncol 25:5287;2007). We agree with these recommendations. ER, PR, and HER2 overexpression should be evaluated on every primary invasive breast cancer. Hormone receptor expression should be used to guide endocrine therapy decisions. HER2 expression should be used to select patients for whom HER2-directed therapy use is appropriate in the metastatic and adjuvant setting. The available data are insufficient to recommend that HER2 overexpression be used for determining prognosis in patients with early breast cancer. Furthermore, there is insufficient evidence to support the clinical use of serum HER2 extracellular domain (ECD) testing.
  - Multiparameter gene expression analysis (i.e., the Oncotype DX™ assay) should be used to predict the risk of recurrence in women with newly diagnosed, node-negative, ER+ breast cancer who will be receiving tamoxifen. A low recurrence score should be used to identify patients who are predicted to obtain the most therapeutic benefit from tamoxifen and who may not require adjuvant chemotherapy. Patients with a high recurrence score appear to derive relatively more benefit from chemotherapy (specifically the CMF regimen) than from tamoxifen.

1.1 Introduction

Cancer of the breast is one of the most common neoplasms and the leading cause of carcinoma deaths in women. In 2012, 226,870 women are diagnosed with and 39,510 women will die of breast cancer (SEER). Breast cancer is associated with western lifestyle with incidence rates being highest in countries with advanced economic development.

Most breast malignancies arise from epithelial tissue and are categorized as carcinomas. Breast carcinomas are a diverse group of lesions that differ in microscopic appearance and biologic behavior, although these diverse disorders are often discussed as a single disease.

The in situ carcinomas of the breast are either ductal (also known as intraductal carcinoma) or lobular. This distinction is primarily based upon the growth pattern and cytologic features of the lesions, rather than their anatomic location within the mammary ductal-lobular system.

The invasive breast carcinomas consist of several histologic subtypes; the estimated percentages are taken from a contemporary population-based series of

135,157 women with breast cancer reported to the Surveillance Epidemiology and End Results (SEER) database of the National Cancer Institute between 1992 and 2001 [1].

Other subtypes, including metaplastic breast cancer and invasive micropapillary breast cancer, all account for fewer than 5 % of cases [2].

This section will review the histology of ductal carcinoma in situ and invasive breast carcinoma.

1.2 Ductal Carcinoma In Situ

The term ductal carcinoma in situ (DCIS) encompasses a heterogeneous group of lesions that differ in their clinical presentation, histologic appearance, and biologic potential. DCIS is characterized by proliferation of presumably malignant epithelial cells within the mammary ductal system, with no evidence of invasion into the surrounding stroma as determined by routine light microscopic examination [3]. Ductal carcinoma in situ differs from lobular carcinoma in situ with regard to radiologic features, morphology, biologic behavior, and anatomic distribution in the breast.

Classification schemes that divide DCIS histologically into a variety of subtypes emphasize architectural features or growth patterns of the neoplastic cells, cytologic features, and cell necrosis, both singly and in combination. The traditional method for classifying DCIS lesions is primarily based on the growth pattern (architectural features) of the tumor and recognizes five major types [4–7]:

- The comedo type is characterized by prominent necrosis in the center of the involved spaces (Fig. 1.1). The necrotic material frequently becomes calcified;

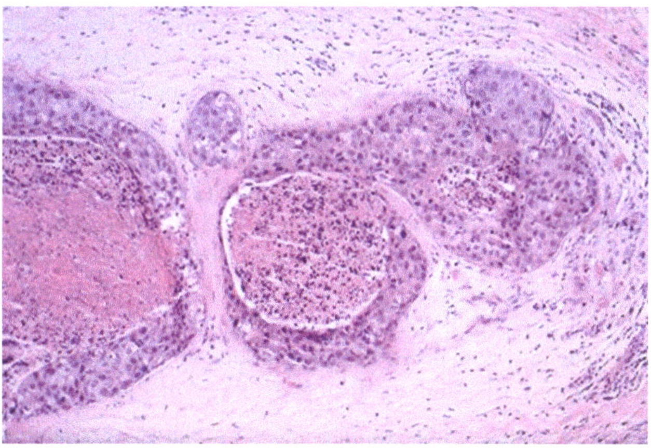

Fig. 1.1 Ductal carcinoma in situ (DCIS) of comedo type. This space shows rim of viable cells with high-grade nuclei and central comedo-type necrosis with focal calcification

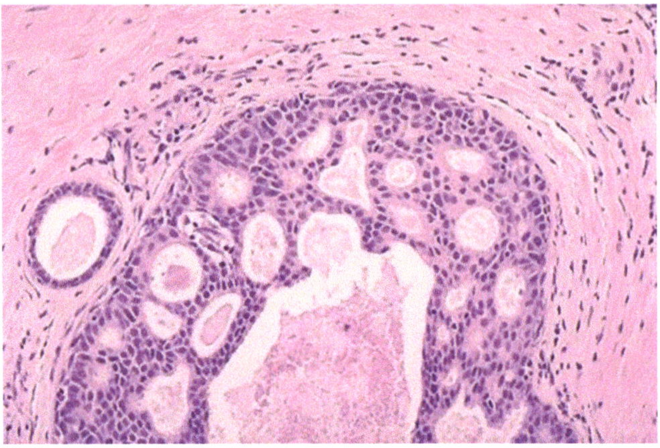

Fig. 1.2 Ductal carcinoma in situ (DCIS) of cribriform pattern. Multiple spaces within the proliferation of monotonous (low-grade) cells are rounded and distributed in an organized fashion

the calcifications may be detected mammographically, characteristically as linear, branching ("casting") calcifications. The tumor cells are large and show nuclear pleomorphism; mitotic activity may be prominent (Fig. 1.1). The comedo type is more often associated with invasion [8, 9], and the degree of comedo necrosis in patients with DCIS appears to be a strong predictor for the risk of ipsilateral breast recurrence after treatment [10].

- The cribriform type is characterized by the formation of back to back glands without intervening stroma. The cells comprising this subtype are typically small to medium sized and have relatively uniform hyperchromatic nuclei. Mitoses are infrequent and necrosis is limited to single cells or small cell clusters (Fig. 1.2).
- The micropapillary type features small tufts of cells that are oriented perpendicular to the basement membrane of the involved spaces and project into the lumina. The apical region of these small papillations is frequently broader than the base, imparting a club-shaped appearance. The micropapillae lack fibrovascular cores. The cells comprising this type of DCIS are usually small to medium in size, and the nuclei show diffuse hyperchromasia; mitoses are infrequent (Fig. 1.3).
- The papillary type shows intraluminal projections of tumor cells that, in contrast to the micropapillary variant, demonstrate fibrovascular cores and thereby constitute true papillations. A variant of papillary DCIS, intracystic papillary carcinoma, is characterized by tumor cells that are primarily or exclusively present in a single cystically dilated space [11, 12].
- The solid type is not as well defined as the other subtypes. It features tumor cells that fill and distend the involved spaces and lack significant necrosis, fenestrations, or papillations. The tumor cells may be large, medium, or small (Fig. 1.3).

Less common variants of DCIS include the "clinging" carcinoma [4, 13], intraductal signet ring cell carcinoma [14], and cystic hypersecretory duct carcinoma [15, 16]. Similar to the comedo type, these variants may show calcifications that can

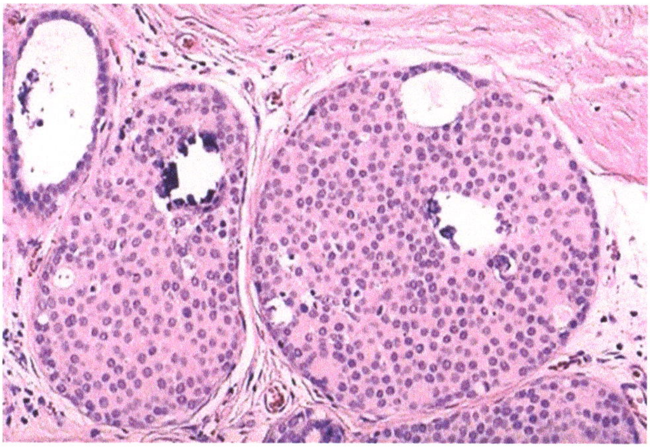

Fig. 1.3 Ductal carcinoma in situ, *solid pattern*. Cells show some variation in nuclear size and nuclei exhibit variably prominent nucleoli

be detected mammographically. However, the mammographic appearance of these microcalcifications is less distinctive than the pattern seen in comedo lesions and can resemble a number of benign processes.

A number of authors have proposed alternative classification systems for DCIS [17–20]. Although they use different terminology, all are primarily based on nuclear grade and/or the presence or absence of necrosis and have in common the recognition of three main categories of DCIS (e.g., high, intermediate, and low grade).

- High-grade lesions typically exhibit aneuploidy, lack estrogen and progesterone receptors, and have a high proliferative rate, overexpression of the HER2 oncogene, mutations of the p53 tumor suppressor gene with accumulation of its protein product, and angiogenesis in the surrounding stroma.
- Low-grade lesions are typically diploid and estrogen and progesterone receptor positive, have a low proliferative rate, and rarely (if ever) show abnormalities of the HER2/neu or p53 oncogenes.
- Lesions categorized histologically as intermediate grade are also intermediate between the high-grade and low-grade lesions with regard to the frequency of alterations in these biologic markers.

These classification systems appear to correlate with biologic prognostic markers and predict groups of patients who are likely to have a recurrence of cancer following breast conservation therapy [17, 20–32].

In 1997, a consensus conference was convened in an attempt to reach agreement on the classification of DCIS [33]. Although the panel did not endorse any single classification system, they recommended that certain features be routinely documented in the pathology report for DCIS lesions, including nuclear grade, the presence of necrosis, cell polarization, and architectural pattern(s).

1.3 Infiltrating Ductal Carcinoma

Infiltrating ductal carcinoma is the most common type of invasive breast cancer, accounting for 70–80 % of invasive lesions. It is also termed infiltrating carcinoma of no special type or infiltrating carcinoma not otherwise specified (NOS).

On gross pathologic evaluation, these lesions are typically hard, gray–white, gritty masses which invade the surrounding tissue in a haphazard fashion to create the characteristic irregular, stellate shape (Fig. 1.4). They are characterized microscopically by cords and nests of tumor cells with varying amounts of gland formation and cytologic features that range from bland to highly malignant. The malignant cells induce a fibrous response as they infiltrate the breast parenchyma, and this reaction is, in large part, responsible for the clinically and grossly palpable mass, the radiologic density, and solid sonographic characteristics of typical invasive carcinomas.

Infiltrating ductal carcinomas are divided into three grades based on a combination of architectural and cytologic features, usually assessed utilizing a scoring system based on three parameters [34]:

- Well differentiated (grade 1)—Well-differentiated tumors have cells that infiltrate the stroma as solid nests of glands. The nuclei are relatively uniform with little or no evidence of mitotic activity (Fig. 1.5).
- Moderately differentiated (grade 2)—Moderately differentiated tumors have cells that infiltrate as solid nests with some glandular differentiation. There is some nuclear pleomorphism and a moderate mitotic rate (Fig. 1.6).
- Poorly differentiated (grade 3)—Poorly differentiated tumors are composed of solid nests of neoplastic cells without evidence of gland formation. There is marked nuclear atypia and considerable mitotic activity (Fig. 1.7).

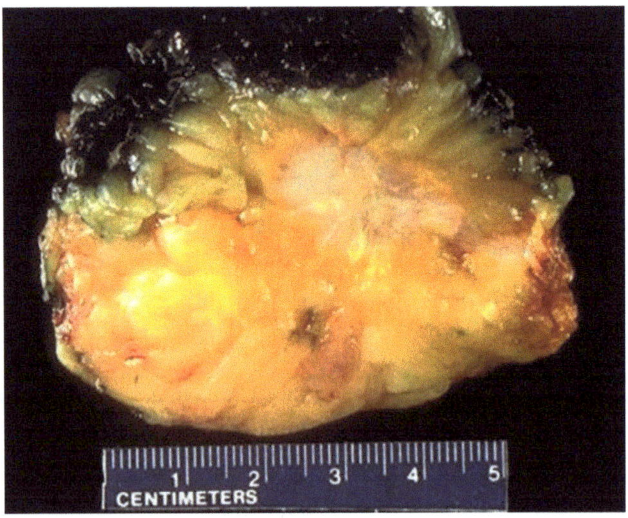

Fig. 1.4 Mastectomy specimen. The cut surface of the tumor is *gray white* and has an irregular stellate outline

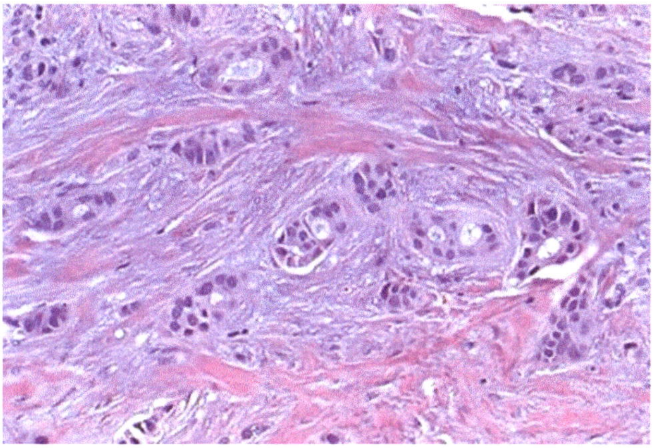

Fig. 1.5 Invasive ductal carcinoma, well differentiated (grade I)

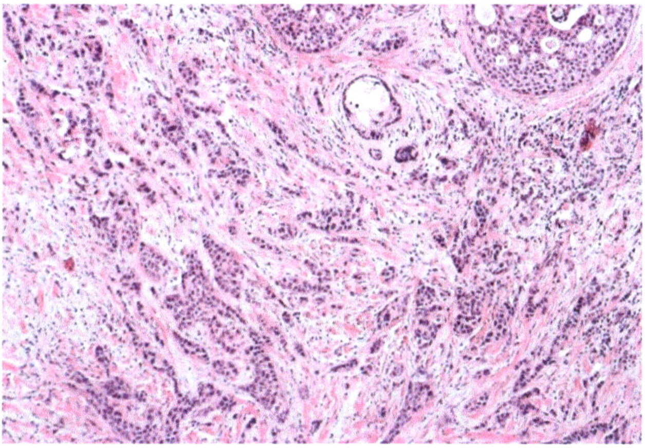

Fig. 1.6 Invasive ductal carcinoma, moderately differentiated (grade II)

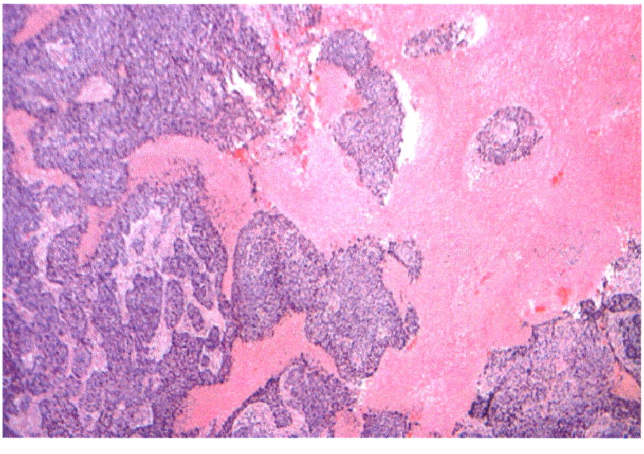

Fig. 1.7 Invasive ductal carcinoma, poorly differentiated (grade III). No evidence of glandular differentiation

A variable amount of associated ductal carcinoma in situ (DCIS) is present in most cases; the extent of DCIS but not lobular carcinoma in situ (LCIS) is an important prognostic factor in patients treated with breast-conserving therapy in which the surgical goal is complete excision of both intraductal and invasive carcinomas [35].

1.4 Infiltrating Lobular Carcinoma

Infiltrating lobular carcinomas are the second most common type of invasive breast cancer, accounting for about 5–10 % of invasive lesions.

Incidence rates of lobular cancer are rising faster than the rates of ductal carcinoma in the USA, and postmenopausal hormone therapy may be more strongly related to lobular cancer risk than to ductal cancer risk.

Some infiltrating lobular carcinomas have a macroscopic appearance identical to that of infiltrating ductal cancers. However, in many cases no mass lesion is grossly evident, and the excised breast tissue may have a normal or only slightly firm consistency. Thus, the microscopic size of invasive lobular carcinoma may be significantly greater than that measured grossly. Some pathologists have used lack of immunohistochemical staining for E-cadherin to distinguish invasive lobular carcinoma from invasive duct carcinoma. While it appears to be a reasonably accurate test, it is for the most part unnecessary in practice.

These tumors are characterized microscopically by small cells that insidiously infiltrate the mammary stroma and adipose tissue individually and in a single-file pattern, often growing in a target-like configuration around normal breast ducts, frequently inducing only minimal fibrous reaction (Fig. 1.8). Associated lobular carcinoma in situ (LCIS) is present in approximately two-thirds of cases; however, DCIS may also accompany invasive lobular carcinoma.

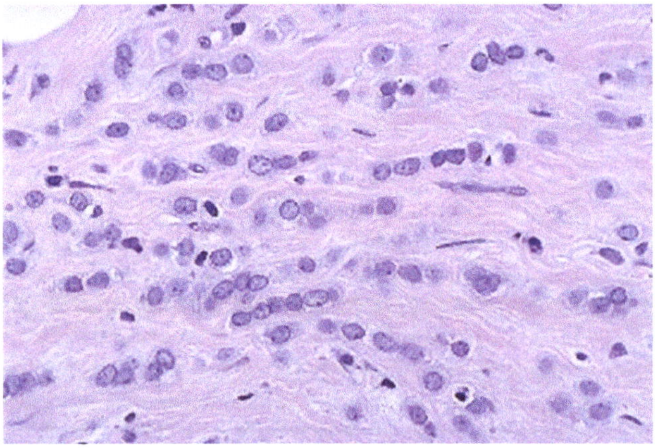

Fig. 1.8 Invasive lobular carcinoma with classic uniform, single-cell files

In addition to their different histologic appearance and mammographic characteristics, there are distinct prognostic and biologic differences between infiltrating lobular and ductal cancers:

Infiltrating lobular carcinomas have a higher frequency of bilaterality and multicentricity than infiltrating ductal carcinomas [36, 37].

- Infiltrating lobular carcinomas arise in older women and are larger and better differentiated tumors [36, 38]. As a rule, invasive lobular carcinomas are ER positive, with variant lesions showing occasional variable expression.
- While older series report a similar prognosis for infiltrating lobular cancers and invasive ductal lesions, more recent reports suggest that outcomes (at least in the short term) may be more favorable for lobular cancers and improving over time [39, 40]. However, variants of infiltrating lobular carcinoma exist, some of which have a poorer prognosis [36].
- As a group, invasive lobular carcinomas tend to metastasize later than invasive duct carcinomas and spread to unusual locations such as peritoneum, meninges, and the gastrointestinal tract [41].

There is an association between mutations in the cadherin (CDH1) gene and invasive lobular breast cancers. Lobular breast cancers have been observed to occur in 20–54 % of women from families with hereditary diffuse gastric cancer who carry germline mutations in the CDH1 gene. However, germline CDH1 mutations can also be cosegregated with invasive lobular breast cancer in the absence of diffuse gastric cancer, suggesting that gastric cancer is not an obligatory hallmark of families with CDH1 mutations. Furthermore, approximately 50 % of sporadic lobular breast cancers contain E-cadherin mutations [42, 43].

1.5 Other Histologic Types

A number of other histologic types account for the remaining invasive breast cancers. These include tubular carcinoma, mucinous carcinoma, medullary carcinoma, invasive micropapillary carcinoma, metaplastic carcinoma, adenoid cystic carcinoma, and others.

1.5.1 Tubular Carcinoma

Tubular carcinomas were relatively infrequent in the pre-mammography era, accounting for 2 % or less of invasive breast cancers. However, in some series of mammographically screened populations, the incidence is higher, accounting for 10–20 % of invasive cancers.

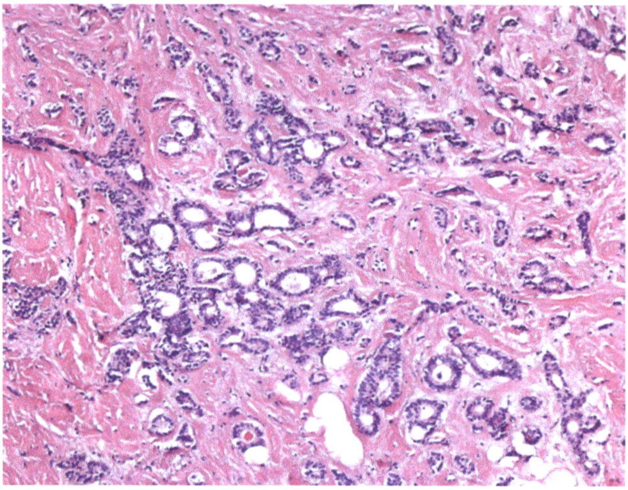

Fig. 1.9 Tubular carcinoma. There is a haphazard distribution of rounded and angulated tubules with open lumina, lined by a single layer of epithelial cells separated by abundant reactive fibroblastic stroma

Tubular carcinoma is characterized by the presence of well-formed tubular or glandular structures infiltrating the stroma (Fig. 1.9):

- The tubules tend to be elongated, and many have pointed ends.
- The cells composing the tubules are cuboidal to columnar and often have apical cytoplasmic protrusions or "snouts."
- The tumor cells are cytologically low grade.
- Associated DCIS, typically of the low-grade type, is present in about three-quarters of the cases.

These lesions have a relatively favorable prognosis compared with infiltrating ductal carcinomas; the natural history is favorable, and metastases are rare [1, 39, 44–46].

1.5.2 Mucinous (Colloid) Carcinoma

Mucinous carcinomas account for between 1 and 2 % of invasive breast cancers and appear to be more common in older patients. These lesions usually have a soft gelatinous appearance on gross examination, and they tend to be well circumscribed. Mucinous carcinomas are characterized microscopically by nests of tumor cells dispersed in large pools of extracellular mucus; the cells tend to have uniform, low-grade nuclei (Fig. 1.10). Similar to tubular carcinomas, these lesions also represent a prognostically favorable variant of invasive breast carcinoma [1, 39, 45, 47].

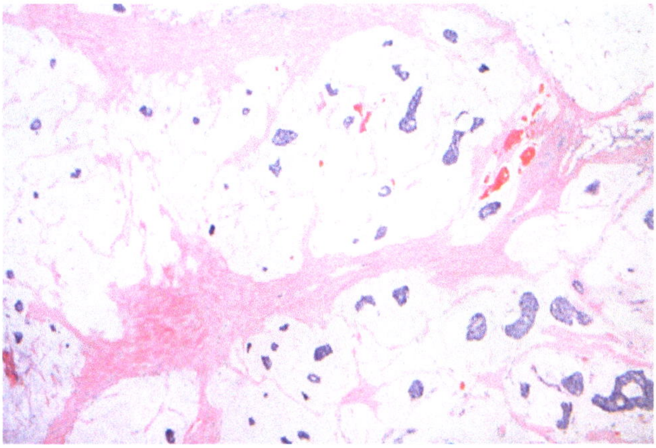

Fig. 1.10 Mucinous carcinoma. Lakes of mucus are separated by fibrous stroma. A few clusters of carcinoma cells are floating in the mucus lakes

1.5.3 Medullary Carcinoma

Medullary carcinomas account for anywhere from 1 to 10 % of invasive breast cancers. However, there is considerable interobserver variability in the diagnosis of this type of breast cancer which is, at least in part, dependent upon the classification system employed [48–50].

Medullary carcinomas are well circumscribed on macroscopic examination and are often soft and tan brown with areas of hemorrhage or necrosis. Circumscription of the lesion is also evident microscopically. The tumor cells are poorly differentiated (high grade), grow in a syncytial pattern, and have an intense associated lymphoplasmacytic infiltrate (Fig. 1.11), and this tumor is actually quite rare when strict diagnostic criteria are followed.

Medullary and medullary-like carcinomas occur more frequently in younger patients than other types of breast cancer. They are also more frequent in women who inherited mutations of the BRCA-1 gene (10 % of breast cancers are medullary in this population, as compared with <1 % of non-BRCA1-related breast cancers). However, the majority of breast cancers in patients with BRCA-1 gene mutations (90 %) are not medullary [51].

The prognosis for pure medullary carcinomas appears to be somewhat more favorable than that of infiltrating ductal carcinomas, despite their aggressive histologic appearance [1, 39, 45, 52].

1.5.4 Tubulolobular Carcinoma

Tubulolobular carcinoma is an often unrecognized breast cancer variant that, as the name implies, has hybrid histologic characteristics of tubular and invasive lobular

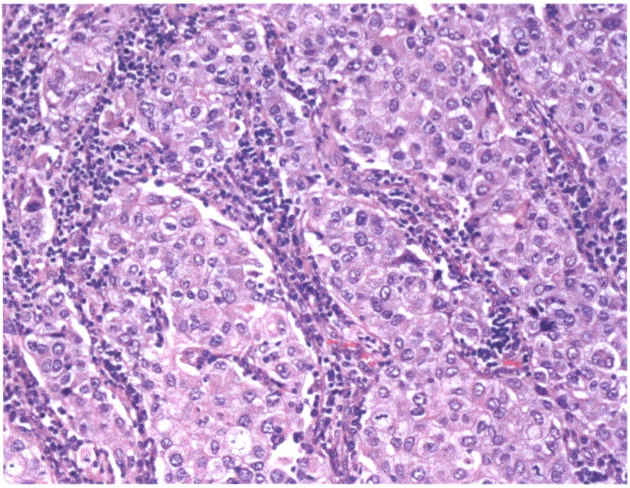

Fig. 1.11 Medullary carcinoma. Syncytial high-grade tumor cell accompanied by lymphoplasmacytic infiltrates

carcinoma with the same cells comprising well-formed glands contiguous with single-file infiltration of stroma. While immunohistochemical studies imply a ductal phenotype [53], from a radiologic and clinical point of view, the tumor is more akin to invasive lobular carcinoma in that its imaging characteristics are identical to lobular breast cancer, and there is the same tendency to multifocality and multicentricity. In terms of staging, however, the tumors behave more like invasive moderately differentiated ductal carcinoma in that they have the same likelihood of nodal metastases when matched by size. Often these tumors are misclassified as invasive carcinoma with mixed ductal and lobular features.

1.5.5 Micropapillary Carcinoma (Fig. 1.12)

Invasive micropapillary carcinoma is a particularly aggressive form of cancer that has a proclivity for lymph node metastasis even when small in size [54].

1.5.6 Metaplastic Carcinoma

Metaplastic carcinoma is a well-circumscribed tumor that consists of various combinations of poorly differentiated ductal adenocarcinoma, mesenchymal (sarcomatous), and other epithelial (e.g., squamous cell) components (Fig. 1.13) [55, 56].

Whether these tumors have a worse prognosis than ordinary invasive ductal cancers is unclear. Some studies suggest that tumors in which the squamous cell

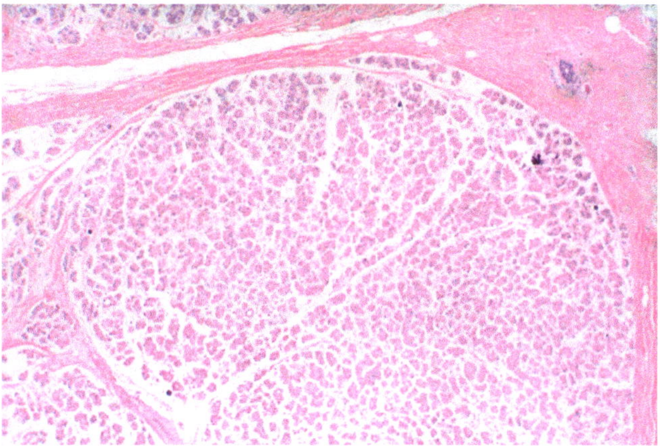

Fig. 1.12 Invasive micropapillary carcinoma. Tumor cell clusters with irregular central spaces proliferate within empty stromal spaces

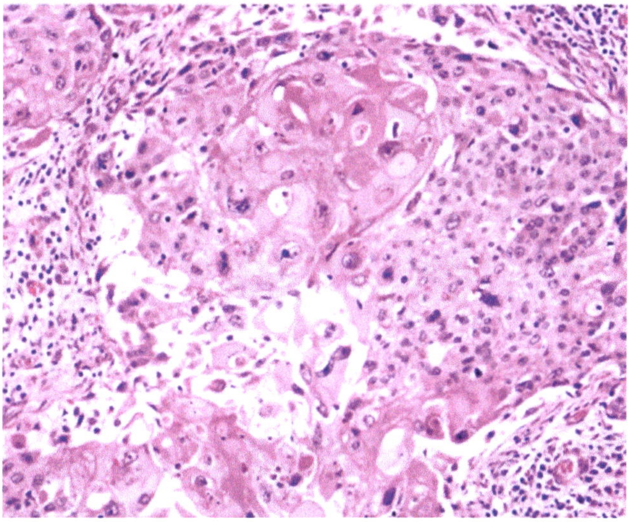

Fig. 1.13 Metaplastic carcinoma with squamous differentiation. Variously shaped spaces with squamous epithelium are characteristic

component predominates (more than 90 % of the malignant cells are of squamous type) are more aggressive and frequently treatment refractory when compared with other infiltrating ductal cancers [57, 58]. However, because metaplastic breast cancer was not officially recognized as a distinct pathologic diagnosis until 2000, knowledge about treatment patterns and outcomes is limited.

The characteristics of 892 metaplastic breast cancers reported to the National Cancer Database between 2001 and 2003 were compared with those of 255,164

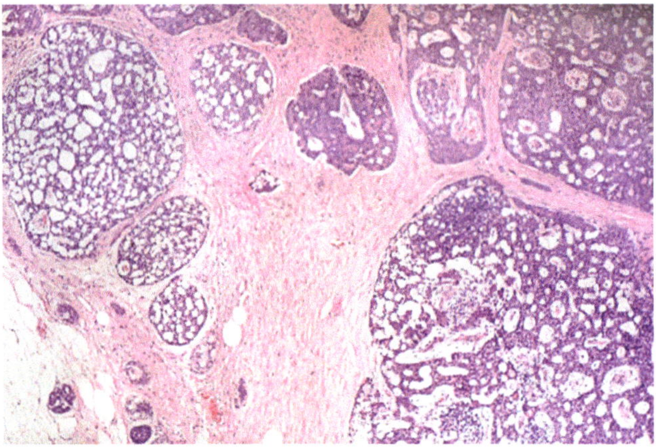

Fig. 1.14 Adenoid cystic carcinoma. Tumor is composed of proliferating glands (adenoid component) and stromal components distributed in varying proportions

typical infiltrating ductal carcinomas [56]. In contrast to patients with infiltrating ductal cancers, the following significant differences were noted in the group with metaplastic tumors:

- Fewer T1 tumors (30 versus 65 %)
- More node-negative tumors (78 versus 66 %)
- More poorly differentiated or undifferentiated tumors (68 versus 39 %)
- Fewer estrogen receptor-positive tumors (11 versus 74 %)

Treatment outcomes were not reported. Despite the perception of a worse prognosis, all metaplastic breast cancers are treated similarly to other invasive breast cancers [59–61].

1.5.7 Adenoid Cystic Carcinoma (Fig. 1.14)

The rare adenoid cystic carcinoma of the breast has a distinctive histologic pattern that is morphologically identical to adenoid cystic carcinomas found in the salivary glands (and other sites). This tumor tends to be associated with a favorable prognosis, even when tumor size is large; the reported incidence of axillary metastases in most series is less than 5 % [62, 63].

Histologic grading based upon the percentage of solid areas (as is used for salivary gland tumors) has been suggested as being prognostically useful [64], although others disagree [63]. At least two series in which outcomes were not as favorable as in most reports were predominated by patients with higher-grade tumors (i.e., the solid variant) [65, 66].

1.6 Tests on Breast Tissue

1.6.1 Hormone Receptors

Assay of hormone receptors has become a routine part of the evaluation of breast cancers, since the results predict the clinical response to hormone therapy, both in the adjuvant setting and for those with metastatic disease. The assays are discussed in detail elsewhere.

The prognostic importance of ER (Fig. 1.15) and PR expression has been a matter of debate for many years. However, taken together, the available evidence suggests that ER-/PR-negative tumors have a worse prognosis, at least in the first 5–10 years after treatment. On the other hand, other data suggest that as a result of sequential improvements in adjuvant chemotherapy (which disproportionately benefits those with hormone receptor-negative tumors) over time, the prognosis of individuals with ER-/PR-negative breast cancer now approaches that of patients with hormone receptor-positive disease.

In keeping with ASCO guidelines, ER/PR analysis should be performed routinely in all invasive breast cancers using either immunohistochemistry (IHC) or the now rarely used ligand binding assay [67]. The information should be used to select patients who are most likely to respond to hormone therapy. The prognostic implications of hormone receptor expression are discussed in more detail elsewhere.

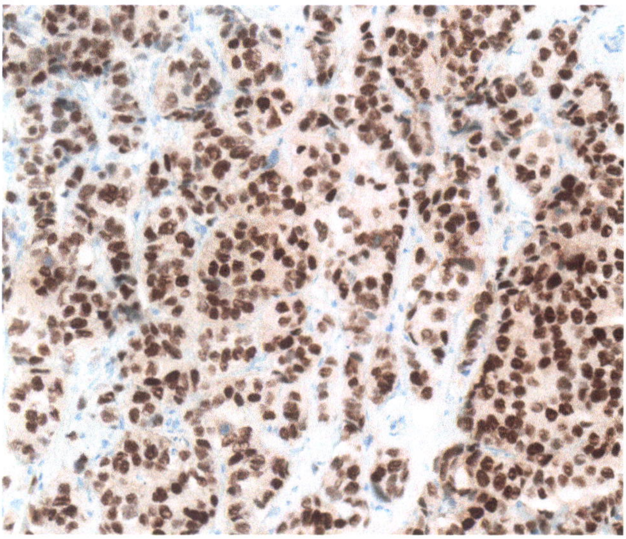

Fig. 1.15 Expression of estrogen receptor (ER) in invasive breast cancers as determined by immunohistochemistry. Approximately 100 % tumor cell nuclei are positive

1.6.2 Markers of Proliferation

In general, markers of an elevated proliferative rate correlate with a worse prognosis in untreated patients [68]. The proliferative rate of breast tumors may be assessed by a variety of methods, including mitotic counts, thymidine labeling index, bromodeoxyuridine (BrdU) labeling, S-phase fractions as determined by flow cytometry, IHC using monoclonal antibodies (MoAbs) to antigens found in proliferating cells (e.g., Ki-67 or proliferating cell nuclear antigen [PCNA/cyclin]), and the assessment of argyrophilic nucleolar organizer regions (AgNOR).

Guidelines from an expert panel on tumor markers proposed by ASCO in 2007 do not recommend the routine use of any markers of proliferation to assign patients to prognostic groups [67].

Dozens of studies have been undertaken to explore the relationship between Ki-67 status and prognosis in breast cancer [69, 70]. A meta-analysis of 46 studies (over 12,000 patients) came to the following conclusions [70]:

- Using cutoff values for Ki-67 positivity from the original studies (range 3.5–34 %), Ki-67/MIB-1 positivity was associated with a significantly higher risk of relapse in both node-positive (hazard ratio 1.59, 95 % CI 1.35–1.87) and node-negative diseases (HR 2.31, 95 % CI 1.83–2.92).
- There was also a significant relationship between Ki-67/MIB-1 positivity and worse breast cancer survival in node-positive (HR for death 2.33, 95 % CI 1.83–2.95) and node-negative disease (HR 2.54, 95 % CI 1.65–3.91).

However, an exceptionally thorough review of the literature on IHC-based markers of proliferation in breast cancer emphasized the lack of prospective studies and the difficulty with literature interpretation due to lack of standardization of assay reagents, procedures, and scoring, reinforcing the position taken by the ASCO expert tumor marker committee in its 2007 updated guidelines [71].

1.6.3 HER2 Overexpression

Amplification or overexpression of the HER2 oncogene is present in approximately 20 % of primary invasive breast cancers. Assay for HER2 overexpression and/or amplification is recommended as a routine part of the diagnostic work-up on all primary breast cancers [67, 72].

The main benefit of HER2 testing is its predictive value (Fig. 1.16). High levels of HER2 expression (3+ by IHC or an amplified HER2 gene copy number by FISH [72]) identify those women who might benefit from trastuzumab in both the adjuvant and metastatic disease settings. High levels of HER2 expression (3+ by IHC or an amplified HER2 gene copy number by fluorescence in situ hybridization [FISH]) represent an important predictive factor, identifying those patients who might

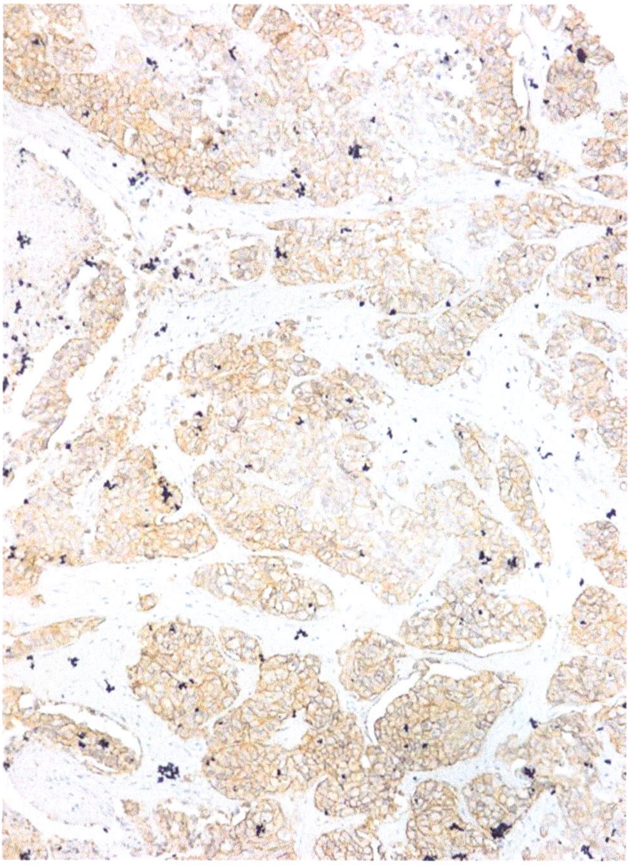

Fig. 1.16 Positive Her2 testing of invasive breast cancer by immunohistochemistry. More than 30 % of tumor cells show strong complete membrane staining

benefit from treatments that target HER2, such as trastuzumab, in both the adjuvant and metastatic disease settings. Guidelines for HER2 testing from a joint ASCO/CAP consensus panel [72], the data on HER2 status and treatment as well as recommendations for optimal performance, interpretation, and reporting of individual assays are discussed in detail elsewhere.

The weight of evidence suggests that HER2 overexpression is a negative prognostic factor. In most, but not all studies, HER2 overexpression in primary tumor tissue (as determined by IHC) is associated with a worse prognosis in untreated patients [73–76]. In some, HER2 overexpression correlates with other factors associated with a poor prognosis (such as tumor grade, size, nodal status) [77].

The independent contribution of HER2 overexpression in tumor tissue as a marker of poor prognosis in patients with node-positive disease is fairly consistent

[78–81]. Although early studies were unclear about the prognostic role of HER2 in node-negative untreated patients, more recent studies have provided more clarity. In a cohort of 2026 patients with node-negative breast cancer, of whom 70 % did not receive adjuvant systemic therapy, HER2 expression (as assessed by IHC on tissue microarrays, with FISH confirmation for those with 2+ IHC staining) was independently associated with significantly worse 10-year relapse-free (66 versus 76 %) and breast cancer-specific survival (76 versus 86 %) [82]. Almost 90 % of the tumors in this series were >1.0 cm in size.

However, the added value of this information in clinical practice is questionable, and the ASCO expert panel on tumor markers in breast cancer did not recommend the use of HER2 for determining prognosis, largely because outcomes are heavily influenced by subsequent therapy [67].

1.6.4 Multigene Predictors of Clinical Outcome

Gene expression profiling has identified molecular signatures, such as the 21-gene recurrence score (RS, Oncotype Dx®), the Amsterdam 70-gene prognostic profile (MammaPrint®), and the Rotterdam/Veridex 76-gene signature, that augment conventional prognostic indicators in their ability to predict breast cancer outcome and response to treatment.

The 21-gene recurrence score can be used to predict the risk of recurrence in patients with newly diagnosed, node-negative, estrogen receptor (ER)-positive disease and to identify patients who are likely to benefit from chemotherapy added to adjuvant endocrine therapy. Preliminary data also suggest utility of the 21-gene recurrence score in patients with node-positive, ER-positive breast cancer. Although use of the test has been cleared by the US FDA for women with ER-positive early breast cancer and one to three positive nodes, further prospective validation in this group is awaited.

Gene expression profiling has also been instrumental in developing the molecular classification of breast cancer.

1.6.5 Urokinase Plasminogen Activator System

Urokinase plasminogen activator (uPA) is a serine protease with an important role in cancer invasion and metastases [83]. When bound to its receptor (uPAR), uPA converts plasminogen into plasmin and mediates degradation of the ECM during tumor cell invasion. Specific inhibitors of uPA (plasminogen activator inhibitors [PAI] types 1 and 2) have been identified. PAI-1 levels are high in tumor tissue and plasma, and PAI-1 is inactivated when bound to uPA. In contrast, PAI-2 is usually present in low levels except for some conditions such as pregnancy or myeloid leukemia [83].

The ASCO expert panel recommended that measurement of uPA and PAI-1 by ELISA on at least 300 mg of fresh or frozen breast cancer tissue may be used for determination of prognosis in patients with newly diagnosed, node-negative breast cancer [67]. While preliminary reports are promising for immunohistochemical staining to determine uPA and PAI-1 status, validation is needed.

- In retrospective reports, high levels of uPA, uPAR, and PAI-1 have been associated with shorter survival in women with breast cancer, while high levels of PAI-2 were associated with better outcomes [83–86]. Further support for the prognostic value of these molecules was derived from a pooled analysis of individual patient data from 8,377 women treated in clinical trials sponsored by the EORTC, in which tumor uPA and PAI-1 levels were determined in primary tumor tissue extracts [87]. In multivariate analysis, uPA and PAI-1 levels were the strongest predictors of disease-free and overall survival, after nodal status, for all patients combined. Although higher uPA and PAI-1 levels were associated with poorer outcomes in women with both node-positive and node-negative diseases, expression of uPA or PAI-1 was the strongest significant predictor of relapse-free survival (hazard ratio [HR] 2.3 for uPA and 1.9 for PAI-1) in women with node-negative disease, particularly when expression of both markers was considered together.
- Much of the data on uPA and PAI-1 concern its ability to stratify women with node-negative early breast cancer into prognostically relevant subgroups. In an initial report of 269 women with node-negative disease (none of whom received chemotherapy) whose primary tumors were assayed for uPA and PAI-1, high levels of both predicted a significant 3.9-fold higher risk of relapse and a 2.8-fold higher risk of death [88].

  A low level of both markers is associated with a sufficiently low risk of recurrence, particularly in women with ER-/PR-positive disease who will receive adjuvant hormone therapy, that the additional benefit derived from chemotherapy is minimal. CMF-based chemotherapy provides substantial benefit, compared with observation alone, in patients with high levels of uPA and PAI-1.
- Enthusiasm for the use of uPA and PAI-1 as prognostic indicators in the USA is tempered by problems with assay methodology. Nearly all of the prognostic data have been derived by measuring these factors in relatively large frozen tissue sections removed at the time of resection by ELISA. With the widespread adoption of IHC techniques to determine ER and PR status, routine tissue freezing of breast cancer specimens has been discontinued, obviating the need for expensive equipment and supplies for freezing and sample storage. Returning to such a system would require a major paradigm shift in this country. Although preliminary data on IHC staining for uPA and PAI-1 in cryostat specimens support the utility of this technique, studies correlating these results with outcome are lacking [89]. As a result, assessment of these markers is still considered investigational in the USA.

1.6.6 P53 Gene Analysis

Mutations in the p53 tumor suppressor gene or accumulation of p53 protein (a mutated p53 gene produces a protein that is not degraded as quickly as the wild-type protein leading to protein accumulation which can be detected by IHC) are reported in 20–50 % of human breast cancers. These abnormalities are more often seen in patients with hereditary breast cancer syndromes (such as the familial breast and ovarian cancer and Li-Fraumeni syndromes) than in those with sporadic breast cancer [90].

A number of studies, including a 1999 meta-analysis, suggest that high tissue p53 protein levels (as measured by IHC) or mutations or deletions in the p53 gene represent an independent predictor of decreased disease-free and overall survival in both node-positive and node-negative patients [91–98]. However, other studies have failed to find an association between p53 abnormalities and clinical outcomes [91, 99, 100].

The likely explanation for the variable results regarding p53 and prognosis is the confounding effect of adjuvant systemic therapy [101]. While p53 abnormalities might be associated with a worse prognosis in untreated patients, they might also be associated with sensitivity to some therapeutic agents and resistance to others. Thus, p53 mutations or deletions might confer a favorable or unfavorable prognosis depending on the specific type of treatment.

The ASCO panel on tumor markers concluded that IHC for p53 protein was unlikely to provide sufficiently accurate results to be clinically useful [67]. At present, there is no role for p53 gene analysis in women with breast cancer.

1.6.7 Other Markers of Invasion and Metastasis

Many other markers of invasion and metastatic potential have been proposed and/or studied in retrospective reports. These include nm23, E-cadherin, catenins, tissue inhibitors of metalloproteinases (TIMPs), prostate-specific antigen, tissue factor, and osteopontin [102–110]. Allelic loss, microsatellite instability, or methylation silencing of tumor suppressor genes may also provide prognostic information [111–114].

All of these potential indicators of prognosis require further evaluation and validation. None should be considered routine in the evaluation of breast cancer specimens.

1.7 Summary

- DCIS is characterized by a proliferation of abnormal cells confined within the mammary ductal system (See Sect. 1.2 above).
  - DCIS is commonly classified according to architectural and cytologic features and cell necrosis as low and intermediate grade (papillary, cribriform, and solid) and high grade (comedo).

- DCIS represents a precursor to invasive breast cancer.

- The invasive breast carcinomas consist of several histologic subtypes.
  - Infiltrating ductal carcinoma is the most common type of invasive breast cancer, accounting for 70–80 % of invasive cancers (See Sect. 1.3).
  - Infiltrating lobular carcinoma is the second most common invasive breast cancer, accounting for 5–10 % of invasive cancers (See Sect. 1.4).
  - As compared with infiltrating ductal carcinomas, infiltrating lobular carcinomas tend to be multicentric and/or bilateral, more differentiated, and hormone receptor positive, arise in older women, metastasize later, and spread to unusual locations, such as meninges, peritoneum, or gastrointestinal tract (See Sect. 1.4).
  - Other less common invasive breast carcinoma histologies include tubular, mucinous, and medullary carcinomas (See Sect. 1.5).

- The only uniformly accepted prognostic markers that provide critical information necessary for treatment decisions are TNM stage, axillary lymph node status, tumor size and grade, hormone receptor status, and HER2 receptor status.
  - Guidelines for measurement and use of prognostic markers are available from a 2007 ASCO expert panel on tumor markers in breast cancer [67]. We agree with these recommendations. ER, PR, and HER2 overexpression should be evaluated on every primary invasive breast cancer. Hormone receptor expression should be used to guide endocrine therapy decisions. HER2 expression should be used to select patients for whom HER2-directed therapy use is appropriate in the metastatic and adjuvant setting. The available data are insufficient to recommend that HER2 overexpression be used for determining prognosis in patients with early breast cancer. Furthermore, there is insufficient evidence to support the clinical use of serum HER2 extracellular domain (ECD) testing.
  - Multiparameter gene expression analysis (i.e., the Oncotype DX™ assay) should be used to predict the risk of recurrence in women with newly diagnosed, node-negative, ER+ breast cancer who will be receiving tamoxifen. A low recurrence score should be used to identify patients who are predicted to obtain the most therapeutic benefit from tamoxifen and who may not require adjuvant chemotherapy. Patients with a high recurrence score appear to derive relatively more benefit from chemotherapy (specifically the CMF regimen) than from tamoxifen. (See "Prognostic molecular profiles of breast cancer.").
  - Measurement of urokinase plasminogen activator (uPA) and plasminogen activator inhibitor-1 (PAI-1) by ELISA on fresh or frozen tissue may be used for determination of prognosis in patients with newly diagnosed, node-negative breast cancer. Low levels of both markers are associated with a sufficiently low risk of recurrence, particularly in women with ER-/PR-positive disease who will receive adjuvant hormone therapy that the additional benefit of chemotherapy is minimal (see Sect. 1.6.5).

– The data are insufficient to recommend the use of p53 or immunohistochemically based markers of proliferation to assign patients to prognostic groups (see Sects. 1.6.2 and 1.6.6).

References

1. Li CI, Uribe DJ, Daling JR. Clinical characteristics of different histologic types of breast cancer. Br J Cancer. 2005;93:1046.
2. Schnitt SJ, Guidi AJ. Pathology of invasive breast cancer. In: Harris JR, Lippman ME, Morrow M, Osborne CK, editors. Diseases of the breast. 3rd ed. Philadelphia: Lippincott Williams and Wilkins; 2004. p. 393.
3. Allred DC. Ductal carcinoma in situ: terminology, classification, and natural history. J Natl Cancer Inst Monogr. 2010;2010:134.
4. Azzopardi JG. Problems in breast pathology. Philadelphia: WB Saunders; 1963. p. 244.
5. Page DL, Anderson TJ. Diagnostic histopathology of the breast. Edinburgh: Churchill Livingstone; 1987. p. 157.
6. Rosen PP, Oberman H. Tumors of the mammary gland. Washington, DC: Armed Forces Institute of Pathology; 1993. p. 143.
7. Morrow M, Schnitt SJ, Harris JR. Ductal carcinoma in situ. In: Harris JR, Lippman ME, Morrow M, Hellman S, editors. Diseases of the breast. Philadelphia: Lippincott-Raven; 1995. p. 355.
8. Schwartz GF, Patchefsky AS, Finklestein SD, et al. Nonpalpable in situ ductal carcinoma of the breast. Predictors of multicentricity and microinvasion and implications for treatment. Arch Surg. 1989;124:29.
9. Silverstein MJ, Waisman JR, Gamagami P, et al. Intraductal carcinoma of the breast (208 cases). Clinical factors influencing treatment choice. Cancer. 1990;66:102.
10. Fisher ER, Dignam J, Tan-Chiu E, et al. Pathologic findings from the National Surgical Adjuvant Breast Project (NSABP) eight-year update of Protocol B-17: intraductal carcinoma. Cancer. 1999;86:429.
11. Grabowski J, Salzstein SL, Sadler GR, Blair S. Intracystic papillary carcinoma: a review of 917 cases. Cancer. 2008;113:916.
12. Lefkowitz M, Lefkowitz W, Wargotz ES. Intraductal (intracystic) papillary carcinoma of the breast and its variants: a clinicopathological study of 77 cases. Hum Pathol. 1994;25:802.
13. Eusebi V, Foschini MP, Cook MG, et al. Long-term follow-up of in situ carcinoma of the breast with special emphasis on clinging carcinoma. Semin Diagn Pathol. 1989;6:165.
14. Fisher ER, Brown R. Intraductal signet ring carcinoma. A hitherto undescribed form of intraductal carcinoma of the breast. Cancer. 1985;55:2533.
15. Rosen PP, Scott M. Cystic hypersecretory duct carcinoma of the breast. Am J Surg Pathol. 1984;8:31.
16. Guerry P, Erlandson RA, Rosen PP. Cystic hypersecretory hyperplasia and cystic hypersecretory duct carcinoma of the breast. Pathology, therapy, and follow-up of 39 patients. Cancer. 1988;61:1611.
17. Lagios MD, Margolin FR, Westdahl PR, Rose MR. Mammographically detected duct carcinoma in situ. Frequency of local recurrence following tylectomy and prognostic effect of nuclear grade on local recurrence. Cancer. 1989;63:618.
18. Poller DN, Silverstein MJ, Galea M, et al. Ideas in pathology. Ductal carcinoma in situ of the breast: a proposal for a new simplified histological classification association between cellular proliferation and c-erbB-2 protein expression. Mod Pathol. 1994;7:257.
19. Holland R, Peterse JL, Millis RR, et al. Ductal carcinoma in situ: a proposal for a new classification. Semin Diagn Pathol. 1994;11:167.

20. Silverstein MJ, Poller DN, Waisman JR, et al. Prognostic classification of breast ductal carcinoma-in-situ. Lancet. 1995;345:1154.
21. Bur ME, Zimarowski MJ, Schnitt SJ, et al. Estrogen receptor immunohistochemistry in carcinoma in situ of the breast. Cancer. 1992;69:1174.
22. Meyer JS. Cell kinetics of histologic variants of in situ breast carcinoma. Breast Cancer Res Treat. 1986;7:171.
23. Killeen JL, Namiki H. DNA analysis of ductal carcinoma in situ of the breast. A comparison with histologic features. Cancer. 1991;68:2602.
24. van de Vijver MJ, Peterse JL, Mooi WJ, et al. Neu-protein overexpression in breast cancer. Association with comedo-type ductal carcinoma in situ and limited prognostic value in stage II breast cancer. N Engl J Med. 1988;319:1239.
25. Bartkova J, Barnes DM, Millis RR, Gullick WJ. Immunohistochemical demonstration of c-erbB-2 protein in mammary ductal carcinoma in situ. Hum Pathol. 1990;21:1164.
26. Lodato RF, Maguire Jr HC, Greene MI, et al. Immunohistochemical evaluation of c-erbB-2 oncogene expression in ductal carcinoma in situ and atypical ductal hyperplasia of the breast. Mod Pathol. 1990;3:449.
27. Poller DN, Roberts EC, Bell JA, et al. p53 protein expression in mammary ductal carcinoma in situ: relationship to immunohistochemical expression of estrogen receptor and c-erbB-2 protein. Hum Pathol. 1993;24:463.
28. O'Malley FP, Vnencak-Jones CL, Dupont WD, et al. p53 mutations are confined to the comedo type ductal carcinoma in situ of the breast. Immunohistochemical and sequencing data. Lab Invest. 1994;71:67.
29. Guidi AJ, Fischer L, Harris JR, Schnitt SJ. Microvessel density and distribution in ductal carcinoma in situ of the breast. J Natl Cancer Inst. 1994;86:614.
30. Bobrow LG, Happerfield LC, Gregory WM, et al. The classification of ductal carcinoma in situ and its association with biological markers. Semin Diagn Pathol. 1994;11:199.
31. Zafrani B, Leroyer A, Fourquet A, et al. Mammographically-detected ductal in situ carcinoma of the breast analyzed with a new classification. A study of 127 cases: correlation with estrogen and progesterone receptors, p53 and c-erbB-2 proteins, and proliferative activity. Semin Diagn Pathol. 1994;11:208.
32. Wärnberg F, Nordgren H, Bergh J, Holmberg L. Ductal carcinoma in situ of the breast from a population-defined cohort: an evaluation of new histopathological classification systems. Eur J Cancer. 1999;35:714.
33. Consensus Conference on the classification of ductal carcinoma in situ. The Consensus Conference Committee. Cancer 1997; 80:1798.
34. Elston CW, Ellis IO. Pathological prognostic factors in breast cancer. I. The value of histological grade in breast cancer: experience from a large study with long-term follow-up. Histopathology. 1991;19:403.
35. Abner AL, Connolly JL, Recht A, et al. The relation between the presence and extent of lobular carcinoma in situ and the risk of local recurrence for patients with infiltrating carcinoma of the breast treated with conservative surgery and radiation therapy. Cancer. 2000;88:1072.
36. Orvieto E, Maiorano E, Bottiglieri L, et al. Clinicopathologic characteristics of invasive lobular carcinoma of the breast: results of an analysis of 530 cases from a single institution. Cancer. 2008;113:1511.
37. Winchester DJ, Chang HR, Graves TA, et al. A comparative analysis of lobular and ductal carcinoma of the breast: presentation, treatment, and outcomes. J Am Coll Surg. 1998;186:416.
38. Pestalozzi BC, Zahrieh D, Mallon E, et al. Distinct clinical and prognostic features of infiltrating lobular carcinoma of the breast: combined results of 15 International Breast Cancer Study Group clinical trials. J Clin Oncol. 2008;26:3006.
39. Li CI, Moe RE, Daling JR. Risk of mortality by histologic type of breast cancer among women aged 50 to 79 years. Arch Intern Med. 2003;163:2149.
40. Cristofanilli M, Gonzalez-Angulo A, Sneige N, et al. Invasive lobular carcinoma classic type: response to primary chemotherapy and survival outcomes. J Clin Oncol. 2005;23:41.

41. Ferlicot S, Vincent-Salomon A, Médioni J, et al. Wide metastatic spreading in infiltrating lobular carcinoma of the breast. Eur J Cancer. 2004;40:336.
42. Berx G, Cleton-Jansen AM, Strumane K, et al. E-cadherin is inactivated in a majority of invasive human lobular breast cancers by truncation mutations throughout its extracellular domain. Oncogene. 1996;13:1919.
43. De Leeuw WJ, Berx G, Vos CB, et al. Simultaneous loss of E-cadherin and catenins in invasive lobular breast cancer and lobular carcinoma in situ. J Pathol. 1997;183:404.
44. Liu GF, Yang Q, Haffty BG, Moran MS. Clinical-pathologic features and long-term outcomes of tubular carcinoma of the breast compared with invasive ductal carcinoma treated with breast conservation therapy. Int J Radiat Oncol Biol Phys. 2009;75:1304.
45. Thurman SA, Schnitt SJ, Connolly JL, et al. Outcome after breast-conserving therapy for patients with stage I or II mucinous, medullary, or tubular breast carcinoma. Int J Radiat Oncol Biol Phys. 2004;59:152.
46. Sullivan T, Raad RA, Goldberg S, et al. Tubular carcinoma of the breast: a retrospective analysis and review of the literature. Breast Cancer Res Treat. 2005;93:199.
47. DiSaverio S, Gutierrez J, Avisar E. A retrospective review with long term follow up of 11,400 cases of pure mucinous breast carcinoma. Breast Cancer Res Treat. 2008;111:541.
48. Gaffey MJ, Mills SE, Frierson Jr HF, et al. Medullary carcinoma of the breast: inter observer variability in histopathologic diagnosis. Mod Pathol. 1995;8:31.
49. Ridolfi RL, Rosen PP, Port A, et al. Medullary carcinoma of the breast: a clinicopathologic study with 10 year follow-up. Cancer. 1977;40:1365.
50. Wargotz ES, Silverberg SG. Medullary carcinoma of the breast: a clinicopathologic study with appraisal of current diagnostic criteria. Hum Pathol. 1988;19:1340.
51. Armes JE, Venter DJ. The pathology of inherited breast cancer. Pathology. 2002;34:309.
52. Vu-Nishino H, Tavassoli FA, Ahrens WA, Haffty BG. Clinicopathologic features and long-term outcome of patients with medullary breast carcinoma managed with breast-conserving therapy (BCT). Int J Radiat Oncol Biol Phys. 2005;62:1040.
53. Esposito NN, Chivukula M, Dabbs DJ. The ductal phenotypic expression of the E-cadherin/catenin complex in tubulolobular carcinoma of the breast: an immunohistochemical and clinicopathologic study. Mod Pathol. 2007;20:130.
54. Walsh MM, Bleiweiss IJ. Invasive micropapillary carcinoma of the breast: eighty cases of an underrecognized entity. Hum Pathol. 2001;32:583.
55. Tavassoli FA. Classification of metaplastic carcinomas of the breast. Pathol Annu. 1992;27(Pt 2):89.
56. Pezzi CM, Patel-Parekh L, Cole K, et al. Characteristics and treatment of metaplastic breast cancer: analysis of 892 cases from the National Cancer Data Base. Ann Surg Oncol. 2007;14:166.
57. Hennessy BT, Krishnamurthy S, Giordano S, et al. Squamous cell carcinoma of the breast. J Clin Oncol. 2005;23:7827.
58. Behranwala KA, Nasiri N, Abdullah N, et al. Squamous cell carcinoma of the breast: clinicopathologic implications and outcome. Eur J Surg Oncol. 2003;29:386.
59. Dave G, Cosmatos H, Do T, et al. Metaplastic carcinoma of the breast: a retrospective review. Int J Radiat Oncol Biol Phys. 2006;64:771.
60. Rayson D, Adjei AA, Suman VJ, et al. Metaplastic breast cancer: prognosis and response to systemic therapy. Ann Oncol. 1999;10:413.
61. Chao TC, Wang CS, Chen SC, Chen MF. Metaplastic carcinomas of the breast. J Surg Oncol. 1999;71:220.
62. Arpino G, Clark GM, Mohsin S, et al. Adenoid cystic carcinoma of the breast: molecular markers, treatment, and clinical outcome. Cancer. 2002;94:2119.
63. Kleer CG, Oberman HA. Adenoid cystic carcinoma of the breast: value of histologic grading and proliferative activity. Am J Surg Pathol. 1998;22:569.
64. Ro JY, Silva EG, Gallager HS. Adenoid cystic carcinoma of the breast. Hum Pathol. 1987;18:1276.

65. Millar BA, Kerba M, Youngson B, et al. The potential role of breast conservation surgery and adjuvant breast radiation for adenoid cystic carcinoma of the breast. Breast Cancer Res Treat. 2004;87:225.
66. Shin SJ, Rosen PP. Solid variant of mammary adenoid cystic carcinoma with basaloid features: a study of nine cases. Am J Surg Pathol. 2002;26:413.
67. Harris L, Fritsche H, Mennel R, et al. American Society of Clinical Oncology 2007 update of recommendations for the use of tumor markers in breast cancer. J Clin Oncol. 2007;25:5287.
68. Colozza M, Azambuja E, Cardoso F, et al. Proliferative markers as prognostic and predictive tools in early breast cancer: where are we now? Ann Oncol. 2005;16:1723.
69. Urruticoechea A, Smith IE, Dowsett M. Proliferation marker Ki-67 in early breast cancer. J Clin Oncol. 2005;23:7212.
70. deAzambuja E, Cardoso F, De Castro Jr G, et al. Ki-67 as prognostic marker in early breast cancer: a meta-analysis of published studies involving 12,155 patients. Br J Cancer. 2007;96:1504.
71. Yerushalmi R, Woods R, Ravdin PM, et al. Ki67 in breast cancer: prognostic and predictive potential. Lancet Oncol. 2010;11:174.
72. Wolff AC, Hammond ME, Schwartz JN, et al. American Society of Clinical Oncology/College of American Pathologists guideline recommendations for human epidermal growth factor receptor 2 testing in breast cancer. J Clin Oncol. 2007;25:118.
73. Slamon DJ, Clark GM, Wong SG, et al. Human breast cancer: correlation of relapse and survival with amplification of the HER-2/neu oncogene. Science. 1987;235:177.
74. Paik S, Hazan R, Fisher ER, et al. Pathologic findings from the National Surgical Adjuvant Breast and Bowel Project: prognostic significance of erbB-2 protein overexpression in primary breast cancer. J Clin Oncol. 1990;8:103.
75. van de Vijver MJ, Mooi WJ, Wisman P, et al. Immunohistochemical detection of the neu protein in tissue sections of human breast tumors with amplified neu DNA. Oncogene. 1988;2:175.
76. Gilcrease MZ, Woodward WA, Nicolas MM, et al. Even low-level HER2 expression may be associated with worse outcome in node-positive breast cancer. Am J Surg Pathol. 2009;33:759.
77. Taucher S, Rudas M, Mader RM, et al. Do we need HER-2/neu testing for all patients with primary breast carcinoma? Cancer. 2003;98:2547.
78. Tandon AK, Clark GM, Chamness GC, et al. HER-2/neu oncogene protein and prognosis in breast cancer. J Clin Oncol. 1989;7:1120.
79. Gusterson BA, Gelber RD, Goldhirsch A, et al. Prognostic importance of c-erbB-2 expression in breast cancer. International (Ludwig) Breast Cancer Study Group. J Clin Oncol. 1992;10:1049.
80. Pauletti G, Dandekar S, Rong H, et al. Assessment of methods for tissue-based detection of the HER-2/neu alteration in human breast cancer: a direct comparison of fluorescence in situ hybridization and immunohistochemistry. J Clin Oncol. 2000;18:3651.
81. Kröger N, Milde-Langosch K, Riethdorf S, et al. Prognostic and predictive effects of immunohistochemical factors in high-risk primary breast cancer patients. Clin Cancer Res. 2006;12:159.
82. Chia S, Norris B, Speers C, et al. Human epidermal growth factor receptor 2 overexpression as a prognostic factor in a large tissue microarray series of node-negative breast cancers. J Clin Oncol. 2008;26:5697.
83. Stephens RW, Brünner N, Jänicke F, Schmitt M. The urokinase plasminogen activator system as a target for prognostic studies in breast cancer. Breast Cancer Res Treat. 1998;52:99.
84. Malmström P, Bendahl PO, Boiesen P, et al. S-phase fraction and urokinase plasminogen activator are better markers for distant recurrences than Nottingham Prognostic Index and histologic grade in a prospective study of premenopausal lymph node-negative breast cancer. J Clin Oncol. 2001;19:2010.
85. Foekens JA, Peters HA, Look MP, et al. The urokinase system of plasminogen activation and prognosis in 2780 breast cancer patients. Cancer Res. 2000;60:636.

86. Chappuis PO, Dieterich B, Sciretta V, et al. Functional evaluation of plasmin formation in primary breast cancer. J Clin Oncol. 2001;19:2731.
87. Look MP, van Putten WL, Duffy MJ, et al. Pooled analysis of prognostic impact of urokinase-type plasminogen activator and its inhibitor PAI-1 in 8377 breast cancer patients. J Natl Cancer Inst. 2002;94:116.
88. Harbeck N, Kates RE, Schmitt M. Clinical relevance of invasion factors urokinase-type plasminogen activator and plasminogen activator inhibitor type 1 for individualized therapy decisions in primary breast cancer is greatest when used in combination. J Clin Oncol. 2002;20:1000.
89. Visscher DW, Sarkar F, LoRusso P, et al. Immunohistologic evaluation of invasion-associated proteases in breast carcinoma. Mod Pathol. 1993;6:302.
90. Patocs A, Zhang L, Xu Y, et al. Breast-cancer stromal cells with TP53 mutations and nodal metastases. N Engl J Med. 2007;357:2543.
91. Mansour EG, Ravdin PM, Dressler L. Prognostic factors in early breast carcinoma. Cancer. 1994;74:381.
92. Thor AD, Moore II DH, Edgerton SM, et al. Accumulation of p53 tumor suppressor gene protein: an independent marker of prognosis in breast cancers. J Natl Cancer Inst. 1992;84:845.
93. Allred DC, Clark GM, Elledge R, et al. Association of p53 protein expression with tumor cell proliferation rate and clinical outcome in node-negative breast cancer. J Natl Cancer Inst. 1993;85:200.
94. Isola J, Visakorpi T, Holli K, Kallioniemi OP. Association of overexpression of tumor suppressor protein p53 with rapid cell proliferation and poor prognosis in node-negative breast cancer patients. J Natl Cancer Inst. 1992;84:1109.
95. Olivier M, Langerød A, Carrieri P, et al. The clinical value of somatic TP53 gene mutations in 1,794 patients with breast cancer. Clin Cancer Res. 2006;12:1157.
96. Overgaard J, Yilmaz M, Guldberg P, et al. TP53 mutation is an independent prognostic marker for poor outcome in both node-negative and node-positive breast cancer. Acta Oncol. 2000;39:327.
97. Pharoah PD, Day NE, Caldas C. Somatic mutations in the p53 gene and prognosis in breast cancer: a meta-analysis. Br J Cancer. 1999;80:1968.
98. Bonnefoi H, Piccart M, Bogaerts J, et al. TP53 status for prediction of sensitivity to taxane versus non-taxane neoadjuvant chemotherapy in breast cancer (EORTC 10994/BIG 1-00): a randomised phase 3 trial. Lancet Oncol. 2011;12:527.
99. Rosen PP, Lesser ML, Arroyo CD, et al. p53 in node-negative breast carcinoma: an immunohistochemical study of epidemiologic risk factors, histologic features, and prognosis. J Clin Oncol. 1995;13:821.
100. Bull SB, Ozcelik H, Pinnaduwage D, et al. The combination of p53 mutation and neu/erbB-2 amplification is associated with poor survival in node-negative breast cancer. J Clin Oncol. 2004;22:86.
101. Bergh J, Norberg T, Sjögren S, et al. Complete sequencing of the p53 gene provides prognostic information in breast cancer patients, particularly in relation to adjuvant systemic therapy and radiotherapy. Nat Med. 1995;1:1029.
102. Tuck AB, O'Malley FP, Singhal H, et al. Osteopontin expression in a group of lymph node negative breast cancer patients. Int J Cancer. 1998;79:502.
103. Heimann R, Hellman S. Individual characterisation of the metastatic capacity of human breast carcinoma. Eur J Cancer. 2000;36:1631.
104. Heimann R, Lan F, McBride R, Hellman S. Separating favorable from unfavorable prognostic markers in breast cancer: the role of E-cadherin. Cancer Res. 2000;60:298.
105. Heimann R, Ferguson DJ, Hellman S. The relationship between nm23, angiogenesis, and the metastatic proclivity of node-negative breast cancer. Cancer Res. 1998;58:2766.
106. Yu H, Levesque MA, Clark GM, Diamandis EP. Enhanced prediction of breast cancer prognosis by evaluating expression of p53 and prostate-specific antigen in combination. Br J Cancer. 1999;81:490.
107. Remacle A, McCarthy K, Noël A, et al. High levels of TIMP-2 correlate with adverse prognosis in breast cancer. Int J Cancer. 2000;89:118.

108. Ueno T, Toi M, Koike M, et al. Tissue factor expression in breast cancer tissues: its correlation with prognosis and plasma concentration. Br J Cancer. 2000;83:164.
109. Yoshida R, Kimura N, Harada Y, Ohuchi N. The loss of E-cadherin, alpha- and beta-catenin expression is associated with metastasis and poor prognosis in invasive breast cancer. Int J Oncol. 2001;18:513.
110. Lipton A, Ali SM, Leitzel K, et al. Elevated plasma tissue inhibitor of metalloproteinase-1 level predicts decreased response and survival in metastatic breast cancer. Cancer. 2007; 109:1933.
111. Emi M, Yoshimoto M, Sato T, et al. Allelic loss at 1p34, 13q12, 17p13.3, and 17q21.1 correlates with poor postoperative prognosis in breast cancer. Genes Chromosomes Cancer. 1999;26:134.
112. Utada Y, Emi M, Yoshimoto M, et al. Allelic loss at 1p34-36 predicts poor prognosis in node-negative breast cancer. Clin Cancer Res. 2000;6:3193.
113. Hui R, Macmillan RD, Kenny FS, et al. INK4a gene expression and methylation in primary breast cancer: overexpression of p16INK4a messenger RNA is a marker of poor prognosis. Clin Cancer Res. 2000;6:2777.
114. Yang X, Yan L, Davidson NE. DNA methylation in breast cancer. Endocr Relat Cancer. 2001;8:115.

Chapter 2
Multicentric/Multifocal Breast Cancer: Overview, Biology, and Therapy

Federico Buggi, Annalisa Curcio, Fabio Falcini, and Secondo Folli

Abstract Multiple breast cancers may present with different clinical and biological characteristics as compared with unicentric disease, and in certain instances this may have implications as far as treatment is concerned. Multiple tumors may have increased lymph node involvement compared with unifocal tumors, and some available data suggest that multifocal/multicentric breast cancer is actually more aggressive and carries worse overall outcomes than unifocal disease. In other studies, multifocality itself does not appear to be a contributing factor for worse outcome; more aggressive systemic disease or decreased response to systemic therapies, instead, seem to play a role. It has been suggested that multi- and unifocal tumors do not share the same biology since factors other than those currently employed for staging and prognostic purposes have been shown to affect behavior. In fact, the prognostic impact of multiple breast cancer has been poorly studied, and the necessity for specific adjuvant treatment in order to counteract the potentially unfavorable effect of multifocality is still subject to investigation.

2.1 Introduction

The presence of multiple foci of disease in the cancer-containing breast has been reported during most of the twentieth century [1]. Its significance has been lively discussed over the last 30 years [2], and sequential revisions of the biological and clinical implications of multiple breast cancer foci have followed one another over time; at least in part, the interpretation of multicentric tumors has been paralleled by the contemporary shift in paradigms for the treatment of breast cancer, with mutual influence.

F. Buggi • A. Curcio • S. Folli (✉)
Breast Unit, Morgagni-Pierantoni Hospital, Forl, Italy
e-mail: federicobuggi@hotmail.com; s.folli@ausl.fo.it

F. Falcini
Romagna Cancer Institute (IRST), Meldola, Italy

H. Schatten (ed.), *Cell and Molecular Biology of Breast Cancer*,
DOI 10.1007/978-1-62703-634-4_2, © Springer Science+Business Media New York 2013

2.2 Multifocality: What Definition?

In our opinion, the peculiarity of multifocal/multicentric breast cancer resides in the question whether the intrinsic profile of a certain multiple breast cancer can be extrapolated from the characterization of a single lesion (namely, the major, as is currently assumed) or, otherwise, whether each focus carries its own independent contribution to the overall behavior of the disease. An additional issue is whether multifocality itself may represent an independent biological feature to take into account when outlining a tumor profile.

The early findings by Holland et al. [3] (see below) in terms of high prevalence of multiple cancer foci as well as his depiction of the distribution of multiple lesions over a sort of topographic gradient around the biggest focus were fully confirmed in more recent studies: for example, Vaidya et al. reported [4] that 63 % of patients in their series harbored multiple foci in addition to the primary tumor and, in particular, only 53 % of patients had all foci contained within 2 cm, while 80 % of lesions could be found within 4 cm and 90 % within 5 cm of the main tumor edge. The importance of such a distribution of cancer within the breast has come to be well valorized after breast-conserving surgery became a standard procedure in order to optimize breast conservation [5], and, possibly, it has even been enhanced by the most extreme conservative efforts adopted in most recent times, such as oncoplastic surgery.

However, results in the literature regarding the significance of multiple breast cancer foci are still conflicting, and the impact of multifocality on overall survival is still controversial, mainly due to the use of various definitions lacking a wide international consensus and to methodological limitations [6].

Several definitions have in fact been used in the literature addressing multifocal breast cancer. Some studies required that tumors should be separated from each other by some arbitrarily selected distance or be located in different quadrants of the breast to be considered multifocal. Others included carcinoma in situ in the definition of multifocality [7] or used the term multicentricity as well: e.g., Katz et al. [8] defined "multicentricity" as more than two areas of carcinoma in more than one quadrant of the breast and separated by at least 4 cm, while "multifocality" was defined as two or more separate areas of carcinoma within the same quadrant and/or separated by less than 4 cm. Often multicentricity implies more than one primary tumor, whereas multifocality indicates multiple foci of the same tumor; some researchers operate with a minimum distance between the foci of 2 cm, and others require normal breast tissue between tumors but no minimal distance between the tumor foci [9].

Some authors, on the other hand, stated that the dichotomous classification of multifocal and multicentric disease is arbitrary and could simply represent different degrees of spatial separation rather than any biological difference between categories; breast quadrants are indeed typically defined by tracing lines which radiate perpendicularly from the nipple, but the lack of anatomical boundaries within the breast can result in variability of these "quadrants" with respect to patient position and the modality of assessment (clinical or radiological); therefore, lesions in different quadrants can in fact have closer proximity than lesions occupying remote locations within the same quadrant [10].

It could thus be argued that despite being quite common in practice, categorizing multifocality solely on metric parameters may not be the most effective approach, and, actually, for some authors the definition of multifocality based on the distance between the individual foci appears to impede the ability to obtain comparable and significant results, first, because the biological importance of a distance is hard to test but also because it is impossible to achieve consensus regarding the required amount of millimeters [6].

In one study of cytogenetic changes, macroscopically distinct tumors in 9 of 12 mastectomy specimens were monoclonal, suggesting a common origin [11], and in these monoclonal cases, multiple foci were in closer proximity. Similar findings resulted from a study where a panel of immunohistochemical markers was utilized [12]. These data would need confirmation with contemporary profiling techniques but would suggest that when multiple lesions are in close proximity (multifocal), they are biologically similar, but not so when they are far apart (multicentric) [2]. However, on the practical side, even with molecular studies it can be difficult to differentiate between multifocal tumors, defined as the presence of multiple foci of the same tumor, and multicentric carcinomas, defined as multiple primary carcinomas in the same breast [13], hence many recent studies did not attempt to separate them.

In this vein, it has also been suggested that a classification based on quadrant location and distance from the nipple and rooted in the notion that the "quadrants" of the breast have some anatomic and biological meaning needs to be reconsidered [2]. A trend in this direction has been included among the changes from the sixth to the seventh edition of the AJCC Cancer Staging Manual, where it has been acknowledged that it is not necessary for tumors to be in separate quadrants to be classified as multiple simultaneous ipsilateral carcinomas, providing that they can be unambiguously demonstrated to be macroscopically distinct and measurable using available clinical and pathological techniques [14].

2.3 Multifocality and Breast Conservative Treatment

From approximately 1890 to 1970, radical mastectomy as introduced by William Halsted was generally accepted as the standard treatment for breast cancer [1], and the many studies that were carried out over such period in order to investigate multicentricity in breast cancer were conducted mainly with the purpose of understanding the process of *origin* of multiple cancers: due to the lack of therapeutic alternatives to destructive surgery, the relevance of multifocality in terms of treatment is yet to come. In the mid-1970s, Fisher et al. effectively summarized the findings yielded so far by the studies in the domain of breast cancer multicentricity when they stated that "the detection of multicentric cancers in mammary quadrants other than that harboring the primary cancer in mastectomy specimens represents a conservative estimate since the probability of identifying such a lesion appears to increase as the number of [tissue] samples [examined] per patient increases" [15].

Actually, the incidence of multicentric foci in the breast has been reported to vary from 18 %, when 1–2 random samples from each quadrant were examined [16], to 69 % when 5 mm sections of whole breast were examined [17].

Some studies that addressed the topic of multiple breast cancer [18] used radiography, but it was Egan [17] who standardized the "correlated pathological-radiological" method of whole-organ analysis, which provides optimum sampling of breast tissue. Many studies conducted with Egan's technique on this matter over the years reported rates of multiple lesions ranging from 56 % [19] to 69 % [17].

In the 1970s, the major debate in the local therapy of breast cancer was the safety of the switch from radical mastectomy to modified radical mastectomy. In that environment, the National Surgical Adjuvant Breast and Bowel Project (NSABP)-B04 trial [15] studied the contribution to survival deriving from the removal of the axillary nodes in clinically node-negative women: its results eventually led to the direct repudiation of the Halstedian concept of breast cancer biology and opened the door to studies that tested the effectiveness of breast conservation treatment.

Over the course of the studies on breast conservation, the new option of partial removal of the breast coupled with the long-standing awareness of high frequency of additional cancer foci beyond an index lesion focused concerns upon the eventuality of local recurrences and brought into focus the importance of assessing the extent and the topographical distribution of tumor foci in the surroundings of an overt cancer.

In a landmark paper on multifocality, Holland et al. [3] examined with Egan's method a consecution series of breast cancers that appeared clinically and radiologically unifocal and that constituted virtual candidates to conservative treatments; the actual presence of additional undetected cancer foci beyond the index lesion was studied in mastectomy specimens of these cases with the purpose to estimate the frequency with which tumor would remain in the breast after a breast-conserving surgical intervention.

Their results showed that only a minority of tumors had their clinical unifocality confirmed in the surgical specimen, and actually a 63 % rate of multicentricity was reported; moreover, the authors addressed the additional issue of the spatial distribution of clinically undetected foci in terms of their distances from the primary tumor, and the results showed that the likelihood of finding additional tumor foci in breast tissue decreased as the distance from the index lesion increased, despite the size of the index tumor.

In other words, assuming the distance of additional tumor foci from the index lesion as a surrogate for the surgical margin during breast-conserving surgery for infiltrating carcinoma, the percentage of patients still harboring tumor foci following excision would be higher with a margin of 2 cm rather than 4 cm (42 % and 10 %, respectively, according to Holland). Therefore, the emerging scenario was that an average quadrantectomy on an average-sized breast was unlikely to remove all cancer foci: earlier concerns "that local excision may ignore residual tumors, particularly those which may occur as clinically and pathologically undetected de novo cancers at sites within the breast quite remote from the dominant mass" [15] still represented a deterrent to the acceptability of breast-conserving surgery.

At this stage it was difficult to scientifically accept the possibility that a cancer might not be a cancer of clinical significance and there was a need for information regarding the kinetics of the multicentric foci. In 1969 a randomized study to compare radical mastectomy with breast-conserving surgery, which was termed "quadrantectomy," was approved by the World Health Organization Committee of Investigators for Evaluation of Methods of Diagnosis and Treatment of Breast Cancer; after the new procedure was standardized, the recruitment of patients began at the Milan Cancer Institute in 1973, and preliminary data showed that survival rates were equal after radical and breast-conserving surgery [20]. A few years later, another randomized controlled clinical trial—NSABP-06 [21]—was conducted in order to evaluate the efficacy of breast-conserving surgery and the biological importance of tumor multicentricity: after 20 years of follow-up, the absence of a significant difference in overall survival among women who underwent mastectomy and those who underwent conservation treatment was confirmed. Besides, the rate of ipsilateral breast tumor recurrence, as well, did not differ significantly among the two groups [22]. Over time, the association of breast-conserving surgery with radiotherapy has proved to grant patients equivalent survival with respect to mastectomy, as pointed out by six prospective randomized trials with long-term follow-up, some more than 20 years [22].

Moreover, adjuvant treatments have been extensively employed and refined with increasing success: in a report of 3,799 node-negative women participating in five NSABP trials of adjuvant systemic therapy, the cumulative incidence of in-breast recurrence at 12 years for those receiving adjuvant therapy was only 6.6 % [23].

In summary, the persistence of tumor foci after breast conservation treatment was known for a fact, but, nevertheless, the observation of increasing survival rates and a parallel decrease in local recurrence due to improvements in adjuvant treatments and refinements of diagnostic tools led to the replacement of concerns about the mere presence of remnant cancer cells by issues concerning the specific biological feature of the (remaining) disease.

The importance of biology and targeted therapy has been supported by the emerging literature on the impact of tumor subtypes on local recurrence after BCT or mastectomy. Both Millar et al. [24] and Nguyen et al. [25] demonstrated that the rate of local recurrence after BCT varies among the intrinsic subtypes of breast cancer as approximated by the ER, progesterone receptor (PgR), and human epidermal growth factor receptor (HER)-2 status.

2.4 Multifocality and Outcome

When it comes to considering outcomes of multiple tumors, overall survival and other prognostic factors (that may in turn indirectly affect overall survival) should deserve distinct consideration.

An insight of the appropriateness in this separation comes from a study published by Pedersen et al. [9], which also offers an effective example of mismatches in results due to differences in definitions and methodologies. Pedersen et al. reported that in a study from 1982 [17], Egan found that the presence of multifocality signaled a worse prognosis, while in their own investigation, it was not found to have an independent effect on overall survival, when controlling for known prognostic factors. The findings Pedersen reported were in accordance with the data published by Rakowsky et al. [26], who also found that multifocality had no influence on disease-free survival, and those of Vlastos et al. [27], who found in a set of patients with early-stage cancer that patients with unicentric tumors had a 10-year disease-free survival of 84 % and the patients with multicentric tumors had a 10-year disease-free survival of 83 %. The differences in the conclusions reported by Egan and Pedersen could be explained by differences in the criteria adopted for defining multifocality (Egan had a broader definition of multifocality including lobular carcinoma in situ) and by the fact that the former did not use a multivariate analysis to find the prognostic influence of multifocality; in fact, when comparing the prognostic influence of multifocality against other prognostic factors in a Cox multivariate analysis, multifocality lost its independent prognostic influence. Patients with multifocal disease often had more positive lymph nodes and larger tumors than patients with unifocal cancers: these two factors are known to be strong prognostic factors, and this can explain why multifocality appeared as a prognostic factor in the univariate analysis and why it had no significant effect on overall survival in the Cox multivariate analysis.

Actually, on a more general level, multiple carcinomas have been repeatedly reported to carry a higher frequency of lymph node metastases and a less favorable patient outcome when compared with unifocal lesions; unfortunately, it is still unclear whether this difference reflects a different biological behavior (which could be responsible for the multifocality as well) or merely larger tumor burden [13].

Other studies led to depict a multifaceted scenario as far as outcome of multiple breast cancer is concerned. Litton [28] studied a subset of young women with breast cancer (<35 years old), and multifocality by itself did not prove to worsen the initially poor prognosis of young breast cancer patients. Multifocal disease was associated with an increase in the risk of death (hazard ratio, 1.57) and decrease in the risk of recurrence (hazard ratio, 0.87), but did not reach statistical significance. Multifocality was instead statistically associated with an increase in the risk of death after recurrence (HR, 3.71). There were, however, statistically significant differences when looking at specific biological features, including pathological tumor grade, hormone receptor status, DCIS, and oral contraceptive use. For those women who did recur a more aggressive systemic disease, a decreased response to systemic therapies was hypothesized, and other biological factors or genetic profiling that might help explain these differences beyond multifocality itself were claimed to play a role.

Similarly, Cabioglu [29] reported that no difference could be found between patients with unifocal or multiple breast cancer in terms of rates of systemic metastases, local recurrences, and disease-free or overall survival when patients were tabulated according to the stage, but their findings led the authors to conclude that breast tumors with multiple foci have a different biology, with an increased metastatic potential to axillary lymph nodes, regardless of tumor size (that reflects an advanced stage).

Theoretically, the difference in behavior may be a reflection of either the increased intrinsic aggressiveness of multiple tumors or their larger tumor size, in keeping with a consistent biological relation between tumor burden and angiolymphatic dissemination. In fact, the prognostic relevance of the T category has been extensively established, but survival of patients with breast cancer depends on two different types of prognostic factors: tumor size as a marker of tumor biology (as a time-dependent phenomenon) and biological factors (i.e., histological grade, the estrogen and progesterone receptor status, as well as the number of mitotic figures per ten high-power fields), which represent tumor aggressiveness [30].

2.5 Assessing Tumor Burden in Multiple Foci

Unfortunately, evaluating the burden of breast carcinomas is subject to several problems; routinely, the diameter of tumor nodules is used, primarily for practical reasons, but tumors are in fact variably and irregularly shaped tridimensional objects, and therefore diameters inaccurately reflect their real size. Nevertheless, the largest tumor diameter is currently used as an approximation (or surrogate) of the tumor volume for staging purposes in cases of unifocal breast carcinoma, and, in the same line, the American Joint Committee on Cancer and the International Union Against Cancer (AJCC/UICC) recommend the usage of the diameter of the largest tumor (only) also for the staging of multifocal/multicentric breast carcinomas [31]. Therefore, in practice, for the purpose of obtaining a simple and consistent measurement, the actual tumor burden is underestimated because secondary tumor foci, which are often sizable, are not included [32].

Andea et al. [32] hypothesized that the propensity of multifocal/multicentric tumors for metastasis is best described as a function of aggregate tumor size. This prompted the authors to explore the relation between tumor size and lymph node involvement in multifocal/multicentric tumors by using aggregate tumor size estimates, and the findings were used to investigate whether the current staging criteria optimally reflect the metastatic behavior of multifocal breast carcinomas.

Two different methods for estimating tumor size in multifocal/multicentric carcinomas (i.e., diameter of the largest nodule and combined diameters) were used, and the two methods resulted in statistically significant differences in both size and

T classification distribution when compared with unifocal cases. Using the diameter of the largest nodule as a size estimate produced a lower mean tumor size for multifocal than for unifocal cases (2.53 vs. 3.47 cm, respectively; $P=0.026$), and, conversely, combining the diameters of multifocal/multicentric tumors resulted in a larger mean tumor size compared with unifocal lesions (4.2 vs. 3.47 cm, respectively; $P=0.003$). Consequently, the multifocal tumors had a different distribution within T classifications depending on the method of tumor size estimation.

However, as far as the relation between T-stage and lymph node involvement is concerned, when utilizing the standard tumor size estimate, multifocal/multicentric tumors were demonstrated to have a significantly higher incidence of axillary lymph node metastases than unifocal tumors of similar size (odds ratio, 2.8; $P=0.0001$), but if combined diameters of all tumor nodules were utilized for the T-stage assessment, the metastatic behavior of multifocal carcinomas was not significantly different from that of unifocal tumors (odds ratio, 1.4; $P=0.13$). A multivariate logistic regression model was analyzed assessing the impact of multifocality versus unifocality on lymph node status when controlling for tumor size, and results showed that multifocality did not significantly influence lymph node status for the same tumor size when a combined diameter is used as a tumor size estimate.

In other words, results by Andea et al. [32] confirmed that, within similar T classification groups, the currently used measurement methods for staging multifocal carcinomas (diameter of the largest nodule) resulted in a significantly higher incidence of positive lymph node status in multiple tumors as compared to unifocal tumors; on the other hand, using the combined diameters as a size estimate resulted in frequencies of positive lymph nodes that did not significantly differ from the unifocal control series. In particular, the most prominent change was observed in T1 classification, where the incidence of lymph node positivity for multifocal and unifocal series became equal.

2.6 Tumor Burden Versus Lymphatic Metastases

As the authors acknowledge, one potential criticism in this study is that it uses tumor diameters when, more likely, the propensity for metastases is a function of tumor volume or surface area. In multiple tumors, adding diameters of nodules in an attempt to estimate total tumor bulk would result in a consistent error because volumes and areas are proportional to the third and second power of the diameter, respectively, and therefore summing diameter of tumor nodules would overestimate total tumor volume and, to a lesser extent, area. Thus, the same author pushed further his research [13] and quantified the relationship between tumor volume and area in multifocal tumors as compared to single cancers. The results confirmed former findings and showed that multifocal tumors have a significantly larger aggregate diameter, but they have a lower median volume and a similar distribution of tumor surface area than unifocal tumors of similar stage. So it was concluded that aggregate diameter measurements would actually overestimate the total volume of multiple tumors, but such an approach was sustained because, even though summing diameters of multifocal tumors "overcorrects" for volume, this inadvertently accounts for their increased biological aggressiveness.

Based on these data, the authors suggested that alternative T classification algorithms for multiple breast carcinomas should be investigated; in their opinion the utilization of aggregate diameter measurements would allow multifocal/multicentric and unifocal tumors to be staged uniformly, although this may not reflect accurately the total tumor burden. Alternatively, multiple tumors could be designated in a separate T classification to convey the increased risk for metastatic dissemination.

This sort of concern was addressed by the College of American Pathologists and in the protocol for the examination of specimens from patients with invasive carcinoma of the breast, based on the 7th edition of the AJCC cancer staging manual [33], where the size of the largest invasive carcinoma is still used for T classification but, if multiple carcinomas are present, the modifier "m" is used in the assessment of the "T" stage to indicate that multiple foci are present.

Others have addressed the issue that, currently, in multicentric/multifocal disease the size of the tumor is assessed by measuring the largest tumor focus only. The aim of a study of Weissenbacher et al. [30] was to compare the prognosis of multicentric/multifocal tumors with unifocal tumors with apparently identical tumor size according to TNM staging. A total of 288 pairs, each consisting of one patient with unifocal disease and one patient with multifocal or multicentric breast cancer, were created by matched-pair analysis to achieve statistical balance of the major prognostic factors between both groups. All match criteria (tumor size, grading, and hormone receptor status) were equally distributed in both sets, and, furthermore, no significant difference was found between the two groups in terms of systemic therapy and primary operation. The Cox multivariate regression analysis regarding breast cancer-specific survival and local or systemic relapse showed that multicentricity/multifocality is a significant independent predictor of reduced breast cancer-specific survival as well as of reduced relapse-free survival. Tumor size, grading, and lymph node status were also significant independent predictors. As far as overall survival is concerned, hormonal therapy resulted in significance ($P=0.002$), while both chemotherapy and radiation showed no statistical significance by multivariate analysis. Concerning disease recurrence, neither chemotherapy, hormonal therapy, nor radiation showed any significance.

In order to explain their findings, the authors stated that the currently used algorithms, which employs the diameter of the largest nodule, result in the downplaying of multifocal breast carcinomas due to the underestimation of actual tumor size. They concluded that failure to measure the additional tumor burden provided by multiple small foci may underestimate the disease, and, besides, ignoring the contribution of the smaller foci to the incidence of node positivity and survival may deny patients the opportunity of adjuvant therapies.

In summary it appears that, even when biases due to obsolete topographic definitions are left aside, foreseeing multiple breast cancer prognosis still poses peculiar difficulties that may reside in the incomplete adequacy of the presently used staging systems when applied to multiple disease and in the practice of neglecting smaller tumor foci, each one carrying its own biological features.

2.7 Future Perspectives: Biological Determinants of Aggressiveness

The sensitivity of some breast cancers to hormones and the possibility of reducing the growth of these tumors by removal of circulating estrogens have been known for over 100 years; quantitatively, it is reported that 50–80 % of breast cancers are ER positive, thus hormone therapy has possibly prognostic relevance for a significant proportion of patients [34]. However, as mentioned above, in cases of multiple breast cancer current guidelines recommend that the highest T category tumor should be the one selected for classification and staging. The reported grade corresponds to the largest area of invasion and ER, PgR, and HER2 status are determined solely on the largest invasive carcinoma; biological tests on smaller invasive carcinomas are recommended only if these cancers are of different histological type or of higher grade [33]. Smaller cancers thus tend to be ignored, and homogeneity among different cancer foci is assumed.

Among papers that created grounds for such an attitude, a pilot study [34] involving 18 patients showed that ER status was the same in all foci of multiple breast cancers and therefore concluded that the current practice of establishing ER status in the primary focus is adequate in relation to hormone therapy. Yet, as acknowledged by the authors themselves, the variability of ER and PgR status between individual foci in multiple breast cancer has not been widely studied, and, besides, it has been previously demonstrated by allelotyping that multiple breast lesions may exhibit distinct clonal patterns at the molecular level [35, 36].

In fact, some authors described quite a different picture. Poulsen et al. [37] presented a case report in 1981 of a patient with two foci of invasive ductal breast cancer in the same breast: one tumor was ER positive, while the other was ER negative. Panahy et al. [38] studied hormone receptor distribution in normal and cancerous breast tissue from nine patients harboring ER-positive cancers and identified—in four patients—multifocal tumors with a varying phenotype for ER and PgR status; multiple tumors were reported to be somehow different in regard to histological criteria, too, but no further detail on the histology of foci was reported. In one case, all three cancer foci were ER positive/PgR positive, but other cases showed multifocal tumors of more than one soluble receptor phenotype. One case had both ER-positive/PgR-positive and ER-positive/PgR-negative tumors, while another case had an ER-positive/PgR-negative and an ER-negative/PgR-positive tumor. A third instance combined two distinct ER-positive/PgR-positive tumors with, remarkably, a second phenotype (ER negative/PgR negative) on a further focus. Interestingly, in addition, variability was also seen in different regions of large tumors.

In order to analyze whether biological features that play a role in the choice of adjuvant treatment of breast cancer are differently expressed in distinct foci of invasive multiple breast cancers with a single histological feature, we prospectively studied the expression of biological markers connected with adjuvant therapy and prognosis over a series of 113 cases [39]. In particular, ER and PgR status, proliferative index Ki-67 (measured as Mib-1 staining), and the amplification of HER2 were

Table 2.1 Overview of diverging expression of biological parameters among different foci in 113 multiple breast cancers (modified from Buggi et al. [39])

	ER-positive minor focus/ ER-negative main focus	ER-negative minor focus/ ER-positive main focus	Total
Divergent ER status	2 cases (1.7 %)	3 cases (2.5 %)	5 (4.4 %)
	PgR-positive minor focus/ PgR-negative main focus	PgR-negative minor focus/ PgR-positive main focus	
Divergent PgR status	10 cases (8.8 %)	8 cases (7.7 %)	18 (15.9 %)
	High-grade minor focus/ low-grade main focus	Low-grade minor focus/ high-grade main focus	
Divergent grading	3 cases (2.6 %)	18 cases (15,9 %)	21 (18.6 %)
	"High" in minor focus/ "low" in main focus	"Low" in minor focus/ "high" in main focus	
Divergent Ki-67 staining	8 cases (7 %)	9 cases (7.9 %)	17 (15 %)
	Amplified in minor focus Not amplified in main focus	Not amplified in minor focus Amplified in main focus	
Divergent HER2 expression	4 cases (3.5 %)	7 cases (6.2 %)	11 (9.7 %)

ER estrogen receptor, *PgR* progesterone receptor

assessed. The expression of all these features was prospectively assessed in each, and every tumor focus of multiple lesions and mismatches among foci were present in 4.4–18.6 % of cases, according to the parameter considered (see Table 2.1).

In detail, mismatches in ER status were present in 4.4 % of cases: in two cases (1.7 %) out of five, a smaller focus was positive and the main focus was negative, while in the remaining three cases (2.5 %), a smaller focus was negative and the main focus was positive. For PgR status, mismatches were present in 18 cases (15.9 %). For tumor grading, 21 (18.6 %) mismatches were found, with a minor focus displaying a higher grade in comparison to the main nodule in 3 cases (2.6 %). Proliferative index Ki-67 differed in 17 (15 %) cases, with eight cases (7 %) in which a "high" index was reported in minor foci only. A mismatch in HER2 amplification was present in 11 (9.7 %) cases showing amplification exclusively in a minor focus in four cases (3.5 %).

The specific biological parameters were chosen because they are known as key elements in planning adjuvant treatments and their specific combination is an issue in the indication to hormonal therapy, chemotherapy, target therapy, or associated treatments [40, 41]; actually, in our series, extending the biological characterization to each tumor focus led to identifying heterogeneous characteristics over the foci and therefore to issuing different indications to adjuvant treatment in 14 (12.4 %) patients out of 113 as compared with what would have been prescribed if the status of the main focus only were taken into account.

2.8 Closing Remarks

Findings such as those reported in the experiences cited above need consistent confirmation, but nevertheless they allow to obtain some insight into biological heterogeneity among foci of multiple breast cancer. As previously claimed, the design of a rational therapeutic strategy for breast cancer should begin with a clear understanding of the biological basis of multicentricity and multifocality; given this, questions of definition and therapeutic strategy would follow logically [2].

Future efforts will be required in order to confirm preliminary data, but if present findings will prove consistent, a thorough assessment over all tumor foci of biological features that play a role in the adjuvant treatment decision-making process may eventually lead to optimal therapy tailoring.

References

1. Fisher B. Role of science in the treatment of breast cancer when tumor multicentricity is present. J Natl Cancer Inst. 2011;103:1292–8. doi:10.1093/jnci/djr240.
2. Khan S. The many questions that surround multicentric and multifocal breast cancer. Breast J. 2010;16(3):219–21. doi:10.1111/j.1524-4741.2010.00929.x.
3. Holland R, et al. Histologic multifocality of Tis, T1-2 breast carcinomas. Cancer. 1985;56:979–90.
4. Vaidya JS, Vyas JJ, Chinoy RF, Merchant N, Sharma OP, Mittra I. Multicentricity of breast cancer: whole-organ analysis and clinical implications. Br J Cancer. 1996;74:820–4.
5. Connolly JL, Harris JR, Schnitt SJ. Understanding the distribution of cancer within the breast is important for optimizing breast-conserving treatment [editorial]. Cancer. 1995;76:1–3.
6. Tot T. Clinical relevance of the distribution of the lesions in 500 consecutive breast cancer cases documented in large-format histologic sections. Cancer. 2007;110(11):2551–60. doi:10.1002/cncr.23052.
7. Luttges J, Kalbfleisch H, Prinz P. Nipple involvement and multicentricity in breast cancer. J Cancer Res Clin Oncol. 1987;113:481–7.
8. Katz A, Strom E, Buchholz T, Theriault R, Singletary E, McNeese M. The influence of pathologic tumor characteristics on loco regional recurrence rates following mastectomy. Int J Radiat Oncol Biol Phys. 2001;50(3):735–42.
9. Pedersen L, Gunnarsdottir KA, Rasmussen BB, Moeller A, Lanng C. The prognostic influence of multifocality in breast cancer patients. Breast. 2004;13:188–93. doi:10.1016/j.breast.2003.11.004.
10. Patani N, Carpenter R. Oncological and aesthetic considerations of conservational surgery for multifocal/multicentric breast cancer. Breast J. 2010;16(3):222–32. doi:10.1111/j.1524-4741.2010.00917.x.
11. Teixeira MR, Pandis N, Bardi G, et al. Discrimination between multicentric and multifocal breast carcinoma by cytogenetic investigation of macroscopically distinct ipsilateral lesions. Genes Chromosomes Cancer. 1997;18:170–4.
12. Dawson PJ, Baekey PA, Clark RA. Mechanisms of multifocal breast cancer: an immunocytochemical study. Hum Pathol. 1995;26:965–9.
13. Andea AA, Bouwman D, Wallis T, Visscher DW. Correlation of tumor volume and surface area with lymph node status in patients with multifocal/multicentric breast carcinoma. Cancer. 2004;100:20–7. doi:10.1002/cncr.11880.

14. American Joint Committee on Cancer. Summary of Changes: Understanding the Changes from the Sixth to the Seventh Edition of the AJCC Cancer Staging Manual. http://www.cancerstaging.org/staging/changes2010.pdf. Accessed 30 April 2012.
15. Fisher ER, Gregorio R, Redmond C, et al. Pathologic findings from the National Surgical Adjuvant Breast Project (Protocol No. 4). Observations concerning the multicentricity of mammary cancer. Cancer. 1975;35(1):247–54.
16. Spinelli C, Berti P, Ricci E, Miccoli P. Multicentric breast tumour: an anatomical-clinical study of 100 cases. Eur J Surg Oncol. 1992;18:23–6.
17. Egan RL. Multicentric breast carcinomas: clinical-radiographic-pathologic whole organ studies and 10-year survival. Cancer. 1982;49:1123–30.
18. Gallager HS, Martin JE. The study of mammary carcinoma by mammography and whole organ sectioning. Early observations. Cancer. 1969;23:855–73.
19. Lagios MD. Multicentricity of breast carcinoma demonstrated by routine correlated serial subgross and radiographic examination. Cancer. 1977;40:1726–34.
20. Veronesi U, Cascinelli N, Mariani L, et al. Twenty-year follow-up of a randomized study comparing breast-conserving surgery with radical mastectomy for early breast cancer. N Engl J Med. 2002;347(16):1227–32.
21. Fisher B, Anderson S, Bryant J, et al. Twenty-year follow-up of a randomized trial comparing total mastectomy, lumpectomy, and lumpectomy plus irradiation for the treatment of invasive breast cancer. N Engl J Med. 2002;347(16):1233–41.
22. Ho A, Morrow M. The evolution of the locoregional therapy of breast cancer. Oncologist. 2011;16:1367–79.
23. Anderson SJ, Wapnir I, Dignam JJ, et al. Prognosis after ipsilateral breast tumor recurrence and locoregional recurrences in patients treated by breast-conserving therapy in five national surgical adjuvant breast and bowel Project protocols of node-negative breast cancer. J Clin Oncol. 2009;27:2466–73. doi:10.1200/JCO.2008.19.8424.
24. Millar EK, Graham PH, O'Toole SA, et al. Prediction of local recurrence, distant metastases, and death after breast-conserving therapy in early-stage invasive breast cancer using a five-biomarker panel. J Clin Oncol. 2009;27:4701–8.
25. Nguyen PL, Taghian AG, Katz MS, et al. Breast cancer subtype approximated by estrogen receptor, progesterone receptor, and HER-2 is associated with local and distant recurrence after breast-conserving therapy. J Clin Oncol. 2008;26:2373–8.
26. Rakowsky E, Klein E, Kahan E, Derazne E, Lurie H. Prognostic factors in node-positive breast cancer patients receiving adjuvant chemotherapy. Breast Cancer Res Treat. 1992;21(2):121–31.
27. Vlastos G, Rubio IT, Mirza NQ, Newman A, Aurora R, Alderfer J. Impact of multicentricity in patients with T1-2, N0-1, M0 breast cancer. Ann Surg Oncol. 2000;7(8):581–7.
28. Litton JK, Eralp Y, Gonzalez-Angulo AM, Broglio K, Uyei A, Hortobagyi GN, Banu Arun B. Multifocal breast cancer in women <35 years old. Cancer. 2007;110:1445–50. doi:10.1002/cncr.22928.
29. Cabioglu N, Ozmen V, Kaya H, Tuzlali S, Igci A, Muslumanoglu M, Kecer M, Dagoglu T. Increased lymph node positivity in multifocal and multicentric breast cancer. J Am Coll Surg. 2009;208:67–74. doi:10.1016/j.jamcollsurg.2008.09.001.
30. Weissenbacher TM, Zschage M, Janni W, Jeschke U, Dimpfl T, Mayr D, Rack B, Schindlbeck C, Friese K, Dia D. Multicentric and multifocal versus unifocal breast cancer: is the tumor-node-metastasis classification justified? Breast Cancer Res Treat. 2010;122:27–34. doi:10.1007/s10549-010-0917-9.
31. Sobin LH, Gospodarowitz MK, Wittekind CH, editors. Union for International Cancer Control. TNM classification of malignant tumours. 7th ed. New York, NY, USA: Wiley-Blackwell; 2009.
32. Andea AA, Wallis T, Newman LA, Bouwman D, Dey J, Visscher DW. Pathologic analysis of tumor size and lymph node status in multifocal/multicentric breast carcinoma. Cancer. 2002;94(5):1383–90. doi:10.1002/cncr.10331.
33. Lester SC, Shikha Bose S, Chen YY, et al. Protocol for the examination of specimens from patients with invasive carcinoma of the breast. Arch Pathol Lab Med. 2009;133:1515–38.

34. Garimella V, Long ED, O'Kane SL, Drew PJ, Cawkwell L. Oestrogen and progesterone receptor status of individual foci in multifocal invasive ductal breast cancer. Acta Oncol. 2007;46:204–7. doi:10.1080/02841860600897884.
35. Tse GM, Kung FY, Chan AB, Law BK, Chang AR, Lo KW. Clonal analysis of bilateral mammary carcinomas by clinical evaluation and partial allelotyping. Am J Clin Pathol. 2003;120:168–74.
36. Smeds J, Warngerg F, Norberg T, Nordgren H, Holmberg L, Bergh J. Ductal carcinoma in situ of the breast with different histopathological grades and corresponding new breast tumour events: analysis of loss of heterozygosity. Acta Oncol. 2005;44:41–9.
37. Poulsen HS, Jensen J, Hermansen C. Human breast cancer: heterogeneity of estrogen binding sites. Cancer. 1981;48:1791–3.
38. Panahy C, Puddefoot JR, Anderson E, Vinson GP, Berry CL, Turner MJ, Brown CL, Goode AW. Oestrogen and progesterone receptor distribution in the cancerous breast. Br J Cancer. 1987;55:459–62.
39. Buggi F, Folli S, Curcio A, Casadei-Giunchi D, Rocca A, Pietri E, Medri L, Serra L. Multicentric/multifocal breast cancer with a single histotype: is the biological characterization of all individual foci justified? Ann Oncol. 2012;23:2042–46. doi:10.1093/annonc/mdr570.
40. Goldhirsch A, Ingle JN, Gelber RD, et al. Thresholds for therapies: highlights of the St Gallen International Expert Consensus on the primary therapy of early breast cancer 2009. Ann Oncol. 2009;20:1319–29. doi:10.1093/annonc/mdp322.
41. Goldhirsch A, Wood WC, Coates AS, Gelber RD, Thurlimann B, Senn HJ, Panel members. Strategies for subtypes—dealing with the diversity of breast cancer: highlights of the St Gallen International Expert Consensus on the Primary Therapy of Early Breast Cancer 2011. Ann Oncol. 2011;22:1736–47. doi:10.1093/annonc/mdr304.

Chapter 3
The Immune System in Breast Cancer Initiation and Progression: Role of Epithelial to Mesenchymal Transition

Keith L. Knutson and Derek C. Radisky

Abstract The mammary gland is a complex organ necessary for providing nutrition and immunity to the newborn. Understanding the normal physiology and architecture of the mammary gland is essential because it is an anatomic site that is often diseased in humans. Cancer remains one of the dominant diseases detected in breast tissue. Worldwide, breast cancer is responsible for over 400,000 deaths annually. Despite our tremendous efforts, the molecular and cellular pathways to initiation and progression of breast cancer remain poorly understood. What has become increasingly clear is the microenvironment in the normal and malignant breast can have a profound influence on malignancy. While there are data that suggest that one component of the microenvironment, the immune system, protects against breast cancer, other data support a pathological role. In this review the authors take a comprehensive approach in defining this paradoxical, double-edged, role of the immune system.

K.L. Knutson (✉)
Vaccine and Gene Therapy Institute of Florida, 9801 SW Discovery Way,
Port St. Lucie, FL 34987, USA

Department of Immunology, Mayo Clinic, Rochester, MN 55906, USA
e-mail: kknutson@vgtifl.org

D.C. Radisky
Department of Biochemistry and Molecular Biology, Mayo Clinic,
4500 San Pablo Road, Jacksonville, FL 32224, USA
e-mail: radisky.derek@mayo.edu

H. Schatten (ed.), *Cell and Molecular Biology of Breast Cancer*,
DOI 10.1007/978-1-62703-634-4_3, © Springer Science+Business Media New York 2013

3.1 Introduction

The human breast is composed of a bilayered epithelial structure in which luminal epithelial cells are surrounded by basal myoepithelial cell layer. It is thought that the vast majority (~80 %) of breast cancers, whether $ER\alpha^+$ or $ER\alpha^-$, develops from luminal epithelium of the terminal duct lobular units (TDLUs) [1–3]. This hypothesis is consistent with recent work from Keller and colleagues who showed that transformation of luminal $EpCAM^+$ progenitors resulted in the formation of the most common forms of breast cancer including $ER\alpha^+$ or $ER\alpha^-$ with luminal- and basal-like characteristics, respectively [4]. In contrast, transformation of basal (i.e., myoepithelial) progenitors resulted in the development of rare metaplastic tumors with the claudin-low phenotype. Rather than sudden oncogenic transformation of the epithelium, the prevalent model of breast cancer development involves a progression through a continuum of tissue events that begin as benign breast lesions before eventually developing into malignancy [5]. This progressive model suggests that in situ carcinoma and subsequent breast cancer are preceded by at least two stages of proliferative changes: benign or usual proliferation and atypical ductal hyperplasia. This is based on molecular pathology studies that showed that benign proliferative disease and atypical ductal hyperplasia (ADH) are more similar to normal healthy tissue than to either in situ disease or invasive breast cancer, suggesting that unknown key molecular events occur with the development of in situ disease leading to abnormal cell phenotypes that look identical to invasive breast cancer but have not yet assumed the full malignant phenotype. With recent comparative genomics and expression profiling studies, this model has been challenged in recent years. For example, Lopez-Garcia and colleagues suggest that only about 50 % of $ER\alpha^+$ breast cancers originate from low-grade benign lesions and that the other 50 % originate as de novo cancers (assuming that in situ disease represents cancer), and moreover that the HER-2/neu and basal subtypes of breast cancer also develop as de novo cancers rather than progressing from low-grade proliferative intermediates, given that these breast cancer subtypes preferentially present as high-grade (III) lesions [6]. Whether forming from apparently normal epithelium as de novo cancers or progressing through precursor benign stages, it is clear that breast cancer encompasses a variety of diseases consisting of invasive tumors that exhibit a wide spectrum of histologic and molecular subtypes. This heterogeneity in subtypes suggests that there are multiple steps along the pathway to malignancy. In recent years it has become clear that the nonneoplastic components of the tumor microenvironment can play a significant role in cancer initiation and progression. These components include the extracellular matrix (ECM), cells that comprise the vasculature, stromal fibroblasts, and cells that comprise the immune system (myeloid and lymphoid). It was originally thought that many of these nonneoplastic microenvironmental components were passive enablers, but in recent years it has become clear that they can be disease "inducers" and "progressers" [7]. Thus, in addition to genetic abnormalities in the neoplastic cell itself, one could envision that the cells of the microenvironment, given their diversity, could be responsible for the heterogeneity of this frequent disease. In the present review, we focus on the role of the immune system in breast cancer initiation and progression.

3.2 Inflammation in Initiation and Progression of Breast Cancer

3.2.1 As a Mucosal Tissue, the Mammary Gland and the Immune System Are Naturally Linked

The mammary gland is a modified sweat gland and unlike other prominent organs has no capsule or sheath. The gland lies in the superficial fascia, anterior to the thorax, and the deeper aspects of the breast are separated from the pectoral muscles by the deep fascia. Mammary glands are ubiquitous throughout mammals. With the notable exception of the human and some ruminants, the glandular mass is typically present within subcutaneous fat and diffusely spread over the fascia of the pectoral muscles, making observation difficult. In many species, including the mouse, a useful experimental model, the glandular mass can spread over the abdominal muscles as well as to the sides and back of the animal. The mammary glands of humans are contained entirely within the hemispheres of the pectoral fatty tissue. The amount of the glandular tissue is fairly homogenous, in contrast to the fatty tissue which is the primary determinant of breast size.

The mammary glands are accessory organs of the female reproductive system but are also present in males. Although varying in size and complexity throughout the life of an individual, the gland develops in the embryo and is first observed in the first trimester of life. The primary function of the gland is lactation, the secretion of milk for the nourishment of the infant; lactation also provide for maternal transfer of immune effectors such as immunoglobulins which confer immunity against pathogenic organisms. Each gland consists of 15–20 lobules of glandular tissue, each equipped with a lactiferous duct that extends to the nipples. Each lobule consists of many alveoli, hollow cavities, which are several millimeters in thickness. Alveoli are saclike structures where milk is synthesized. The lumen of the alveolus is lined by a single layer of secretory epithelial cells, which is surrounded by contractile myoepithelial cells, which aid in the injection of milk into the ducts. This bilayered structure is wrapped in a specialized extracellular matrix structure called the basement membrane, which separates the epithelial glandular tissue from the surrounding stroma, which is composed of stromal fibroblasts that support capillary beds which provide nutrients and oxygen to the tissue, adipocytes that provide lipids for milk production, and vasculature. Because milk production is only required at defined times during the life of the individual, the amount of glandular tissue can be highly variable being divided into four physiologic stages, resting, pregnant, lactating, and involution (pregnancy associated).

Although often overlooked, in addition to the glandular, stroma, and vascular cells, the mammary gland also contains a significant number of cellular immune effectors making it similar to other secretory organs, such as the gastrointestinal tract. Integration of the immune system into the mammary gland is essential for at least three reasons. First, secretory IgA from breast milk exhibits specificity for an array of common intestinal, as well as respiratory pathogens [8, 9]. The secretory antibodies in breast milk are thus highly targeted against infectious agents likely to be encountered by the infant during its first weeks of life when both innate and adaptive immunity remain underdeveloped. Secondly, the mucosal immunity (both antibody and T cells) protects the gland from spontaneous and breast-feeding-associated microbial (usually *S. aureus*) colonization (i.e., mastitis) [10]. Lastly, there is some evidence suggesting that immune cells and effectors (e.g., cytokines, macrophages, neutrophils) are involved in mammary gland development, post-lactational involution, and maintenance of glandular structure and function [11–13]. Dramatic changes in cellular composition occur in the mammary gland between stages of pregnancy, lactation, and post-lactation involution and menopause-associated involution. While the observation that immunodeficient mice are able to develop mammary glands and to nurse mice suggests that the adaptive immune system is not absolutely required, the fact that immunodeficient mice develop breast adenomas and cancers more frequently than their immunocompetent counterparts suggests that the immune system plays a critical neoplastic role [14].

There are two immune compartments in mucosal tissues, an initiation compartment and an effector compartment [11]. The initiation compartment consists of structures not unlike normal lymph nodes. There are also lymph node-like structures that are unique in some mucosal tissues, for example, the Peyer's patches of the gastrointestinal (GI) tract [15]. Mammary glands are equipped with very few initiation sites but do have at least one internal lymph node and in most cases more than one. However, there is no data to support that the mammary gland contains unique immune-initiation structures, like the Peyer's patches observed in the GI tract. There is however strong evidence of linkage between the GI mucosal and the mammary gland such that there may not be an evolutionary reason for formation of novel initiation sites in the breast, which by design, makes sense given that a primary role of the immune system in the gland is to confer immunity to gut pathogens to the newborn [16].

The main effector compartment is the lamina propria (region immediately below the epithelial plane basement membrane) to which numerous T cells migrate after exposure in the initiation compartment [17]. Although there is an abundance of both T and B cells, T cells seem to be the predominant population in the lamina propria in higher organisms, including in the human [18]. Lamina propria T lymphocytes (LPLs) are much like conventional peripheral T cells in other immune tissues and organs [19, 20]. For example, they express the αβ-T cell receptor and typically either CD4 and/or CD8αβ. There are, however, notable differences as compared to peripheral T cells including oligoclonality and an enrichment in an antigen-experienced phenotype which suggest that the mucosa initiates and accumulates selected clones based on local antigen exposure [21, 22]. In addition to the lamina propria T cells, other T cells, known as intraepithelial lymphocytes (IELs), accumulate directly in the epithelial plane [21, 22]. IELs are more frequently CD8+ T cells that exhibit immediate cytolytic activity [23]. There are two mains categories of IELs, conventional CD4 or CD8αβ αβ-T cells and the CD8αα single-positive IELs. The latter CD8αα T cells are further divided based on expression of either the αβ-TCR or the γδ-TCR. Conventional CD4 or CD8αβ αβ-TCR T cells appear to be predominantly nonself-antigen-specific effector T cells primed within the mucosal node-like structures (e.g., Peyer's patches). In contrast, CD8αα αβ-TCR and γδ-TCR T cells are self-antigen specific and can adopt either effector function, regulatory function, or tissue repair function [24, 25]. Although MHC-restricted (both conventional and nonconventional MHCs), studies have shown that CD8αα expression is not associated with costimulation or MHC class I-restriction, but rather is associated with T cell survival and mucosa-specific properties [25]. Studies have shown that CD8αα αβ-TCR and γδ-TCR T cells demonstrate minimal pathogen-specific activities. However, in *Toxoplasma gondii*-infected mice, CD8αα+ T cells are required for CD8αβ αβ-TCR-dependent microbial clearance. Based on these and other results, there is some consensus that the primary responsibility of CD8αα T cells is in maintaining the integrity of the epithelial barrier by surveying for stress epithelium and controlling immune responses [26]. Indeed, this is further supported by the observation that CD8αα T cells arise very early in neonatal development and express autoreactive TCRs. Further research into the differentiation and selection of these autoreactive T cells has led to a revision in the thymic education theory to include "agonist selection" in addition to "negative selection." Under the original theory, αβ-TCR T cells that migrated from the thymus into the periphery were thymically selected based on moderate affinity binding of the TCR to self-peptide and MHC, whereas T cells with a TCR that bind with high affinity to self-peptide in association with MHC, as well as those that do not bind, are deleted. By contrast, CD8αα αβ-TCR T cells preferentially accumulate in mucosal tissues under conditions in which conventional αβ-TCR T cells are deleted. Murine TCR transgenic mouse studies have shown that the accumulation of CD8αα αβ-TCR T cells is thymus dependent and occurs most effectively in the presence of high-affinity interactions between the T cells and the agonistic peptide. While it is likely that the major role of T cells in the gland is protection against infection and to support development and maintenance of an antibody response during lactation, T cells may also have a critical role in alveolar development. Studies done in IL-4/IL-13 double knockout mice reveal delayed alveolar development [27]. One caveat with interpretation of these studies, however, is that it is also known that mammary gland epithelial cells have a unique capability of producing both Th1 and Th2 cytokines [27]. Lastly, T cells (as well as B cells) in the gland are released into the milk during lactation in many species, including humans [18]. The role of the milk borne lymphocytes is unclear but it has been suggested that, like IgA, they may survive passage through the harsh gastric environment to confer immunity to the infant [18].

The B cells that infiltrate into the mucosa are the primary cells that release secretory IgA an antibody that is unique compared to other antibody subsets, such as IgG, IgE, and IgM. Given the vast surface area constituted by the mucosal immune systems of the body, IgA is the most abundant antibody (75 % of the total Ig), and in humans between 3 and 5 g are produced daily [17]. IgA is secreted into the lumens of mucosal tissues typically as a dimer complexed to two additional proteins chains. One of these is the J chain (joining chain), which is a protein of molecular mass 15 kDa that joins two IgA monomers [28]. Secretory IgA also contain a protein of a much larger molecular mass (70 kDa) called the secretory component that is produced by epithelial cells and originates from the poly-Ig receptor (130 kDa) that is responsible for the uptake and transcellular transport of oligomeric (but not monomeric) IgA across the epithelial cells and into secretions such as tears, saliva, sweat, milk, and gut fluid. Plasma B cells responsible for productions of secretory IgA are induced in the mucosa-associated lymphoid structures such as lymph nodes and Peyer's patches of the gut. In addition to producing immunoglobulins, plasma B cells may also be involved in post-lactational involution, although their role is unclear [29].

In addition to adaptive immune effectors, innate immune cells in the effector compartment have been studied extensively in mouse models as regulators of mammary gland development and function [30]. Primary ductal branching is associated with the presence of eosinophils at TEBs, and increased numbers of mast cells are found at the points of TEB bifurcation; genetic depletion of either eosinophils or mast cells significantly affects primary branching and ductal complexity [31, 32]. A role for immune cells in the secondary branching process is less clear, although macrophages are known to move along the length of the ducts in the developing gland [33], and macrophage-deficient mice also have defective branching and exhibit unusual shaping of TEBs due to reduced collagen deposition [31, 34].

At present, little is known of roles for innate immune effectors in lactation. Macrophage infiltration into developing alveolar has been observed during pregnancy. Macrophages exclusively express the colony-stimulating factor receptor in the mammary gland, and in CSF knockout mice, the absence of macrophages in the developing alveolar is associated with an inability of dams to nurse their pups [35]. Macrophages, particularly of the M2 wound repair phenotype, are also found in involuting glands where they are postulated to have a role in phagocytosing apoptotic cells [36].

3.2.2 Is Chronic Inflammation Linked to Breast Cancer Incidence or Disease Progression?

The fact that the immune system integrates into the normal healthy gland suggests that immunity or immune deregulation could have a role in breast cancer incidence or progression. Although a speculative idea, the concept is well supported by parallel observations in which deregulation of the immune system in the gut predisposes mice and humans to the development of colorectal cancer [37]. The role of the immune system in breast cancer in humans has been somewhat controversial. In mice, some breast cancers are promoted by the immune system [38]. Inflammation results in oxidative damage that could initiate the carcinogenic process by causing inactivating mutations in tumor suppressor genes or posttranslational modifications in proteins involved in apoptosis and DNA repair [39]. Additionally, cytokines produced during the inflammatory process directly promote growth and proliferation of cells, normal and malignant [40]. If such immune promotion occurs in humans, epidemiologic studies should show lower risk patterns in women who are immunosuppressed following transplant, who regularly use anti-inflammatory agents (e.g., aspirin), or who have reduced systemic inflammatory markers. In the transplant setting, solid organ recipients are treated long term with various immunosuppressive drugs that act to prevent T and B cell-mediated immunity, such as tacrolimus, prednisone, and sirolimus. Stewart and colleagues reported in a study of 25,914 women that there were 86 cases of breast cancer compared to an expected number of 114 cases. Relative risk in this transplant group at 1 year was 0.49 rising to 0.84 in subsequent years [41]. This reduction was recently confirmed by Engels and colleagues who showed a relative risk of 0.84 in a larger cohort of 175,732 transplant patients [42]. Although these findings would suggest a role for the immune system in promotion of breast cancer, other independent studies have not shown lower risk, although the cohorts in the latter studies are substantially smaller [43–45]. A case for an inflammatory cause of breast cancer can also be made from population-based studies of the use of nonsteroidal anti-inflammatory drugs such as aspirin and ibuprofen. While these compounds are proposed to have several mechanisms of action, one major effect is blockade of cyclooxygenase-2, the enzyme responsible for catalyzing the production of inflammatory mediators such as prostaglandins [46]. Case–control studies have repeatedly shown that both aspirin and ibuprofen are associated with a reduced risk of breast cancer, and the analysis of each study in the context of this chapter is prohibitive. Algra and Rothwell recently published the results of their intensive study of the published literature on the effects of aspirin use on long-term incidence [47]. Their pooled analysis reveals that aspirin use is associated with an overall 19 % reduction in breast cancer. Alternative to immune suppression, it is well known that the risk of developing breast cancer is 2.5-fold higher in women who exhibit chronic mastitis [48]. Although frank mastitis is not usually associated with the vast majority of breast cancer cases, the observations of increased incidence of breast cancer in immunosuppressed patients might suggest that inflammation is subclinical, yet chronic.

The suggestion that initiation of breast cancer is mediated by subclinical inflammation is largely supported by studies evaluating systemic markers of inflammation such as the C-reactive protein (CRP). CRP is a protein found in the blood, the levels of which rise rapidly in response to inflammation. Its physiological role is to bind to phosphocholine expressed on the surface of dead or dying cells in order to activate the complement system. CRP levels rise in the blood in response to IL-6 released during local inflammatory processes, and thus it is not a disease-specific marker [49]. CRP protein levels in the blood are associated with a wide range of diseases, such as diabetes and atherosclerosis [50]. Siemes and colleagues reported, using data collected in the Rotterdam study, that in individuals with higher CRP levels, there was a 70 % increased risk of breast cancer incidence [51]. Excluding the first 5 years as a latency period, risk remained elevated, further supporting a role for inflammation as causative. Despite these compelling, other studies of similar or greater sample size have failed to show strong correlations between CRP levels and incident breast cancer [52, 53].

Genes encoding various cytokines related to inflammation and immunity have been linked with the development of breast cancer. This includes the identification several SNPs contained with various immune cytokines [54–56]. Dunning et al. found that the T+29C (Leu10Pro) signal peptide polymorphism embedded in the signal peptide region of TGF-β is associated with enhanced secretion [54]. Individuals with the homozygous ProPro phenotype have a modestly elevated risk of developing breast cancer (RR ~1.21). While TGF-β is typically anti-inflammatory, it is constitutively expressed in mucosal microenvironments to regulate immunity, particularly by mediating immunoglobulin class-switching to IgA. It is unlikely that TGF-β alone induces cancer, particularly because it is well known to block proliferation. However, in the presence of other inflammatory cytokines, it may drive pathogenesis. For example, our group has observed that immortalized breast epithelial cells (e.g., MCF10) exposed to TGF-β and TNF-α undergo epithelial to mesenchymal transition and acquire a breast cancer stem cell phenotype with anchorage independent growth [57].

The IL-12B gene encodes for IL-12 subunit p40. IL-12 p40 associates with IL-12 subunit p35, to form the potent inflammatory cytokine, IL-12, which is involved in activating cellular immunity. IL-12 production is highly restricted to dendritic cells and macrophages, both normal cellular components of mucosal immune systems. Macrophages, recruited to the terminal end buds of the mammary gland, are essential to branching morphogenesis [58]. Although it remains unclear whether gland-associated macrophages secrete IL-12, a recent study from Kaarvatn showed that the rs3212227 SNP of IL-12b is associated with the development of breast cancer. At the genotype level, AA homozygosity portends a 68 % increase in risk of breast cancer incidence [55]. The A allele is a high producer of IL-12 p40 mRNA. It should also be noted that IL-12 p40 is also a subunit of the IL-23 cytokine. IL-23 is central to development of autoimmune disease and chronic inflammation associated Th17 helper T cells [59]. As the name implies, a major cytokine produced by Th17 T cell is IL-17. IL-17 refers to IL-17A, the namesake cytokine of a family of structurally related cytokines, IL-17A-F. IL-17A is a well-studied cytokine and has a spectrum of proinflammatory activities. While essential to host defense, IL-17 is linked to a wide variety of inflammatory conditions such as asthma, inflammatory bowel disease, and rheumatoid arthritis [60]. While there is no evidence in case–control studies that the cytokines IL-17A and IL-23 directly confer an elevated risk of breast cancer, various studies do provide indirect evidence. For example, obesity induces a chronic state of inflammation and numerous studies show that obese women are at an increased risk for postmenopausal breast cancer [61, 62]. Interestingly, the IL-17/IL-23 axis is highly upregulated in obesity as shown by Sumarac-Dumanovic and others [61, 63]. Further supporting a link between IL-17A and breast cancer incidence, Wang showed that IL-17 SNPs are associated with a significantly increased risk of breast cancer in a cohort of Han Chinese women [64].

A completely different kind of inflammation follows tumor development [65]. Rather than initiating malignant transformation, chronic inflammation fosters and supports various cellular events resulting in induction and/or selection of aggressive cancer cells. A growing body of evidence has shown that cytokines released by the immune response either can directly alter tumor cell biology or can activate other cells in the tumor microenvironment such as tumor-associated macrophages, fibroblasts, and vascular endothelial cells [66]. Alternatively, systemic inflammation, regardless of its source, may promote the development of metastatic niches throughout the body [67]. A potential role for inflammation in disease progression has been established by several lines of investigation. In contrast to the weaker findings with disease incidence, correlations between elevated CRP and disease progression are fairly strong in both early and advanced disease [68–70]. In addition to CRP, reduced survival in breast cancer is also associated with elevated levels of another acute phase liver-derived protein, serum amyloid A (SAA) [70]. SAA is involved in extracellular matrix degradation, cholesterol transport, and recruitment of leukocytes [71, 72]. Both CRP and SAA are released in response to IL-6 secretion by various immune effectors as well as tumor cells themselves. IL-6 is a pleiotropic cytokine that supports the development of both innate and adaptive immune responses and notably suppresses regulatory T cell development and fosters inflammation-associated Th17 cell expansion [73]. Consistent with the CRP and SAA findings, elevated levels of IL-6 are associated with poor outcomes in patients diagnosed with metastatic breast cancer [74, 75]. Although there are likely many factors regulating production of IL-6, levels appear strongly linked to the genetics of the individual. At the promoter position –174 of the IL-6 gene, a G to C variation is linked to reduced serum levels of IL-6 [76], and in two separate studies, G to C variation is associated with a better prognosis [77, 78].

In summary, there is some support of the notion that chronic inflammation in the mammary gland is involved in breast cancer development. It is largely accepted that the potential for inflammation in the breast drives carcinogenesis given direct evidence of a higher rate of cancer diagnoses in individuals with clinically recognizable mastitis. Unfortunately, no studies to the best of our knowledge have been reported which directly compare in either prospective study or case–control studies that smoldering subclinical inflammation drives the development of neoplasia of the breast. In fact, primary early breast cancer removed from patients rarely show any evidence of significant inflammation [66]. The best evidence in our view is from the studies demonstrating reduced risk of breast cancer in the setting of immune suppression in the transplant and aspirin use settings. However, there are several other mechanisms that could explain the effects of drugs used in these settings. Furthermore, unlike for colorectal cancers, there is no appropriate animal model data to support the claim (including the cow, an animal which does not get breast cancer) [79]. Lastly, there is a possibility that simple evaluation of systemic markers of inflammation may not reveal a role for the immune system and inflammation in driving tumorigenesis. Although speculative, future studies evaluating modifications of the mucosal immune microenvironment may reveal patterns or abnormalities that may better define a role for immunity and subclinical inflammation in driving breast cancer initiation.

3.3 The Tumor Immune Microenvironment

3.3.1 The Role of Infiltrating Immune Effectors in Disease Pathogenesis and Survival

Immunohistochemical and immunofluorescent studies examining the infiltration of various immune effectors into primary operable breast cancers have pointed to involvement of the immune system in the clinical course of established disease.

An authoritative overview of the multiple studies completed over the last decade has been recently summarized by Mohammed and colleagues [80]. Here, the focus is on immune cells that can have a pro-tumor phenotype.

The two immune effectors which appear to have the dominant pro-tumor impact on breast cancer are the macrophage and the regulatory T cell (Treg). Data for a pro-tumor role for infiltrating macrophages is strong and conclusive. Derived from blood monocytes, macrophages are specialized phagocytic immune effectors that have a broad role in both in innate and adaptive immune responses. The plastic nature of macrophages results in distinct phenotypes that are dependent on the microenvironment in which they are localized [81]. Rather uniformly, macrophage infiltration in breast cancer portends a poor outcome [82]. In tumors, macrophages are known to release a variety of mediators involved in angiogenesis and production of extracellular matrices. Treg refers to a diverse group of either thymically derived or peripherally induced, suppressor CD4 or CD8 T cells that control peripheral activation and function of both self- and foreign-antigen reactive T cells [83–90]. Tregs infiltrate into tumors and essentially behave as shields by suppressing T and B cell immune responses through a variety of mechanisms. Several studies show that elevated Tregs in the tumor are associated with poor survival in breast cancer [91–96].

Some studies have suggested a role for cytotoxic T cells by associating CD8 staining with disease-free and overall survival. Cytotoxic T cells are the primary cell type staining with CD8 antibodies. Cytotoxic T cells are crucial components of the tumor-specific adaptive immune response. CD8 T cells recognize tumor-associated antigens presented in the context of HLA-A, HLA-B, and HLA-C (major histocompatibility complex, MHC in mice) molecules. Ligation of the T cell receptor by peptide-bound HLA leads to release of several toxic molecules associated with the induction of apoptosis in the target cells [97]. Strong associations of CD8 T cell infiltration with improved survival has been observed in several cancers such as ovarian and colorectal cancers [98–100]. The role of CD8 T cells in breast cancer pathogenesis appears to be complex and not easily resolvable. Recently, Mahmoud and colleagues showed that stromal, but not intratumoral, CD8 T cells were associated with improved survival among breast cancer patients [101]. However, other studies have failed to demonstrate such an association [102]. Other studies suggest a potential for CD8 T cells to be involved in increased tumor aggressiveness. For example, Matkowski found that CD8 T cell infiltration was associated with poor survival and node-positive disease, albeit in a small group of 88 patients [103]. While murine data remains scarce, work from our group has suggested a role for CD8 T cells in both tumorigenesis and the genesis of breast cancer stem cells in the FBV/N-based neu-transgenic mouse model of breast cancer [104, 105]. Despite those findings, mechanisms remain elusive. Further work is required to understand a potentially dual-role for CD8 T cells in breast cancer. Notwithstanding, it is notable from the studies described above that normal and malignant epithelium in the breast is associated with a complex immune microenvironment that could promote tumor initiation, growth, and metastasis under conditions which remain obscure. While there are likely to be multiple mechanisms involved in the pro-tumorigenic activities of macrophages, Tregs, and possibly cytotoxic T cells, below we consider the possibility that one of the major mechanisms may be induction of epithelial to mesenchymal transition (EMT). This is considered, largely because the dominant pro-tumorigenic immune cells that populated the normal and malignant breast produce a number of factors (e.g., TNF-α, TGF-β, IL-6, and MMP-3) which are critically involved in the activation of EMT.

3.3.2 *Epithelial to Mesenchymal Transition*

EMT is a cellular transdifferentiation process wherein epithelial cells take on the characteristics of mesenchymal cells, lose cell–cell interactions, alter cell–extracellular matrix (ECM) interactions, acquire increased migratory capability, increased resistance to cell death, and substantially altered cellular metabolism. The EMT process has been most studied in the context of embryonic development, where it has been found to be essential for the development of many tissues and organs during development [106–108]. In early developmental processes, EMT drives gastrulation, which leads to formation of mesoderm, as well as neural crest formation, which is responsible for the release of mesenchymal cells that migrate through the body, ultimately generating the vertebrate head and a wide variety of tissue types, including glial and neuronal cells, adrenal glandular tissues, melanocytes, and skeletal and connective tissues [107, 109]. EMT of embryonic endocardial cells into the endocardial cushion creates precursors of the valvular and septal structures [110]. When activated during development, EMT is a highly regulated process that follows a defined series of events controlled by signals from neighboring cells as well as soluble molecules [107]; this is necessary to retain the permeability function of the epithelial sheet from which the mesenchymal cells will emerge. First, where EMT occurs becomes specified through identification of the cells that will undergo EMT concomitant with a morphogenic rearrangement that moves those cells to the site of EMT. Following this, the interaction between epithelial cells and the basement membrane (BM) must become disrupted, through regulation of the integrins that mediate cell–BM contacts or through proteolytic degradation of the BM itself. Third, the cells undergoing EMT detach from the epithelial sheet so as to maintain overall epithelial tissue integrity, generally involving coordination of elongation of the cells undergoing EMT as they leave the epithelial sheet in combination with contraction of the remaining epithelial cells so as to close the new gap. After all this has occurred, the ingressed cells differentiate into the mesenchymal cellular phenotype. Thus, developmental EMT is much more than just the acquisition of motility but is a process involving the entire epithelial tissue.

While physiological EMT is largely viewed as driving developmental processes in the embryo, EMT has been implicated in postnatal mammary gland development. During puberty, the rudimentary mammary gland develops into the fat pad through extension and branching morphogenesis of the ductal tree [111–113]. This process has been studied extensively using mouse models, which has revealed that matrix metalloproteinases (MMPs) are key mediators of the developmental processes. In the mouse mammary gland, ductal extension in the developing gland occurs at terminal end buds (TEBs), invasive structures that express high levels of EMT-associated transcription factors, including Snail and Twist [114], as well as matrix metalloproteinase-2 (MMP-2) and the membrane-spanning MMP-14 [115]. Development of the mammary ductal tree through branching morphogenesis occurs through primary branching or end bud bifurcation, and secondary branching, in which differentiated, ductal epithelium dedifferentiates, detaches from the adjacent epithelial cells, penetrates the basement membrane, and invades into the surrounding tissue. The key mediator of secondary branch formation is matrix metalloproteinase-3 (MMP-3): transgenic mice lacking MMP-3 expression have significantly reduced secondary branching, while the overexpression of MMP-3 leads to increased secondary branching and ductal complexity [115, 116]. Inhibition of retinoic acid signaling pathways, another key mediator of side branching, also increases MMP-3 expression and also increases side branching [117]. As we will discuss below, MMP-3 has also emerged as a key inducer of EMT and activator of invasive, metastatic behavior in the breast cancer context.

At the cellular level, the process of EMT is defined as the transdifferentiation of the epithelial phenotype. Epithelial cells are tightly interconnected by cell–cell junctions and are attached to the BM through specialized cell–ECM contacts. Whereas some cell rearrangement can occur in the context of tissue morphogenesis, movement of epithelial cells is generally constrained by cell–cell junctions to the epithelial sheet. One of the key components of epithelial junctions is E-cadherin, a transmembrane molecule that connects adjacent cells, and β-catenin, an intracellular molecule that is part of the protein complex that connects cadherins to the actin cytoskeleton. In addition to restraining epithelial cell movement, E-cadherin/β-catenin complexes function to transmit information from the microenvironment to the cells as well as inhibit transition to the mesenchymal state [118–120]. Accordingly, reduced expression/function of E-cadherin is viewed as one of the key steps in completion of EMT. By contrast, cells which have fully acquired the mesenchymal phenotype lack cell–cell junctions and are capable of free movement in three dimensions, traveling along collagen strands and moving within the interstitial ECM. While cell–ECM contacts are still critical for mesenchymal cells, these contacts become specialized for interaction with the components of the interstitial ECM [121]. During the EMT process most cell structures and functions are altered. In addition to the loss of cell–cell adhesion through downregulation of E-cadherin, there is upregulation of key transcription factors with the expression of associated mesenchymal genes, such as the cytoskeletal intermediate filament vimentin and the ECM molecules fibronectin and collagen [122]. EMT is regulated by soluble growth factors or cytokines, including epidermal growth factor (EGF), hepatocyte growth factor (HGF), fibroblast growth factors (FGFs), and transforming growth factor (TGF)-β [123], as well as the composition and structure of the ECM components and ECM-remodeling matrix metalloproteinases (MMPs) [124–126].

3.3.3 Immune-Induced EMT in Cancer Progression

EMT in embryonic development generally occurs in an immunologically privileged setting with little or no inflammatory responses. In the adult organism, transition between epithelial and mesenchymal cell types occurs under conditions in which the immune system can be highly activated, such as during wound healing, as well as during remodeling of tissues that develop postnatally, such as the mammary gland, and in the pathological context, such as inflammation, fibrosis, or breast cancer development [127, 128]. By contrast to normal embryonic development, the disrupted, chaotic context present in developing or progressing breast cancers can stimulate EMT in a disorganized and cell-autonomous fashion, leading to disruption of epithelial integrity and disorganization of epithelial tissue as well as production of new mesenchymal cells which can perpetuate the disease process. Most importantly, activation of EMT in premalignant cells can confer the highly tumorigenic, malignant cancer stem cell (CSC) phenotype, producing cells which are highly metastatic, resistant to many chemotherapeutic treatments, and which possess the ability to reconstitute the entire tumor from only a few cells (or even a single cell).

TGF-β, the product of several subsets of infiltrating immune cells (e.g., Tregs), is one of the most potent activators of EMT [129]. During gastrulation, EMT induced by TGF-β leads to the formation of the mesoderm. TGF-β is also implicated in developmental EMT in the kidney, endocardial cushion, and subsequent atrioventricular valve formation [130]. Virtually all cell types are responsive to TGF-β, which regulates proliferation, migration, differentiation, and survival and, in cancer, can act either to suppress tumor growth or to activate tumor progression, depending upon cellular context [131–133]. Induction of EMT by TGF-β involves a combination of Smad-dependent and Smad-independent events on cell junction complexes [134]. While TGF-β is a central mediator of normal inflammatory responses, its sustained expression stimulates fibrogenic processes and tumor promotion. It could be speculated that poor outcomes associated with infiltration of Tregs may in fact be due to direct actions of Treg-produced TGF-β on the tumor cells rather than on the infiltrating immune effectors. The cooperation between TGF-β1 and other proinflammatory cytokines (e.g., TNF-α) produced in a chronically inflamed microenvironment by macrophages, CD8 T cells, and CD4 T cells activates autoregulatory loops which further reinforce the EMT program [57, 135]. In addition to providing TNF-α, macrophages may also contribute to EMT by providing MMP-3 [136]. MMP-3 induces EMT in both non-transformed and transformed mammary epithelia through a cascade of events involving production of reactive oxygen species (also produced by macrophages) mediated by MMP-3-induced production of alternatively spliced forms of Rac1b [137].

Several transcription factors have been identified in EMT, involved either in transcriptional inactivation of epithelial genes or activation of mesenchymal genes [122]. Among these, Snail has emerged as a central regulator of both developmental and pathological EMT [138]. Snail is critical for gastrulation in normal development of mice; homozygous knockout of Snail is lethal as embryos fail to produce mesoderm [139]. Snail has also been associated with pathological conditions, such as fibrosis and cancer, in which its detrimental role is determined by its ability to induce EMT-like processes. Snail has been found to directly suppress expression of E-cadherin, and other epithelial cell–cell adhesion molecules, and to promote expression of mesenchymal proteins, such as fibronectin and MMP-9 [108]. Snail is expressed in premalignant breast epithelial cells and in fibroblasts associated with damaged or neoplastic tissues [140], and its increased expression is commonly observed in cultured cells treated with agents that stimulate the EMT program, for example, by treatment with TGF-β, interleukin 6, and growth factors such as EGF or FGF or by exposure to MMPs [57, 141, 142]. Expression of Snail is further regulated by an integrated and complex signaling network at the transcriptional and posttranscriptional level that includes integrin-linked kinase (ILK), phosphatidylinositol 3-kinase (P13-K), mitogen-activated protein kinases (MAPKs), glycogen synthase kinase 3-beta (GSK-3β), and NF-κB pathways. The increase of the transcription of Snail comes about through the direct binding of NF-κB to the Snail promoter [143], as well as activation of NF-κB by GSK-3β inhibition [134, 141]. Lastly, T cells and macrophages are the body's major producers of IL-6, and it has been shown that ectopic expression of IL-6 in tumor cells can lead to expression of Twist, an EMT regulator and direct transcriptional repressor of E-cadherin [144].

3.3.4 Immune-Induced EMT Can Lead to Cancer Stem Cell Phenotype

Activation of the EMT program in cancer cells facilitates tumor progression through several distinct mechanisms (a) EMT in the tumor cells can trigger invasive and anti-apoptotic mechanisms that drive cancer metastasis, (b) EMT can generate activated stromal cells that drive cancer progression through biochemical and structural alterations of the tumor microenvironment, and (c) EMT mediators can stimulate the increased malignancy associated with the CSC phenotype.

In normal tissues, stem cells are characterized as having low proliferative rates, existing as minority populations within tissues in defined tissue compartments, or niches, and having responses to extracellular stimuli that are distinct from those of the more differentiated cells within the organ [145]. Stem cells exhibit self-renewal and can divide symmetrically, to produce additional stem cells, or asymmetrically, to generate progenitor cells that can subsequently differentiate into the many different cell types within the organ. More recently, cells with stem/progenitor characteristics have been found to play critical roles in tumor formation and progression [146]. These CSCs were first identified in human acute myeloid leukemia [147], although subsequent studies have identified CSCs in solid tumors, including breast, brain, colon, and pancreas (reviewed in [148]). The identification of CSCs has resulted in increased efforts to identify treatments that selectively target this subpopulation of cells, as traditional treatments for breast cancer that target the bulk of the tumor such as surgery, conventional chemotherapy, and radiation may be less effective toward the CSCs, due to their decreased proliferation rate and resistance to apoptosis [149]. While identification of therapies that selectively target CSCs is a potentially important goal, a parallel and perhaps equally effective approach would be to target the mechanisms that generate and maintain the CSC population in tumors. A recent series of studies showed that induction of EMT in human or mouse mammary epithelial or epithelial tumor cells can directly result in the generation of CSCs [105, 150, 151]. Evaluation of such model systems provides an experimental platform to explore how to target EMT-induced CSC development.

Immune responses can eliminate premalignant and malignant cells in a process called immunosurveillance, which has been identified as a key factor in reducing the earliest stages of tumor development [152]. However, aberrant activation of the immune system can also promote tumor development, as chronic activation of inflammatory responses has been directly linked to development of cancer [153, 154]. The interaction of tumors with the immune system progresses through an initial stage in which the immune system can effectively target and kill tumor cells, but as the tumor immune interaction evolves, tumors acquire the ability to evade immune targeting entirely, by becoming less immunogenic or more immune suppressive [152, 155]. Evaluation of immunoediting in a mouse model of breast cancer relapse identified EMT as a mechanism for escape from immunosurveillance, as well as activation of the CSC phenotype [156]. Transgenic mice that express the cell surface rat HER2/neu oncogene under control of the mouse mammary tumor virus (MMTV) promoter, which drives expression in mammary epithelial cells, develop tumors with a highly epithelial morphology; transplantation of these tumors into non-transgenic syngeneic mice stimulated a T cell-dependent rejection, followed by relapse of phenotypically mesenchymal tumors enriched in neu-negative variant cells [157, 158]. Activation of the EMT program led to downregulation of the epithelial-specific MMTV promoter and loss of expression of the neu oncogene; the resultant tumor was able to grow without provoking an immune reaction. Further studies identified T cells of the CD8 subtype as required for outgrowth of the neu-negative mesenchymal variants [105], and the induction of EMT in the neu-expressing epithelial cells was associated with TGF-β and TNF-α signaling pathways [57]. Further evaluation of tumor cells isolated from relapsed mice demonstrated that these tumors had adopted CSC characteristics, including altered cell surface marker profiles, enhanced mammosphere formation, and substantially increased tumorigenicity [105]. CD8-induced mesenchymal tumor cells also have elevated expression of drug transporters, DNA repair enzymes, and resistance to chemotherapy and radiation [57, 105]. Moreover, the neu-negative CSCs were fully capable of differentiating into epithelial cells; when neu-negative cells were injected into MMTV-HER2/neu mice (where selection against neu was absent) showed development of tumors with a predominantly epithelial phenotype associated the reacquisition of neu expression [105]. These studies suggest that migration or metastasis of CSCs to distant regions of the body may circumvent immunoediting processes.

3.4 Conclusions

Like the stromal and vascular system, the immune system is an integral component of the normal healthy mammary gland. When deregulated, however, the immune system could precipitate breast cancer and aid in its progression. While the mechanisms by which the immune system fosters disease progression are unclear, EMT is a likely target given the abundance of immune effectors that induce and regulate EMT. Thus, it seems reasonable that future approaches to disease prevention and therapy include modulators of one or more targets along this pathway.

Acknowledgments This author's work was supported by a generous gift from Martha and Bruce Atwater (KLK), support from the Susan B. Komen Foundation (FAS0703855, DCR), and the National Cancer Institute (Howard Temin Award [K01-CA100764, KLK]; R01-CA122086, DCR; P50-CA116201, KLK; and P50-CA136393, KLK).

References

1. Wellings SR, Jensen HM. On the origin and progression of ductal carcinoma in the human breast. J Natl Cancer Inst. 1973;50(5):1111–8.
2. Wellings SR, Jensen HM, Marcum RG. An atlas of subgross pathology of the human breast with special reference to possible precancerous lesions. J Natl Cancer Inst. 1975;55(2): 231–73.
3. Wellings SR, Jentoft VL. Organ cultures of normal, dysplastic, hyperplastic, and neoplastic human mammary tissues. J Natl Cancer Inst. 1972;49(2):329–38.
4. Keller PJ, Arendt LM, Skibinski A, et al. Defining the cellular precursors to human breast cancer. Proc Natl Acad Sci USA. 2012;109(8):2772–7.
5. Burstein HJ, Polyak K, Wong JS, et al. Ductal carcinoma in situ of the breast. N Engl J Med. 2004;350(14):1430–41.
6. Lopez-Garcia MA, Geyer FC, Lacroix-Triki M, et al. Breast cancer precursors revisited: molecular features and progression pathways. Histopathology. 2010;57(2):171–92.
7. Bhowmick NA, Neilson EG, Moses HL. Stromal fibroblasts in cancer initiation and progression. Nature. 2004;432(7015):332–7.
8. Brandtzaeg P. The mucosal immune system and its integration with the mammary glands. J Pediatr. 2010;156(2 Suppl):S8–15.
9. Goldman AS. The immune system of human milk: antimicrobial, antiinflammatory and immunomodulating properties. Pediatr Infect Dis J. 1993;12(8):664–71.
10. Spencer JP. Management of mastitis in breastfeeding women. Am Fam Physician. 2008;78(6):727–31.
11. Cheroutre H, Madakamutil L. Acquired and natural memory T cells join forces at the mucosal front line. Nat Rev Immunol. 2004;4(4):290–300.
12. Ghajar CM. On leukocytes in mammary development and cancer. Cold Spring Harb Perspect Biol. 2012;4(5):a013276.
13. Reed JR, Schwertfeger KL. Immune cell location and function during post-natal mammary gland development. J Mammary Gland Biol Neoplasia. 2010;15(3):329–39.
14. Shankaran V, Ikeda H, Bruce AT, et al. IFNgamma and lymphocytes prevent primary tumour development and shape tumour immunogenicity. Nature. 2001;410(6832):1107–11.
15. Pabst R, Rothkotter HJ. Structure and function of the gut mucosal immune system. Adv Exp Med Biol. 2006;579:1–14.

16. Goldblum RM, Ahlstedt S, Carlsson B, et al. Antibody-forming cells in human colostrum after oral immunisation. Nature. 1975;257(5529):797–8.
17. Brandtzaeg P. Mucosal immunity: induction, dissemination, and effector functions. Scand J Immunol. 2009;70(6):505–15.
18. Wirt DP, Adkins LT, Palkowetz KH, et al. Activated and memory T lymphocytes in human milk. Cytometry. 1992;13(3):282–90.
19. Schieferdecker HL, Ullrich R, Hirseland H, et al. T cell differentiation antigens on lymphocytes in the human intestinal lamina propria. J Immunol. 1992;149(8):2816–22.
20. Zeitz M, Schieferdecker HL, Ullrich R, et al. Phenotype and function of lamina propria T lymphocytes. Immunol Res. 1991;10(3–4):199–206.
21. Regnault A, Kourilsky P, Cumano A. The TCR-beta chain repertoire of gut-derived T lymphocytes. Semin Immunol. 1995;7(5):307–19.
22. Blumberg RS, Yockey CE, Gross GG, et al. Human intestinal intraepithelial lymphocytes are derived from a limited number of T cell clones that utilize multiple V beta T cell receptor genes. J Immunol. 1993;150(11):5144–53.
23. Montufar-Solis D, Garza T, Klein JR. T-cell activation in the intestinal mucosa. Immunol Rev. 2007;215:189–201.
24. Rocha B, Vassalli P, Guy-Grand D. The V beta repertoire of mouse gut homodimeric alpha CD8+ intraepithelial T cell receptor alpha/beta + lymphocytes reveals a major extrathymic pathway of T cell differentiation. J Exp Med. 1991;173(2):483–6.
25. Leishman AJ, Gapin L, Capone M, et al. Precursors of functional MHC class I- or class II-restricted CD8alphaalpha(+) T cells are positively selected in the thymus by agonist self-peptides. Immunity. 2002;16(3):355–64.
26. Taylor BC, Dellinger JD, Cullor JS, et al. Bovine milk lymphocytes display the phenotype of memory T cells and are predominantly CD8+. Cell Immunol. 1994;156(1):245–53.
27. Khaled WT, Read EK, Nicholson SE, et al. The IL-4/IL-13/Stat6 signalling pathway promotes luminal mammary epithelial cell development. Development. 2007;134(15):2739–50.
28. Suzuki K, Kawamoto S, Maruya M, et al. GALT: organization and dynamics leading to IgA synthesis. Adv Immunol. 2010;107:153–85.
29. Stein T, Morris JS, Davies CR, et al. Involution of the mouse mammary gland is associated with an immune cascade and an acute-phase response, involving LBP, CD14 and STAT3. Breast Cancer Res. 2004;6(2):R75–91.
30. Coussens LM, Pollard JW. Leukocytes in mammary development and cancer. Cold Spring Harb Perspect Biol. 2011;3(3):a003285.
31. Gouon-Evans V, Rothenberg ME, Pollard JW. Postnatal mammary gland development requires macrophages and eosinophils. Development. 2000;127(11):2269–82.
32. Lilla JN, Werb Z. Mast cells contribute to the stromal microenvironment in mammary gland branching morphogenesis. Dev Biol. 2010;337(1):124–33.
33. Ingman WV, Wyckoff J, Gouon-Evans V, et al. Macrophages promote collagen fibrillogenesis around terminal end buds of the developing mammary gland. Dev Dyn. 2006;235(12): 3222–9.
34. Lin EY, Gouon-Evans V, Nguyen AV, et al. The macrophage growth factor CSF-1 in mammary gland development and tumor progression. J Mammary Gland Biol Neoplasia. 2002;7(2):147–62.
35. Pollard JW, Hennighausen L. Colony stimulating factor 1 is required for mammary gland development during pregnancy. Proc Natl Acad Sci USA. 1994;91(20):9312–6.
36. O'Brien J, Lyons T, Monks J, et al. Alternatively activated macrophages and collagen remodeling characterize the postpartum involuting mammary gland across species. Am J Pathol. 2010;176(3):1241–55.
37. Saleh M, Trinchieri G. Innate immune mechanisms of colitis and colitis-associated colorectal cancer. Nat Rev Immunol. 2011;11(1):9–20.
38. Wei WZ, Gill RF, Wang H. Mouse mammary tumor virus associated antigens and superantigens–immuno-molecular correlates of neoplastic progression. Semin Cancer Biol. 1993;4(3):205–13.

39. Hussain SP, Harris CC. Inflammation and cancer: an ancient link with novel potentials. Int J Cancer. 2007;121(11):2373–80.
40. Philip M, Rowley DA, Schreiber H. Inflammation as a tumor promoter in cancer induction. Semin Cancer Biol. 2004;14(6):433–9.
41. Stewart T, Tsai SC, Grayson H, et al. Incidence of de-novo breast cancer in women chronically immunosuppressed after organ transplantation. Lancet. 1995;346(8978):796–8.
42. Engels EA, Pfeiffer RM, Fraumeni Jr JF, et al. Spectrum of cancer risk among US solid organ transplant recipients. JAMA. 2011;306(17):1891–901.
43. Adami J, Gabel H, Lindelof B, et al. Cancer risk following organ transplantation: a nationwide cohort study in Sweden. Br J Cancer. 2003;89(7):1221–7.
44. Birkeland SA, Storm HH, Lamm LU, et al. Cancer risk after renal transplantation in the Nordic countries, 1964–1986. Int J Cancer. 1995;60(2):183–9.
45. Kelly DM, Emre S, Guy SR, et al. Liver transplant recipients are not at increased risk for nonlymphoid solid organ tumors. Cancer. 1998;83(6):1237–43.
46. Xu XC. COX-2 inhibitors in cancer treatment and prevention, a recent development. Anticancer Drugs. 2002;13(2):127–37.
47. Algra AM, Rothwell PM. Effects of regular aspirin on long-term cancer incidence and metastasis: a systematic comparison of evidence from observational studies versus randomised trials. Lancet Oncol. 2012;13(5):518–27.
48. Monson RR, Yen S, MacMahon B. Chronic mastitis and carcinoma of the breast. Lancet. 1976;2(7979):224–6.
49. Gabay C, Kushner I. Acute-phase proteins and other systemic responses to inflammation. N Engl J Med. 1999;340(6):448–54.
50. Ridker PM. Inflammatory biomarkers and risks of myocardial infarction, stroke, diabetes, and total mortality: implications for longevity. Nutr Rev. 2007;65(12 Pt 2):S253–9.
51. Siemes C, Visser LE, Coebergh JW, et al. C-reactive protein levels, variation in the C-reactive protein gene, and cancer risk: the Rotterdam Study. J Clin Oncol. 2006;24(33):5216–22.
52. Zhang SM, Lin J, Cook NR, et al. C-reactive protein and risk of breast cancer. J Natl Cancer Inst. 2007;99(11):890–4.
53. Heikkila K, Harris R, Lowe G, et al. Associations of circulating C-reactive protein and interleukin-6 with cancer risk: findings from two prospective cohorts and a meta-analysis. Cancer Causes Control. 2009;20(1):15–26.
54. Dunning AM, Ellis PD, McBride S, et al. A transforming growth factor beta1 signal peptide variant increases secretion in vitro and is associated with increased incidence of invasive breast cancer. Cancer Res. 2003;63(10):2610–5.
55. Kaarvatn MH, Vrbanec J, Kulic A, et al. Single nucleotide polymorphism in the interleukin 12B gene is associated with risk for breast cancer development. Scand J Immunol. 2012;76(3):329–35.
56. Slattery ML, Curtin K, Baumgartner R, et al. IL6, aspirin, nonsteroidal anti-inflammatory drugs, and breast cancer risk in women living in the southwestern United States. Cancer Epidemiol Biomarkers Prev. 2007;16(4):747–55.
57. Asiedu MK, Ingle JN, Behrens MD, et al. TGFbeta/TNF(alpha)-mediated epithelial-mesenchymal transition generates breast cancer stem cells with a claudin-low phenotype. Cancer Res. 2011;71(13):4707–19.
58. Pollard JW. Tumour-educated macrophages promote tumour progression and metastasis. Nat Rev Cancer. 2004;4(1):71–8.
59. Langrish CL, McKenzie BS, Wilson NJ, et al. IL-12 and IL-23: master regulators of innate and adaptive immunity. Immunol Rev. 2004;202:96–105.
60. Korn T, Bettelli E, Oukka M, et al. IL-17 and Th17 Cells. Annu Rev Immunol. 2009;27:485–517.
61. Sumarac-Dumanovic M, Stevanovic D, Ljubic A, et al. Increased activity of interleukin-23/interleukin-17 proinflammatory axis in obese women. Int J Obes. 2009;33(1):151–6.
62. Anderson GL, Neuhouser ML. Obesity and the risk for premenopausal and postmenopausal breast cancer. Cancer Prev Res. 2012;5(4):515–21.

63. Winer S, Paltser G, Chan Y, et al. Obesity predisposes to Th17 bias. Eur J Immunol. 2009;39(9):2629–35.
64. Wang L, Jiang Y, Zhang Y, et al. Association analysis of IL-17A and IL-17F polymorphisms in Chinese Han women with breast cancer. PLoS One. 2012;7(3):e34400.
65. Grivennikov SI, Greten FR, Karin M. Immunity, inflammation, and cancer. Cell. 2010;140(6):883–99.
66. Cole SW. Chronic inflammation and breast cancer recurrence. J Clin Oncol. 2009;27(21):3418–9.
67. Fidler IJ. The pathogenesis of cancer metastasis: the 'seed and soil' hypothesis revisited. Nat Rev Cancer. 2003;3(6):453–8.
68. Albuquerque KV, Price MR, Badley RA, et al. Pre-treatment serum levels of tumour markers in metastatic breast cancer: a prospective assessment of their role in predicting response to therapy and survival. Eur J Surg Oncol. 1995;21(5):504–9.
69. Al Murri AM, Bartlett JM, Canney PA, et al. Evaluation of an inflammation-based prognostic score (GPS) in patients with metastatic breast cancer. Br J Cancer. 2006;94(2):227–30.
70. Pierce BL, Ballard-Barbash R, Bernstein L, et al. Elevated biomarkers of inflammation are associated with reduced survival among breast cancer patients. J Clin Oncol. 2009;27(21):3437–44.
71. Schultz DR, Arnold PI. Properties of four acute phase proteins: C-reactive protein, serum amyloid A protein, alpha 1-acid glycoprotein, and fibrinogen. Semin Arthritis Rheum. 1990;20(3):129–47.
72. Malle E, Sodin-Semrl S, Kovacevic A. Serum amyloid A: an acute-phase protein involved in tumour pathogenesis. Cell Mol Life Sci. 2009;66(1):9–26.
73. Kimura A, Kishimoto T. IL-6: regulator of Treg/Th17 balance. Eur J Immunol. 2010;40(7):1830–5.
74. Bachelot T, Ray-Coquard I, Menetrier-Caux C, et al. Prognostic value of serum levels of interleukin 6 and of serum and plasma levels of vascular endothelial growth factor in hormone-refractory metastatic breast cancer patients. Br J Cancer. 2003;88(11):1721–6.
75. Salgado R, Junius S, Benoy I, et al. Circulating interleukin-6 predicts survival in patients with metastatic breast cancer. Int J Cancer. 2003;103(5):642–6.
76. Hulkkonen J, Pertovaara M, Antonen J, et al. Elevated interleukin-6 plasma levels are regulated by the promoter region polymorphism of the IL6 gene in primary Sjogren's syndrome and correlate with the clinical manifestations of the disease. Rheumatology. 2001;40(6):656–61.
77. Gonzalez-Zuloeta Ladd AM, Arias Vasquez A, Witteman J, et al. Interleukin 6 G-174 C polymorphism and breast cancer risk. Eur J Epidemiol. 2006;21(5):373–6.
78. Madeleine MM, Johnson LG, Malkki M, et al. Genetic variation in proinflammatory cytokines IL6, IL6R, TNF-region, and TNFRSF1A and risk of breast cancer. Breast Cancer Res Treat. 2011;129(3):887–99.
79. Kraus S, Arber N. Inflammation and colorectal cancer. Curr Opin Pharmacol. 2009;9(4):405–10.
80. Mohammed ZM, Going JJ, Edwards J, et al. The role of the tumour inflammatory cell infiltrate in predicting recurrence and survival in patients with primary operable breast cancer. Cancer Treat Rev. 2012;38(8):943–55.
81. Geissmann F, Manz MG, Jung S, et al. Development of monocytes, macrophages, and dendritic cells. Science. 2010;327(5966):656–61.
82. Pollard JW. Macrophages define the invasive microenvironment in breast cancer. J Leukoc Biol. 2008;84(3):623–30.
83. Buckner JH, Ziegler SF. Regulating the immune system: the induction of regulatory T cells in the periphery. Arthritis Res Ther. 2004;6(5):215–22.
84. Kronenberg M, Rudensky A. Regulation of immunity by self-reactive T cells. Nature. 2005;435(7042):598–604.
85. Mills KH, McGuirk P. Antigen-specific regulatory T cells–their induction and role in infection. Semin Immunol. 2004;16(2):107–17.
86. Wing K, Suri-Payer E, Rudin A. CD4+CD25+-regulatory T cells from mouse to man. Scand J Immunol. 2005;62(1):1–15.

87. Gregori S, Bacchetta R, Hauben E, et al. Regulatory T cells: prospective for clinical application in hematopoietic stem cell transplantation. Curr Opin Hematol. 2005;12(6):451–6.
88. Bacchetta R, Gregori S, Roncarolo MG. CD4+ regulatory T cells: mechanisms of induction and effector function. Autoimmun Rev. 2005;4(8):491–6.
89. Maggi E, Cosmi L, Liotta F, et al. Thymic regulatory T cells. Autoimmun Rev. 2005;4(8):579–86.
90. Lan RY, Ansari AA, Lian ZX, et al. Regulatory T cells: development, function and role in autoimmunity. Autoimmun Rev. 2005;4(6):351–63.
91. Mahmoud SM, Paish EC, Powe DG, et al. An evaluation of the clinical significance of FOXP3+ infiltrating cells in human breast cancer. Breast Cancer Res Treat. 2011;127(1):99–108.
92. Ladoire S, Mignot G, Dabakuyo S, et al. In situ immune response after neoadjuvant chemotherapy for breast cancer predicts survival. J Pathol. 2011;224(3):389–400.
93. Bates GJ, Fox SB, Han C, et al. Quantification of regulatory T cells enables the identification of high-risk breast cancer patients and those at risk of late relapse. J Clin Oncol. 2006;24(34):5373–80.
94. de Kruijf EM, van Nes JG, Sajet A, et al. The predictive value of HLA class I tumor cell expression and presence of intratumoral Tregs for chemotherapy in patients with early breast cancer. Clin Cancer Res. 2010;16(4):1272–80.
95. Gobert M, Treilleux I, Bendriss-Vermare N, et al. Regulatory T cells recruited through CCL22/CCR4 are selectively activated in lymphoid infiltrates surrounding primary breast tumors and lead to an adverse clinical outcome. Cancer Res. 2009;69(5):2000–9.
96. Liu F, Lang R, Zhao J, et al. CD8 cytotoxic T cell and FOXP3 regulatory T cell infiltration in relation to breast cancer survival and molecular subtypes. Breast Cancer Res Treat. 2011;130(2):645–55.
97. Russell JH, Ley TJ. Lymphocyte-mediated cytotoxicity. Annu Rev Immunol. 2002;20:323–70.
98. Naito Y, Saito K, Shiiba K, et al. CD8+ T cells infiltrated within cancer cell nests as a prognostic factor in human colorectal cancer. Cancer Res. 1998;58(16):3491–4.
99. Galon J, Costes A, Sanchez-Cabo F, et al. Type, density, and location of immune cells within human colorectal tumors predict clinical outcome. Science. 2006;313(5795):1960–4.
100. Sato E, Olson SH, Ahn J, et al. Intraepithelial CD8+ tumor-infiltrating lymphocytes and a high CD8+/regulatory T cell ratio are associated with favorable prognosis in ovarian cancer. Proc Natl Acad Sci USA. 2005;102(51):18538–43.
101. Mahmoud SM, Paish EC, Powe DG, et al. Tumor-infiltrating CD8+ lymphocytes predict clinical outcome in breast cancer. J Clin Oncol. 2011;29(15):1949–55.
102. Murri AM, Hilmy M, Bell J, et al. The relationship between the systemic inflammatory response, tumour proliferative activity, T-lymphocytic and macrophage infiltration, microvessel density and survival in patients with primary operable breast cancer. Br J Cancer. 2008;99(7):1013–9.
103. Matkowski R, Gisterek I, Halon A, et al. The prognostic role of tumor-infiltrating CD4 and CD8 T lymphocytes in breast cancer. Anticancer Res. 2009;29(7):2445–51.
104. Behrens MD, Wagner WM, Krco CJ, et al. The endogenous danger signal, crystalline uric acid, signals for enhanced antibody immunity. Blood. 2007;111(3):1472–9.
105. Santisteban M, Reiman JM, Asiedu MK, et al. Immune-induced epithelial to mesenchymal transition in vivo generates breast cancer stem cells. Cancer Res. 2009;69(7):2887–95.
106. Radisky DC. Epithelial-mesenchymal transition. J Cell Sci. 2005;118(Pt 19):4325–6.
107. Shook D, Keller R. Mechanisms, mechanics and function of epithelial-mesenchymal transitions in early development. Mech Dev. 2003;120(11):1351–83.
108. Thiery JP, Acloque H, Huang RY, et al. Epithelial-mesenchymal transitions in development and disease. Cell. 2009;139(5):871–90.
109. Duband JL. Neural crest delamination and migration: integrating regulations of cell interactions, locomotion, survival and fate. Adv Exp Med Biol. 2006;589:45–77.
110. Runyan RB, Heimark RL, Camenisch TD, et al. Epithelial-mesenchymal transformation in the embryonic heart. In: Savagner P, editor. Rise and fall of epithelial phenotype. New York: Springer; 2005. p. 40–55.

111. Fata JE, Werb Z, Bissell MJ. Regulation of mammary gland branching morphogenesis by the extracellular matrix and its remodeling enzymes. Breast Cancer Res. 2004;6(1):1–11.
112. Sternlicht MD, Kouros-Mehr H, Lu P, et al. Hormonal and local control of mammary branching morphogenesis. Differentiation. 2006;74(7):365–81.
113. Wiseman BS, Werb Z. Stromal effects on mammary gland development and breast cancer. Science. 2002;296(5570):1046–9.
114. Kouros-Mehr H, Werb Z. Candidate regulators of mammary branching morphogenesis identified by genome-wide transcript analysis. Dev Dyn. 2006;235(12):3404–12.
115. Wiseman BS, Sternlicht MD, Lund LR, et al. Site-specific inductive and inhibitory activities of MMP-2 and MMP-3 orchestrate mammary gland branching morphogenesis. J Cell Biol. 2003;162(6):1123–33.
116. Sympson CJ, Talhouk RS, Alexander CM, et al. Targeted expression of stromelysin-1 in mammary gland provides evidence for a role of proteinases in branching morphogenesis and the requirement for an intact basement membrane for tissue-specific gene expression. J Cell Biol. 1994;125(3):681–93.
117. Wang YA, Shen K, Wang Y, et al. Retinoic acid signaling is required for proper morphogenesis of mammary gland. Dev Dyn. 2005;234(4):892–9.
118. Perez-Moreno M, Fuchs E. Catenins: keeping cells from getting their signals crossed. Dev Cell. 2006;11(5):601–12.
119. Chanson L, Brownfield D, Garbe JC, et al. Self-organization is a dynamic and lineage-intrinsic property of mammary epithelial cells. Proc Natl Acad Sci USA. 2011;108(8):3264–9.
120. LaBarge MA, Nelson CM, Villadsen R, et al. Human mammary progenitor cell fate decisions are products of interactions with combinatorial microenvironments. Integr Biol. 2009;1(1): 70–9.
121. Cukierman E, Pankov R, Stevens DR, et al. Taking cell-matrix adhesions to the third dimension. Science. 2001;294(5547):1708–12.
122. Yang J, Weinberg RA. Epithelial-mesenchymal transition: at the crossroads of development and tumor metastasis. Dev Cell. 2008;14(6):818–29.
123. Shirakihara T, Horiguchi K, Miyazawa K, et al. TGF-beta regulates isoform switching of FGF receptors and epithelial-mesenchymal transition. EMBO J. 2011;30(4):783–95.
124. Radisky ES, Radisky DC. Matrix metalloproteinase-induced epithelial-mesenchymal transition in breast cancer. J Mammary Gland Biol Neoplasia. 2010;15(2):201–12.
125. Lochter A, Galosy S, Muschler J, et al. Matrix metalloproteinase stromelysin-1 triggers a cascade of molecular alterations that leads to stable epithelial-to-mesenchymal conversion and a premalignant phenotype in mammary epithelial cells. J Cell Biol. 1997;139(7):1861–72.
126. Lochter A, Srebrow A, Sympson CJ, et al. Misregulation of stromelysin-1 expression in mouse mammary tumor cells accompanies acquisition of stromelysin-1-dependent invasive properties. J Biol Chem. 1997;272(8):5007–15.
127. Lopez-Novoa JM, Nieto MA. Inflammation and EMT: an alliance towards organ fibrosis and cancer progression. EMBO Mol Med. 2009;1(6–7):303–14.
128. Nieto MA. The ins and outs of the epithelial to mesenchymal transition in health and disease. Annu Rev Cell Dev Biol. 2011;27:347–76.
129. Sanjabi S, Zenewicz LA, Kamanaka M, et al. Anti-inflammatory and pro-inflammatory roles of TGF-beta, IL-10, and IL-22 in immunity and autoimmunity. Curr Opin Pharmacol. 2009;9(4):447–53.
130. Acloque H, Adams MS, Fishwick K, et al. Epithelial-mesenchymal transitions: the importance of changing cell state in development and disease. J Clin Invest. 2009;119(6): 1438–49.
131. Massague J. TGFbeta in cancer. Cell. 2008;134(2):215–30.
132. Sieweke MH, Stoker AW, Bissell MJ. Evaluation of the cocarcinogenic effect of wounding in Rous sarcoma virus tumorigenesis. Cancer Res. 1989;49(22):6419–24.
133. Sieweke MH, Thompson NL, Sporn MB, et al. Mediation of wound-related Rous sarcoma virus tumorigenesis by TGF-beta. Science. 1990;248(4963):1656–60.

134. Fuxe J, Vincent T, Garcia de Herreros A. Transcriptional crosstalk between TGFbeta and stem cell pathways in tumor cell invasion: role of EMT promoting Smad complexes. Cell Cycle. 2010;9(12):2363–74.
135. Bates RC, Mercurio AM. Tumor necrosis factor-alpha stimulates the epithelial-to-mesenchymal transition of human colonic organoids. Mol Biol Cell. 2003;14(5):1790–800.
136. Souslova V, Townsend PA, Mann J, et al. Allele-specific regulation of matrix metalloproteinase-3 gene by transcription factor NFkappaB. PLoS One. 2010;5(3):e9902.
137. Radisky DC, Levy DD, Littlepage LE, et al. Rac1b and reactive oxygen species mediate MMP-3-induced EMT and genomic instability. Nature. 2005;436(7047):123–7.
138. Barrallo-Gimeno A, Nieto MA. The Snail genes as inducers of cell movement and survival: implications in development and cancer. Development. 2005;132(14):3151–61.
139. Carver EA, Jiang R, Lan Y, et al. The mouse snail gene encodes a key regulator of the epithelial-mesenchymal transition. Mol Cell Biol. 2001;21(23):8184–8.
140. Rowe RG, Li XY, Hu Y, et al. Mesenchymal cells reactivate Snail1 expression to drive three-dimensional invasion programs. J Cell Biol. 2009;184(3):399–408.
141. de Herreros AG, Peiro S, Nassour M, et al. Snail family regulation and epithelial mesenchymal transitions in breast cancer progression. J Mammary Gland Biol Neoplasia. 2010;15(2):135–47.
142. Przybylo JA, Radisky DC. Matrix metalloproteinase-induced epithelial-mesenchymal transition: tumor progression at Snail's pace. Int J Biochem Cell Biol. 2007;39(6):1082–8.
143. Franco DL, Mainez J, Vega S, et al. Snail1 suppresses TGF-beta-induced apoptosis and is sufficient to trigger EMT in hepatocytes. J Cell Sci. 2010;123(Pt 20):3467–77.
144. Sullivan NJ, Sasser AK, Axel AE, et al. Interleukin-6 induces an epithelial-mesenchymal transition phenotype in human breast cancer cells. Oncogene. 2009;28(33):2940–7.
145. Fuchs E, Tumbar T, Guasch G. Socializing with the neighbors: stem cells and their niche. Cell. 2004;116(6):769–78.
146. Kakarala M, Wicha MS. Implications of the cancer stem-cell hypothesis for breast cancer prevention and therapy. J Clin Oncol. 2008;26(17):2813–20.
147. Lapidot T, Sirard C, Vormoor J, et al. A cell initiating human acute myeloid leukaemia after transplantation into SCID mice. Nature. 1994;367(6464):645–8.
148. Hollier BG, Evans K, Mani SA. The epithelial-to-mesenchymal transition and cancer stem cells: a coalition against cancer therapies. J Mammary Gland Biol Neoplasia. 2009;14(1):29–43.
149. Creighton CJ, Li X, Landis M, et al. Residual breast cancers after conventional therapy display mesenchymal as well as tumor-initiating features. Proc Natl Acad Sci USA. 2009;106(33):13820–5.
150. Mani SA, Guo W, Liao MJ, et al. The epithelial-mesenchymal transition generates cells with properties of stem cells. Cell. 2008;133(4):704–15.
151. Morel AP, Lievre M, Thomas C, et al. Generation of breast cancer stem cells through epithelial-mesenchymal transition. PLoS One. 2008;3(8):e2888.
152. Dunn GP, Old LJ, Schreiber RD. The immunobiology of cancer immunosurveillance and immunoediting. Immunity. 2004;21(2):137–48.
153. Mantovani A, Allavena P, Sica A, et al. Cancer-related inflammation. Nature. 2008;454(7203):436–44.
154. Tan TT, Coussens LM. Humoral immunity, inflammation and cancer. Curr Opin Immunol. 2007;19(2):209–16.
155. Reiman JM, Kmieciak M, Manjili MH, et al. Tumor immunoediting and immunosculpting pathways to cancer progression. Semin Cancer Biol. 2007;17(4):275–87.
156. Reiman JM, Knutson KL, Radisky DC. Immune promotion of epithelial-mesenchymal transition and generation of breast cancer stem cells. Cancer Res. 2010;70(8):3005–8.
157. Kmieciak M, Knutson KL, Dumur CI, et al. HER-2/neu antigen loss and relapse of mammary carcinoma are actively induced by T cell-mediated anti-tumor immune responses. Eur J Immunol. 2007;37(3):675–85.
158. Knutson KL, Lu H, Stone B, et al. Immunoediting of cancers may lead to epithelial to mesenchymal transition. J Immunol. 2006;177(3):1526–33.

Chapter 4
Remodelling of the Extracellular Matrix: Implications for Cancer

Thomas R. Cox and Janine T. Erler

Abstract Remodelling of the extracellular matrix (ECM) is essential for development, wound healing and normal organ homeostasis. The tight control of biochemical and biomechanical properties of the ECM and the ongoing remodelling of the ECM is critical in maintaining correct organ function. When this remodelling becomes excessive or uncontrolled, life-threatening pathological conditions such as cancer can result. In this chapter, we focus on the importance and complexity of cell–ECM interactions in determining correct cell behaviour under normal conditions and what happens when it all goes wrong in situations of cancer. We discuss the pathogenesis of breast cancer initiation, progression and metastatic dissemination in the context of the ECM and illustrate how dynamic ECM remodelling contributes to all of these stages. We discuss how improving our understanding of ECM remodelling in all stages of carcinogenesis is crucial for uncovering novel therapeutic targets and treatment strategies for a disease which currently presents challenging obstacles with respect to clinical treatment.

4.1 What Is the Extracellular Matrix?

The extracellular matrix (ECM) is defined as the diverse collection of biochemically distinct components including proteins, glycoproteins, proteoglycans and polysaccharides that surrounds cells in all solid tissues. Each of these individual components exhibits markedly different physical and biochemical properties and their exquisite networks of interaction are responsible for determining the structure and function of all organs in the body [1–3].

T.R. Cox • J.T. Erler (✉)
Biotech Research and Innovation Centre (BRIC), University of Copenhagen,
Ole Maaløes Vej 5, Copenhagen, Denmark
e-mail: janine.erler@bric.ku.dk

H. Schatten (ed.), *Cell and Molecular Biology of Breast Cancer*,
DOI 10.1007/978-1-62703-634-4_4, © Springer Science+Business Media New York 2013

Historically, the ECM was thought to provide structural support to tissues by maintaining a complex insoluble scaffold, which is in turn responsible for defining the characteristic shape and dimensions of organs and tissues. The intricate interlocking mesh of fibrillar and non-fibrillar collagens, elastin fibres, glycosaminoglycan-containing non-collagenous glycoproteins, hyaluronic acid and proteoglycans contributes most of the structural support provided by the ECM. Whilst the ECM indeed fulfils a primarily structural and hence physical role, two important considerations are now widely accepted: firstly, that the matrix is far from static, instead being one of the most dynamic compartments of the body; and secondly, that the matrix plays an intricate role in providing contextual information responsible for controlling both individual and collective cell behaviour.

4.2 Laying Down the ECM

Following intracellular synthesis, ECM components are secreted and incorporated into the surrounding environment that encompasses and supports the cells. The exquisite structure and hence function of the ECM manifests in the manner in which these individual ECM components are assembled. Because of their remarkable structural and biochemical diversity, and their functional versatility, the precise composition and organisation of the ECM lends itself to the generation of unique physical (i.e. rigidity, porosity, insolubility, topography); biochemical (i.e. composition, spatial ligand distribution); and biomechanical properties (i.e. stiffness, elasticity, viscosity) that are essential for regulating cell behaviour. All of these properties are capable of exerting both positive and negative effects on cell behaviour, and together the complex summation of inputs is critical in determining the net result of environmental cues and leads to highly tuned properties of the ECM critical to supporting the diverse wealth of tissue architecture, integrity and function observed in the body [4].

Structurally, the individual components of the ECM make up both the basement membrane (BM) (produced by epithelial, endothelial and stromal cells) and interstitial matrix (produced primarily by stromal cells). BM is a highly specialised form of ECM, forming sheet-like structures, which are more compact and less porous than interstitial matrix and important in separating epithelium (or endothelium) from stroma. BM has a distinctive composition containing high amounts of type IV collagen, laminins, fibronectin, heparan sulphate proteoglycans and linker proteins such as nidogen and entactin, which connect the collagens with other protein components [5]. BMs play a key role in epithelial cell function, providing orientational cues that help establish and maintain apico-basal polarity and cellular differentiation. In contrast, interstitial matrix is highly rich in fibrillar collagens, proteoglycans and various glycoproteins such as tenascin C, decorin, fibromodulin, SPARC, lumican, osteopontin and fibronectin. The interstitial matrix is highly charged, hydrated and as a result provides a significant contribution to the tensile strength of tissues

[4, 6, 7]. Although collagens are collectively the most abundant component of the ECM, the differential expression of individual subtypes as well as other ECM components underpins the specific functions of many organs and tissues.

4.3 Functional Properties of the ECM

As we will explore throughout this chapter, the ECM serves many functions in addition to the classical role in providing structural support. Indeed macroscopically, the ECM physically segregates cells and organs, acting as a barrier, anchorage site, movement track for cells, but also as a protective cushion—for example, by regulating hydrostatic pressure within tissues and organs during exposure to deforming stresses. At a microscopic level, the precise organisation and orientation of ECM components creates a highly organised topology that contributes to the functional properties of the matrix [8, 9]. As we shall see in this chapter, through direct or indirect means, the ECM regulates almost all cellular behaviour and is indispensable for the major developmental processes [10–13]. Similarly, the dynamic nature and remodelling of the ECM (discussed later) is an equally important contributor to tissue function as the more static structural properties. At this microscopic level, the highly dynamic molecular network is capable of regulating cellular behaviour through modulation of, among other things, proliferation, cytoskeletal organisation, cellular differentiation and receptor signalling [8, 9]. Functionally discrete tissues and organs have markedly distinct biomechanical and biochemical properties that are critical to defining their precise structure and function, and which are subject to change during the course of development and more importantly, as we will discuss later, during pathogenesis [14].

4.4 The Dynamic Nature of the ECM

More often than not, the ECM is wrongly viewed as a static structure that plays a predominantly scaffolding role in maintaining tissue morphology and changes only in response to growth or repair. In fact, the ECM is an essential part of the milieu of a cell that is surprisingly dynamic and versatile, and influences all fundamental aspects of cell biology [15]. The highly dynamic nature of the ECM means that it is constantly being remodelled at various stages of life, beginning during embryonic development and continuing postnatally. It is worth noting that, whilst the microscopic matrix topology is determined by continuous dynamic ECM remodelling, often macroscopic topology remains mostly unchanged, i.e. except in pathological cases, tissues and organs rarely undergo changes in size, shape or organisation. An excellent exception to this rule, which will be discussed shortly, is the case of mammary gland morphogenesis during involution.

Important, the concept of "ECM dynamics" covers a broad range of activities, which may include changes in the absolute amount of ECM components, or indeed the relative amounts of ECM components, as a result of altered synthesis or degradation. It is becoming increasingly important to distinguish between increased amounts and increased concentrations of an ECM component in disease progression. This was first highlighted in a study by Cattell and co-workers [16]. In their study, aortic collagen and elastin concentrations increased with age, whereas the absolute amounts decreased. The authors attributed this discrepancy to the age-related differential loss (through the activity of matrix-degrading enzymes, discussed later) of other tissue components. Alternatively, and perhaps equally as important, the ECM may show no compositional change of its components at all, either absolute or relative, but instead, in such situations, "ECM dynamics" refers to how individual ECM components are laid down, cross-linked and spatially arranged together via covalent and non-covalent modifications (discussed below). Typically, a combination of both manifests, for example, in aged breast, there is reduced collagen deposition and increased MMP activity over time [14]. However, this is, perhaps counter-intuitively, accompanied by increases in tissue stiffness (i.e. loss of elasticity), and is most likely associated with a disproportionate increase in inappropriate posttranslational cross-linking and modification of ECM proteins.

4.5 Regulation of ECM Dynamics

A carefully controlled balance between matrix synthesis, secretion, modification and enzymatic degradation regulates ECM dynamics. Such regulation is critically important during development and is primarily accomplished by controlling the expression or activities of both ECM components and remodelling enzymes at multiple levels, including transcriptional, translational and posttranslational [17, 18].

ECM components are degraded by matrix-degrading enzymes, including heparanase, cathepsins, hyaluronidases, matriptases and various serine and threonine proteases [19] and the large superfamily of metzincins, including ADAMs, ADAMTSs and the matrix metalloproteinases (MMPs) and their inhibitors, tissue inhibitor of MMPs (TIMPs) [20]. Cleavage and covalent cross-linking of ECM components (posttranslational modification) is also important in determining higher order structure of the ECM, its solubility and ability to be degraded. Such events are key to determining both physical and biomechanical properties of the ECM and will be discussed shortly. The tightly controlled ECM homeostasis is sensitive to altered expression of these proteases, which, if present over prolonged periods of time, will result in excessive ECM remodelling as is frequently observed in both fibrotic diseases and cancer [6]. Changes in matrix homeostasis affect not only the biochemical properties of the matrix but also the resulting physical and biomechanical properties, both of which are crucial for development and correct tissue function.

4.6 Posttranslational Cross-Linking of the ECM

One of the major factors controlling the properties of the ECM is posttranslational modification of the constituent components. One of the biggest posttranslational modifications of the interstitial matrix is cross-linking of collagens, which is critical in determining higher order collagen fibre formation and ECM structure. Collagen cross-linking can occur in both a regulated and nonregulated manner, typically via enzyme-mediated or nonenzyme-mediated processes, respectively. Regulated collagen cross-linking is almost exclusively mediated by lysyl oxidase (LOX) and LOX family members [6, 21, 22], and primarily occurs during developmental processes and wound healing. The LOX family of secreted amine oxidases catalyses the cross-linking of collagens (and elastin) through oxidative deamination of lysine residues. The importance of correct LOX expression and the resulting collagen cross-linking is exemplified by the fact that the LOX-knockout mouse dies at birth due to collapse of a lethally fragile diaphragm and cardiovascular system [23, 24]. The significant importance of aberrant LOX family expression and activity in breast cancer is discussed later in the chapter.

Nonenzymatic collagen cross-linking typically occurs through glycation [25] and transglutamination [26, 27] or increased biglycan and proteoglycan levels [28]. In some cases, fructosylation will also lead to in vivo collagen cross-linking, although this is observed less frequently. Such collagen cross-linking acts to stiffen the ECM, although this process occurs much more slowly than its enzymatic alternative [29]. However, due to the remarkable longevity of ECM proteins in vivo, often measured in years as opposed to hours for intracellular proteins, glycation-mediated collagen cross-linking becomes important and is thought to play a key role in many age-associated diseases [30–34].

4.7 The Importance of Correct Cell–ECM Interactions

The interaction of cells with their surrounding ECM is by no means a unidirectional process and the most prominent feature of cell–ECM interactions is that they are inherently reciprocal. Whilst the cells are continuously secreting, degrading, rearranging and realigning components of the ECM, resulting in changes in ECM properties, these changes are continuously feeding back onto the cell influencing the behaviour in a cyclical manner [6, 14]. Such reciprocal interaction leads to the establishment of a regulatory feedback mechanisms with cell behaviour controlling ECM dynamics and vice versa. Such an intimate relationship under normal conditions ultimately maintains correct tissue structure, function and behaviour allowing cells to adapt to ongoing stresses imparted upon them [35]. However, when this delicate balance is upset, as in the case of many diseases, essential cellular functions become disrupted, significantly altering the way in which cells can respond to their environment. Since cellular responses are tissue and context dependent [36], understanding the complex processes surrounding ECM dynamics and the cues provided is key to determining how ECM changes control or influence cellular responses in disease.

4.8 Biochemical Cues

Biochemical cues are typically provided by ECM composition, and the importance of correct ECM composition and organisation is apparent when we consider orientational cues provided by the BM to cells. BMs are responsible for maintaining apico-basal polarity in epithelial cells and have organ-specific compositions. Changes in BM composition can lead to changes in the physical properties of the BM and hence to changes in cellular shape and behaviour, which, in turn are capable of driving proliferation and tumorigenesis through altered binding or spatial distribution of cell surface receptors [37–39]. Although for many years it was thought that loss of apico-basal polarity due to BM disruption was a secondary consequence of oncogenic transformation, recent investigations of *Drosophila* mutants have shown that loss of polarity determinants might in fact act as a driver of tumorigenesis [40]. As more than 80 % of human cancers, including breast, are derived from the epithelium, the contextual environmental components of tumour initiation and progression are becoming increasingly important in finding commonalities between seemingly distinct tumour types.

In addition to a precise composition of individual components, the ECM also sequesters and hence acts as a "local depot" for a wide range of growth factors and cytokines. As a consequence, the ECM can both induce and suppress signal transduction cascades due to its highly charged nature (primarily due to the presence of polysaccharide modifications) affecting sequestration, release, diffusive capacity and availability of potent growth factors such as tissue growth factor-β (TGF-β), bone morphogenic proteins (BMPs), fibroblast growth factors (FGFs), Wnts and Hedgehogs. At the same time, when local changes in physiological conditions and/or tissue injury occur, they trigger protease activities leading to rapid release of signalling molecules which in turn allows the swift and local growth factor-mediated activation of cellular functions without de novo synthesis.

4.9 Biomechanical Cues

Arguably one of the most rapidly expanding areas of ECM biology is how the biomechanical properties of the ECM, including properties such as ECM elasticity, contribute to development and disease [41, 42]. The biomechanical properties of the ECM are tightly controlled by the specific composition and concentration of matrix components and also by posttranslational modifications, such as glycosylation and cross-linking [5]. These biomechanical properties of the ECM regulate mechanosensory pathways that prompt cells to detect and respond to changes in tissue stiffness. As a result, during the last two decades, mechanical forces have emerged as important regulators in the complex interplay between cells and their environment [43, 44] with many cellular processes including cell survival [45], cell fate determination, angiogenesis [46], differentiation and tissue function [47–49] depending on the correct adhesive geometry and stiffness of the matrix.

Recent research has also focused on how perturbations in ECM stiffness can affect the behaviour of tissue cells, and as such, the elasticity of the three-dimensional ECM is considered to be critical in determining how a cell senses and perceives external forces [50] and thus provides a major environmental cue that determines cellular behaviour [51–56]. For example, cell adhesion and organisation has been shown to depend strongly on matrix stiffness, with most cells cultured on soft elastic substrates failing to form mature stress fibres and adhesions whereas the same cells cultured on stiff plastic of glass substrates tend to develop large focal adhesion and stress fibres [57]. This has been shown to act via mechanisms whereby increasing matrix stiffness enhances Rho-generated cytoskeletal tension to promote focal adhesion (FA) assembly and increase growth factor-dependent ERK activation [50]. Furthermore, force has been shown to be crucial in the assembly and stabilisation of FAs as they build and remain stable under tension, but tend to disintegrate as force is reduced [58] which holds important implications for cell migration.

Gene expression studies of stem cells plated onto matrices of different compliances have shown that matrix stiffness can drive cellular differentiation down alternative lineages [47]. Furthermore, microarray studies showed that gene expression changes that are typically associated with cardiovascular disease occur in cells of the ascending aorta in response to changes in arterial stiffening [59]. The range of stiffness typically encountered by cells in their physiological environments can vary from 100 Pa in the brain to 10 kPa in connective tissue and 50 kPa in the bone [43, 60], but it is important to note that cellular response is not determined by single mechanosensor but by a much more complex network of integrated mechanosensory elements.

4.10 The Importance of the ECM in Development: Mammary Gland Morphogenesis

ECM composition, remodelling and ECM stiffness play a major role in coordinating the function of a variety of tissue types by providing informational cues to many cell types during embryogenesis and organism development. Two excellent examples of which are the development of branching morphogenesis in the lung and the mammary gland, which we will discuss below. In both, studies have shown that the ECM is essential for the establishment and maintenance of tissue polarity and architecture leading to correct organ development. For example, β1-integrin interaction with the ECM is critical in maintaining tissue polarity in the mammary gland [61]. Branching morphogenesis of the mammary gland also occurs as a result of progressive end bud enlargement and expansion to form the highly organised ductal tree. The ductal tree terminates in alveolar acini units (lobules), and these acini undergo further differentiation to produce milk in the presence of lactogenic hormones [62].

All of these processes are accompanied by significant ECM remodelling and the topology and composition of the ECM supporting these ductal trees affect the matrix stiffness and play a key role in modulating cell differentiation [14]. Compliant basement membrane surrounds the mammary epithelial cell bilayer of the ductal tree and provides a mechanical shield that is crucial to maintain its functional integrity [63, 64]. Adjacent to the basement membrane and surrounding the ductal tree is the stiffer intralobular matrix, which is comprised of structural proteins such as collagen type I and III and elastin. This intralobular matrix, together with the basement membrane, defines the form and the function of the breast by providing both mechanical and biochemical cues.

Through attempting to understand the role of the ECM in development, it has allowed us to develop models to interrogate how pathological changes in the ECM promote disease progression. An excellent example of this phenomenon is the growth of mammary epithelial cells (MECs) within reconstituted BM (Matrigel). MECs grown in such 3D in vitro model systems form polarised mammary acini and differentiate in response to lactogenic hormones in a manner that recapitulates the in vivo situation mentioned above [65]. This provides an ideal platform to model how transformation events trigger cells to invade the acini lumen [66] as is typically seen during breast tumorigenesis. Importantly, these studies have also shown that oncogenic behaviour can be induced solely by increasing matrix stiffness [from 170 Pascal's (Pa) to 1,200 Pa], without altering biochemical composition. These mechanical alterations lead to an increase in cell growth and a compromised cell–cell junction integrity that ultimately impedes lumen formation [50]. These studies have also shown that stiff matrices drive the formation of continuously growing, nonpolarised, disorganised and invasive colonies that lack detectable cell–cell junction proteins, exhibiting irregular cell shape, increased focal adhesion contacts and activated forms of the cell-adhesion proteins focal adhesion kinase (FAK), vinculin and p130CAS [50]. Conversely, soft compliant matrix leads to emerging but not mature focal contacts and maintenance of intact, organised acini structures [50]. Such examples of ECM influence on development can also be seen in ECM-vascular cell interactions for blood vessel formation [67] and ECM-neural cell interactions during brain development [68].

Thus, the ECM dictates organ development through a variety of mechanisms, and these consistent, multiple regulatory mechanisms exist to ensure that ECM dynamics, collectively measured by its production, degradation and remodelling, are normal during organ development and function [18]. The disruption of these control mechanisms deregulates and disorganizes the ECM, leading to abnormal behaviours of cells and ultimately failure of organ homeostasis and function. However, despite the presence of multiple control mechanisms, ECM dynamics often become deregulated with age and under disease conditions. Consequently, ECM dynamics become abnormal as the amount, composition or topography of the ECM become aberrant, leading to disorganisation and changes in the essential properties of the ECM. Altered ECM dynamics are one of the most ostensible clinical outcomes in diseases such as tissue fibrosis and cancer [6].

4.11 ECM Dynamics in Adult Mammary Gland Tissue

ECM remodelling and subsequent changes in ECM stiffness are critical in regulating and maintaining normal tissue homeostasis in the adult. For example, the establishment, and more importantly maintenance, of cell polarity relies on correct matrix stiffness [69–71]. Typically, ECM turnover is mostly observed during injury and repair (discussed later). However, matrix stiffness and ECM remodelling are also critical for the regulation of normal function of certain tissues, such as the mammary gland during pregnancy. Increases in ECM protein concentration, increased matrix cross-linking or parallel reorientation of matrix fibrils within a stromal matrix, can stiffen a tissue locally to alter cell growth or direct cell migration [14]. Interstitial fibrillar collagens are major contributors to this tissue stiffness as they are subjected to multiple posttranslational modifications including protease cleavage, glycosylation and cross-linking, which modify their tensile strength and stiffness.

ECM remodelling and stiffness also regulate cell behaviour and tissue phenotype in mammary gland remodelling after pregnancy (discussed above). During milk production, the matrix around the differentiated acini remains soft and compliant. However, extensive remodelling of the ECM occurs during gland involution. This dramatically alters the composition and architecture of the stroma and as a result drives epithelial cell proliferation and differentiation [72].

Repair of damaged tissues is a fundamental biological process that is critically important for tissue and organism survival. Tissue damage can arise from mechanical injury, infection and autoimmune reactions, which may be acute or chronic. Wound healing is critically dependent on ECM remodelling [73]. Disruption of the carefully controlled wound healing processes is often pathological leading to fibrotic disease.

During wound healing, the goal is to replace injured cells and tissue as swiftly as possible. Early in this process, inflammatory mediators are released from the damaged tissue that initiates the coagulation cascade; resulting in blood clot formation (primarily composed of fibrins) and early ECM deposition, typically type III collagen [73, 74]. Fibroblasts transform into collagen secreting α-SMA+ myofibroblasts as they migrate into the wound and also produce MMPs, which degrade the basement membrane and allow immune cell infiltration. Endothelial and epithelial cells also secrete MMPs together with many different growth factors, cytokines and chemokines that recruit leukocytes and promote their proliferation. Granulation tissue is laid down but then, the initial, weaker collagen III-rich ECM is degraded and replaced with a stronger predominantly collagen I-rich ECM with activated myofibroblasts promoting wound contraction and epithelial/endothelial cell division and migration which regenerates the damaged tissue. In the case of scarring, excess collagen I-rich ECM is continuously deposited resulting in the remodelling and loss of normal tissue

[75] and is usually the result of a persistent irritation and a chronic inflammatory response, often as a consequence of repeated injury. These chronic responses arise from sustained expression of proteolytic enzymes, fibrogenic cytokines, growth factors and angiogenic factors [73] and deposition of ECM components.

4.12 When It All Goes Wrong: The ECM and the Tumour Microenvironment

The concept that the local microenvironment plays an important role in regulating cell behaviour is far from a novel one and has always been the central theme in classical embryology for decades. However, only recently, and perhaps embarrassingly, has the importance become appreciated in the cancer biology field [36, 76, 77]. Over two decades ago Dvorak published a paper in the New England Journal of Medicine proposing that tumours behaved as wounds that could not heal, highlighting the striking similarity between the generation of tumour stroma and wound healing [78]. Over the past decade, solid tumours have increasingly been recognised as discrete "organs" that show a complexity that approaches and may even exceed that of normal healthy tissue [79] and the extrinsic contribution of the tumour microenvironment (TME) is now considered to be equally as important as tumour cell intrinsic factors. Indeed, despite being a disorganised organ, tumours still develop using many of the same cellular and developmental processes regulated by environmental cues essential for organogenesis [4, 7, 80].

Although tightly controlled during embryonic development and organ homeostasis, the ECM is commonly deregulated and becomes disorganised in diseases such as cancer. Importantly, the unique characteristic properties of the ECM, responsible for providing the exquisite variation in tissues, also contribute enormous importance in the development of cancer. Abnormal ECM affects tumour progression directly by acting on tumour cells; for example, abnormal ECM dynamics can compromise basement membrane as a physical barrier and promote epithelial–mesenchymal transition, which together can facilitate tissue invasion by cancer cells [81, 82]. Abnormal ECM can also affect tumorigenesis indirectly by deregulating the behaviour of stromal cells, in turn facilitating tumour-associated angiogenesis and inflammation leading to the generation of the TME. With this in mind, it is important to emphasise that solid tumours comprise not only the malignant cancer cells but also several other non-malignant cell types which are equally as important and together constitute the stromal tissue including fibroblasts [83], also often referred to as cancer-associated fibroblasts (CAFs); resident non-malignant epithelial cells; pericytes; myofibroblasts; vascular and lymphovascular endothelial cells; and infiltrating cells of the immune system, myeloid-derived suppressor cells (MDSCs) [84], tumour-associated macrophages (TAMs) and MAST cells.

These stromal-derived cells act to stimulate tumour cell proliferation and invasion via the secretion of high amounts of ECM components such as collagen types I, II, III, V and IX [85, 86] and tenascin C (TNC) [87] as well as both paracrine and autocrine growth factors, along with the secretion of matrix remodelling enzymes, such as urokinase plasminogen activator (uPA), cathepsins B and D [88], MMP2 and MMP9 [89], tryptases and chymases [90]. Indeed, increased ECM deposition is a classical hallmark of breast cancer formation and progression [91], and breast tumours (as well as many other solid tumours) are initially detected as a palpable stiffening of the local tissue. As such, several approaches such as magnetic resonance and ultrasound elastography have been developed to exploit this characteristic and enhance cancer detection [92, 93].

All together, the tumour stroma acts to contribute to the altered ECM microenvironment, providing additional oncogenic signals that can lead to an acceleration of cancer progression [83]. Unfortunately, a more detailed discussion of the contribution of the diverse stromal cells to tumour progression is beyond the scope of this chapter. However, in the context of cancer, it is exciting to note that the therapeutic disruption of these microenvironmental, ECM-derived cues, provided by stromal cells cues could retard breast tumour growth and prove to be a promising strategy in difficult to treat cancer subtypes.

Classically, the majority of work has aimed to determine how cellular components of the TME initiate and promote cancer development [94]. However, recent progress has also highlighted the significant importance of the noncellular components of the niche, namely the ECM and ECM dynamics, during cancer progression [50, 95–97]. It is important to remember that the vastly diverse properties of the ECM are not independent; rather, they are intertwined. For example, biochemical changes in ECM composition under pathological conditions also lead to changes in ECM biomechanical properties and other physical ECM properties, which combine to exert effects on the cells interaction with the ECM.

4.13 Deregulated ECM Dynamics Play a Critical Role in Influencing the Tumour Microenvironment

Tumour development is a complex, dynamic and progressive process that involves both cellular and environmental cues, and now it is clear that the ECM is much more than a passive bystander in regulating cellular behaviour and phenotype during development and in particular during the progression of cancers. Instead there is a network of bidirectional, dynamic and intricately complex interactions between the cells and the ECM that is regulated by both tumour and non-malignant stromal cells. Non-malignant cells of stromal tissue are capable of producing a unique microenvironment that can modify the neoplastic properties of the tumour cells. In turn, during the course of multistep tumorigenesis, the tumour cells contribute to the generation and modification of the local microenvironment to further enhance their survival thereby creating a positive tumourigenic feedback loop. Therefore, the TME is a dynamic network that includes the cancer cells and the stromal tissue, as well as the all-surrounding ECM, and it has been proposed that, once established, tumours should be considered functionally discrete organs [36].

In addition to changes in its biochemical properties (compositional makeup), the architecture and other physical properties of tumour-associated ECM are fundamentally different from that of the normal tissue ECM. For example, rather than relaxed non-orientated fibrils, the collagen I in breast tumours is often highly linearised and either orientated adjacent to the epithelium or projecting perpendicularly into the tissue at the invasive front of the tumour [96, 98]. Such ECM organisation is a classic representation of metastatic dissemination, and often, transformed mammary epithelial cells are found on bundles of linear collagen fibres adjacent to blood vessels [99]. Consistent with this, intravital imaging has shown tumour cells are capable of travelling along these realigned collagen fibres to facilitate invasion through tissue and extravasation into the bloodstream [100]. These changes are also reflected in the biomechanical properties, where reports have demonstrated that tumour stroma is typically significantly stiffer than normal stroma (~400 Pa compared with 150 Pa) [8, 14, 96]. More recently however, increased matrix stiffness and ECM remodelling has been shown to occur in premalignant tissue and significantly contribute to malignant transformation in the breast (~350 Pa vs. 150 Pa) [96]. In this case it has been shown that part of this increase can be attributed to excess activities of lysyl oxidase (LOX), which cross-links collagen fibres and other ECM components [101]. Indeed such upregulation of LOX has been seen in several cancers including breast, head and neck and colorectal and is seen as a poor prognostic marker [95, 102–104]. Importantly, a study using mouse genetics has shown that overexpression of LOX increases ECM stiffness and promotes tumour cell invasion and progression. In contrast, inhibition of LOX reduces tissue fibrosis and tumour incidence in the *Neu* breast cancer model [96]. Together these data demonstrate that deregulation of collagen cross-linking and ECM stiffness is more than just a secondary outcome but instead plays a causative role in breast cancer pathogenesis. Interestingly though, the same study showed that overexpression of LOX alone is insufficient to cause tumours to form suggesting that deregulation of ECM remodelling is a co-conspirator rather than a primary inducer of tumorigenesis in the breast [6].

A large majority of increased tumour and adjacent tissue stiffness also occurs in part as a result of increased ECM deposition, as well as from increased ECM remodelling. The deposition of type I collagen and fibronectin are the most common and abundant ECM components deposited in cancer [105]. This predominantly results from desmoplasia; the uncontrolled fibrosis driven by signals from the primary tumour and mediated by myofibroblasts. Desmoplasia results in a pervasive dense fibrous tissue which typically surrounds the tumour and increases local tissue stiffness and is generally only associated with malignant tumours as it is rarely observed in benign tumours [14]. The mechanisms and key players implicated in tumour-associated fibrosis are much the same as pathological fibrosis. For example, CCL2, CXC4, CCR2, CCR7, TGFβ, CTGF and IL-6 have all been reported to play a role in tumour progression and its associated fibrosis [106–110].

Remodelling of the basement membrane (BM) is also commonly associated with cancer and, moreover, malignant progression and metastatic dissemination. Disruption of the BM abrogates apico-basal polarity and also allows tumour cells to escape the primary tumour. Similarly, many other ECM components and their receptors such as heparin sulphate proteoglycans and CD44 that facilitate growth factor signalling are frequently overproduced in cancer [111–113].

4.14 Disrupting ECM Dynamics and Turnover

ECM dynamics play a critical role in tumour progression, and a careful balance of matrix remodelling enzymes governs this, and indeed expression of many of these ECM remodelling enzymes is often deregulated in human cancers. Lysyl oxidases, heparanases, 6-*O*-sulfatases, cysteine cathepsins, urokinase and most notable, the MMPs are frequently overexpressed in breast and other cancers [114, 115]. LOX (as mentioned above) is required to cross-link newly synthesised collagen and its expression/activity is elevated in response to increased collagen deposition. Elevated LOX and LOX-Like (LOXL) family member expression is significantly correlated with metastasis and decreased survival in cancer patients and mouse models of breast cancer [95, 116]. Increased LOX activity results in increased matrix stiffness and ECM remodelling in premalignant tissue [96] and has been shown to increase the invasiveness of many cancer cell types [95, 117, 118]. Whilst the exact mechanisms by which this occurs remain to be elucidated, it is thought to occur through increased cell-matrix adhesion, increased matrix stiffness and integrin activation leading to increased SRC and focal adhesion kinase (FAK) activation [95, 102, 103, 119, 120]. In particular, the increase in collagen deposition and subsequently matrix stiffness in and around tumours leads to increased integrin clustering [50, 121] resulting in an enhancement of mechanotransduction into the cell via ERK–ROCK signalling, consequently promoting cell survival, proliferation and migration. Interestingly, work by Samuel et al. [35] has also shown that constitutive ROCK activation in skin leads to elevated tissue stiffness via upregulated collagen deposition resulting in increased tumour number, growth and progression in the K14-ROCK:ER mouse. Finally, increased collagen cross-linking and ECM stiffness as a result of LOX overexpression also promotes focal adhesion assembly and ERK and PI3 kinase signalling and facilitates *Neu*-mediated oncogenic transformation [96]. Thus it still remains unclear whether ROCK signalling drives, or is as a consequence of aberrant ECM dynamics, but importantly both are inexplicably linked in ECM-mediated tumour progression.

A distinct but closely related member of the lysyl oxidase family is the so-called lysyl oxidase-like 2 (LOXL2). Whilst possessing a conserved catalytic domain, leading to a proposed similar function of ECM cross-linking to that of LOX, the presence of multiple scavenger receptor cysteine-rich (SRCR) domains in the N-terminal are thought to tailor LOXL2-specific protein–protein interactions. As such, this second family member has recently become the focus of increased research activity, and upregulation of LOXL2 has been observed in a number of human cancers including breast [116, 118, 122–129]. In humans, LOXL2 expression has been shown to be closely associated with regions of collagenous ECM, activated fibroblasts, tumour–stroma boundaries and tumour-associated vasculature [122].

In particular, high LOXL2 expression is associated with a more aggressive phenotype in breast cancer and has also been correlated with metastasis and the reduced survival of patients with estrogen receptor (ER)-negative breast cancer [116]. Whilst the precise mechanisms are not clear, most reports suggest that increased LOXL2 expression leads to tumour progression and metastasis, through promoting tumour cell invasion via the remodelling of the tumour microenvironment. LOXL2 activity has also been shown to regulate the expression of claudin 1 (CLDN1) and lethal giant larvae homologue 2 (LGL2), which are genes that encode components of tight junctions and cell polarity complexes, respectively [124]. In breast cancer, secreted LOXL2 has been shown to drive breast cancer invasion and has also been associated with the activation of tissue inhibitor of metalloproteinases 1 (TIMP1) and MMP9 [116, 122, 125]. Such dynamic ECM remodelling is likely to increase integrin activation and subsequent intracellular signalling involving mediators such as focal adhesion kinase (FAK) and MAPK67. Importantly, Akiri et al. [130] showed that upregulation of LOXL2 in non-invasive breast cancer cells implanted orthotopically in nude mice induced fibrotic foci formation and increased the invasiveness of these cells. As well as affecting tumour cells directly, LOXL2 has been shown to be upregulated in tumours as a result of hypoxia, leading to increases in ECM cross-linking and stiffening which leads to sprouting angiogenesis and vascularisation of the tumour [131]. Importantly, the inhibition of secreted LOXL2 and LOX for that matter leads to the significant reduction in breast tumour progression in both xenograft and transgenic mouse models [95, 96, 122, 127].

At this point, it is especially interesting to note the control the ECM may exert over tumour cells. Elegant studies have shown that cells with a tumorigenic genotype can become phenotypically normal if the environmental context is appropriately manipulated, and there is increasing evidence that it might be possible to restore aggressive breast cancer cell lines to a near-normal phenotype by manipulating environmental cues and simultaneously inhibiting multiple signalling pathways [36]. Thus, oncogenic transformation is by no means the be-all and end-all of tumour progression and as mentioned previously, the ECM may be critical in determining the penetrance of oncogenic transformations. One could propose the theory that the ECM and in particular ECM dynamics may cause irreversible changes to the normal cellular niche which at a pivotal point switch it from a tumour-suppressing to tumour-promoting environment.

4.15 The ECM in Metastatic Dissemination and Colonisation: The Century Old Seed and Soil Hypothesis

Even with the wealth of information available to us, it remains unclear as to why some solid tumours preferentially metastasise to a particular organ and others exhibit different or lesser degrees of specificity. For example, prostate cancer will, in most cases, exhibit metastasis specifically to the bone [132], whereas breast cancer will metastasise to bone, liver, lung and brain and colorectal predominantly to liver. One of the current bottlenecks in understanding this is due to the fact that metastasis as a process is inherently difficult to observe and study, since macro-metastases are the clinical end point of the process. In the 1920s, Ewing [133] made the proposal that when a solid tumour metastasises, the final resting organ is determined as a function of mechanical and anatomical features of the vascular system. Whilst the anatomy of the circulatory and lymphatic systems certainly helps to explain the delivery of tumour cells to distant organs, it cannot, in reality, fully account for the site-specific bias of some tumours. Instead, it appears that the successful colonisation of macro-metastases is dependent on a receptive microenvironment and that the mere entry into and exit of tumour cells from the circulation is not enough to give rise to clinically detectable metastases. It has been estimated from experimental models that there is somewhere in the region of one million cells per gram of solid tumour tissue released on a daily basis [134]. These enormous numbers of cells circulate the body and lodge in various organs, yet only a miniscule percentage of these cells will propagate into overt metastases. The conclusion is that one of the defining steps in tumour metastasis, is the ability of the lodging tumour cells to successfully colonise either the same and perhaps more frighteningly, seemingly distinct organs with which they have previously had no connection. It is without doubt that oncogenic transformation is considered the most important event in initiation of tumorigenesis, however, it is not sufficient for metastatic competence, as has been evidenced by many in vivo models of oncogene-driven tumorigenesis that fail to show establishment of distant metastases [135], or indeed the fact that in some patients disseminated tumour cells are detectable, yet overt metastases fail to form [136]. Thus this model assumes that tumour cells do not solely dictate their own fate, but that presence of a hospitable microenvironment is essential—not just permissive—for disseminating tumour cells to spawn a secondary tumour growth.

Over a century ago in 1889, the English surgeon Stephen Paget [137] first proposed the "seed and soil" hypothesis to explain the seemingly predictable spread of solid tumours. By analysing autopsy records of 735 cases of advanced breast cancer, Paget discovered predictable patterns of bone and visceral metastasis. He introduced the concept of a receptive milieu and his hypothesis put forth that for a tumour cell (seed) to grow it requires the appropriate local microenvironment (soil). Seminal work by Fidler and colleagues [138, 139] to support this showed that although circulating tumour cells are found in the tumour vasculature of multiple organs, they do not give rise to metastatic disease; however, other selective sites consistently develop metastatic tumour deposits and as such these sites must be more conducive to tumour cell colonisation. More recently, exciting work has been published by Cox et al. [140] in which the authors present critical evidence linking tissue fibrosis to enhanced metastasis at secondary organs. The authors show that fibrotic ECM remodelling in organs, independent of primary tumours, is capable of generating pro-metastatic milieu within tissues that subsequently enhances tumour cell colonization and outgrowth. This work elegantly highlights how ECM remodelling is important in regulating tumour cell ability to colonise secondary organs.

4.16 ECM Dynamics and the Pre-metastatic Niche

Perhaps one of the most frightening concepts of metastasis is that some solid tumours appear capable of actively appropriating distant secondary sites in advance of their arrival [6, 141–143]. The idea that tumours are capable of predefining future sites of metastasis is both exciting and terrifying as we try to understand the dynamic networks associated with solid tumour metastasis. Exactly how a tumour cell can alter the distant metastatic microenvironment is of great importance and will unlock novel strategies for successfully targeting these processes.

In 2002, a paper published by Hiratsuka [144] and colleagues reignited an interest in the metastatic milieu of primary tumours and that of tumour cells at metastatic sites. In this paper, they presented compelling evidence in support of Paget's original "seed and soil" hypothesis showing that primary tumours are capable of appropriating secondary sites in advance of tumour cell arrival by modulating the local environment. A paper closely followed this in 2005 by Kaplan et al. in which they coined the term "pre-metastatic niche" [145, 146]. The authors showed that deposition of fibronectin and recruitment of VEGF receptor 1^{+} hematopoietic progenitor cells to these sites of future metastasis was critical for tumour cell colonisation. In 2009, Erler and colleagues published a paper showing for the first time that ECM remodelling was critical to the generation of the pre-metastatic niche in breast cancer metastasis [117], reaffirming the notion that the ECM is a pivotal player in metastatic dissemination. The authors showed that as in the case of the primary tumour microenvironment, LOX activities are upregulated at pre-metastatic sites [117], resulting in increased ECM dynamics and leading to the recruitment of host immune cells and further remodelling the ECM thus facilitating colonisation of circulating mammary carcinoma cells. Fibronectin deposition is a critical factor in the regulation of the formation of the pre-metastatic niches, and fibronectin matrices have been shown to provide specific microenvironments for the regulation of LOX catalytic activity [147]. Recently, Yaqoob et al. have shown the expression of neuropilin-1 (NRP-1) promotes integrin-dependent fibronectin fibril assembly promoting matrix stiffness and tumour growth [148]. Therefore it is likely that the initial deposition of fibronectin and the function of LOX during pre-metastatic niche formation generates an ECM which acts to facilitate the recruitment of bone-marrow-derived cells and other mesenchymal cells to create permissive niches for circulating tumour cell colonisation.

The contribution of other ECM components to pre-metastatic niche formation has also been highlighted in other studies where hyaluronan (an anionic, nonsulfated glycosaminoglycan) and its receptor CD44 facilitate CXCR4/CXCL12 signalling essential for tumour cell homing to the lung and bone marrow [149–151]. Further still, the role of ECM engagement of the tumour cells at the pre-metastatic niche has been further underlined in more recent studies by Oskarsson et al. [87] and O'Connell et al. [152] in which they demonstrated that TNC engages the Notch and Wnt signalling pathways to support the colonisation of initiating breast cancer cells during the establishment of lung metastases. They further went on to show that interference with cancer cell-derived TNC production results in suppression of the survival and expansion of micrometastatic colonies. Similar work has also recently shown that periostin exerts similar effects acting as a key component in metastatic niches for tumour-initiating cells invading into the lungs [153]. In fact, both tenascin C and periostin have been shown to bind tightly to one another, with periostin further binding collagen I and fibronectin anchoring them both to these more ubiquitous ECM components [154, 155].

During the generation of the pre-metastatic niche, factors produced by disseminated tumour cells, trapped in the peripheral organs, or even released from the primary tumour itself serve to downregulate fibulin 5 (FBLN5), a 66 kDa glycoprotein that belongs to the group of modulators of cell–ECM interactions known as the matricellular proteins. These matricellular proteins are essential for the formation of elastic fibres. Møller and collaborators [156] demonstrated that FBLN5 expression in fibroblasts suppresses metastasis formation by inhibition of the production of MMP9 and by reducing the invasive behaviour of fibroblasts at metastatic sites. Therefore, the downregulation of FBLN5 in stromal fibroblasts driven by tumour-derived factors results in the upregulation of MMP9, ECM remodelling and invasion of fibroblasts facilitating tumour cell metastatic colonisation.

However it is defined, the fact remains that the TME undergoes extensive changes during the establishment, evolution and progression of metastases and that this is a major factor in the determination of the survival and growth of disseminated tumour cells at potential metastatic sites. As such, the underlying message regarding the metastatic and pre-metastatic niche is that all of these events lead towards the creation of a fertile milieu within which tumour cells can develop metastases.

4.17 Influence of the Tumour Microenvironment on Response to Therapy

The importance of the ECM in the tumour context is becoming increasingly clear from the results of clinical drug trials, in which compounds show antitumour activity in vitro, but subsequently fail to show efficacy in a clinical setting. It is becoming increasingly accepted that the response of cancer cells to drugs is determined in part by the 3D TME. Importantly, the precise location of a cell within a tumour may also be an important contributing factor. It has been shown that the ECM-mediated organisational cues can act as survival signals, as is seen in the case of breast carcinoma studies, in which upregulated expression of β4 integrin suppresses apoptosis by inducing Nek activity [157]. Similarly, upregulated expression of drug transporters such as P-glycoprotein, which is observed in 3D but not 2D culture conditions, might explain apparent drug resistance [158]. Furthermore, recent reports demonstrating that stromal cells or stromal cell-derived matrix can modulate the responses of tumour cells to chemotherapeutic agents also underscore the importance of considering the role of the TME in drug discovery and drug sensitivity testing [159].

4.18 Targeting the Tumour Microenvironment, Clinical Implications and Perspective

The unravelling of the relationships between tumour cells and their microenvironment represents an important issue for the development of new therapeutic agents that can fight both initiation and recurrence of cancer. An important area of research is the targeting of ECM remodelling enzymes to disrupt aberrant ECM dynamics and this is becoming an increasingly attractive therapeutic approach for preventing cancer progression. However, this is not always straightforward. For example, although there is much evidence that MMPs play multiple roles in metastases, clinical trials with MMP inhibitors have failed to show significant efficacy because they failed to increase survival rates in patients, prompting a re-examination of MMP function and its complex roles [160]. This has largely been due to unexpected normal tissue toxicities and conflicting roles in both promoting and reducing metastatic dissemination and a lack of robust preclinical models. In addition, most MMP inhibitors are not specific and, more importantly, MMPs are implicated in a wide variety of functions both pro- and antitumorigenic. Finally, there is thought to be functional redundancy between MMP family members. Although a new generation of highly specific MMP inhibitors and other inhibitors of matrix-degrading proteases hold promise [161], targeting the specific enzymes involved in ECM remodelling whilst avoiding unwanted side effects currently remains challenging.

Conversely, targeting enzymes that play a role in increasing matrix stiffness might be a more successful approach. Indeed, inhibiting LOX or LOXL2 should decrease tissue desmoplasia and tumour incidence [101, 162]. Inhibition of LOX reduces primary tumour growth and mechanotransduction in the mammary epithelium [96]. Furthermore, LOX inhibition prevents the formation of invasive branching structures of breast cancer cells in collagen in vitro, the invasion of tumours in vivo, and abrogates immune cell recruitment and the establishment of pre-metastatic niches and consequently metastases in vivo [95, 117]. Indeed, LOX seems an excellent therapeutic target and inhibitors are now in development for use in the clinic.

In terms of the LOX family member LOXL2, therapeutic development is rapidly progressing. Work by Barker et al. showed that inhibition of LOXL2 catalytic activity demonstrated a marked reduction in metastatic spread of primary breast tumours [116]. Consistent with this, Barry-Hamilton et al. [122] showed that mammary tumours treated with a LOXL2-specific monoclonal antibody (AB0023) were significantly smaller, demonstrated a significant reduction in activated fibroblasts and growth factor signalling usually present in activated stroma and also displayed less cross-linked collagenous matrix. Most importantly, treatment with AB0023 was able to reduce metastatic burden in mammary and ovarian metastatic mouse models. As a therapeutic, LOXL2 is a highly attractive target which should have relatively few deleterious side effects owing to the very low expression levels of LOXL2 in normal tissues [163]. As such, a humanised version (AB0024) has recently cleared phase I and entered phase II clinical trials.

These data on LOX and LOXL2 support the idea that antagonising matrix modifications is a promising therapeutic cancer prevention strategy [101]. However, although LOX/LOXL2 are encouraging targets, lessons from the clinic have taught us that targeting a single molecule in a disease network often results in "network compensation" and subsequent drug resistance. The TME contains many overlapping mechanisms that help to maintain its functional disorder; therefore, there is a need to better understand where key individual and collective players sit in cellular networks, how these nodes respond to the dynamics of the network and, more importantly, how environmental cues modulate these signalling networks. Such studies will provide insight into the most effective drug combination therapies and timing of treatments, and these network biology studies can only be done through extensive, multidisciplinary collaborative modelling of tissue disease [164, 165]. Similarly, a greater understanding of ECM dynamics as a driver of tumorigenesis will pave the way to the targeting of both tumour and its environment and as such may offer new insights into combinatorial drug treatment.

4.19 Concluding Remarks

First and foremost, it is essential to reiterate the notion that cancers are highly heterogeneous and that the TME is mechanically and biologically active, and perhaps, more importantly, dynamic, as is highlighted by the fact that it is continuously and progressively remodelled [166]. Thus, abnormal changes in the amount and composition of ECM components can greatly alter ECM biochemical properties, disrupting and deregulating cell behaviour and potentiating the oncogenic effects of various growth factor signalling pathways during malignant transformation. The lynchpin of the model is that both intrinsic properties of metastatic cells (i.e. the genetic and epigenetic alterations) and the extrinsic properties of the tumour microenvironment are important in determining disease progression.

The tight control of ECM production, turnover and remodelling, and the resulting biomechanical properties, allows for correct organism development, wound healing and normal tissue homeostasis. When matrix homeostasis is perturbed, aberrant ECM can contribute to pathological conditions, including fibrotic disease, tumour progression and metastasis. Such changes can result from excessive deposition of ECM components in response to chronic chemokine, cytokine and growth factor signalling or unbalanced ECM dynamics. A fundamentally important area of future cancer research will be understanding how ECM composition and topography are maintained and how their deregulation influences cancer progression. Such work holds the promise for developing new therapeutic interventions against abnormal ECM in a multitude of pathological conditions, including cancer. Indeed, further studies on the mechanisms of the crosstalk between tumour cells and the TME will allow a better understanding of the nature of these interactions and will help determine the roles of these cells from the TME in cancer progression and metastasis formation.

The hope is that targeting and monitoring ECM abnormalities associated with pathological conditions will soon become standard clinical practice. The challenge remains to identify effective ways to spatially and temporally monitor ECM changes at both the biochemical and biomechanical levels in a non-invasive and quantitative manner. Researchers should always be aware that both cells and their surrounding matrix together make up tissues and organs, and that the issue to address—especially in the case of cancer—is that of complex tissue diseases in which all components should be studied simultaneously.

References

1. Ozbek S, Balasubramanian PG, et al. The evolution of extracellular matrix. Mol Biol Cell. 2010;21(24):4300–5.
2. Radisky ES, Radisky DC. Matrix metalloproteinase-induced epithelial-mesenchymal transition in breast cancer. J Mammary Gland Biol Neoplasia. 2010;15(2):201–12.
3. Whittaker CA, Bergeron KF, et al. The echinoderm adhesome. Dev Biol. 2006;300(1):252–66.
4. Lu P, Weaver VM, et al. The extracellular matrix: a dynamic niche in cancer progression. J Cell Biol. 2012;196(4):395–406.
5. Erler JT, Weaver VM. Three-dimensional context regulation of metastasis. Clin Exp Metastasis. 2009;26(1):35–49.
6. Cox TR, Erler JT. Remodeling and homeostasis of the extracellular matrix: implications for fibrotic diseases and cancer. Dis Model Mech. 2011;4(2):165–78.
7. Egeblad M, Rasch MG, et al. Dynamic interplay between the collagen scaffold and tumor evolution. Curr Opin Cell Biol. 2010;22(5):697–706.
8. Kass L, Erler JT, et al. Mammary epithelial cell: influence of extracellular matrix composition and organization during development and tumorigenesis. Int J Biochem Cell Biol. 2007;39(11):1987–94.
9. Paszek MJ, Weaver VM. The tension mounts: mechanics meets morphogenesis and malignancy. J Mammary Gland Biol Neoplasia. 2004;9(4):325–42.
10. Lu P, Takai K, et al. Extracellular matrix degradation and remodeling in development and disease. Cold Spring Harb Perspect Biol. 2011;3(12).
11. Rebustini IT, Myers C, et al. MT2-MMP-dependent release of collagen IV NC1 domains regulates submandibular gland branching morphogenesis. Dev Cell. 2009;17(4):482–93.
12. Stickens D, Behonick DJ, et al. Altered endochondral bone development in matrix metalloproteinase 13-deficient mice. Development. 2004;131(23):5883–95.
13. Wiseman BS, Sternlicht MD, et al. Site-specific inductive and inhibitory activities of MMP-2 and MMP-3 orchestrate mammary gland branching morphogenesis. J Cell Biol. 2003;162(6):1123–33.
14. Butcher DT, Alliston T, et al. A tense situation: forcing tumour progression. Nat Rev Cancer. 2009;9(2):108–22.
15. Hynes RO. The extracellular matrix: not just pretty fibrils. Science. 2009;326(5957):1216–9.
16. Cattell MA, Anderson JC, et al. Age-related changes in amounts and concentrations of collagen and elastin in normotensive human thoracic aorta. Clin Chim Acta. 1996;245(1): 73–84.
17. Aitken KJ, Bagli DJ. The bladder extracellular matrix. Part I: architecture, development and disease. Nat Rev Urol. 2009;6(11):596–611.
18. Page-McCaw A, Ewald AJ, et al. Matrix metalloproteinases and the regulation of tissue remodelling. Nat Rev Mol Cell Biol. 2007;8(3):221–33.

19. Roycik MD, Fang X, et al. A fresh prospect of extracellular matrix hydrolytic enzymes and their substrates. Curr Pharm Des. 2009;15(12):1295–308.
20. Mott JD, Werb Z. Regulation of matrix biology by matrix metalloproteinases. Curr Opin Cell Biol. 2004;16(5):558–64.
21. Csiszar K. Lysyl oxidases: a novel multifunctional amine oxidase family. Prog Nucleic Acid Res Mol Biol. 2001;70:1–32.
22. Kagan HM, Li W. Lysyl oxidase: properties, specificity, and biological roles inside and outside of the cell. J Cell Biochem. 2003;88(4):660–72.
23. Hornstra IK, Birge S, et al. Lysyl oxidase is required for vascular and diaphragmatic development in mice. J Biol Chem. 2003;278(16):14387–93.
24. Maki JM, Rasanen J, et al. Inactivation of the lysyl oxidase gene Lox leads to aortic aneurysms, cardiovascular dysfunction, and perinatal death in mice. Circulation. 2002;106(19):2503–9.
25. Schnider SL, Kohn RR. Glucosylation of human collagen in aging and diabetes mellitus. J Clin Invest. 1980;66(5):1179–81.
26. Mosher DF, Schad PE. Cross-linking of fibronectin to collagen by blood coagulation Factor XIIIa. J Clin Invest. 1979;64(3):781–7.
27. Mosher DF, Schad PE, et al. Inhibition of blood coagulation factor XIIIa-mediated cross-linking between fibronectin and collagen by polyamines. J Supramol Struct. 1979;11(2):227–35.
28. Wiberg C, Klatt AR, et al. Complexes of matrilin-1 and biglycan or decorin connect collagen VI microfibrils to both collagen II and aggrecan. J Biol Chem. 2003;278(39):37698–704.
29. Avery NC, Bailey AJ. The effects of the Maillard reaction on the physical properties and cell interactions of collagen. Pathol Biol (Paris). 2006;54(7):387–95.
30. Bunn HF, Gabbay KH, et al. The glycosylation of hemoglobin: relevance to diabetes mellitus. Science. 1978;200(4337):21–7.
31. Frank RN. On the pathogenesis of diabetic retinopathy. A 1990 update. Ophthalmology. 1991;98(5):586–93.
32. Glenn JV, Stitt AW. The role of advanced glycation end products in retinal ageing and disease. Biochim Biophys Acta. 2009;1790(10):1109–16.
33. Sasaki N, Fukatsu R, et al. Advanced glycation end products in Alzheimer's disease and other neurodegenerative diseases. Am J Pathol. 1998;153(4):1149–55.
34. Vitek MP, Bhattacharya K, et al. Advanced glycation end products contribute to amyloidosis in Alzheimer disease. Proc Natl Acad Sci USA. 1994;91(11):4766–70.
35. Samuel MS, Lopez JI, et al. Actomyosin-mediated cellular tension drives increased tissue stiffness and beta-catenin activation to induce epidermal hyperplasia and tumor growth. Cancer Cell. 2011;19(6):776–91.
36. Bissell MJ, Radisky D. Putting tumours in context. Nat Rev Cancer. 2001;1(1):46–54.
37. Giancotti FG, Ruoslahti E. Integrin signaling. Science. 1999;285(5430):1028–32.
38. Roskelley CD, Srebrow A, et al. A hierarchy of ECM-mediated signalling regulates tissue-specific gene expression. Curr Opin Cell Biol. 1995;7(5):736–47.
39. Schwartz MA, Baron V. Interactions between mitogenic stimuli, or, a thousand and one connections. Curr Opin Cell Biol. 1999;11(2):197–202.
40. Bilder D, Li M, et al. Cooperative regulation of cell polarity and growth by Drosophila tumor suppressors. Science. 2000;289(5476):113–6.
41. McBeath R, Pirone DM, et al. Cell shape, cytoskeletal tension, and RhoA regulate stem cell lineage commitment. Dev Cell. 2004;6(4):483–95.
42. Reilly GC, Engler AJ. Intrinsic extracellular matrix properties regulate stem cell differentiation. J Biomech. 2010;43(1):55–62.
43. Discher DE, Janmey P, et al. Tissue cells feel and respond to the stiffness of their substrate. Science. 2005;310(5751):1139–43.
44. Vogel V, Sheetz M. Local force and geometry sensing regulate cell functions. Nat Rev Mol Cell Biol. 2006;7(4):265–75.

45. Chen CS, Mrksich M, et al. Geometric control of cell life and death. Science. 1997;276(5317):1425–8.
46. Mammoto A, Connor KM, et al. A mechanosensitive transcriptional mechanism that controls angiogenesis. Nature. 2009;457(7233):1103–8.
47. Engler AJ, Sen S, et al. Matrix elasticity directs stem cell lineage specification. Cell. 2006;126(4):677–89.
48. Gilbert PM, Havenstrite KL, et al. Substrate elasticity regulates skeletal muscle stem cell self-renewal in culture. Science. 2010;329(5995):1078–81.
49. Lutolf MP, Gilbert PM, et al. Designing materials to direct stem-cell fate. Nature. 2009;462(7272):433–41.
50. Paszek MJ, Zahir N, et al. Tensional homeostasis and the malignant phenotype. Cancer Cell. 2005;8(3):241–54.
51. DuFort CC, Paszek MJ, et al. Balancing forces: architectural control of mechanotransduction. Nat Rev Mol Cell Biol. 2011;12(5):308–19.
52. Fernandez-Gonzalez R, Simoes Sde M, et al. Myosin II dynamics are regulated by tension in intercalating cells. Dev Cell. 2009;17(5):736–43.
53. Kolsch V, Seher T, et al. Control of Drosophila gastrulation by apical localization of adherens junctions and RhoGEF2. Science. 2007;315(5810):384–6.
54. Montell DJ. Morphogenetic cell movements: diversity from modular mechanical properties. Science. 2008;322(5907):1502–5.
55. Pouille PA, Ahmadi P, et al. Mechanical signals trigger Myosin II redistribution and mesoderm invagination in Drosophila embryos. Sci Signal. 2009;2(66):ra16.
56. Solon J, Kaya-Copur A, et al. Pulsed forces timed by a ratchet-like mechanism drive directed tissue movement during dorsal closure. Cell. 2009;137(7):1331–42.
57. Pelham Jr RJ, Wang Y. Cell locomotion and focal adhesions are regulated by substrate flexibility. Proc Natl Acad Sci USA. 1997;94(25):13661–5.
58. Riveline D, Zamir E, et al. Focal contacts as mechanosensors: externally applied local mechanical force induces growth of focal contacts by an mDia1-dependent and ROCK-independent mechanism. J Cell Biol. 2001;153(6):1175–86.
59. Durier S, Fassot C, et al. Physiological genomics of human arteries: quantitative relationship between gene expression and arterial stiffness. Circulation. 2003;108(15):1845–51.
60. Janmey PA, Winer JP, et al. The hard life of soft cells. Cell Motil Cytoskeleton. 2009;66(8):597–605.
61. Akhtar N, Marlow R, et al. Molecular dissection of integrin signalling proteins in the control of mammary epithelial development and differentiation. Development. 2009;136(6):1019–27.
62. Nelson CM, Vanduijn MM, et al. Tissue geometry determines sites of mammary branching morphogenesis in organotypic cultures. Science. 2006;314(5797):298–300.
63. Griffith LG, Swartz MA. Capturing complex 3D tissue physiology in vitro. Nat Rev Mol Cell Biol. 2006;7(3):211–24.
64. Timpl R. Macromolecular organization of basement membranes. Curr Opin Cell Biol. 1996;8(5):618–24.
65. Barcellos-Hoff MH, Aggeler J, et al. Functional differentiation and alveolar morphogenesis of primary mammary cultures on reconstituted basement membrane. Development. 1989;105(2):223–35.
66. Weaver VM, Fischer AH, et al. The importance of the microenvironment in breast cancer progression: recapitulation of mammary tumorigenesis using a unique human mammary epithelial cell model and a three-dimensional culture assay. Biochem Cell Biol. 1996;74(6):833–51.
67. Eble JA, Niland S. The extracellular matrix of blood vessels. Curr Pharm Des. 2009;15(12):1385–400.
68. Dityatev A, Fellin T. Extracellular matrix in plasticity and epileptogenesis. Neuron Glia Biol. 2008;4(3):235–47.

69. Besser A, Schwarz US. Hysteresis in the cell response to time-dependent substrate stiffness. Biophys J. 2010;99(1):L10–2.
70. Prager-Khoutorsky M, Lichtenstein A, et al. Fibroblast polarization is a matrix-rigidity-dependent process controlled by focal adhesion mechanosensing. Nat Cell Biol. 2011;13(12):1457–65.
71. Zemel A, Rehfeldt F, et al. Optimal matrix rigidity for stress fiber polarization in stem cells. Nat Phys. 2010;6(6):468–73.
72. McDaniel SM, Rumer KK, et al. Remodeling of the mammary microenvironment after lactation promotes breast tumor cell metastasis. Am J Pathol. 2006;168(2):608–20.
73. Wynn TA. Common and unique mechanisms regulate fibrosis in various fibroproliferative diseases. J Clin Invest. 2007;117(3):524–9.
74. Keeley EC, Mehrad B, et al. Fibrocytes: bringing new insights into mechanisms of inflammation and fibrosis. Int J Biochem Cell Biol. 2010;42(4):535–42.
75. Friedman SL. Mechanisms of disease: Mechanisms of hepatic fibrosis and therapeutic implications. Nat Clin Pract Gastroenterol Hepatol. 2004;1(2):98–105.
76. Bissell MJ, Labarge MA. Context, tissue plasticity, and cancer: are tumor stem cells also regulated by the microenvironment? Cancer Cell. 2005;7(1):17–23.
77. Wiseman BS, Werb Z. Stromal effects on mammary gland development and breast cancer. Science. 2002;296(5570):1046–9.
78. Dvorak HF. Tumors: wounds that do not heal. Similarities between tumor stroma generation and wound healing. N Engl J Med. 1986;315(26):1650–9.
79. Egeblad M, Nakasone ES, et al. Tumors as organs: complex tissues that interface with the entire organism. Dev Cell. 2010;18(6):884–901.
80. Ruoslahti E. Specialization of tumour vasculature. Nat Rev Cancer. 2002;2(2):83–90.
81. Duong TD, Erickson CA. MMP-2 plays an essential role in producing epithelial-mesenchymal transformations in the avian embryo. Dev Dyn. 2004;229(1):42–53.
82. Song W, Jackson K, et al. Degradation of type IV collagen by matrix metalloproteinases is an important step in the epithelial-mesenchymal transformation of the endocardial cushions. Dev Biol. 2000;227(2):606–17.
83. Kalluri R, Zeisberg M. Fibroblasts in cancer. Nat Rev Cancer. 2006;6(5):392–401.
84. Gabrilovich DI, Nagaraj S. Myeloid-derived suppressor cells as regulators of the immune system. Nat Rev Immunol. 2009;9(3):162–74.
85. Huijbers IJ, Iravani M, et al. A role for fibrillar collagen deposition and the collagen internalization receptor endo180 in glioma invasion. PLoS One. 2010;5(3):e9808.
86. Zhu GG, Risteli L, et al. Immunohistochemical study of type I collagen and type I pN-collagen in benign and malignant ovarian neoplasms. Cancer. 1995;75(4):1010–7.
87. Oskarsson T, Acharyya S, et al. Breast cancer cells produce tenascin C as a metastatic niche component to colonize the lungs. Nat Med. 2011;17(7):867–74.
88. Vasiljeva O, Papazoglou A, et al. Tumor cell-derived and macrophage-derived cathepsin B promotes progression and lung metastasis of mammary cancer. Cancer Res. 2006;66(10):5242–50.
89. Hagemann T, Robinson SC, et al. Enhanced invasiveness of breast cancer cell lines upon co-cultivation with macrophages is due to TNF-alpha dependent up-regulation of matrix metalloproteases. Carcinogenesis. 2004;25(8):1543–9.
90. Caughey GH. Mast cell tryptases and chymases in inflammation and host defense. Immunol Rev. 2007;217:141–54.
91. Kauppila S, Stenback F, et al. Aberrant type I and type III collagen gene expression in human breast cancer in vivo. J Pathol. 1998;186(3):262–8.
92. Garra BS. Imaging and estimation of tissue elasticity by ultrasound. Ultrasound Q. 2007;23(4):255–68.
93. Glaser KJ, Felmlee JP, et al. Stiffness-weighted magnetic resonance imaging. Magn Reson Med. 2006;55(1):59–67.

94. Bhowmick NA, Neilson EG, et al. Stromal fibroblasts in cancer initiation and progression. Nature. 2004;432(7015):332–7.
95. Erler JT, Bennewith KL, et al. Lysyl oxidase is essential for hypoxia-induced metastasis. Nature. 2006;440(7088):1222–6.
96. Levental KR, Yu H, et al. Matrix crosslinking forces tumor progression by enhancing integrin signaling. Cell. 2009;139(5):891–906.
97. Sternlicht MD, Lochter A, et al. The stromal proteinase MMP3/stromelysin-1 promotes mammary carcinogenesis. Cell. 1999;98(2):137–46.
98. Provenzano PP, Eliceiri KW, et al. Collagen reorganization at the tumor-stromal interface facilitates local invasion. BMC Med. 2006;4(1):38.
99. Ingman WV, Wyckoff J, et al. Macrophages promote collagen fibrillogenesis around terminal end buds of the developing mammary gland. Dev Dyn. 2006;235(12):3222–9.
100. Condeelis J, Pollard JW. Macrophages: obligate partners for tumor cell migration, invasion, and metastasis. Cell. 2006;124(2):263–6.
101. Barker HE, Cox TR, et al. The rationale for targeting the LOX family in cancer. Nat Rev Cancer. 2012;19(12):540–52.
102. Baker AM, Bird D, et al. Lysyl oxidase enzymatic function increases stiffness to drive colorectal cancer progression through FAK. Oncogene. 2013;32(14):1863–8.
103. Baker AM, Cox TR, et al. The Role of Lysyl Oxidase in SRC-Dependent Proliferation and Metastasis of Colorectal Cancer. J Natl Cancer Inst. 2011;103(5):407–24.
104. Le QT, Harris J, et al. Validation of lysyl oxidase as a prognostic marker for metastasis and survival in head and neck squamous cell carcinoma: Radiation Therapy Oncology Group trial 90-03. J Clin Oncol. 2009;27(26):4281–6.
105. Provenzano PP, Inman DR, et al. Collagen density promotes mammary tumor initiation and progression. BMC Med. 2008;6:11.
106. Aggarwal BB, Gehlot P. Inflammation and cancer: how friendly is the relationship for cancer patients? Curr Opin Pharmacol. 2009;9(4):351–69.
107. Ben-Baruch A. The multifaceted roles of chemokines in malignancy. Cancer Metastasis Rev. 2006;25(3):357–71.
108. Bennewith KL, Huang X, et al. The role of tumor cell-derived connective tissue growth factor (CTGF/CCN2) in pancreatic tumor growth. Cancer Res. 2009;69(3):775–84.
109. Massague J. TGFbeta in Cancer. Cell. 2008;134(2):215–30.
110. Stover DG, Bierie B, et al. A delicate balance: TGF-beta and the tumor microenvironment. J Cell Biochem. 2007;101(4):851–61.
111. Kainz C, Kohlberger P, et al. Prognostic value of CD44 splice variants in human stage III cervical cancer. Eur J Cancer. 1995;31A(10):1706–9.
112. Nasser NJ. Heparanase involvement in physiology and disease. Cell Mol Life Sci: CMLS. 2008;65(11):1706–15.
113. Stauder R, Eisterer W, et al. CD44 variant isoforms in non-Hodgkin's lymphoma: a new independent prognostic factor. Blood. 1995;85(10):2885–99.
114. Ilan N, Elkin M, et al. Regulation, function and clinical significance of heparanase in cancer metastasis and angiogenesis. Int J Biochem Cell Biol. 2006;38(12):2018–39.
115. Kessenbrock K, Plaks V, et al. Matrix metalloproteinases: regulators of the tumor microenvironment. Cell. 2010;141(1):52–67.
116. Barker HE, Chang J, et al. LOXL2-mediated matrix remodeling in metastasis and mammary gland involution. Cancer Res. 2011;71(5):1561–72.
117. Erler JT, Bennewith KL, et al. Hypoxia-induced lysyl oxidase is a critical mediator of bone marrow cell recruitment to form the premetastatic niche. Cancer Cell. 2009;15(1):35–44.
118. Kirschmann DA, Seftor EA, et al. A molecular role for lysyl oxidase in breast cancer invasion. Cancer Res. 2002;62(15):4478–83.
119. Payne SL, Fogelgren B, et al. Lysyl oxidase regulates breast cancer cell migration and adhesion through a hydrogen peroxide-mediated mechanism. Cancer Res. 2005;65(24):11429–36.
120. Payne SL, Hendrix MJ, et al. Lysyl oxidase regulates actin filament formation through the p130(Cas)/Crk/DOCK180 signaling complex. J Cell Biochem. 2006;98(4):827–37.

121. Wozniak MA, Desai R, et al. ROCK-generated contractility regulates breast epithelial cell differentiation in response to the physical properties of a three-dimensional collagen matrix. J Cell Biol. 2003;163(3):583–95.
122. Barry-Hamilton V, Spangler R, et al. Allosteric inhibition of lysyl oxidase-like-2 impedes the development of a pathologic microenvironment. Nat Med. 2010;16(9):1009–17.
123. Fong SF, Dietzsch E, et al. Lysyl oxidase-like 2 expression is increased in colon and esophageal tumors and associated with less differentiated colon tumors. Genes Chromosomes Cancer. 2007;46(7):644–55.
124. Moreno-Bueno G, Salvador F, et al. Lysyl oxidase-like 2 (LOXL2), a new regulator of cell polarity required for metastatic dissemination of basal-like breast carcinomas. EMBO Mol Med. 2011;3(9):528–44.
125. Offenberg H, Brunner N, et al. TIMP-1 expression in human colorectal cancer is associated with TGF-B1, LOXL2, INHBA1, TNF-AIP6 and TIMP-2 transcript profiles. Mol Oncol. 2008;2(3):233–40.
126. Peinado H, Moreno-Bueno G, et al. Lysyl oxidase-like 2 as a new poor prognosis marker of squamous cell carcinomas. Cancer Res. 2008;68(12):4541–50.
127. Peng L, Ran YL, et al. Secreted LOXL2 is a novel therapeutic target that promotes gastric cancer metastasis via the Src/FAK pathway. Carcinogenesis. 2009;30(10):1660–9.
128. Ruckert F, Joensson P, et al. Functional analysis of LOXL2 in pancreatic carcinoma. Int J Colorectal Dis. 2010;25(3):303–11.
129. Sano M, Aoyagi K, et al. Forkhead box A1 transcriptional pathway in KRT7-expressing esophageal squamous cell carcinomas with extensive lymph node metastasis. Int J Oncol. 2010;36(2):321–30.
130. Akiri G, Sabo E, et al. Lysyl oxidase-related protein-1 promotes tumor fibrosis and tumor progression in vivo. Cancer Res. 2003;63(7):1657–66.
131. Bignon M, Pichol-Thievend C, et al. Lysyl oxidase-like protein-2 regulates sprouting angiogenesis and type IV collagen assembly in the endothelial basement membrane. Blood. 2011;118(14):3979–89.
132. Edlund M, Sung SY, et al. Modulation of prostate cancer growth in bone microenvironments. J Cell Biochem. 2004;91(4):686–705.
133. Ewing J. Neoplastic diseases: a treatise on tumors. Philadelphia & London: W.B. Saunders Co.; 1928.
134. Chang YS, di Tomaso E, et al. Mosaic blood vessels in tumors: frequency of cancer cells in contact with flowing blood. Proc Natl Acad Sci USA. 2000;97(26):14608–13.
135. Minna JD, Kurie JM, et al. A big step in the study of small cell lung cancer. Cancer Cell. 2003;4(3):163–6.
136. Klein CA. The systemic progression of human cancer: a focus on the individual disseminated cancer cell–the unit of selection. Adv Cancer Res. 2003;89:35–67.
137. Paget S. The distribution of secondary growths in cancer of the breast. Lancet. 1889;1: 571–3.
138. Hart IR, Fidler IJ. Role of organ selectivity in the determination of metastatic patterns of B16 melanoma. Cancer Res. 1980;40(7):2281–7.
139. Poste G, Fidler IJ. The pathogenesis of cancer metastasis. Nature. 1980;283(5743):139–46.
140. Cox TR, Bird D, Baker AM, Barker HE, Ho MW-Y, Lang G, Erler JT. LOX-mediated collagen crosslinking is responsible for fibrosis-enhanced metastasis. Cancer Research Mar 2013;15;73(6):1721–32.
141. Bateman A. Growing a tumor stroma: a role for granulin and the bone marrow. J Clin Invest. 2011;121(2):516–9.
142. Cox TR, Gartland A, et al. The pre-metastatic niche: is metastasis random? BoneKEy Rep. 2012;1(5).
143. McAllister SS, Weinberg RA. Tumor-host interactions: a far-reaching relationship. J Clin Oncol: Official J Am Soc Clin Oncol. 2010;28(26):4022–8.
144. Hiratsuka S, Nakamura K, et al. MMP9 induction by vascular endothelial growth factor receptor-1 is involved in lung-specific metastasis. Cancer Cell. 2002;2(4):289–300.

145. Kaplan RN, Riba RD, et al. VEGFR1-positive haematopoietic bone marrow progenitors initiate the pre-metastatic niche. Nature. 2005;438(7069):820–7.
146. Psaila B, Lyden D. The metastatic niche: adapting the foreign soil. Nat Rev Cancer. 2009;9(4):285–93.
147. Fogelgren B, Polgar N, et al. Cellular fibronectin binds to lysyl oxidase with high affinity and is critical for its proteolytic activation. J Biol Chem. 2005;280(26):24690–7.
148. Yaqoob U, Cao S, et al. Neuropilin-1 stimulates tumor growth by increasing fibronectin fibril assembly in the tumor microenvironment. Cancer Res. 2012;72(16):4047–59.
149. Avigdor A, Goichberg P, et al. CD44 and hyaluronic acid cooperate with SDF-1 in the trafficking of human CD34+ stem/progenitor cells to bone marrow. Blood. 2004;103(8):2981–9.
150. Jones DH, Nakashima T, et al. Regulation of cancer cell migration and bone metastasis by RANKL. Nature. 2006;440(7084):692–6.
151. Netelenbos T, Zuijderduijn S, et al. Proteoglycans guide SDF-1-induced migration of hematopoietic progenitor cells. J Leukoc Biol. 2002;72(2):353–62.
152. O'Connell JT, Sugimoto H, et al. VEGF-A and Tenascin-C produced by S100A4+ stromal cells are important for metastatic colonization. Proc Natl Acad Sci USA. 2011;108(38): 16002–7.
153. Malanchi I, Santamaria-Martinez A, et al. Interactions between cancer stem cells and their niche govern metastatic colonization. Nature. 2012;481(7379):85–9.
154. Kii I, Nishiyama T, et al. Incorporation of tenascin-C into the extracellular matrix by periostin underlies an extracellular meshwork architecture. J Biol Chem. 2010;285(3):2028–39.
155. Oskarsson T, Massague J. Extracellular matrix players in metastatic niches. EMBO J. 2012;31(2):254–6.
156. Moller HD, Ralfkjaer U, et al. Role of fibulin-5 in metastatic organ colonization. Mol Cancer Res: MCR. 2011;9(5):553–63.
157. Weaver VM, Lelievre S, et al. beta4 integrin-dependent formation of polarized three-dimensional architecture confers resistance to apoptosis in normal and malignant mammary epithelium. Cancer Cell. 2002;2(3):205–16.
158. Wartenberg M, Frey C, et al. Development of an intrinsic P-glycoprotein-mediated doxorubicin resistance in quiescent cell layers of large, multicellular prostate tumor spheroids. Int J Cancer. 1998;75(6):855–63.
159. Serebriiskii I, Castello-Cros R, et al. Fibroblast-derived 3D matrix differentially regulates the growth and drug-responsiveness of human cancer cells. Matrix Biol. 2008;27(6):573–85.
160. Coussens LM, Fingleton B, et al. Matrix metalloproteinase inhibitors and cancer: trials and tribulations. Science. 2002;295(5564):2387–92.
161. Mack GS, Marshall A. Lost in migration. Nat Biotechnol. 2010;28(3):214–29.
162. Barker HE, Erler JT. The potential for LOXL2 as a target for future cancer treatment. Future Oncol. 2011;7(6):707–10.
163. Jourdan-Le Saux C, Tronecker H, et al. The LOXL2 gene encodes a new lysyl oxidase-like protein and is expressed at high levels in reproductive tissues. J Biol Chem. 1999;274(18): 12939–44.
164. Erler JT, Linding R. Network-based drugs and biomarkers. J Pathol. 2010;220(2):290–6.
165. Pawson T, Linding R. Network medicine. FEBS Lett. 2008;582(8):1266–70.
166. Yu H, Mouw JK, et al. Forcing form and function: biomechanical regulation of tumor evolution. Trends Cell Biol. 2011;21(1):47–56.

Chapter 5
Biology and Treatment of Basal-Like Breast Cancer

Bingchen Han, William Audeh, Yanli Jin, Sanjay P. Bagaria, and Xiaojiang Cui

Abstract Breast cancer is one of the most common cancers identified among women worldwide. It is a heterogeneous disease, with each sub-disease displaying unique clinical and histopathologic characteristics. Based on genomic analysis and gene expression profiling, breast cancer has been classified into several distinct subgroups. Estrogen receptor (ER)-positive breast cancers can be treated with antiestrogen drugs such as tamoxifen, and human epidermal growth factor receptor 2 (HER2)-positive breast cancers can be treated with HER2-targeted drugs such as trastuzumab. Because of the low expression of ER and HER2, patients with basal-like breast cancer cannot benefit from these targeted therapies. Thus, to identify and validate pivotal theranostic biomarkers for basal-like breast cancer is of paramount importance. This chapter describes the molecular and histological features of

B. Han • Y. Jin
Department of Surgery and Obstetrics and Gynecology, Women's Cancer Program, Samuel Oschin Comprehensive Cancer Institute, Cedars-Sinai Medical Center, Los Angeles, CA 90048, USA

W. Audeh
Samuel Oschin Comprehensive Cancer Institute, Cedars-Sinai Medical Center, Los Angeles, CA 90048, USA

S.P. Bagaria
Department of Surgery, Mayo Clinic Florida, Jacksonville, FL 32250, USA

X. Cui (✉)
Department of Surgery and Obstetrics and Gynecology, Women's Cancer Program, Samuel Oschin Comprehensive Cancer Institute, Cedars-Sinai Medical Center, Los Angeles, CA 90048, USA

Cedars-Sinai Medical Center, W Tower, Suite 290W, 8700 Beverly Blvd, Los Angeles, CA 90048, USA
e-mail: Xiaojiang.cui@cshs.org

H. Schatten (ed.), *Cell and Molecular Biology of Breast Cancer*,
DOI 10.1007/978-1-62703-634-4_5,

basal-like breast cancer as well as the differences between basal-like breast cancer and the other breast cancer subtypes. It presents current progress in identifying the biomarkers for basal-like breast cancer. It concludes with a description of the current therapeutic strategies to treat basal-like breast cancer and argues that inhibition of basal-like tumor-associated signaling pathways, in conjunction with standard therapies, may enhance the treatment efficacy.

5.1 Introduction

Breast cancer, the most common cancer among women, is a heterogeneous disease that consists of diverse disease subtypes that each has prognostic and predictive value. High-throughput screening technologies—in particular microarray analysis—has classified breast cancer into five major biologically distinct intrinsic subtypes: luminal A, luminal B, HER2-overexpressing, normal-like, and basal-like [1, 2]. Luminal A tumors typically express ER and have a low-grade histology. Luminal B-type tumors tend to have lower ER expression and a higher grade histology compared to luminal A. $HER2^{+}$ tumors are characterized by the amplification of the *HER2* gene and have high-grade histology, and basal-like breast cancers express undetectable or low levels of ER, PR, and HER2 and have high-grade histology.

Based on the expression of various receptors in breast cancer cells, therapy is selected for individual patients. ER-positive or progesterone receptor (PR)-positive tumors can be treated with endocrine therapy, whereas tumors that have HER2 gene amplification may be targeted with antibodies (such as trastuzumab) or small molecule tyrosine kinase inhibitors (such as lapatinib). These targeted therapies have dramatically changed the outcome of patients with ER-positive and/or HER2-positive breast cancer. However, there still remains a daunting challenge to develop targeted therapies against highly invasive, metastatic ER-negative/HER2-negative "basal-like" breast cancers as effective as those against other subtypes of breast cancer.

5.2 What Is Basal-Like Breast Cancer?

Basal-like tumors account for about 15–20 % of all invasive breast cancers and are usually associated with younger patient age and high histologic grade. They display a specific pattern of distant metastasis, a high recurrence rate, short recurrence-free survival, and poor outcome. Basal-like tumors are more prevalent among premenopausal African American women than in postmenopausal African American women and non-African American women of any age [3, 4]. To date, there is no internationally accepted definition for basal-like cancers for use in clinical classification or research [5]. Some have used microarray-based gene expression profiling to define basal-like breast cancers, whereas others have used immunohistochemical markers.

Basal-like breast cancers derived their name from the finding that these tumors express the genes usually found in basal/myoepithelial cells of the normal breast. In the 1980s, it was found that a small subgroup of breast cancers, similar to normal myoepithelial (basal) epithelium, expressed high-molecular-weight basal cytokeratins [6–9]. Cytokeratins are proteins of the intermediate filament family, usually found in the cytoplasmic cytoskeleton of epithelial tissue. In the mammary gland, myoepithelial cells usually express high-molecular-weight cytokeratins 5/6, 14, and 17, whereas luminal cells usually express low-molecular-weight cytokeratins 8 and 18 [10]; however, expression of basal cytokeratins are not restricted to myoepithelial cells.

Immunohistochemical markers that have been proposed to define basal-like breast cancer include ER^-, PR^-, $HER2^-$, the epidermal growth factor receptor (EGFR), the receptor tyrosine kinase c-Kit, P-cadherin, and cytokeratin 5/6 [11–15]. Compared with the triple-negative definition (ER^-, PR^-, $HER2^-$), inclusion of additional biomarkers such as EGFR not only provides a more specific identification of basal-like breast cancer as defined by gene expression profiling analysis, but also better stratifies breast cancer survival among triple-negative breast cancer patients. However, there is a lack of a consensus definition of basal-like breast cancer to date.

Basal-like breast cancers are well known to have aggressive clinical features, poor prognosis and lower survival rates [16, 17], and tend to metastasize to the brain and lung, while luminal subtypes tend to metastasize to the bone [18, 19]. However, it is important to mention here that basal-like breast cancer is likely to be a heterogeneous group of tumors that may have unique outcomes, based on the differential gene expression profiles of basal-like cases in the hierarchical clustering analysis and characterization of basal-like tumors by different immunohistochemical biomarker panels [20, 21]. Interestingly, compared with other subtypes, basal-like tumors are more likely to present as interval breast cancers (i.e., tumors that develop between scheduled mammography screenings) [22, 23], probably as a result of their high proliferation rate [14, 15]. The majority of basal-like tumors are ductal and are occasionally tubular, mixed, metaplastic, or medullary cancers [5]. Microarray-based comparative genomic hybridization (CGH) has demonstrated that basal-like tumors have the highest frequency of DNA losses and gains compared with other subtypes, suggesting a genomic instability [24, 25]. Notably, the basal-like tumors were relatively enriched for low-level copy number gains and losses, while high-level amplification at any locus was infrequent in these tumors [25]. Similarly, genome-wide single nucleotide polymorphism arrays have shown the highest overall rate of loss of heterozygosity in basal-like tumors [26]. Basal-like breast cancers have more frequent TP53 mutations, a higher mitotic index, greater nuclear pleomorphism, and higher grade [4, 27]. They also contain geographic areas of necrosis, pushing borders of invasion, and a stromal lymphocytic response [14].

There is controversy with regard to the true origin of basal-like breast tumors. Although gene expression profiling studies have demonstrated a basal-like genotype in these breast cancers, accumulating evidence argues against a myoepithelial origin. Livasy et al. found that in basal-like breast cancers, the frequency of expression of myoepithelial markers SMA, p63, and CD10 is low, and cytokeratin 8/18, a marker typically expressed in the luminal epithelial cells of the breast, was strongly expressed in 83 % of basal-like tumors [14]. Furthermore, gene expression profiling revealed that basal-like breast tumors were more similar to normal luminal progenitor cells than any other epithelial subsets, including the stem cell-enriched population [28]. In agreement, deleting the BRCA1 gene in mouse mammary epithelial luminal progenitors produces tumors that phenocopy human BRCA1-mutant breast cancers and the majority of sporadic basal-like breast cancer. As is well known, BRCA1 mutation-associated breast tumors frequently display a distinctive basal-like phenotype [29]. These data suggest that basal-like breast cancers arise from luminal progenitors but not from normal basal stem cells [30]. Consistent with their progenitor cell origin, several studies showed that basal-like breast cancers have stem cell-like properties. Breast cancer cells with a $CD44^{+}/CD24^{-}$ phenotype—a cell population characterized by the CD44 positive and CD24 negative or low cell surface antigen expression profile—are found to possess tumor-initiating properties with stem cell-like features [31] and are enriched in basal-like breast tumors [32]. Furthermore, gene expression profiling analysis showed that basal-like tumors have an embryonic stem cell-like gene expression signature [33].

5.3 The Difference Between Basal-Like Breast Cancer and Triple-Negative Breast Cancer

Triple-negative breast cancers are defined immunohistochemically as tumors that lack ER, PR, and HER2 expression. Like basal-like breast cancers, triple-negative breast tumors also have high histologic grade and occur more frequently in young black and Hispanic women than in young women of other racial or ethnic groups [3]. Many cancers meet the definitions of both triple-negative breast cancers and basal-like breast cancers. Most basal-like breast cancers (55–85 %) are of a triple-negative staining pattern, and the majority of triple-negative breast cancers (70–90 %) exhibit a basal-like phenotype [5, 12, 34]. Notably, some researchers claim that the basal-like breast cancers are composed almost entirely of triple-negative tumors [35]. Kreike and colleagues analyzed 97 triple-negative tumors and they found that all triple-negative tumors were classified as basal-like tumors, based upon their overall gene expression profile. These authors concluded that triple-negative tumors are synonymous with basal-like tumors, and thus both can be identified by the three standard immunohistochemical markers ER, PR, and HER2 [36]. Notably, this study confirmed that basal-like tumors are heterogeneous and can be subdivided into at least five distinct subgroups [36].

Although there are numerous similarities between basal-like and triple-negative breast cancers, there is evidence that these two terms are not synonymous. Not all basal-like cancers determined by gene expression profiling lack ER, PR, and/or HER2; on the other hand, not all triple negative cancers show a basal-like phenotype by expression array analysis [12]. Bertucci and colleagues used immunohistochemical markers to define triple-negative breast cancers and gene expression profiling to define basal-like breast cancers. They found that in 172 triple-negative breast cancers, 28 % of the tumors did not have the basal-like gene expression signature. Meanwhile, in 160 basal-like breast cancers, 23 % of the tumors did not have triple-negative phenotype [37]. Several other studies reported similar results: 8–29 % of triple-negative tumors did not show basal-like subtype, while 14–40 % of basal-like tumors did not show the triple-negative phenotype [34, 38, 39], suggesting an intrinsic difference between basal-like breast cancers and some triple-negative breast cancers.

Careful analysis of microarray-based expression profiles suggests that the triple-negative group also encompasses another molecular subgroup of breast cancer—normal-like, which not only have a better prognosis than basal-like tumors but also do not respond to neoadjuvant chemotherapy in the same fashion as basal-like cancers [1, 34, 40]. It is now widely accepted that equating triple-negative tumors with basal-like breast cancer is inaccurate and misleading [41] and that the term "basal like" is not interchangeable with "triple negative."

5.4 BRCA1 and Basal-Like Breast Cancer

The BRCA1 tumor suppressor gene is localized on chromosome 17q and encodes a 1,863-amino acid protein with a C3HC4-type zinc finger domain [42]. BRCA1 has multiple roles within cells, including those related to transcriptional regulation and repair of double-stranded DNA breaks to protect the genome during DNA replication. Cells that lack BRCA1 function have a deficiency in the repair of double-stranded DNA breaks by homologous recombination, which is likely one of the mechanisms behind their association with increased cancer predisposition [43]. Accumulating data show that there is a close relationship between the BRCA1 mutation and basal-like breast cancer. Breast cancers arising in patients with BRCA1 deficiency are of triple-negative/basal-like subtype in the majority of cases [1, 44, 45]. Immunohistochemistry (IHC)-based studies classify 80–90 % of BRCA1-associated tumors as being triple negative or basal-like [44, 46, 47]. Furthermore, microarray results suggested that mutations in the BRCA1 gene predispose to the basal tumor subtype [1], as BRCA1-mutant tumors display the basal-like gene expression profile in hierarchical clustering analysis [1, 48]. This may be partially due to the fact that BRCA1 can transcriptionally regulate genes associated with the basal-like phenotype in breast cancer [49]. For example, wild-type BRCA1 represses the expression of the FOXC1 transcription factor in basal-like breast cancer [50]. FOXC1 has been proposed as a critical biomarker for basal-like breast cancer [51]. Loss of BRCA1 function or expression may induce FOXC1 expression and thereby elicit a basal phenotype. Expression of the cytokeratin CK5/6, a commonly accepted immunohistochemical marker of basal-like breast cancer, is also significantly associated with BRCA1-related breast cancers [46].

Although 5–10 % of breast cancer may be due to the inheritance of autosomal dominant breast cancer susceptibility alleles, the alteration in the expression or function of BRCA1 may be important in the development of sporadic basal-like breast cancer [52]. Hypermethylation of the BRCA1 promoter, which leads to loss of BRCA1 transcription, has been reported to be present in 11–13 % sporadic breast cancers [53, 54]. Interestingly, the ID4 gene, which was identified as a negative regulator of BRCA1 [55], was found to be expressed at 9.1-fold higher levels in sporadic basal-like breast cancer [56], suggesting a potential ID4-mediated mechanism of BRCA1 dysfunction or deficiency in sporadic breast cancer.

Another breast cancer susceptibility gene, BRCA2, also plays an important role in genome protection. BRCA2 is localized on chromosome 13q and encodes a 3,418-amino acid protein [42]. It is a mediator of the core mechanism of homologous recombination [43]. However, unlike BRCA1, it has been shown that BRCA2 is not involved in the biology of basal-like breast cancer as most BRCA2-mutant tumors are ER/PR positive [45, 57] and do not express basal cytokeratins such as CK5/14 [58]. BRCA2-related tumors also have different CGH profiles (which indicate chromosomal gains and losses) and gene expression patterns compared with BRCA1-associated tumors [59, 60]. In addition, compared with basal-like and BRCA1-mutated tumors, BRCA2-mutated tumors express higher levels of p27 and p16 and lower levels of skp2, cyclin E, and caspase 3 [61].

5.5 Potential Targets or Predictive Biomarkers for Basal-Like Breast Cancer Treatment

5.5.1 PI3K/Akt

The phosphatidylinositol 3 kinase (PI3K) pathway is involved in many cellular functions, including cell proliferation, survival, and migration [62]. Deregulated PI3K/AKT activation promotes tumorigenesis, and this signaling pathway has been identified as a putative therapeutic target in a range of human cancers [63]. The activation of PI3K pathway was significantly associated with ER-negative and PR-negative status, high tumor grade, and a basal-like phenotype [64–66], and this activation was associated with the loss of PTEN. Lower levels of the tumor suppressor PTEN were significantly negatively correlated with Akt activity [64], which has been shown to be enriched in triple-negative or basal-like breast cancer [67, 68]. Further study showed that this activation of PI3K pathway in basal-like breast cancer may also be attributed to the loss of inositol polyphosphate 4-phosphatase II (INPP4B), which hydrolyzes phosphatidylinositol (3,4,5)-triphosphate PtdIns(3,4)P2 generated from the PI3K product PtdIns(3,4,5)P3. INPP4B functions as a tumor suppressor by negatively regulating normal and malignant mammary epithelial cell proliferation through regulation of the PI3K/Akt signaling pathway [69]. Interestingly, among the three Akt isoforms, only Akt3 is overexpressed in ER-negative breast tumors and represents the major active Akt in ER-negative breast cancer cells [70]. Consistent with this finding, a recurrent membrane-associated guanylate kinase (MAGI3)—Akt3 fusion, which leads to constitutively active Akt3 activity—was reported to be enriched in triple-negative breast cancer [67]. These findings suggest that the use of isoform-specific Akt inhibitors should be evaluated in clinical trials for the treatment of MAGI3-Akt3 fusion-positive triple-negative breast cancers.

5.5.2 *MAPK/ERK*

MAPK/ERK signaling pathway plays an essential role in regulating the growth and survival of neoplastic cells [71]. A previous study found that cell lines and xenografts with a basal-like gene expression signature were sensitive to the small molecule inhibitor of MAPK/ERK kinase (MEK) as opposed to cell lines and xenograft models of other breast cancer subtypes [72]. The study also found that loss of PTEN is a negative predictor of response to MEK inhibition, and treatment with a selective MEK inhibitor caused upregulation of PI3K pathway signaling, suggesting that single-agent MEK inhibition may be a promising therapeutic modality for basal-like breast cancers with intact PTEN. Similar results showed that basal-like breast cancer cells were particularly susceptible to growth inhibition by small molecule MEK inhibitors [73]. In agreement with these reported results, it was recently found that dual specificity protein phosphatase 4 (DUSP4), an ERK phosphatase, correlated with breast cancer resistance to neoadjuvant chemotherapy and with basal-like breast cancer status [74]. This study also showed that gene expression patterns with respect to Ras-ERK pathway activation, probably due to DUSP4 downregulation, are associated with basal-like tumors.

5.5.3 *FOXC1*

FOXC1 is a transcription factor playing an important role in the development of the brain and the eye during the embryonic stage [75–78]. Loss of FOXC1 functions can cause hydrocephalus, abnormal eyelids, and Axenfeld-Rieger syndrome. Our previous studies found that FOXC1 is a theranostic biomarker that is specific for basal-like breast cancer [51]. Ectopic overexpression of FOXC1 led to increased migration, invasion, and proliferation of breast cancer cells, and knockdown of FOXC1 led to opposite results. Elevated FOXC1 expression predicted poor overall survival in basal-like breast cancer, a higher incidence of brain metastasis, and a shorter brain metastasis-free survival in lymph node-positive or lymph node-negative patients, offering FOXC1 as not only a potential prognostic predictor but also a potential molecular therapeutic target in this breast cancer subtype [51]. Kolacinska et al. also showed that triple-negative breast cancer subtype was associated with higher expression of FOXC1 [79]. In addition, it was reported that basal-like breast cancer defined by triple negative plus FOXC1 demonstrated superior prognostic relevance compared to basal-like tumors defined by triple negative alone or triple negative plus basal CKs [80]. Recent work further suggests that the activation of NF-κB signaling pathway mediates the function of FOXC1 in human basal-like breast cancer [66]. Sustained NF-κB activation exists mostly in human basal-like breast cancer cell lines and ER-negative breast cancer, with its highest activity found in triple-negative tumors [81–83]. FOXC1 may serve as a critical regulator of NF-κB regulator in basal-like breast cancer. Consistent with this result, a recent study showed that matrix metalloprotease 7 (MMP7), which is known to be regulated by NF-κB, mediates the invasion-promoting function of FOXC1 in basal-like breast cancer [84]. Moreover, FOXC1 levels have been shown to predict resistance to chemotherapy in human breast cancers and cell models [50, 85], reflecting another potential mechanism underlying the role of FOXC1 in basal-like tumor progression.

5.5.4 αB-Crystallin

α-Basic-crystallin (αB-crystallin) is a member of the small heat shock protein family and plays an important role in the regulation of apoptosis [86]. Accumulating evidence suggests that αB-crystallin is a potential biomarker for basal-like breast cancer [87–92]. The expression of αB-crystallin was found to be positively correlated with other established basal-like markers and histological subtypes associated with basal-like breast cancer [87], and the expression of αB-crystallin independently predicts shorter survival in patients with basal-like breast cancer [89], indicating that αB-crystallin is a diagnostic and prognostic indicator for this type of tumors. In vitro studies have shown that it is upregulated by the transcription factor Ets1, which promotes cell proliferation, and is also overexpressed in basal-like breast cancer [92]. Mechanistically, αB-crystallin induces breast cancer cell growth, chemoresistance, migration, and invasion by activating MEK/ERK pathway [89], providing a biological basis of why αB-crystallin is associated with poor clinical outcomes in breast cancer.

5.5.5 P-Cadherin

P-cadherin is a calcium-dependent cell-cell adhesion glycoprotein. In normal adult breast tissue, P-cadherin is restricted to myoepithelial cells and is involved in architecture and differentiation functions [93]. Studies have shown that P-cadherin is a reliable biomarker for basal-like breast cancer [15, 94–97] and is used in a IHC marker panel to identify these breast cancers. The expression of P-cadherin in invasive ductal tumors is negatively correlated with ER and PR and positively correlated with recurrence and distant metastasis [97]. In vitro studies have shown that P-cadherin is induced by FOXC1 [51], while suppressed by functional BRCA1 [49]. Moreover, it has been shown that P-cadherin is associated with high-grade tumor subtypes and is a marker for poor prognosis [98], but there are also reports showing that P-cadherin is not an adverse prognostic factor and may not be associated with EGFR [94, 95]. Interestingly, P-cadherin is also associated with the expression of the breast stem cell markers CD44, CD49f, and aldehyde dehydrogenase 1 in the basal-like tumors [99]. It mediates stem cell properties such as the mammosphere-forming capacity and radiation resistance, reflecting its potential role in regulating aggressive behavior of basal-like breast cancer.

5.5.6 *Integrin*

Integrins comprise a large family of cell–cell adhesion receptors that mediate interactions between the extracellular environment and the cytoplasm. These transmembrane proteins play a crucial role in cell growth, survival, and invasion [100]. Both gene expression profiling and immunohistochemical analysis revealed that integrin β4 is significantly associated with basal-like tumors and that the integrin β4 "signature" genes, i.e., the top-ranked 65 genes in correlation with β4 expression, predict poor prognosis in breast cancer [101]. Interestingly, the expression level of integrin β4 in basal-like breast cancer cell lines can be upregulated by the basal-like tumor-associated transcription factor FOXC1 [51]. Integrin α9β1, another member of the integrin family, is a receptor for extracellular matrix (ECM), its expression showed a significant association with reduced overall patient survival and reduced distant-metastasis-free survival, and acts as a potential marker for basal-like breast cancer [102]. Src, a non-receptor tyrosine kinase commonly activated by integrin/focal adhesion kinase signaling, has also been linked with basal-like tumors [103, 104].

5.6 Treatment Strategies for Basal-Like Breast Cancer

Because most of the basal-like breast cancers have low or undetectable expression of ER, PR, and HER2, the majority of these cancers cannot be effectively treated with the existing targeted therapies, such as antiestrogen and anti-HER2 therapies. So far, chemotherapy remains the only modality of systemic therapy for basal-like tumors, and considerable efforts have been made to establish molecular targets and thereby targeted therapy for this subtype of breast cancer. Although some basal-like tumors express ER and HER2 [105, 106], whether antiestrogen or anti-HER2 therapies can improve the outcome of these cancers awaits to be determined, as basal phenotype-associated signaling pathways may induce de novo resistance to the treatment [105, 107].

5.6.1 *Chemotherapy*

There is currently no standard chemotherapeutic regimen for basal-like breast cancer. Anthracycline-containing regimens remain a common modality for treatment of basal-like breast cancer. Three-drug combinations of cyclophosphamide, methotrexate or epirubicin, and fluorouracil are also being evaluated and compared in clinical trials [108]. Recent studies showed that the combination of docetaxel and carboplatin is promising for the treatment of triple-negative breast cancer [109]. In addition, meta-analysis indicates that taxanes may have added benefits in anthracycline chemotherapy treatment of breast cancer regardless of ER levels [110]. Notably, preclinical and clinical studies have shown increased sensitivity of BRCA1-associated breast cancer to the DNA-damaging agents such as platinum drugs [111, 112], due to defective double-strand repair in BRCA1-mutant cells. It merits mentioning that these clinical trials did not stratify heterogeneous triple-negative patient cohorts into basal-like and non-basal-like groups, which may compound the challenges to compare the efficacies of different chemotherapies.

Clinical studies show that basal-like tumors, compared with the other subtypes, are more sensitive to paclitaxel- and doxorubicin-containing preoperative chemotherapy, as well as to high-dose chemotherapy [34, 113], probably due to their higher proliferation rate [14, 114]. Basal-like and HER2 subtypes are associated with the highest rates of pathologic complete response (45 %), in comparison to luminal tumors where the response rate is only 6 % [34]. However, the basal-like subtype itself is not sufficient to predict the likelihood of chemotherapy response because these tumors are frequently of a high nuclear grade and hormone receptor negative, both of which are known to be associated with higher probability of pathologic complete response to preoperative chemotherapy [34]. Although basal-like breast cancers showed a better response rate to chemotherapy, these tumors had an increased likelihood of distant recurrence and shorter survival compared with ER+ tumors if pathologic complete response is not achieved by neoadjuvant therapy [35, 115–117]. One plausible explanation pertains to the notion that chemoresistant cancer stemlike cell populations are enriched in basal-like breast cancer. Contrary to the above observations, studies also showed that basal-like tumors had a relatively low response rate to neoadjuvant chemotherapy with docetaxel, doxorubicin, and cyclophosphamide [118], maybe because of a higher proportion of ER+ (albeit low levels of ER) and a lower proportion of high-grade tumors in that cohort of basal-like tumors.

5.6.2 PARP Inhibitors

DNA repair pathways involved in single-strand and double-strand breaks are critical for maintaining stability and integrity of the genome. As discussed before, the BRCA1 gene plays an essential role in the repair of double-strand breaks by the homologous recombination mechanism [119, 120]. The majority of BRCA1 mutation-associated breast cancers display the basal-like phenotype, and some sporadic basal-like breast cancers, i.e., those with BRCA1 downregulation, also phenotypically resemble BRCA1-mutant breast cancers (a concept called BRCAness) [56]. It is postulated that these BRCA1 mutation or BRCAness-associated tumors, which have defective homologous recombination repair, would be susceptible to the drugs, blocking single-strand break repair mechanism. This approach, referred to as "synthetic lethal," will render tumor cells to rely on the error-prone nonhomologous end-joining mechanism to repair DNA breaks, which eventually leads to genomic instability and apoptosis [121]. Poly(ADP-ribose) polymerases (PARPs) are a family of nuclear proteins (enzymes) involved in base excision repair, a key pathway in the repair of single-strand breaks [122, 123]. In vitro studies showed that BRCA1 dysfunction sensitizes cells to the inhibition of PARP enzymatic activity, resulting in chromosomal instability, cell cycle arrest, and subsequent apoptosis [124], and this is due to DNA repair deficiency [125].

A phase II clinical study involving patients with metastatic triple-negative breast cancer found that addition of iniparib, a compound initially thought to be a PARP inhibitor, to gemcitabine and carboplatin improved the rate of overall response and prolonged the median progression-free survival from 3.6 months to 5.9 months and the median overall survival from 7.7 months to 12.3 months [126]. However, yet to be published data from the phase III trial with iniparib failed to meet its primary objective of improvement in the combined end point of progression-free and overall survival [108]. Surprisingly, it was subsequently reported that iniparib does not inhibit PARP1 or PARP2 activity in vitro [127], negating the relevance of this trial for testing the utility of PARP inhibitors as a therapeutic strategy in breast cancer. Positive clinical benefit in patients with BRCA-mutated breast cancer has been reported with a confirmed inhibitor of PARP 1 and 2, olaparib, in a phase II study, with a reported response rate of 41 % [128]. A possible confounding issue related to this synthetic lethal approach is that the activity of most existing PARP inhibitors against PARP 1 and 2 has been reported with IC50s in the nanomolar range, yet activity against other members of the PARP enzyme family is unknown, raising the issue of a lack of selectivity [129]. There are 17 known PARP family proteins with PARP1 as the most abundant [122]. PARP1 and PARP2 function in base excision repair. Specificity for PARP1 inhibition is a critical issue which needs to be addressed for improving the efficacy of these inhibitors. It also merits mentioning that these clinical studies were designed on the assumption that all studied cases were PARP-positive, but in fact 18.1 % of BRCA1-associated cancers had low expression of or did not express nuclear PARP1 protein [6]. In addition, it is not known whether PARP levels correlated with response to PARP inhibitors in BRCA-mutated cancers. Therefore, the expression of PARP protein in tumor cells should be taken into account in the future trials of PARP inhibitors [130]. Although PARP inhibitors may eventually prove to be effective as monotherapy or in combination with chemotherapy for BRCA-associated breast cancer, there is an urgent need to identify those patients without BRCA mutations who may benefit from this potential therapy and account for the majority of the triple-negative patient population. Two recent reviews summarize the current status and our understanding of PARP inhibitors in breast cancer-related clinical trials [108, 131].

5.6.3 EGFR Inhibitors

Epidermal growth factor receptor (EGFR) is a receptor tyrosine kinase of the ErbB family and is abnormally activated in many epithelial tumors. EGFR mRNA and protein are detected more frequently and at higher levels in triple-negative breast cancers [12, 132, 133], and the *EGFR* gene is amplified in some basal-like breast cancer [134]. In addition, EGFR is a predictor of poor prognosis in triple-negative breast cancer, independent of nodal status and size [135]. As such, EGFR-directed monoclonal antibodies or tyrosine kinase inhibitors, as a single agent or in combination with chemotherapy drugs, have been pursued as treatment modalities for basal-like breast cancer [136]. So far, targeting EGFR seems to only have limited success and has not been approved as a therapy for basal-like breast cancer [137, 138]. These results showed that anti-EGFR drugs were not effective in treatment of unselected triple-negative breast cancer patients. As EGFR levels by themselves may not be a reliable indicator for the activation of EGFR signaling, there is a need to exploit new predictive biomarkers for EGFR-targeted therapy to identify those patients with "EGFR-addicted" basal-like breast cancers which harbor hyperactive EGFR signaling and depend on this signaling for growth and progression.

5.6.4 Other Targeted Therapy Approaches

In addition to the targets mentioned above, several other potential therapeutic targets have been evaluated. Dasatinib, a Src inhibitor, selectively inhibits growth of basal-like/triple-negative breast cancer cell lines [139]. Single-agent dasatinib has limited activity in patients with triple-negative breast cancer [140]. Although abnormal activation or amplification of Src has been detected in various tumors [141], no studies have convincingly shown that Src is a critical basal marker. c-Kit expression, which can be inhibited by the multitargeted anticancer drug imatinib mesylate [5], is more common in basal-like tumors [12, 142, 143]. One caveat is that c-Kit levels do not correlate with prognosis, which argues against a critical role of c-Kit in these breast cancers. Moreover, other targets such as Hsp90, mTOR, VEGF, and androgen receptor are also being investigated as potential therapeutic targets in basal-like breast cancer [144]. However, the relevance or therapeutic potential of targeting these proteins in human basal-like tumors may be diminished if they turn out not to be specific markers or regulators of human basal-like breast cancer.

In summary, lessons from treatment of ER+ and HER2+ breast cancer emphasize the notion that the success and ultimate benefit of targeted therapy depend on the expression level and essential role of the target protein in tumor cells. Thus, elucidation of the functions of these proteins and selection of patients with cancer characterized by these proteins may be the key to an effective targeted therapy.

5.7 Conclusions

Basal-like breast cancer represents an aggressive and biologically heterogeneous subtype of breast cancer, with poor prognosis, a specific pattern of distant metastasis, and a high recurrence rate with standard chemotherapy. Unfortunately, there are no existing targeted treatment strategies for basal-like breast cancer. Additionally, a consensus on the definition for basal-like breast cancer is lacking. Microarray-based gene expression profiling has its limitations in the clinical identification of basal-like tumors, because microarray assays cannot be readily applied to formalin-fixed, paraffin-embedded tissues and its complexity makes it expensive. A suitable, pragmatic, objective solution would be to use IHC surrogates to define the subtype. However, the immunohistochemical markers used to detect basal-like breast cancers, which include ER, PR, HER2, EGFR, and cytokeratin 5/6, also present issues of sensitivity and specificity to detect basal-like tumors defined by expression profiling methods, due to scoring and interpretation of immunohistochemical staining [145]. In addition, these markers alone may be inadequate to identify the clinically relevant biological diversity within this group of cancers. A consensus definition is far from being achieved. Several genes, such as FOXC1, have been shown to be consistently and exclusively overexpressed in basal-like breast cancers. These genes also play essential roles in basal-like breast cancer cell function and thus may serve as theranostic marker and therapeutic target for treatment of basal-like breast cancer. It is also important to design prospective clinical studies to evaluate these markers in comparison with previously established markers such as EGFR and CK5/6 in the diagnosis and prognostication of basal-like breast cancer and to study the impact of these individual protein markers in response to specific chemotherapy. Further understanding of the biology of basal-like breast cancer will lead to more effective treatment and prevention of this devastating disease.

Acknowledgments We thank Alice Chung, Shikha Bose, Jian Huang, and Armando Giuliano for thoughtful comments. We thank the support of National Institutes of Health (CA151610 and UL1TR000124), QVC and the Fashion Footwear Association of New York Charitable Foundation, and the Avon Foundation (02-2010-068).

References

1. Sorlie T, et al. Repeated observation of breast tumor subtypes in independent gene expression data sets. Proc Natl Acad Sci USA. 2003;100:8418–23.
2. Perou CM, et al. Molecular portraits of human breast tumours. Nature. 2000;406:747–52.
3. Foulkes WD, Smith IE, Reis-Filho JS. Triple-negative breast cancer. N Engl J Med. 2010;363:1938–48.
4. Carey LA, et al. Race, breast cancer subtypes, and survival in the Carolina Breast Cancer Study. JAMA. 2006;295:2492–502.
5. Rakha EA, Reis-Filho JS, Ellis IO. Basal-like breast cancer: a critical review. J Clin Oncol. 2008;26:2568–81.
6. Domagala P, Huzarski T, Lubinski J, Gugala K, Domagala W. Immunophenotypic predictive profiling of BRCA1-associated breast cancer. Virchows Arch. 2011;458:55–64.
7. Moll R, Krepler R, Franke WW. Complex cytokeratin polypeptide patterns observed in certain human carcinomas. Differentiation. 1983;23:256–69.
8. Moll R, Franke WW, Schiller DL, Geiger B, Krepler R. The catalog of human cytokeratins: patterns of expression in normal epithelia, tumors and cultured cells. Cell. 1982;31:11–24.
9. Dairkee SH, Puett L, Hackett AJ. Expression of basal and luminal epithelium-specific keratins in normal, benign, and malignant breast tissue. J Natl Cancer Inst. 1988;80:691–5.

10. Gusterson BA, Ross DT, Heath VJ, Stein T. Basal cytokeratins and their relationship to the cellular origin and functional classification of breast cancer. Breast Cancer Res. 2005;7: 143–8.
11. Cheang MC, et al. Basal-like breast cancer defined by five biomarkers has superior prognostic value than triple-negative phenotype. Clin Cancer Res. 2008;14:1368–76.
12. Nielsen TO, et al. Immunohistochemical and clinical characterization of the basal-like subtype of invasive breast carcinoma. Clin Cancer Res. 2004;10:5367–74.
13. Rakha EA, et al. Breast carcinoma with basal differentiation: a proposal for pathology definition based on basal cytokeratin expression. Histopathology. 2007;50:434–8.
14. Livasy CA, et al. Phenotypic evaluation of the basal-like subtype of invasive breast carcinoma. Mod Pathol. 2006;19:264–71.
15. Matos I, Dufloth R, Alvarenga M, Zeferino LC, Schmitt F. p63, cytokeratin 5, and P-cadherin: three molecular markers to distinguish basal phenotype in breast carcinomas. Virchows Arch. 2005;447:688–94.
16. Garcia S, et al. Poor prognosis in breast carcinomas correlates with increased expression of targetable CD146 and c-Met and with proteomic basal-like phenotype. Hum Pathol. 2007;38:830–41.
17. Banerjee S, et al. Basal-like breast carcinomas: clinical outcome and response to chemotherapy. J Clin Pathol. 2006;59:729–35.
18. Smid M, et al. Subtypes of breast cancer show preferential site of relapse. Cancer Res. 2008;68:3108–14.
19. Gaedcke J, et al. Predominance of the basal type and HER-2/neu type in brain metastasis from breast cancer. Mod Pathol. 2007;20:864–70.
20. Rakha EA, Ellis IO. Triple-negative/basal-like breast cancer: review. Pathology. 2009;41:40–7.
21. Rakha EA, El-Sayed ME, Reis-Filho J, Ellis IO. Patho-biological aspects of basal-like breast cancer. Breast Cancer Res Treat. 2009;113:411–22.
22. Seewaldt VL, Scott V. Images in clinical medicine. Rapid progression of basal-type breast cancer. N Engl J Med. 2007;356(e12).
23. Collett K, et al. A basal epithelial phenotype is more frequent in interval breast cancers compared with screen detected tumors. Cancer Epidemiol Biomarkers Prev. 2005;14:1108–12.
24. Bergamaschi A, et al. Distinct patterns of DNA copy number alteration are associated with different clinicopathological features and gene-expression subtypes of breast cancer. Genes Chromosomes Cancer. 2006;45:1033–40.
25. Chin K, et al. Genomic and transcriptional aberrations linked to breast cancer pathophysiologies. Cancer Cell. 2006;10:529–41.
26. Wang ZC, et al. Loss of heterozygosity and its correlation with expression profiles in subclasses of invasive breast cancers. Cancer Res. 2004;64:64–71.
27. Sorlie T, et al. Gene expression patterns of breast carcinomas distinguish tumor subclasses with clinical implications. Proc Natl Acad Sci USA. 2001;98:10869–74.
28. Lim E, et al. Aberrant luminal progenitors as the candidate target population for basal tumor development in BRCA1 mutation carriers. Nat Med. 2009;15:907–13.
29. Molyneux G, et al. BRCA1 basal-like breast cancers originate from luminal epithelial progenitors and not from basal stem cells. Cell Stem Cell. 2010;7:403–17.
30. Stingl J, et al. Purification and unique properties of mammary epithelial stem cells. Nature. 2006;439:993–7.
31. Al-Hajj M, Wicha MS, Benito-Hernandez A, Morrison SJ, Clarke MF. Prospective identification of tumorigenic breast cancer cells. Proc Natl Acad Sci USA. 2003;100:3983–8.
32. Honeth G, et al. The CD44+/CD24- phenotype is enriched in basal-like breast tumors. Breast Cancer Res. 2008;10:R53.
33. Ben-Porath I, et al. An embryonic stem cell-like gene expression signature in poorly differentiated aggressive human tumors. Nat Genet. 2008;40:499–507.
34. Rouzier R, et al. Breast cancer molecular subtypes respond differently to preoperative chemotherapy. Clin Cancer Res. 2005;11:5678–85.

35. Dent R, et al. Triple-negative breast cancer: clinical features and patterns of recurrence. Clin Cancer Res. 2007;13:4429–34.
36. Kreike B, et al. Gene expression profiling and histopathological characterization of triple-negative/basal-like breast carcinomas. Breast Cancer Res. 2007;9:R65.
37. Bertucci F, et al. How basal are triple-negative breast cancers? Int J Cancer. 2008;123: 236–40.
38. de Ronde JJ, et al. Concordance of clinical and molecular breast cancer subtyping in the context of preoperative chemotherapy response. Breast Cancer Res Treat. 2010;119:119–26.
39. Parker JS, et al. Supervised risk predictor of breast cancer based on intrinsic subtypes. J Clin Oncol. 2009;27:1160–7.
40. Sotiriou C, et al. Breast cancer classification and prognosis based on gene expression profiles from a population-based study. Proc Natl Acad Sci USA. 2003;100:10393–8.
41. Rakha EA, et al. Are triple-negative tumours and basal-like breast cancer synonymous? Breast Cancer Res. 2007;9:404 (author reply 405).
42. Murphy CG, Moynahan ME. BRCA gene structure and function in tumor suppression: a repair-centric perspective. Cancer J. 2010;16:39–47.
43. Roy R, Chun J, Powell SN. BRCA1 and BRCA2: different roles in a common pathway of genome protection. Nat Rev Cancer. 2012;12:68–78.
44. Lakhani SR, et al. Prediction of BRCA1 status in patients with breast cancer using estrogen receptor and basal phenotype. Clin Cancer Res. 2005;11:5175–80.
45. Foulkes WD, et al. Estrogen receptor status in BRCA1- and BRCA2-related breast cancer: the influence of age, grade, and histological type. Clin Cancer Res. 2004;10:2029–34.
46. Foulkes WD, et al. Germline BRCA1 mutations and a basal epithelial phenotype in breast cancer. J Natl Cancer Inst. 2003;95:1482–5.
47. Arnes JB, et al. Placental cadherin and the basal epithelial phenotype of BRCA1-related breast cancer. Clin Cancer Res. 2005;11:4003–11.
48. van't Veer LJ, et al. Gene expression profiling predicts clinical outcome of breast cancer. Nature. 2002;415:530–6.
49. Gorski JJ, et al. BRCA1 transcriptionally regulates genes associated with the basal-like phenotype in breast cancer. Breast Cancer Res Treat. 2010;122:721–31.
50. Tkocz D, et al. BRCA1 and GATA3 corepress FOXC1 to inhibit the pathogenesis of basal-like breast cancers. Oncogene. 2011;31(32):3667–78.
51. Ray PS, et al. FOXC1 is a potential prognostic biomarker with functional significance in basal-like breast cancer. Cancer Res. 2010;70:3870–6.
52. Turner N, Tutt A, Ashworth A. Hallmarks of 'BRCAness' in sporadic cancers. Nat Rev Cancer. 2004;4:814–9.
53. Esteller M, et al. Promoter hypermethylation and BRCA1 inactivation in sporadic breast and ovarian tumors. J Natl Cancer Inst. 2000;92:564–9.
54. Catteau A, Harris WH, Xu CF, Solomon E. Methylation of the BRCA1 promoter region in sporadic breast and ovarian cancer: correlation with disease characteristics. Oncogene. 1999;18:1957–65.
55. Beger C, et al. Identification of Id4 as a regulator of BRCA1 expression by using a ribozyme-library-based inverse genomics approach. Proc Natl Acad Sci USA. 2001;98:130–5.
56. Turner NC, et al. BRCA1 dysfunction in sporadic basal-like breast cancer. Oncogene. 2007;26:2126–32.
57. Palacios J, et al. Phenotypic characterization of BRCA1 and BRCA2 tumors based in a tissue microarray study with 37 immunohistochemical markers. Breast Cancer Res Treat. 2005;90:5–14.
58. Laakso M, Loman N, Borg A, Isola J. Cytokeratin 5/14-positive breast cancer: true basal phenotype confined to BRCA1 tumors. Mod Pathol. 2005;18:1321–8.
59. van Beers EH, et al. Comparative genomic hybridization profiles in human BRCA1 and BRCA2 breast tumors highlight differential sets of genomic aberrations. Cancer Res. 2005;65:822–7.
60. Hedenfalk I, et al. Gene-expression profiles in hereditary breast cancer. N Engl J Med. 2001;344:539–48.

61. Turner NC, Reis-Filho JS. Basal-like breast cancer and the BRCA1 phenotype. Oncogene. 2006;25:5846–53.
62. Willems L, et al. PI3K and mTOR signaling pathways in cancer: new data on targeted therapies. Curr Oncol Rep. 2012;14:129–38.
63. Liu P, Cheng H, Roberts TM, Zhao JJ. Targeting the phosphoinositide 3-kinase pathway in cancer. Nat Rev Drug Discov. 2009;8:627–44.
64. Marty B, et al. Frequent PTEN genomic alterations and activated phosphatidylinositol 3-kinase pathway in basal-like breast cancer cells. Breast Cancer Res. 2008;10:R101.
65. Lopez-Knowles E, et al. PI3K pathway activation in breast cancer is associated with the basal-like phenotype and cancer-specific mortality. Int J Cancer. 2010;126:1121–31.
66. Wang J, et al. FOXC1 regulates the functions of human basal-like breast cancer cells by activating NF-kappaB signaling. Oncogene. 2012;31(45):4798–802.
67. Banerji S, et al. Sequence analysis of mutations and translocations across breast cancer subtypes. Nature. 2012;486:405–9.
68. Moulder SL. Does the PI3K pathway play a role in basal breast cancer? Clin Breast Cancer. 2010;10 Suppl 3:S66–71.
69. Fedele CG, et al. Inositol polyphosphate 4-phosphatase II regulates PI3K/Akt signaling and is lost in human basal-like breast cancers. Proc Natl Acad Sci USA. 2010;107:22231–6.
70. Nakatani K, et al. Up-regulation of Akt3 in estrogen receptor-deficient breast cancers and androgen-independent prostate cancer lines. J Biol Chem. 1999;274:21528–32.
71. Santarpia L, Lippman SM, El-Naggar AK. Targeting the MAPK-RAS-RAF signaling pathway in cancer therapy. Expert Opin Ther Targets. 2012;16:103–19.
72. Hoeflich KP, et al. In vivo antitumor activity of MEK and phosphatidylinositol 3-kinase inhibitors in basal-like breast cancer models. Clin Cancer Res. 2009;15:4649–64.
73. Mirzoeva OK, et al. Basal subtype and MAPK/ERK kinase (MEK)-phosphoinositide 3-kinase feedback signaling determine susceptibility of breast cancer cells to MEK inhibition. Cancer Res. 2009;69:565–72.
74. Balko JM, et al. Profiling of residual breast cancers after neoadjuvant chemotherapy identifies DUSP4 deficiency as a mechanism of drug resistance. Nat Med. 2012;18(7):1052–9.
75. Aldinger KA, et al. FOXC1 is required for normal cerebellar development and is a major contributor to chromosome 6p25.3 Dandy-Walker malformation. Nat Genet. 2009;41:1037–42.
76. Maclean K, et al. Axenfeld-Rieger malformation and distinctive facial features: clues to a recognizable 6p25 microdeletion syndrome. Am J Med Genet A. 2005;132:381–5.
77. Mortemousque B, et al. Axenfeld-Rieger anomaly: a novel mutation in the forkhead box C1 (FOXC1) gene in a 4-generation family. Arch Ophthalmol. 2004;122:1527–33.
78. Kume T, et al. The forkhead/winged helix gene Mf1 is disrupted in the pleiotropic mouse mutation congenital hydrocephalus. Cell. 1998;93:985–96.
79. Kolacinska A, et al. Apoptosis-, proliferation, immune function-, and drug resistance- related genes in ER positive, HER2 positive and triple negative breast cancer. Neoplasma. 2012;59:424–32.
80. Ray PS, et al. Basal-like breast cancer defined by FOXC1 expression offers superior prognostic value: a retrospective immunohistochemical study. Ann Surg Oncol. 2011;18(13):3839–47.
81. Gershtein ES, et al. The expression and DNA-binding activity of NF-kappaB nuclear transcription factor in the tumors of patients with breast cancer. Bull Exp Biol Med. 2010;150:71–4.
82. Biswas DK, Iglehart JD. Linkage between EGFR family receptors and nuclear factor kappaB (NF-kappaB) signaling in breast cancer. J Cell Physiol. 2006;209:645–52.
83. Yamaguchi N, et al. Constitutive activation of nuclear factor-kappaB is preferentially involved in the proliferation of basal-like subtype breast cancer cell lines. Cancer Sci. 2009;100:1668–74.
84. Sizemore ST, Keri RA. The forkhead box transcription factor FOXC1 promotes breast cancer invasion by inducing matrix metalloprotease 7 (MMP7) expression. J Biol Chem. 2012;287(29):24631–40.
85. Dejeux E, et al. DNA methylation profiling in doxorubicin treated primary locally advanced breast tumours identifies novel genes associated with survival and treatment response. Mol Cancer. 2010;9:68.

86. Acunzo J, Katsogiannou M, Rocchi P. Small heat shock proteins HSP27 (HspB1), alphaB-crystallin (HspB5) and HSP22 (HspB8) as regulators of cell death. Int J Biochem Cell Biol. 2012;44(10):1622–31.
87. Tsang JY, et al. AlphaB-crystallin is a useful marker for triple negative and basal breast cancers. Histopathology. 2012;61(3):378–86.
88. Sitterding SM, et al. AlphaB-crystallin: a novel marker of invasive basal-like and metaplastic breast carcinomas. Ann Diagn Pathol. 2008;12:33–40.
89. Moyano JV, et al. AlphaB-crystallin is a novel oncoprotein that predicts poor clinical outcome in breast cancer. J Clin Invest. 2006;116:261–70.
90. Ivanov O, et al. AlphaB-crystallin is a novel predictor of resistance to neoadjuvant chemotherapy in breast cancer. Breast Cancer Res Treat. 2008;111:411–7.
91. Chan SK, et al. Increased alpha-B-crystallin expression in mammary metaplastic carcinomas. Histopathology. 2011;59:247–55.
92. Bosman JD, Yehiely F, Evans JR, Cryns VL. Regulation of alphaB-crystallin gene expression by the transcription factor Ets1 in breast cancer. Breast Cancer Res Treat. 2010;119:63–70.
93. Paredes J, et al. P-cadherin expression in breast cancer: a review. Breast Cancer Res. 2007;9:214.
94. Sousa B, et al. P-cadherin, vimentin and CK14 for identification of basal-like phenotype in breast carcinomas: an immunohistochemical study. Histol Histopathol. 2010;25:963–74.
95. Potemski P, et al. Relationship of P-cadherin expression to basal phenotype of breast carcinoma. Pol J Pathol. 2007;58:183–8.
96. Paredes J, Lopes N, Milanezi F, Schmitt FC. P-cadherin and cytokeratin 5: useful adjunct markers to distinguish basal-like ductal carcinomas in situ. Virchows Arch. 2007;450:73–80.
97. Liu N, et al. P-cadherin expression and basal-like subtype in breast cancers. Med Oncol. 2012;29(4):2606–12.
98. Turashvili G, et al. P-cadherin expression as a prognostic biomarker in a 3992 case tissue microarray series of breast cancer. Mod Pathol. 2011;24:64–81.
99. Vieira AF, et al. P-cadherin is coexpressed with CD44 and CD49f and mediates stem cell properties in basal-like breast cancer. Stem Cells. 2012;30:854–64.
100. Lambert AW, Ozturk S, Thiagalingam S. Integrin signaling in mammary epithelial cells and breast cancer. ISRN Oncol. 2012;2012:493283.
101. Lu S, Simin K, Khan A, Mercurio AM. Analysis of integrin beta4 expression in human breast cancer: association with basal-like tumors and prognostic significance. Clin Cancer Res. 2008;14:1050–8.
102. Allen MD, et al. Clinical and functional significance of alpha9beta1 integrin expression in breast cancer: a novel cell-surface marker of the basal phenotype that promotes tumour cell invasion. J Pathol. 2011;223:646–58.
103. Switzer CH, et al. S-nitrosylation of EGFR and Src activates an oncogenic signaling network in human basal-like breast cancer. Mol Cancer Res. 2012;10(9):1203–15.
104. Kurebayashi J, et al. Preferential antitumor effect of the Src inhibitor dasatinib associated with a decreased proportion of aldehyde dehydrogenase 1-positive cells in breast cancer cells of the basal B subtype. BMC Cancer. 2010;10:568.
105. Bagaria SP, et al. Prognostic value of basal phenotype in HER2-overexpressing breast cancer. Ann Surg Oncol. 2012;19:935–40.
106. Liu H, et al. Basal-HER2 phenotype shows poorer survival than basal-like phenotype in hormone receptor-negative invasive breast cancers. Hum Pathol. 2008;39:167–74.
107. Oliveras-Ferraros C, et al. Pathway-focused proteomic signatures in HER2-overexpressing breast cancer with a basal-like phenotype: new insights into de novo resistance to trastuzumab (Herceptin). Int J Oncol. 2010;37:669–78.
108. Metzger-Filho O, et al. Dissecting the heterogeneity of triple-negative breast cancer. J Clin Oncol. 2012;30:1879–87.
109. Chang HR, et al. Differential response of triple-negative breast cancer to a docetaxel and carboplatin-based neoadjuvant treatment. Cancer. 2010;116:4227–37.
110. De Laurentiis M, et al. Taxane-based combinations as adjuvant chemotherapy of early breast cancer: a meta-analysis of randomized trials. J Clin Oncol. 2008;26:44–53.

111. Michalak EM, Jonkers J. Studying therapy response and resistance in mouse models for BRCA1-deficient breast cancer. J Mammary Gland Biol Neoplasia. 2011;16:41–50.
112. Kennedy RD, Quinn JE, Mullan PB, Johnston PG, Harkin DP. The role of BRCA1 in the cellular response to chemotherapy. J Natl Cancer Inst. 2004;96:1659–68.
113. Diallo-Danebrock R, et al. Protein expression profiling in high-risk breast cancer patients treated with high-dose or conventional dose-dense chemotherapy. Clin Cancer Res. 2007;13:488–97.
114. Herschkowitz JI, He X, Fan C, Perou CM. The functional loss of the retinoblastoma tumour suppressor is a common event in basal-like and luminal B breast carcinomas. Breast Cancer Res. 2008;10:R75.
115. Rakha EA, et al. Triple-negative breast cancer: distinguishing between basal and nonbasal subtypes. Clin Cancer Res. 2009;15:2302–10.
116. Rakha EA, et al. Basal phenotype identifies a poor prognostic subgroup of breast cancer of clinical importance. Eur J Cancer. 2006;42:3149–56.
117. Kassam F, et al. Survival outcomes for patients with metastatic triple-negative breast cancer: implications for clinical practice and trial design. Clin Breast Cancer. 2009;9:29–33.
118. Rody A, et al. The erbB2+ cluster of the intrinsic gene set predicts tumor response of breast cancer patients receiving neoadjuvant chemotherapy with docetaxel, doxorubicin and cyclophosphamide within the GEPARTRIO trial. Breast. 2007;16:235–40.
119. Wang Y, et al. BASC, a super complex of BRCA1-associated proteins involved in the recognition and repair of aberrant DNA structures. Genes Dev. 2000;14:927–39.
120. Moynahan ME, Chiu JW, Koller BH, Jasin M. Brca1 controls homology-directed DNA repair. Mol Cell. 1999;4:511–8.
121. Patel AG, Sarkaria JN, Kaufmann SH. Nonhomologous end joining drives poly(ADP-ribose) polymerase (PARP) inhibitor lethality in homologous recombination-deficient cells. Proc Natl Acad Sci USA. 2011;108:3406–11.
122. Ame JC, Spenlehauer C, de Murcia G. The PARP superfamily. Bioessays. 2004;26:882–93.
123. Fong PC, et al. Inhibition of poly(ADP-ribose) polymerase in tumors from BRCA mutation carriers. N Engl J Med. 2009;361:123–34.
124. Farmer H, et al. Targeting the DNA repair defect in BRCA mutant cells as a therapeutic strategy. Nature. 2005;434:917–21.
125. McCabe N, et al. Deficiency in the repair of DNA damage by homologous recombination and sensitivity to poly(ADP-ribose) polymerase inhibition. Cancer Res. 2006;66:8109–15.
126. O'Shaughnessy J, et al. Iniparib plus chemotherapy in metastatic triple-negative breast cancer. N Engl J Med. 2011;364:205–14.
127. Patel AG, De Lorenzo SB, Flatten KS, Poirier GG, Kaufmann SH. Failure of iniparib to inhibit poly(ADP-Ribose) polymerase in vitro. Clin Cancer Res. 2012;18:1655–62.
128. Tutt A, et al. Oral poly(ADP-ribose) polymerase inhibitor olaparib in patients with BRCA1 or BRCA2 mutations and advanced breast cancer: a proof-of-concept trial. Lancet. 2010;376:235–44.
129. Yuan Y, Liao YM, Hsueh CT, Mirshahidi HR. Novel targeted therapeutics: inhibitors of MDM2, ALK and PARP. J Hematol Oncol. 2011;4:16.
130. Domagala P, Lubinski J, Domagala W. Iniparib in metastatic triple-negative breast cancer. N Engl J Med. 2011;364:1780 (author reply 1781).
131. Bayraktar S, Gluck S. Systemic therapy options in BRCA mutation-associated breast cancer. Breast Cancer Res Treat. 2012;135(2):355–66.
132. Reis-Filho JS, et al. Metaplastic breast carcinomas exhibit EGFR, but not HER2, gene amplification and overexpression: immunohistochemical and chromogenic in situ hybridization analysis. Breast Cancer Res. 2005;7:R1028–35.
133. Hoadley KA, et al. EGFR associated expression profiles vary with breast tumor subtype. BMC Genomics. 2007;8:258.
134. Reis-Filho JS, et al. EGFR amplification and lack of activating mutations in metaplastic breast carcinomas. J Pathol. 2006;209:445–53.

135. Corkery B, Crown J, Clynes M, O'Donovan N. Epidermal growth factor receptor as a potential therapeutic target in triple-negative breast cancer. Ann Oncol. 2009;20:862–7.
136. Mendelsohn J, Baselga J. Epidermal growth factor receptor targeting in cancer. Semin Oncol. 2006;33:369–85.
137. Gholam D, Chebib A, Hauteville D, Bralet MP, Jasmin C. Combined paclitaxel and cetuximab achieved a major response on the skin metastases of a patient with epidermal growth factor receptor-positive, estrogen receptor-negative, progesterone receptor-negative and human epidermal growth factor receptor-2-negative (triple-negative) breast cancer. Anticancer Drugs. 2007;18:835–7.
138. Reeder-Hayes KE, Carey LA, Sikov WM. Clinical trials in triple negative breast cancer. Breast Dis. 2010;32:123–36.
139. Finn RS, et al. Dasatinib, an orally active small molecule inhibitor of both the src and abl kinases, selectively inhibits growth of basal-type/"triple-negative" breast cancer cell lines growing in vitro. Breast Cancer Res Treat. 2007;105:319–26.
140. Finn RS, et al. Dasatinib as a single agent in triple-negative breast cancer: results of an open-label phase 2 study. Clin Cancer Res. 2011;17:6905–13.
141. Yeatman TJ. A renaissance for SRC. Nat Rev Cancer. 2004;4:470–80.
142. Nalwoga H, Arnes JB, Wabinga H, Akslen LA. Expression of EGFR and c-kit is associated with the basal-like phenotype in breast carcinomas of African women. APMIS. 2008;116:515–25.
143. Kim MJ, et al. Clinicopathologic significance of the basal-like subtype of breast cancer: a comparison with hormone receptor and Her2/neu-overexpressing phenotypes. Hum Pathol. 2006;37:1217–26.
144. Caldas-Lopes E, et al. Hsp90 inhibitor PU-H71, a multimodal inhibitor of malignancy, induces complete responses in triple-negative breast cancer models. Proc Natl Acad Sci USA. 2009;106:8368–73.
145. Rhodes A, Jasani B, Barnes DM, Bobrow LG, Miller KD. Reliability of immunohistochemical demonstration of oestrogen receptors in routine practice: interlaboratory variance in the sensitivity of detection and evaluation of scoring systems. J Clin Pathol. 2000;53:125–30.

Chapter 6
Re-excision After Lumpectomy for Breast Cancer

Suzanne B. Coopey

Abstract The frequency of positive margins after lumpectomy for breast cancer ranges from 18 to 50 %. Negative margins are necessary in order to minimize the risk of local recurrence after breast-conserving therapy. Many approaches have been described to reduce re-excision rates, including tumor localization techniques, surgical techniques, intraoperative imaging, and intraoperative pathologic assessment. Localization of nonpalpable tumors is possible with single or bracketed wires and with radioactive seeds. Surgical techniques such as shaved cavity margins and intraoperative specimen inking are advocated by some surgeons as a way to reduce margin positivity. Intraoperative imaging with ultrasonography, mammography, micro-computed tomography, and radiofrequency spectroscopy may have a role in decreasing re-excision rates after lumpectomy. Intraoperative pathologic evaluation of margins by frozen section or touch preparation cytology may also be beneficial. This chapter reviews these approaches in detail.

6.1 Introduction

Breast-conserving therapy (BCT) is a safe alternative to mastectomy for many invasive and in situ breast cancers. Evidence from large, prospective, randomized, clinical trials has shown no significant difference in overall or disease-free survival comparing mastectomy to lumpectomy plus radiation for early stage breast cancer [1–3]. The main disadvantage of BCT compared to mastectomy is the potential for local recurrence. Local recurrence rates with breast conservation range from 9 to 22 % with 20 years of follow-up [2–5]. In comparison, the risk of local chest wall recurrence after mastectomy for stage I and II breast cancers ranges from 0 to 10 % (Table 6.1).

S.B. Coopey (✉)
Surgical Oncology, Massachusetts General Hospital, Boston, MA, USA
e-mail: SCOOPEY@PARTNERS.ORG

H. Schatten (ed.), *Cell and Molecular Biology of Breast Cancer*,
DOI 10.1007/978-1-62703-634-4_6, © Springer Science+Business Media New York 2013

Table 6.1 Local recurrence rates for breast conservation therapy and mastectomy

Trial	Number of patients	Breast cancer stage	Length of follow-up (years)	Local recurrence BCT	Local recurrence mastectomy
NSABP B-06 [3]	1,851	I, II	20	14 %	Not reported (8 % at 8 years)
National Cancer Institute, USA [4]	237	I, II	18	22 %	0 %
National Cancer Institute, Milan [2]	701	I	20	9 %	2 %
Institut Gustave-Roussy [5]	179	I, IIA	22	16 %	10 %

NSABP National Surgical Adjuvant Bowel and Breast Project

6.2 Patient Selection

There are few absolute contraindications to BCT. They include first or second trimester of pregnancy, multicentric disease with two or more tumors in separate quadrants of the breast, a history of prior therapeutic breast irradiation including mantle irradiation for Hodgkin's disease, and inability to achieve negative surgical margins [6, 7]. Relative contraindications include certain collagen vascular disorders, such as scleroderma and systemic lupus erythematosus.

Tumor size must also be taken into account when considering breast conservation. While most trials included tumors less than 5 cm in size, there is no specific tumor size which precludes BCT. More important is the ratio of tumor size to breast size. Breast conservation is not ideal for a large tumor within a small breast when excision would result in a poor cosmetic outcome.

6.3 Definition of a Negative Margin

The most important risk factor for local recurrence after BCT is the presence of positive lumpectomy margins [8–11]. While there is no consensus as to what constitutes a negative margin, most surgeons typically re-excise for margins less than 2 mm [12] (Fig. 6.1). Re-excision of an initially close (<2 mm) or positive margin which results in a negative final margin reduces the risk of an in-breast recurrence to that of an initially negative margin [13]. In contrast to surgeons, the majority of radiation oncologists in North America (45.9 %) only require no tumor on the inked margin to deem a margin negative [14].

6.4 Re-excision Rates and Risk Factors

The frequency of involved (positive and/or close) margins after lumpectomy ranges from 18 to 50 % [15–23]. Two of the largest reported series with over 3,000 patients cite rates of 18–23 % [15, 16]. Recognized risk factors for involved margins include

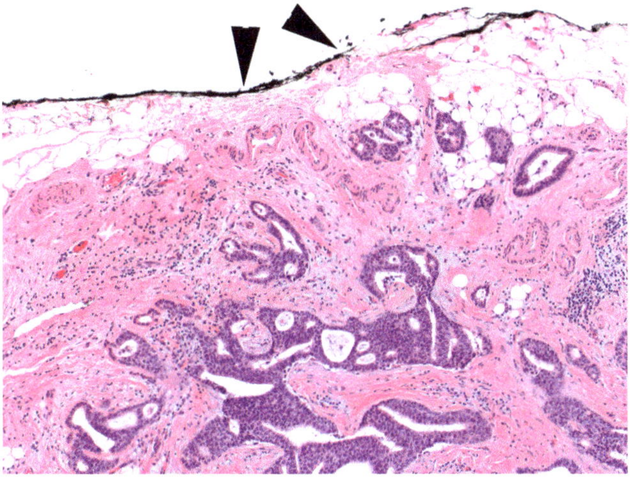

Fig. 6.1 Invasive carcinoma extending to <0.1 cm from the new, inked margin (*arrowheads*). *Courtesy of Elena F. Brachtel MD, Department of Pathology, Massachusetts General Hospital, Boston, MA*

diagnosis by excisional biopsy, younger age, larger tumor size, multifocality, extensive intraductal component, and invasive lobular carcinoma [17–23]. When margins are close or positive, reoperation by re-excision or mastectomy is advised in order to attain negative margins.

6.5 Tumor Localization

6.5.1 Wire Localization

For more than 20 years, wire localization has been used for clinically occult tumors. This technique allows for intraoperative detection of nonpalpable tumors. Wires may be placed by mammographic, sonographic, or magnetic resonance imaging guidance. After excision, the lumpectomy specimen is evaluated mammographically for the presence of the target lesion(s). Target lesions can include microcalcifications, masses, and/or clips. A single wire is most commonly used, but multiple bracketing can also be placed to better define the extent of disease.

6.5.1.1 Multiple Wires

Multiple wires may be placed to bracket large areas of malignant calcifications. Placing multiple wires to delineate the extent of disease has the potential to decrease re-excision rates. However, the effect of multiple wires on the volume of resection is debatable. Kirstein et al. found a significant decrease in re-excision rates after lumpectomy with the use of multiple wires compared with a single wire, 28 % versus 36 %, respectively [24]. Unfortunately, this decrease in re-excision rates with the use of multiple wires was associated with a significant increase in the volume of tissue excised [24]. While Burkholder and colleagues detected a trend toward decreased re-excision rates with multiple wires compared to a single wire (25.6 % versus 17.4 %, respectively), this did not reach significance [25]. In contrast to Kirstein, Burkholder found that multiple wires compared to a single wire led to equivalent lumpectomy volumes for DCIS and significantly smaller lumpectomy volumes for stage 1 and 2 invasive breast cancers [24, 25].

6.5.2 Radioactive Seed Localization

An alternative to wire localization is radioactive seed localization. With this technique a titanium seed containing ^{125}I is placed preoperatively using sonographic or mammographic guidance [26]. Because the half-life of the ^{125}I seed is 60 days, it can be placed days before surgery, an advantage over wire localization. It also eliminates potential complications of wire localization, including wire displacement and transection. A standard intraoperative gamma probe, set to the ^{125}I setting, is used to scan over the breast in order to detect the area of greatest activity. This allows for incision placement directly over the target lesion if desired. The gamma probe is then used throughout the surgery to give audible feedback to the surgeon as to the location and depth of the lesion [26].

Two randomized, controlled trials comparing wire localization with radioactive seed localization have been published [27, 28]. Both compared re-excision rates between groups and results were conflicting. Gray et al. found a significant improvement in re-excision rates with radioactive seed localization (26 %) compared with wire localization (57 %, $p=0.02$) [27]. However, only 42 % of patients had a known preoperative cancer diagnosis, which may have limited the extent of resection. In contrast, Lovric et al. only included patients with a preoperative cancer diagnosis and found no significant difference in reoperation rates with radioactive seed localization (15.1 %) compared with wire localization (19.0 %, $p=0.389$) [28].

6.6 Surgical Techniques

6.6.1 Shaved Cavity Margins

Taking additional shaved cavity margins at the time of lumpectomy is advocated by some surgeons as a way to decrease re-excision rates and/or reduce the volume of breast tissue excised [29–38]. This technique involves excision of all margins adjacent to the lumpectomy cavity. The absolute reduction in re-excision rates with shaved cavity margins compared with lumpectomy alone in retrospective studies ranges from 0 % to 20 % [36–38]. No prospective, randomized trials comparing these two techniques have been reported. At least two studies have found that taking additional shaved cavity margins results in a significant reduction in the total volume of breast tissue excised [37, 38].

Another benefit of taking shaved cavity margins is the possibility of improved pathological assessment of true margins. False-positive lumpectomy margins are thought to occur due to disruption of friable tissue at the edge of lumpectomy specimens. This tissue can inadvertently be removed during the procedure itself, during specimen handling, and/or during specimen radiography. In a study by Dooley, the incidence of false-positive lumpectomy margins was found to be 26.7 % [39]. Taking additional shaved cavity margins for definitive pathological evaluation could help avert this problem.

6.6.2 Intraoperative Specimen Inking

Intraoperative specimen inking by the surgeon at the time of lumpectomy may have a role in decreasing re-excision rates. When Singh et al. performed a retrospective review looking at six-color specimen inking by surgeons intraoperatively versus pathologists in the laboratory, they found a 50 % relative reduction in close/positive margins, from 46 % to 23 % [40]. Although results are promising, they did not reach significance ($p=0.06$), likely due to the small number of patients included in the study ($n=65$). Additional studies are needed to validate these results.

6.7 Intraoperative Imaging

6.7.1 Intraoperative Ultrasound

Intraoperative ultrasound for localization of nonpalpable breast cancers was introduced in 1988 as an alternative to wire localization [41]. Intraoperative ultrasound is successful in identifying the target lesion in 95.7–100 % of cases [42, 43]. Advantages of this technique include flexibility in surgical scheduling by omitting wire placement by radiology, improvement in patient comfort and anxiety, and enhanced intraoperative margin assessment with the potential to lower re-excision rates [42, 43].

Utilizing this technique, re-excision rates of 4–11 % have been cited [43–46]. In a small, prospective, randomized trial, Rahusen and colleagues found a significant improvement in re-excision rates with intraoperative ultrasound compared to wire localization for nonpalpable tumors (11 % versus 45 %, $p=0.007$) [45]. While intraoperative ultrasound has proven useful for localization of invasive cancers, visualization of DCIS has been more difficult [42, 46, 47]. In fact, when James et al.

compared intraoperative ultrasound to wire localization for DCIS, they found no significant difference in re-excision rates between these two modalities [47]. This is likely because the calcifications present in DCIS are not typically seen with ultrasound. When using intraoperative ultrasound for excision of DCIS, there must be a sonographically visible target, such as a hematoma, post-biopsy clip, or biopsy cavity, present [48].

6.7.2 Specimen Mammography

Specimen mammography is the standard of care when excising a nonpalpable breast cancer which has been wire localized to ensure excision of the target lesion. It can also show where the target lesion is in relation to the lumpectomy margins. The sensitivity of specimen mammography for predicting margins is 49–62 %, while the specificity is 77–95 % [49–51]. Prior to the introduction of specimen mammography, Layfield and colleagues relied on macroscopic inspection of the lumpectomy cavity and specimen to ensure an adequate resection; the sensitivity of this technique was much lower at 7.7 % [51].

If specimen mammography shows compromised margins, taking immediate directed cavity shaves has the potential to reduce re-excision rates [52]. While a recent retrospective cohort study by Layfield et al. failed to demonstrate a significant reduction in re-excision rates with specimen mammography, it did show that intraoperative specimen mammography was able to significantly reduce the volume of breast tissue excised during lumpectomy [51]. This may have a positive impact on cosmetic outcome with BCT.

6.7.3 Specimen Micro-Computed Tomography

Micro-computed tomography (micro-CT) is a newer technology for breast tissue evaluation. It provides a three-dimensional imaging of specimens with spatial resolution down to <1 μm [53]. It has the capability to differentiate breast masses from fibrous breast tissue and benign calcifications from malignant ones [54]. Micro-CT provides noninvasive, high-quality imaging of intact small specimens up to 15 cm in diameter, is self-shielded, and is compact enough to be placed near the operating room for convenient access. It takes 7 min to scan the specimen and create two-dimensional cross-sectional images and an additional 7 min to fully reconstruct the images into three-dimensional images [55].

A recent study comparing specimen mammography to specimen micro-CT for evaluation of lumpectomy margins found increased sensitivity for positive margin detection with micro-CT [56]. Micro-CT had a sensitivity of 60 % and a specificity of 93 %, while specimen mammography had a sensitivity of 36 % and a specificity of 96 % [56]. Micro-CT may also be useful for the intraoperative evaluation of shaved cavity margins according to a small pilot study [55].

6.7.4 Radiofrequency Spectroscopy

Radiofrequency spectroscopy with MarginProbe (Dune Medical, Israel) is a novel method for intraoperative margin assessment. The device probe is directly applied to the lumpectomy specimen after excision. Radiofrequency signals are transmitted from the probe to the tissue, reflected back, and collected by the console [57]. The reflected signals are algorithmically analyzed and reported as "positive" or "negative." A positive reading indicates invasive cancer or DCIS detected within 1 mm of the lumpectomy specimen edge. Utilizing this technology, Allweis and colleagues performed a randomized, controlled trial of 300 patients and found a 56 % reduction in re-excision rates, from 12.7 % to 5.6 % [58]. One downside of the device is that it is less successful at assessing margin widths of 2 mm and 5 mm compared with 1 mm [58, 59].

6.8 Intraoperative Pathologic Assessment

6.8.1 Intraoperative Frozen Section

Some institutions routinely perform intraoperative frozen sections to analyze breast cancer lumpectomy margins to decrease the need for re-excision. With this process, the specimen is inked, frozen, and sectioned and then thawed, stained, and evaluated by a pathologist while the patient remains in the operating room [60]. When Esbona and colleagues performed a systematic review of the literature, they found that intraoperative frozen section analysis of lumpectomy margins led to a significant reduction in re-excision rates compared to permanent section analysis, from 35 % (±3 %) to 10 % (±6 %) ($p<0.0001$) [60].

The reported sensitivity of intraoperative frozen section analysis was 83 % (± 13 %), while the reported specificity was 95 % (±8 %) [60]. The variability in sensitivity may be due to pathologist experience [61]. The use of neoadjuvant chemotherapy has also been associated with a lower sensitivity of intraoperative frozen section [62], while lower specificity has been linked to the presence of atypical cells and sclerosing adenosis [60].

One major drawback of frozen section analysis of lumpectomy margins is that it is a labor-intensive process which requires an on-site dedicated pathology team [63]. It also adds significantly to operating room time, around 30–53 min [62, 64]. However, despite the increased operating room time and added costs involved, Osborn and colleagues found that intraoperative frozen section analysis of lumpectomy margins may be cost saving when an institution's re-excision rates without frozen section are greater than 36 % [63]. Opponents of this process do worry that freezing and thawing the specimen could results in loss of tissue which could compromise definitive evaluation of the tumor by the pathologist for histologic information and tumor staging [60, 61].

6.8.2 *Intraoperative Touch Preparation Cytology*

Intraoperative touch preparation cytology or "imprint cytology" has been proposed as a more rapid and simple alternative to intraoperative frozen section [65]. With this technique, the lumpectomy specimen is oriented and pressed onto glass slides making an imprint of all margins [60]. Slides are then fixed, stained, and microscopically evaluated. Malignant cells adhere to the slides, while benign adipose tissue does not [61]. The entire process takes an average of 10 min [66].

Utilizing this technique, re-excision rates of 11 % (± 4 %) have been reported [60]. The sensitivity of intraoperative imprint cytology is 72 % (± 38 %), while the specificity is 97 % (±3 %) [60]. There is a greater degree of variability in the sensitivity of imprint cytology among studies compared to frozen sections [60]. The use of imprint cytology for margin evaluation in patients with invasive lobular carcinoma is limited [67]. Another shortcoming of imprint cytology is that it only detects positive margins (tumor cells on the lumpectomy surface) without taking into account close margins [61].

6.9 Multiple Re-excisions

Performing multiple re-excisions to complete breast conservation therapy is a safe alternative to mastectomy. 61–70 % of patients undergoing two or more re-excisions will achieve negative margins [68, 69]. There is an acceptably low risk of locoregional and systemic failure when negative margins are ultimately achieved in the setting of appropriate radiation and systemic therapy. The 5-year locoregional recurrence rate after multiple re-excisions ranges from 2 to 5.5 % and is not significantly different than that of patients undergoing a single re-excision [68, 70].

References

1. Arriagada R, Le MG, Rochard F, Contesso G, The Institut Gustave-Roussy Breast Cancer Group. Conservative treatment versus mastectomy in early breast cancer: patterns of failure with 15 years of follow-up data. J Clin Oncol. 1996;14:1558–64.
2. Veronesi U, Cascinelli N, Mariani L, et al. Twenty-year follow-up of a randomized study comparing breast-conserving surgery with radical mastectomy for early breast cancer. N Engl J Med. 2002;347:1227–32.
3. Fisher B, Anderson S, Bryant J, et al. Twenty-year follow-up of a randomized trial comparing total mastectomy, lumpectomy, and lumpectomy plus irradiation for the treatment of invasive breast cancer. N Engl J Med. 2002;347:1233–41.
4. Poggi MM, Danforth DN, Sciuto LC, et al. Eighteen-year results in the treatment of early breast carcinoma with mastectomy versus breast conservation therapy. Cancer. 2003;98:697–702.
5. Arriagada R, Lê MG, Guinebretière JM, Dunant A, Rochard F, Tursz T. Late local recurrences in a randomised trial comparing conservative treatment with total mastectomy in early breast cancer patients. Ann Oncol. 2003;14:1617–22.

6. Whelan T. Use of conventional radiation therapy as part of breast-conserving treatment. J Clin Oncol. 2005;23:1718–25.
7. Morrow M, Strom EA, Bassett LW, et al. Standard for breast conservation therapy in the management of invasive breast carcinoma. CA Cancer J Clin. 2002;52:277–300.
8. Park CC, Mitsumori M, Nixon A, et al. Outcome at 8 years after breast-conserving surgery and radiation therapy for invasive breast cancer: influence of margin status and systemic therapy on local recurrence. J Clin Oncol. 2000;18:1668–75.
9. Singletary SE. Surgical margins in patients with early-stage breast cancer treated with breast conservation therapy. Am J Surg. 2002;184:383–93.
10. Kreike B, Hart AAM, Van de Velde T, et al. Continuing risk of ipsilateral breast relapse after breast-conserving therapy at long term follow-up. Int J Radiat Oncol Biol Phys. 2008;71: 1014–21.
11. Smitt MC, Nowels KW, Zdeblick MJ, Jeffrey S, Carlson RW, Stockdale FE, Goffinet DR. The importance of the lumpectomy surgical margin status in long term results of breast conservation. Cancer. 1995;76:259–67.
12. Blair SL, Thompson K, Rococco J, Malcarne V, Beitsch PD, Ollila DW. Attaining negative margins in breast-conservation operations: is there a consensus among breast surgeons? J Am Coll Surg. 2009;209:608–13.
13. Freedman G, Fowble B, Hanlon A, et al. Patients with early stage invasive cancer with close or positive margins treated with conservative surgery and radiation have an increased risk of breast recurrence that is delayed by adjuvant systemic therapy. Int J Radiat Oncol Biol Phys. 1999;44:1005–15.
14. Taghian A, Mohiuddin M, Jagsi R, Goldberg S, Ceilley E, Powell S. Current perceptions regarding surgical margin status after breast-conserving therapy. Ann Surg. 2005;241:629–39.
15. Gupta A, Subhas G, Dubay L, et al. Review of re-excision for narrow or positive margins of invasive and intraductal carcinoma. Am Surg. 2010;76:731–4.
16. Coopey S, Smith BL, Hanson S, et al. The safety of multiple re-excisions after lumpectomy for breast cancer. Ann Surg Oncol. 2011;18:3797–801.
17. Smitt MC, Horst K. Association of clinical and pathologic variables with lumpectomy surgical margin status after preoperative diagnosis or excisional biopsy of invasive cancer. Ann Surg Oncol. 2007;14:1040–4.
18. Sabel MS, Rogers K, Griffith K, et al. Residual disease after re-excision lumpectomy for close margins. J Surg Oncol. 2009;99:99–103.
19. Schiller DE, Le LW, Cho CJ, Youngson BJ, McCready DR. Factors associated with negative margins of lumpectomy specimen: potential use in selecting patients for intraoperative radiotherapy. Ann Surg Oncol. 2008;15:833–42.
20. Waljee JF, Hu ES, Newman LA, Alderman AK. Predictors of re-excision among women undergoing breast-conserving surgery for cancer. Ann Surg Oncol. 2008;15:1297–303.
21. Kurniawan ED, Wong MH, Windle I, et al. Predictors of surgical margin status in breast-conserving surgery within a breast screening program. Ann Surg Oncol. 2008;15:2542–9.
22. Dillon MF, Hill ADK, Quinn CM, McDermott EW, O'Higgins N. A pathologic assessment of adequate margin status in breast conserving therapy. Ann Surg Oncol. 2006;13:333–9.
23. Sanchez C, Brem RF, McSwain AP, Rapelyea JA, Torrente J, Teal CB. Factors associated with re-excision in patients with early-stage breast cancer treated with breast conservation therapy. Am Surg. 2010;76:331–4.
24. Kirstein LJ, Rafferty E, Specht MC, et al. Outcomes of multiple wire localization for larger breast cancers: when can mastectomy be avoided? J Am Coll Surg. 2008;207:342–6.
25. Burkholder HC, Witherspoon LE, Burns RP, Horn JS, Biderman MD. Breast surgery techniques: preoperative bracketing wire localization by surgeons. Am Surg. 2007;73:574–8.
26. Jakub JW, Gray RJ, Degnim AC, Boughey JC, Gardner M, Cox CE. Current status of radioactive seed for localization of nonpalpable breast lesions. Am J Surg. 2010;199:522–8.
27. Gray RJ, Salud C, Nguyen K, et al. Randomized prospective evaluation of a novel technique for biopsy or lumpectomy of nonpalpable breast lesions: radioactive seed versus wire localization. Ann Surg Oncol. 2001;8:711–5.

28. Lovrics PJ, Goldsmith CH, Hodgson N, et al. A multicentered, randomized, controlled trial comparing radioguided seed localization to standard wire localization for nonpalpable, invasive and in situ breast carcinomas. Ann Surg Oncol. 2011;18:3407–14.
29. Jacobson AF, Asad J, Boolbol SK, Osborne MP, Boachie-Adjei K, Feldman SM. Do additional shaved margins at the time of lumpectomy eliminate the need for re-excision? Am J Surg. 2008;196:556–8.
30. Keskek M, Kothari M, Ardehali B, Betambeau N, Nasiri N, Gui GPH. Factors predisposing to cavity margin positivity following conservation surgery for breast cancer. EJSO. 2004;30:1058–64.
31. Cao D, Lin C, Woo S, Vang R, Tsangaris TN, Argani P. Separate cavity margin sampling at the time of initial breast lumpectomy significantly reduces the need for re-excisions. Am J Surg Pathol. 2005;29:1625–32.
32. Hequet D, Bricou A, Delpech Y, Barranger E. Surgical management modifications following systematic additional shaving of cavity margins in breast conservation treatment. Ann Surg Oncol. 2011;18:114–8.
33. Rizzo M, Iyengar R, Gabram SGA, et al. The effects of additional tumor cavity sampling at the time of breast-conserving surgery on final margin status, volume of resection, and pathologist workload. Ann Surg Oncol. 2010;17:228–34.
34. Huston TL, Pigalarga R, Osborne MP, Tousimis E. The influence of additional surgical margins on the total specimen volume excised and the reoperative rate after breast-conserving surgery. Am J Surg. 2006;192:509–12.
35. Patel RR, Li T, Ross EA, Sesa L, Sigurdson ER, Bleicher RJ. The effect of simultaneous peripheral excision in breast conservation upon margin status. Ann Surg Oncol. 2010;17:2933–9.
36. Kobbermann A, Unzeitig A, Xie XJ, et al. Impact of routine cavity shave margins on breast cancer re-excision rates. Ann Surg Oncol. 2011;18:1349–55.
37. Mook J, Klein R, Kobberman A, et al. Volume of excision and cosmesis with routine cavity shave margins technique. Ann Surg Oncol. 2012;19:886–91.
38. Coopey SB, Buckley JM, Smith BL, Hughes KS, Gadd MA, Specht MC. Lumpectomy cavity shaved margins do not impact re-excision rates in breast cancer patients. Ann Surg Oncol. 2011;18:3036–40.
39. Dooley WC, Parker J. Understanding the mechanisms creating false positive lumpectomy margins. Am J Surg. 2005;190:606–8.
40. Singh M, Singh G, Hogan KT, Atkin KA, Schroen AT. The effect of intraoperative specimen inking on lumpectomy re-excision rates. World J Surg Oncol. 2010;8:4.
41. Schwartz GF, Goldberg BB, Rifkin MD, D'Orazio SE. Ultrasonography: an alternative to x-ray-guided needle localization of nonpalpable breast masses. Surgery. 1988;104:870–3.
42. Smith LF, Rubio IT, Henry-Tillman R, Korourian S, Klimberg VS. Intraoperative ultrasound-guided breast biopsy. Am J Surg. 2000;180:419–23.
43. Ngo C, Pollet AG, Laperrelle J, et al. Intraoperative ultrasound localization of nonpalpable breast cancers. Ann Surg Oncol. 2007;14:2485–9.
44. Olsha O, Shemesh D, Carmon M, et al. Resection margins in ultrasound-guided breast-conserving surgery. Ann Surg Oncol. 2011;18:447–52.
45. Rahusen FD, Bremers AJA, Fabry HFJ, van Amerongen AHMT, Boom RPA, Meijer S. Ultrasound-guided lumpectomy of nonpalpable breast cancer versus wire-guided resection: a randomized clinical trial. Ann Surg Oncol. 2002;9:994–8.
46. Harlow SP, Krag DN, Ames SE, Weaver DL. Intraoperative ultrasound localization to guide surgical excision of nonpalpable breast carcinoma. J Am Coll Surg. 1999;189:241–6.
47. James TA, Harlow S, Sheehey-Jones J, et al. Intraoperative ultrasound versus mammographic needle localization for ductal carcinoma in situ. Ann Surg Oncol. 2009;16:1164–9.
48. Fisher CS, Al Mushawah F, Cyr AE, Gao F, Margenthaler JA. Ultrasound-guided lumpectomy for palpable breast cancers. Ann Surg Oncol. 2011;18:3198–203.
49. Graham RA, Homer MJ, Sigler CJ, et al. The efficacy of specimen radiography in evaluating the surgical margins of impalpable breast carcinoma. AJR. 1994;162:33–6.

50. Lee CH, Carter D. Detecting residual tumor after excisional biopsy of impalpable breast carcinoma: efficacy of comparing preoperative mammograms with radiographs of the biopsy specimen. AJR. 1995;164:81–6.
51. Layfield DM, May DJ, Cutress RI, et al. The effect of introducing an in-theatre intra-operative specimen radiography (iosr) system on the management of palpable breast cancer within a single unit. Breast. 2012;21:459–63.
52. McCormick JT, Keleher AJ, Tikhomirov VB, Budway RJ, Caushaj PF. Analysis of the use of specimen mammography in breast conservation therapy. Am J Surg. 2004;188:433–6.
53. Ritman EL. Current status of developments and applications of micro-CT. Annu Rev Biomed Eng. 2011;13:531–52.
54. Gufler H, Franke FE, Wagner S, et al. Fine structure of breast tissue on micro-computed tomography: a feasibility study. Acad Radiol. 2011;18:230–4.
55. Tang R, Coopey SB, Buckley JM, et al. A pilot study evaluating shaved cavity margins with micro-CT tomography – a novel method for predicting lumpectomy margin status intraoperatively. Breast J. 2013. [Epub ahead of print].
56. Saksena M A, Tang R, Buckley J., et al. Micro-computed tomography of breast lumpectomy specimens: comparison with conventional specimen radiograph for margin assessment. Radiological Society of North America meeting abstracts, Dec 2012, Chicago, IL.
57. Karni T, Pappo I, Sandbank J, et al. A device for real-time, intraoperative margin assessment in breast-conservation surgery. Am J Surg. 2007;194:467–73.
58. Allweis TM, Kaufman Z, Lelcuk S, et al. A prospective, randomized, controlled, multicenter study of a real-time, intraoperative probe for positive margin detection in breast-conserving surgery. Am J Surg. 2008;196:483–9.
59. Thill M, Roder K, Diedrich K, Dittmer C. Intraoperative assessment of surgical margins during breast conserving surgery of ductal carcinoma in situ by use of radiofrequency spectroscopy. Breast. 2011;20:579–80.
60. Esbona K, Li Z, Wilke LG. Intraoperative imprint cytology and frozen section pathology for margin assessment in breast conservation surgery: a systematic review. Ann Surg Oncol. 2012;19(10):3236–45.
61. Pleijhuis RG, Graafland M, de Vries J, Bart J, de Jong JS, van Dam GM. Obtaining adequate surgical margins in breast-conserving therapy for patients with early-stage breast cancer: current modalities and future directions. Ann Surg Oncol. 2009;16:2717–30.
62. Riedl O, Fitzal F, Mader N, et al. Intraoperative frozen section analysis for breast-conserving therapy in 1016 patients with breast cancer. EJSO. 2009;35:264–70.
63. Osborn JB, Keeney GL, Jakub JW, Degnim AC, Boughey JC. Cost-effectiveness analysis of routine frozen-section analysis of breast margins compared with reoperation for positive margins. Ann Surg Oncol. 2011;18:3204–9.
64. Fukamachi K, Ishida T, Usami S, et al. Total-circumference intraoperative frozen section analysis reduces margin-positive rate in breast-conservation surgery. Jpn J Clin Oncol. 2010;40:513–20.
65. Klimberg VS, Westbrook KC, Korourian S. Use of touch preps for diagnosis and evaluation of surgical margins in breast cancer. Ann Surg Oncol. 1998;5(220):226.
66. D'Halluin F, Tas P, Rouquette S, et al. Intra-operative touch preparation cytology following lumpectomy for breast cancer: a series of 400 procedures. Breast. 2009;18:248–53.
67. Valdes EK, Boolbol SK, Ali I, Feldman SM, Cohen JM. Intraoperative touch preparation cytology for margin assessment in breast-conservation surgery: does it work for lobular carcinoma? Ann Surg Oncol. 2007;14:2940–5.
68. Coopey SB, Smith BL, Hanson S, et al. The safety of multiple re-excisions after lumpectomy for breast cancer. Ann Surg Oncol. 2011;18:3797–801.
69. Cellini C, Huston TL, Martins D, et al. Multiple re-excisions versus mastectomy in patients with persistent residual disease following breast conserving surgery. Am J Surg. 2005;189:662–6.
70. O'Sullivan MJ, Li T, Freedman G, Morrow M. The effect of multiple re-excisions on the risk of local recurrence after breast conserving surgery. Ann Surg Oncol. 2007;14:3133–40.

Chapter 7
Novel Anti-angiogenic Therapies Using Naturally Occurring and Synthetic Drugs to Combat Progestin-Dependent Breast Cancer

Salman M. Hyder, Benford Mafuvadze, and Cynthia Besch-Williford

Abstract Angiogenesis, the process by which new blood vessels are formed, is essential for both normal and pathological tissue expansion and provides the nourishment necessary for growth. The role of growth factors which promote angiogenesis in pathologic conditions of the breast is now well established. Recently the synthetic progestin component of combination estrogen and progestin hormone replacement therapy (HRT) has been associated with increased risk of breast cancer in postmenopausal women. We demonstrated that progestins induce breast cancer cells to produce vascular endothelial growth factor (VEGF), a potent angiogenic growth factor that promotes angiogenesis and causes tumors to grow. Unfortunately, synthetic antiprogestins are toxic, precluding their use as a means by which to suppress the proangiogenic activity of administered progestins. In this chapter we will discuss our studies aimed at identifying both naturally occurring and synthetic compounds with antiprogestin and anti-angiogenic activities. We will describe our progress using agents which block the production of progestin-induced VEGF from breast cancer cells and which also have the capacity to both treat and prevent progestin-dependent breast disease in animal models. We contend that information gained from such studies could facilitate the development of personalized medicine which might be used to more precisely and selectively target a specific signal transduction pathway essential to angiogenesis, thereby controlling the formation of new blood vessels essential for nourishing the rapid growth of hormone-dependent

S.M. Hyder (✉) • B. Mafuvadze
Dalton Cardiovascular Research Center, University of Missouri, 134 Research Park Drive, Columbia, MO 65211, USA
e-mail: hyders@missouri.edu

C. Besch-Williford
IDEXX RADIL, Columbia, MO 65201, USA

H. Schatten (ed.), *Cell and Molecular Biology of Breast Cancer*,
DOI 10.1007/978-1-62703-634-4_7, © Springer Science+Business Media New York 2013

breast tumors. Our studies could also further the concept of "angio-prevention" of breast cancer in individuals who are particularly susceptible to progestin-dependent disease, for example, women who have mutations in tumor suppressors such as p53 and Brca-1. By maintaining the angiogenic switch "off" within tumor cells, the development of hormone-dependent breast cancers may be prevented.

7.1 Introduction

Developing tumors require a steady supply of nutrients to grow. A primary function of angiogenesis, or new blood vessel formation, is to provide nourishment to the growing tissue and thereby facilitate tumor development. Since the realization that tumor expansion, irrespective of tissue origin, depends on angiogenesis, there has been intense interest in the role played in this process by angiogenic growth factors [1, 2]. Folkman suggested that by inhibiting angiogenesis within tumor tissue, we might treat and perhaps arrest tumor progression [3]. In the United States alone, about 240,000 new cases of breast cancer are detected annually, and approximately 40,000 women die each year of this deadly disease [4]. Consequently, therapies based on the disruption of angiogenesis could be extremely beneficial, particularly to women taking combination HRT. Endocrine therapy is the choice of treatment for those tumors that possess estrogen or progesterone receptors (ER or PR). However, many tumors either do not respond to this form of treatment or, after initially responding to therapy, go on to develop a population of cells which become resistant and therefore unresponsive to antihormones. Alternative methods of treatment include chemotherapy or the use of aromatase inhibitors [5, 6]. It is only recently that convincing scientific evidence has been acquired proving that breast cancer is an angiogenic-dependent disease which is potentially treatable by anti-angiogenic therapy [7–10]. Since the presence of both ER and PR affects disease progression, it has been suggested that these receptors must control angiogenic factors via their respective ligands [11, 12]. Although several angiogenic growth factors have been identified in breast cancers, most studies have had as their goal the elucidation of the mechanism by which VEGF promotes angiogenesis, since VEGF is also considered a molecular therapeutic target by which to treat a number of different types of cancer, including breast cancer [13]. We have examined the role played by ER and PR in the control of VEGF production by breast cancer cells and have undertaken the identification of naturally occurring and synthetic compounds with the capacity to inhibit this process, leading to tumor regression. Since clinical trials in postmenopausal women show that synthetic progestins used in combination HRT are associated with a higher incidence of breast cancer [14, 15], our recent research endeavors have been aimed at controlling the progression of progestin-dependent breast disease by suppressing angiogenesis. In this context we have described angiogenesis as a therapeutic and preventive target with respect to progestin-dependent breast cancer [16–22]. In the following pages we will first discuss the role played by progestins in controlling VEGF production by tumor cells, a process which leads to increased angiogenesis and progestin-dependent in vivo growth of mammary tumors. We will go on to describe studies in which we identify a number of synthetic and naturally occurring compounds which could be used therapeutically to suppress progestin-dependent VEGF induction by tumor cells, inhibit angiogenesis, and thereby promote tumor regression. Studies will also be described that demonstrate the mechanism by which the aforementioned compounds exert their anti-angiogenic effects and which provide a rationale for developing such agents for preventing or delaying the development of hormone-dependent breast cancer in general and progestin-dependent breast cancer in particular. Since monotherapies are generally ineffective in the long term and resistant tumors almost always emerge, we will describe a highly effective combination of a small synthetic molecule that activates the p53 pathway and a tumor-specific vascular disrupting antibody that destroys tumor blood vessels, which could be used to control the progression of breast disease in a nontoxic fashion.

7.2 Angiogenesis in Breast Cancer

It is now well established that breast cancer is an angiogenic-dependent disease and that angiogenesis plays an essential role in breast cancer development, invasion, and metastasis [10, 23–26]. Even though it has been recognized for decades that breast cancer is an endocrine-dependent disease, it is only recently that attention has focused on the role of sex steroids in the process of angiogenesis in breast cancer cells. Strong evidence that angiogenesis is an essential component of breast cancer is afforded by hyperplastic murine breast papillomas [27] and histologically normal lobules adjacent to cancerous breast tissue that possess higher levels of blood vessels [28], suggesting that angiogenesis precedes transformation of mammary hyperplasia to malignancy [10, 29]. In addition, transfection of breast cancer cells with angiogenic stimulatory peptides increases tumor growth, invasiveness, and metastasis [30]. Conversely, transfection of tumor cells with inhibitors of angiogenesis decreases growth and metastasis [29]. It has been suggested that tumor progression occurs once a balance is attained that favors angiogenic activators over inhibitors and a number of factors have been shown to alter from a preinvasive to an invasive status within breast tumors [31, 32]. This could involve either the induction of angiogenic growth factors (e.g., VEGF, FGF, MMPs, etc) or loss of inhibitors of angiogenesis such as TSP-1, sVEGF receptors, and TIMPs, etc. Several factors such as hypoxia or loss of tumor suppressors and inhibitors of angiogenesis can disrupt the angiogenic balance in a tissue leaning toward tumor progression. This is commonly referred to as an angiogenic switch and describes the sudden growth of tissues that otherwise remain quiescent for a long time [32–34], as occurs in many human breast tumors. Some progress has been made in defining the role of sex steroids in the angiogenic switch [9] though much remains to be investigated in this field since modulation of the switch could prove crucial in treating and preventing the progression of breast cancer.

It is our contention that an important growth factor with the ability to induce the angiogenic switch is VEGF, a key feature of which is the capacity to promote tissue permeability, resulting in hyperemia [35, 36]. There is evidence that HRT increases breast tissue density; it is possible that some of these effects occur due to estrogen-induced hyperemia which arises due to VEGF production, which may become more pronounced with the inclusion of progestins [37]. Indeed, evidence suggests that progesterone may have effects on breast vasculature [38]. Thus it is likely that both estrogens and progestins have direct effects on both normal and neoplastic cells as well as resident endothelial cells, via their respective receptors. In addition there is a possibility that a paracrine mechanism exists that could also arise from the production of angiogenic growth factors in breast tissue following hormonal treatment. It has been proposed [39] that such paracrine mechanisms might even facilitate the proliferation of tumor cells that lack the capacity to produce their own angiogenic growth factors in response to steroid hormones but which respond to VEGF produced from other cells, since they may retain the necessary cell-surface receptors [40]. Thus, targeting both angiogenic ligand and receptors would seem to be the most comprehensive strategy for better controlling tumor proliferation.

7.2.1 Progestin Regulation of VEGF and Angiogenesis in Breast Cancer

Given the importance of estrogens and progestins in regulating breast cancer, and the clear evidence that angiogenesis plays an important role in disease onset, progression, and metastasis, a number of studies have been conducted to explain VEGF regulation by sex steroids [11, 41–45]. Herein, we will discuss how progestins induce the production of VEGF in breast cancer cells. Estrogen regulation of VEGF in breast cancer has been reviewed elsewhere [9, 42].

The role of endogenous progesterone and synthetic progestins in the development and progression of breast cancer has been controversial [46]. Until recently, the conventional view that progestins are terminally differentiative in the endometrium was extrapolated to other tissues such as breast [47], and so progestins were, in general, considered antiproliferative and therefore protective against breast cancer. Estrogens on the other hand were primarily blamed for inducing the disease. There is indeed evidence suggesting that in some cases and at certain dosages, progestins may reduce the risk of breast cancer [48]. For example, initial studies showed progestins to be antiproliferative in normal human breast cells [49]. More recently, Rajkumar et al. [50] reported that exogenous estrogen and progesterone reduced the incidence of DMBA-induced mammary tumors in genetically engineered mice overexpressing Her2/neu. There is also evidence showing that administration of exogenous progestins to animals prior to exposure to chemical carcinogens is protective whereas progestin administration shortly after carcinogen exposure exacerbates tumor proliferation and growth [16, 51]. In contrast to those studies describing the negative effects of progestins with respect to tumor growth, recent experimental and clinical evidence suggests that progestins are indeed responsible for inducing mammary carcinogenesis and promoting the maintenance and progression of breast tumors [21, 52]. We first reported that progestins induce the production of VEGF in breast cancer cells and hypothesized that VEGF released from tumor cells might attract new blood vessels, facilitating tumor growth [44, 53]. Our in vitro observations that progestins affect expression of the VEGF gene were confirmed by others [54, 55] and further supported in vivo observations when data from the Women's Health Initiative study showed a higher risk of breast cancer in postmenopausal women taking combined estrogen-progestin HRT compared with those taking estrogen alone or placebo [15, 56]. This report drew attention to the potential dangers of combination HRT and for the first time made the public aware of the possibility that the progestin component of such HRT regimens might be increasing breast cancer incidence in postmenopausal women. Furthermore, as well as raising the incidence of breast cancer, combined HRT was also shown to increase benign breast proliferative lesions, further supporting the notion that a combination of estradiol and progestins may promote mammary cancer growth [57]. Overall, the effects of progestins on breast tissue are somewhat complicated since in experimental systems progestins can elicit either proliferative or antiproliferative effects on breast epithelial cell growth depending on the model system used, cell context, specific progestin, dosage, and duration of treatment [18, 58]. The clinical effects of progestins are, however, more clearly defined, warranting caution when they are used as a component of HRT.

It is now evident from both in vivo and in vitro studies that progestins stimulate the proliferation of both normal and neoplastic mammary gland cells in a variety of animal species [21, 59–61]. As discussed above, epidemiologic studies and clinical trials show there to be a correlation between HRT containing progestins and an elevated risk of breast cancer in postmenopausal women [14, 56, 62]. Given that it takes only a short time between exposure to progestin-containing HRT and occurrence of breast tumors, we initially suggested that progestins promote the growth of already existing latent tumor cells in the breast (Fig. 7.1). Recent studies suggest that progestins promote tumor growth by creating a microenvironment conducive to tumor development and progression [58, 63–66]. Moreover, recent studies from our laboratory show that progestins promote tumor cell metastasis [22], a phenomenon that has also been reported in humans undergoing combined HRT [57]. Numerous mechanisms involving conventional nuclear PR and PR-independent pathways, a variety of different growth factors, neurotransmitters, and polypeptide hormones have been proposed to explain how progestins exert their effects on breast cancer progression [47]. As previously described, we showed in vitro that progestins induce growth factors such as VEGF, an essential component of angiogenesis which is vital for tumor growth [44, 53]. However, studies in the laboratory were confusing and inconclusive since initially we were unable to show in vivo proliferation of human breast cancer cells, though progestin-dependent breast tumor progression was demonstrated in vivo in rodent models [18, 67–69]. In order to overcome this obstacle, we developed our own model, the first in which progestins were shown to promote the in vivo progression of human tumor cell xenografts (see Sect. 7.2.2.2). Progestin-dependent increases in tumor growth in both rodent and human xenografts involved the induction of VEGF together with increased angiogenesis [16, 18, 21]. There is also

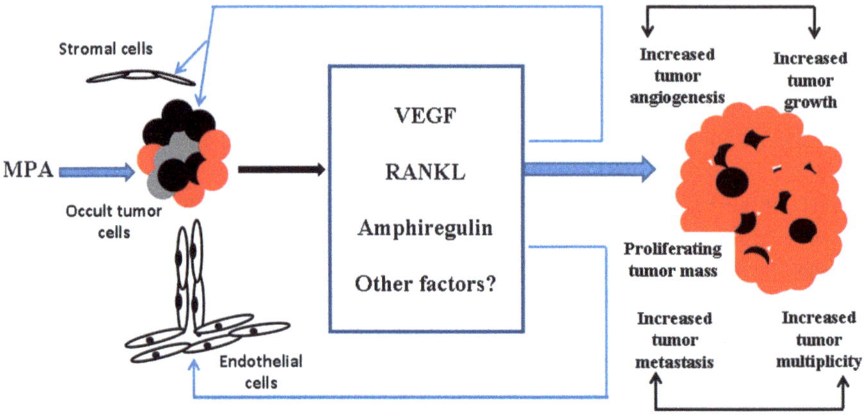

Fig. 7.1 Schematic diagram illustrating potential mechanisms through which progestins promote proliferation and growth of occult tumors. Tumor epithelial cells (*red*) containing PR bind progestins such as medroxyprogesterone acetate (MPA) and respond by secreting various factors with known tumor proliferative effects, which then act in an autocrine and paracrine manner on PR-positive (*red*) and PR-negative (*black*) tumor epithelial cells that express growth factor receptors, stem cells (*gray*), stromal cells, and endothelial cells. The secreted factors include VEGF [20, 44, 53], RANKL [63, 64], and amphiregulin [65]. Growth factors increase angiogenesis, which nourishes proliferating breast tumor cells. Progestins also increase breast cancer metastasis and tumor multiplicity in various animal models referred to in the text

evidence that progestins regulate a number of intracellular factors that are involved in activation and increased expression of several proliferation genes [58]. Recently, Horwitz and Sartorius [70] reported that exposure to medroxyprogesterone acetate (MPA), a progestin commonly used in HRT, increased the stemlike ER-, PR-, and CK5+ subpopulation from 2 % to more than 20 %, lending further support to the notion that progestins play a role in activating and transforming dormant breast cancer stem cells into intermediate subpopulations with the capacity to differentiate into breast cancer cells. Schramek et al. [64] and Gonzalez-Suarez et al. [63] reported that progestin-induced RANKL (receptor activator of nuclear factor kappa-B ligand) may also play a significant role in the development of mammary cancer in rodents, while Kariagina et al. [65] described studies showing that estrogen and progesterone act synergistically in cells expressing ERα, PRA, and PRB to induce robust proliferation of hormone-dependent mammary cancers through induction of amphiregulin, which in turn activates epidermal growth factor receptor (EGFR). Activated EGFR stimulates cell cycle regulatory genes and promotes tumor proliferation through mechanisms involving Akt, ERK, and JNK pathways. Progestins also appear to prime precancerous cells which then respond more vigorously to growth factors [58]. For example, after being primed with MPA, T47D breast cancer cells become highly sensitive to the proliferative effects of epidermal growth factor (EGF) [71].

While it appears that progestins promote breast cancer through various mechanisms, their capacity to increase angiogenesis is clearly central to disease progression [72]. Unfortunately, commonly available antiprogestins such as mifepristone are "promiscuous," in that they also bind to other receptors and thereby have harmful side effects. Such drugs are therefore of limited use in humans. With this in mind we have focused our studies on screening and identifying naturally occurring and synthetic compounds with antiprogestin activity that are safer and can therefore be used to suppress progestin-induced angiogenesis and tumor growth. In subsequent sections, we will describe a number of such compounds identified by us and others in recent years and detail possible mechanisms by which these agents exert anti-angiogenic and anti-proliferative effects when used to combat progestin-dependent breast cancer.

7.2.2 *In Vivo Models Used in Our Laboratory to Study Progestin-Dependent Breast Cancer*

Currently, due to variations in morphology, gene-specific mutations, response to endogenous growth factors, and clinical outcome for breast cancer [73], several in vitro models are used to examine the role of steroid hormones in breast carcinogenesis. However, in vitro studies have inherent limitations when it comes to studying the effects of progestins on breast cancer [74]. For example, progestins are generally antiproliferative in vitro, though why this should be the case when they increase breast cancer cell proliferation in vivo remains a mystery. Consequently, animal models provide a much clearer insight into the pathologic processes underlying progestin-induced breast cancer than in vitro systems. In order to evaluate the anti-progestin and anti-angiogenic activities of synthetic and naturally occurring compounds, we developed two angiogenesis-dependent in vivo models which faithfully recapitulate the in vivo proliferative effects of progestins. These models, which are described below, allow us to identify compounds and agents that can be used therapeutically or which have the potential to prevent the onset of disease. Examples of compounds which have been identified as anti-angiogenic and which have been used to treat and prevent progestin-dependent breast cancer are described next.

7.2.2.1 The Progestin-Accelerated DMBA Rat Mammary Tumor Model

The DMBA-induced rat mammary tumor is a popular in vivo model since rat mammary glands possess comparable ductal-lobular organization to that of human breast [75] and exhibit similar patterns of expression and co-localization of PRA and PRB isoforms [76]. Furthermore, DMBA-induced mammary tumors in rats are ER and PR positive and hormone-dependent [65], making this model an appropriate and valuable means by which to study progestin-dependent breast cancer. Using it we and others have demonstrated that both the natural hormone progesterone and the most widely used synthetic progestin, MPA, accelerate the development of mammary tumors [16, 67] in an angiogenesis-dependent manner [16]. The protocol used and the results obtained to support the statement are described in Fig. 7.2. Briefly, a single oral dose of DMBA (100 mg/kg) given to 45–50-day-old Sprague-Dawley

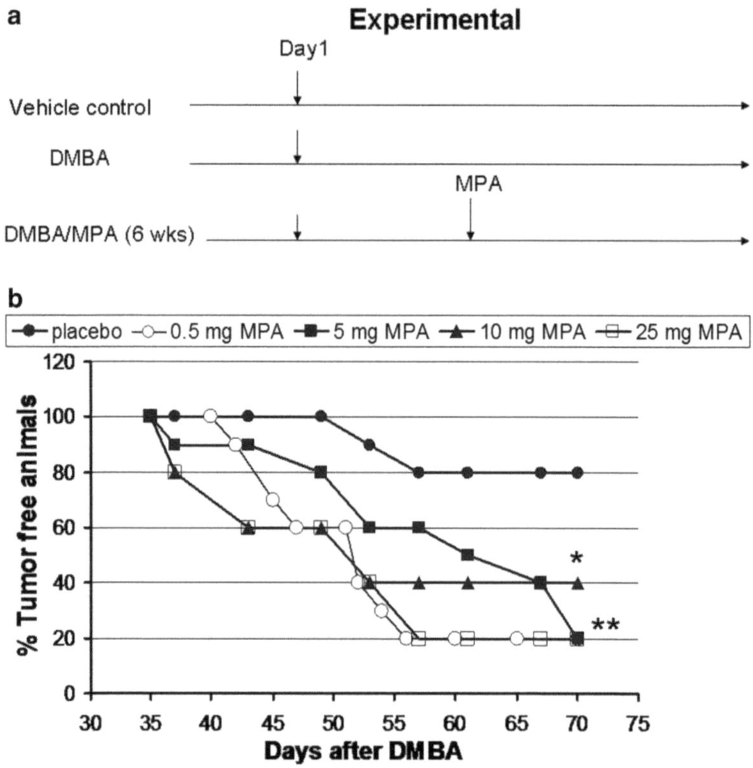

Fig. 7.2 Progestin-induced progression of mammary tumors in vivo. (**a**) Progestins accelerate the development of mammary tumors in DMBA-induced rat model of breast cancer by inducing VEGF. General experimental protocol. 45- to 50-day-old Sprague-Dawley rats were treated with DMBA by oral gavage (20 mg/kg/rat). MPA pellets (25 mg/60-day release) were implanted 4 weeks after DMBA administration. Control animals were implanted with placebo pellets. Following pellet implantation, animals were palpated for tumors every other day until termination of the study. In specific experiments doses of MPA were varied (**b**) and progesterone (**c**, **d**) was administered. (**b**) Dose-dependent effects of MPA on mammary tumor development. MPA pellets (0.5, 5, 10, and 25 mg) were implanted 4 weeks after DMBA treatment. Tumor latency was significantly reduced in animals treated with 5, 10, and 25 mg pellets compared with animals given placebo pellets ($P=0.010$, log-rank test). With 0.5 mg MPA pellets tumor latency was 42 days, which was still significantly different from placebo ($P<0.05$, unpaired *t* test). Tumor incidence varied from 60 to 80 % in MPA-treated groups compared with 20 % in the placebo group. (**c**) Mammary tumors were collected from randomly selected animals at 8, 10, and 12 weeks after DMBA treatment. *Upper panel*: Immunohistochemical staining and analysis showed increased VEGF expression in both MPA- and progesterone (P)-treated animals compared with animals given placebo. *Lower panel*: Factor VIII staining showed more developed blood vessels in MPA and progesterone-treated tumors, suggesting promotion of angiogenesis by progestins in these tumors. (**d**) Multivessel density analysis of tumor tissue with or without exposure to progestins. A number of blood vessels were significantly higher in tumors collected from animals treated with either MPA or progesterone ($*P<0.04$), compared with placebo groups (unpaired *t* test). Taken from Benakanakere et al. *Clin Can Res* [16]

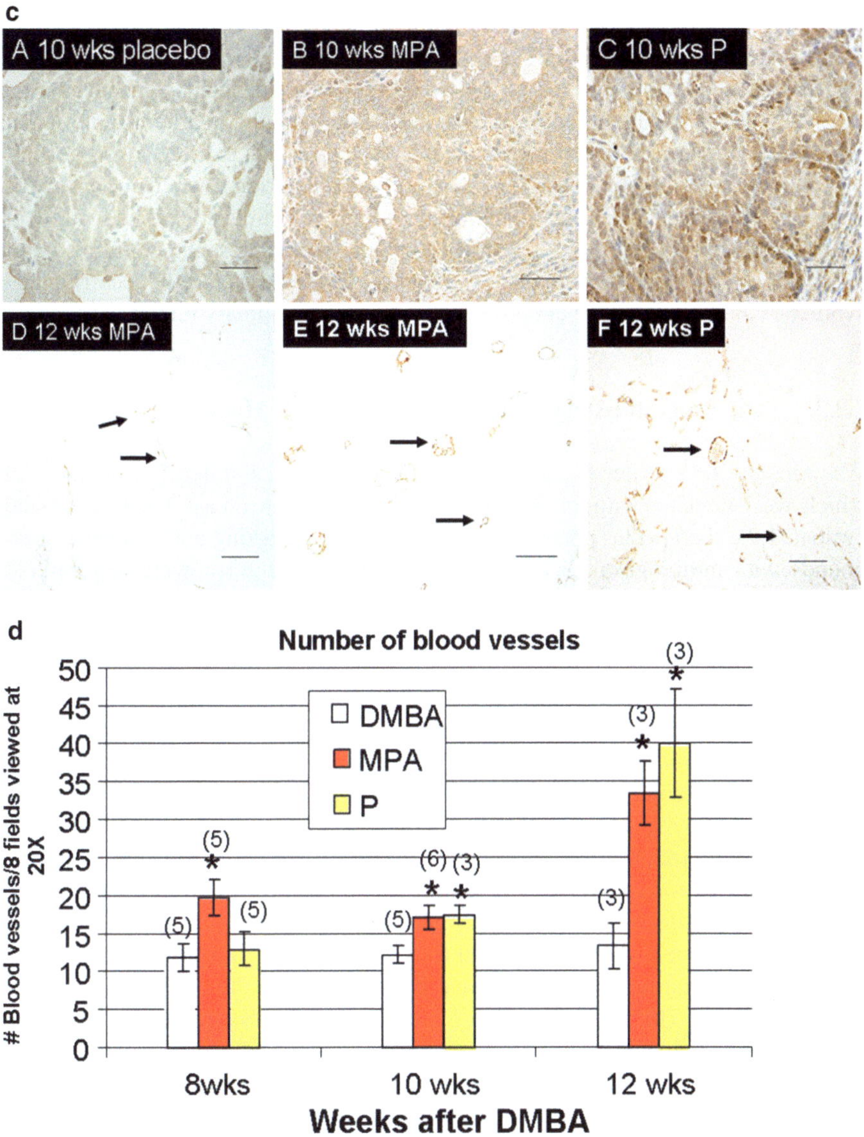

Fig. 7.2 (continued)

rats leads to the development of mammary tumors [16–18]. It is important to note that implantation of MPA (or progesterone) pellets following DMBA treatment accelerates tumor development and increases tumor multiplicity, which correlates with hormone-induced VEGF in mammary tumors [16]. This concurs with our hypothesis that progestins used in HRT promote the growth of occult tumors or lesions in the mammary gland, lesions which most likely require the angiogenic switch to be triggered by progestin-induced VEGF to facilitate their growth. Commonly used antiprogestins such as RU-486 inhibit progestin-driven growth of DMBA-induced mammary tumors, proving that PR-dependent mechanisms are involved [16]. Using this model we have been able to study the micro-tumor environment and determine specific changes in gene expression induced by progestins. Furthermore, the DMBA model has allowed us to determine whether any of the antiprogestin effects are due to inhibition of progestin-dependent induction of angiogenesis in this tissue. In subsequent sections we will present data from studies in which we employed this model to evaluate the ability of naturally occurring and synthetic compounds to prevent progestin-dependent mammary tumors.

7.2.2.2 The Progestin-Dependent Human Xenograft Model

We were the first to develop and employ a novel progestin-dependent model in which we inoculated immunodeficient mice with human breast cancer cells and evaluated the therapeutic potential of both naturally occurring and synthetic compounds with antiprogestin activity [21]. In this model, mice were sequentially implanted with estradiol prior to inoculation with human breast cancer cells and were subsequently implanted with progestin-containing pellets. Most in vivo studies involve inoculation with cells that are incorporated in Matrigel, which contains high levels of growth factors. It is therefore possible that previous in vivo studies which failed to demonstrate progestin-dependent breast cancer cell proliferation did so due to initial induction of maximum proliferation by growth factors present in the Matrigel [77]. In order to avoid this, we injected human breast cancer cells without Matrigel and thereby demonstrated a progestin-dependent enhancement of human tumor cell proliferation as well as increased metastasis [21, 22]. Using this model we found that estradiol supports a short burst of tumor cell growth, followed by regression and tumor cell senescence and/or apoptosis. Progestin supplementation (natural or synthetic) rescues tumor growth, and tumors continue to develop in an angiogenesis-dependent manner that is facilitated by increased production of VEGF from tumor cells [21]. This is illustrated in Fig. 7.3, which shows that MPA-dependent xenograft tumor growth is suppressed significantly by the administration of human-specific anti-VEGF antibodies (2C3), confirming that progestin-induced VEGF derived from human breast cancer cells is largely responsible for tumor development in this model. PRIMA-1-dependent suppression of tumors will be discussed later. It is interesting to note that when progestin pellets are removed, tumor growth is suppressed [21], providing further evidence that this process is progestin-driven. RU-486 suppressed progestin-dependent growth, further supporting a PR-dependent mechanism. Our xenograft tumor model thus offers a practical system for evaluating the role of progestins in breast cancer development as well as for testing different compounds with potential antiprogestin and anti-angiogenic activity. Furthermore, this model provides a means by which to assess progestin-induced metastasis.

7.3 Compounds with Anti-angiogenic Activity and Therapeutic Potential Against Progestin-Dependent Breast Cancer

Progestin-dependent tumor cell proliferation is dependent upon PR activation and a large number of co-modulators, which act in concert within a network of transduction pathways, ultimately promoting proliferation and tumor growth. Consequently, PR and its associated co-activators are legitimate targets through which to treat and perhaps prevent cancer [78]. By targeting transduction mechanisms that are overtly activated by progestins in ways that are both PR-dependent and PR-independent, cancerous cells can be selectively destroyed, while normal cells are spared. The signal transduction pathway leading to increased angiogenesis clearly plays an important role in progestin-mediated tumor growth, and since VEGF is vital to this process, VEGF receptors must also be involved. As discussed earlier, commonly available synthetic PR antagonists such as mifepristone cause progestin-accelerated tumors to regress [16]; however, due to the adverse side effects of such compounds, their extended use in humans is limited. Therefore, we have focused our studies on screening and identifying nontoxic, naturally occurring, and synthetic compounds which possess antiprogestin and/or anti-angiogenic activity. Our ultimate goal is to develop such compounds for use in humans as chemotherapeutic and chemopreventive agents. Due to low mutation rates in endothelial cells, anti-angiogenic compounds are less likely to induce drug resistance than other classes of antitumor agents [79], making their development particularly attractive to cancer researchers. We believe that the use of naturally occurring and synthetic antiprogestins with anti-angiogenic properties, alone or in combination with other chemotherapeutic drugs, is an extremely promising option for both long-term chemoprevention and chemotherapy.

7.3.1 Natural Compounds

Compounds that occur naturally are particularly attractive as chemopreventive and chemotherapeutic drugs because of their wide availability and lower cytotoxicity, which enables chronic consumption without adverse side effects. We initially screened and identified several naturally occurring compounds with potential anti-progestin activity using in vitro assays that specifically evaluated the ability of these compounds to suppress progestin-induced VEGF secretion from breast cancer cells (unpublished data). Subsequently we performed detailed in vitro and in vivo studies using two such compounds, curcumin and apigenin, both of which arrested progestin-dependent growth of breast cancer cells in vivo by suppressing VEGF and angiogenesis. These results are discussed below.

7.3.1.1 Curcumin

Curcumin, the major yellow coloring pigment from the Indian spice turmeric (*Curcuma longa*), exhibits antineoplastic [80–82] and anti-angiogenic properties in a number of cancer cell lines [83]. Although curcumin has several molecular targets [84], it clearly suppresses angiogenesis in part through mechanisms that involve the down-expression of VEGF. Since previous studies showed curcumin to possess low affinity for both estrogen and progesterone receptors [85], we sought to determine whether it would suppress progestin-dependent tumor growth. Initially we showed in vitro that curcumin dose-dependently inhibited MPA-dependent VEGF induction in cultured human T47D cells [86], though interestingly it had no effect on induction of VEGF by endogenous progesterone or other synthetic progestins, such as norethindrone. Since different progestins possess diverse properties with respect to potency, clinical effects, and endocrine function [87], it is possible that the effects of curcumin might be specific to the 17α-hydroxyprogesterone group of progestins alone. Using the DMBA rat model described above (refer to Sect. 7.2.2.1), we tested the ability of curcumin to inhibit MPA-accelerated DMBA-induced mammary tumors and found that while it was unable to delay natural DMBA-induced tumor development, curcumin did interrupt further MPA-accelerated tumor growth [80]. Furthermore, curcumin reduced the average number of tumors per tumor-bearing animal. Immunohistochemical analysis showed that compared with control animals, curcumin suppressed VEGF expression within the mammary glands of those treated with MPA. Interestingly, curcumin treatment did not affect the expression of ERα and ERβ, or PR [80], suggesting that in this model it exerts its anticancer properties with little or no effect on ovarian hormone receptors. This would indicate that agents targeting ER and PR could be combined with curcumin to combat breast cancer, though human clinical studies are required to determine the pharmacokinetic characteristics of curcumin alone or in combination with other drugs, as well as to establish whether curcumin might be used therapeutically to suppress existing progestin-dependent breast tumors.

7.3.1.2 Apigenin

Apigenin is a low molecular weight polyphenolic compound abundantly present in common fruits, vegetables, and beverages, which has been shown to possess anti-mutagenic properties [88]. Apigenin belongs to a group of ubiquitous compounds called flavonoids and is classified under a subgroup known as flavones [88]. Studies have shown that apigenin affects several signaling pathways in different cancer cell lines, via various mechanisms that include antioxidant and anti-inflammatory effects and induction of apoptosis [89]. Since apigenin was shown in our initial screening to possess antiprogestin activity, we conducted studies to determine its ability to suppress progestin-dependent VEGF induction in human breast cancer cells. Apigenin suppressed induction of VEGF by progesterone, MPA, and norethindrone, suggesting that it may act against a wider range of progestins [90] than curcumin,

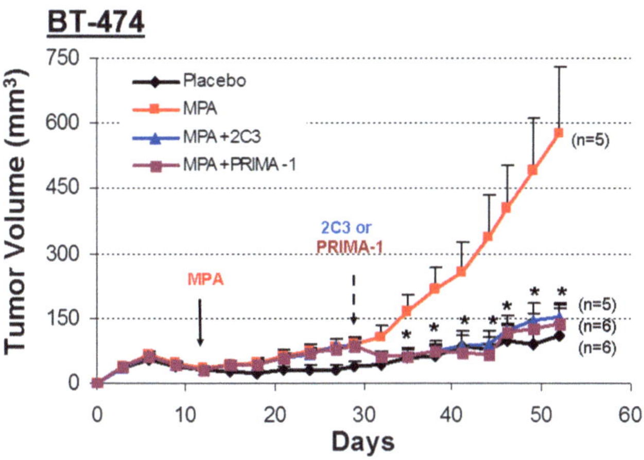

Fig. 7.3 Progestins rescue growth of regressing xenograft tumors in nude mice. BT-474 cells were injected into nude mice implanted with pellets containing 17β-estradiol (1.7 mg/60-day release pellet). Tumor growth was monitored, and once tumors began to regress, MPA pellets (10 mg/60-day release) were implanted (*Solid arrow*). When mean average tumor volume reached approximately 100 mm^3, animals were injected with 2C3, C44 (control IgG), or PRIMA-1 (*broken arrow*). Xenograft tumors from animals treated with 2C3 and PRIMA-1 were significantly different ($P<0.05$, ANOVA) compared with MPA group. Taken from Liang et al. Cancer Research [21]

whose effects were specific to MPA [80]. We further showed, by semiquantitative RT-PCR, that apigenin dose-dependently inhibits progestin-induced VEGF mRNA synthesis. It has been reported that apigenin suppresses expression of VEGF through mechanisms that involve PI3K/AKT/p70S6K1 and HDM2/p53 pathways [91].

Having ascertained that apigenin opposes the effects of progestins in vitro, we used the DMBA rat model to examine its in vivo potential. Intraperitoneal administration of apigenin delayed and decreased significantly the occurrence of MPA-accelerated DMBA-induced mammary tumors in a dose-dependent manner [92]. In contrast to earlier studies with curcumin [80], apigenin did not prevent MPA-induced hyperplasia of mammary gland epithelial cells, suggesting that it might specifically target mammary cancer cells at a particular stage of growth [92]. Disparate effects of apigenin have also been observed between normal prostate cells and cancerous cells [93]. Immunohistochemical analysis of non-tumor mammary tissues demonstrated suppression of VEGF by apigenin, suggesting that the flavonoid exerts its anticancer effects at least in part by inhibiting angiogenesis.

Further studies using the immunodeficient mouse model of carcinogenesis in which human breast cancer cell growth is sustained by implanted progestin pellets [20–22] (Fig. 7.3) showed that apigenin suppressed the growth of MPA-dependent BT-474 xenograft tumors. Histochemical analysis of tumors collected from apigenin-treated animals demonstrated dramatic induction of apoptosis and suppressed VEGF and Her-2/neu expression [94]. While apigenin did not cause a reduction in the number of blood vessels within tumors, it did prevent

MPA-dependent vessel dilation. This suggests that apigenin might arrest tumor growth by restricting tumor blood flow, though perfusion studies are required to confirm whether or not this is the case.

These studies are the first to show that apigenin, a nontoxic, naturally occurring component of many fruits and vegetables, has immense potential as a chemopreventive and chemotherapeutic agent against progestin-dependent breast cancer. Apigenin exerts its anticancer effects via a number of different mechanisms [88, 89], and further studies are required, both in vivo and in vitro, to enhance our understanding of exactly how the flavonoid reduces tumor cell viability and arrests tumor growth. Additional human clinical studies are necessary to establish the optimum dose and route of apigenin administration. However, by virtue of its low intrinsic toxicity and capacity to exert its effects specifically against cancer cells while not affecting normal cells [93], it is abundantly clear that apigenin is a compound with great potential as a chemotherapeutic agent.

7.3.2 Synthetic Compounds

Synthetic antiprogestins have been used to oppose the harmful effects of progestins; however, such compounds are toxic due to their cross-reactivity with steroid receptors other than PR. Consequently, it is essential that safe antiprogestins, which are specific for PR, be developed before clinical studies aimed at arresting breast cancer progression through PR antagonism can occur. In the meantime, we have screened some synthetic compounds with antiprogestin activity based on their in vitro ability to block progestin-dependent VEGF induction [21]. One such compound is PRIMA-1 (p53 reactivation induction of massive apoptosis), which met the criteria of inhibiting progestin-dependent VEGF induction in cultured breast cancer cells [21, 95]. To complement our in vitro studies, we tested PRIMA-1 extensively in vivo for its effectiveness to both treat and prevent progestin-accelerated breast cancer and ascertained, specifically, its anti-angiogenic activity. Our findings support the further investigation of PRIMA-1 as a compound with great potential for controlling progestin-dependent breast cancer, both as a therapeutic and a chemopreventive agent. The salient findings from our studies are described below.

7.3.2.1 PRIMA-1

An estimated 50 % of all breast cancers carry a mutation in the gene-encoding p53 (mtp53) [96–98]. Mutations in wild-type p53 (wtp53) protein undoubtedly lead to cell proliferation and tumor growth due to loss of tumor suppressor activity [99]. Furthermore, mtp53 protein enhances tumor survival and resistance to chemotherapeutic drugs [100] as well as leading to increased angiogenesis with subsequent tumor growth. Based on our observation that loss of p53 function within breast cancer cells facilitates induction of VEGF by progestins via a PR-mediated event [21, 39], we hypothesized that by reintroducing wtp53, or converting mtp53 to an active functional form, we might arrest the progression of progestin-dependent disease. As we predicted, activation of mtp53 or expression of wtp53 blocked progestin-stimulated VEGF expression in breast cancer cells. Thus, restoration of normal p53 function, by converting existing mtp53 to the wild-type conformation, represents a novel strategy by which we might prevent angiogenesis and promote tumor apoptosis. As mentioned above, PRIMA-1 is a molecule with great anticancer potential by virtue of its ability to transform mtp53 to the active conformation [95]. In our studies, PRIMA-1 inhibited progestin-induced expression of VEGF in human breast cancer cells via mechanisms which included in vitro activation of mtp53 [20, 95]. We also found that in vivo, treatment of animals with PRIMA-1 inhibited the growth of both progesterone and MPA-dependent BT-474 xenograft tumors [21]. Two weeks of PRIMA-1 treatment reduced xenograft tumor volume by 75 % compared with tumors collected from animals treated with vehicle medium. Tumors from PRIMA-1-treated animals were generally less vascularized, demonstrating the anti-angiogenic properties of the drug (Fig. 7.4). In addition to blocking VEGF secretion, it is possible that PRIMA-1also inhibits other pathways essential for tumor growth, such as the VEGF-mediated tumor survival pathway [19].

Having determined that administration of PRIMA-1 alone represents an extremely promising avenue for anticancer therapy, we conducted studies aimed at testing its effect when used in combination with other drugs [101]. In particular, we targeted anionic phospholipids on the tumor vasculature surface with a specific monoclonal antibody, 2aG4, which binds specifically to anionic phospholipids in the presence of β2-glycoprotein 1 and thereby disrupts the vascular structure of the tumor. When given together, PRIMA-1 and 2aG4 effectively suppressed the growth of hormone-dependent BT-474 and HCC-1428 xenograft tumors [101], supporting our rationale for using a combination drug regimen which reactivates mtp53 and disrupts tumor vasculature. The administration of PRIMA-1 in combination with other compounds also has the advantage of arresting tumor growth using drug dosages that cause minimal cytotoxic side effects. In future xenograft studies we will use a combination of PRIMA-1 and apigenin since the latter has also been shown to activate and stabilize mtp53 [102]. It is possible that by structural modification, analogues of PRIMA-1 might be produced with increased half-lives, which may aid delivery to mammary tumors.

PRIMA-1 was also effective when used chemotherapeutically to treat DMBA-induced tumors in rats, converting mtp53 to wtp53 and inducing apoptosis in tumor cells [17]. In the same series of experiments, PRIMA-1 reduced levels of VEGF in tumor tissues and also reduced blood flow, as determined by blood perfusion measurements using fluorescent dye [17]. Thus, although PRIMA-1 may not lower actual blood vessel density within tumor tissue (unlike curcumin, which did decrease blood vessel density within progestin-dependent breast tumor tissue; [80]), it can still destroy growing tumors by reducing the nourishment they receive. This is analogous to reduced blood flow in MPA-accelerated tumors in response to apigenin, which likely causes blood vessel constriction [94].

The chemotherapeutic use of PRIMA-1 described above prevented the emergence of new tumors [17]. In this model new tumors become palpable over time as

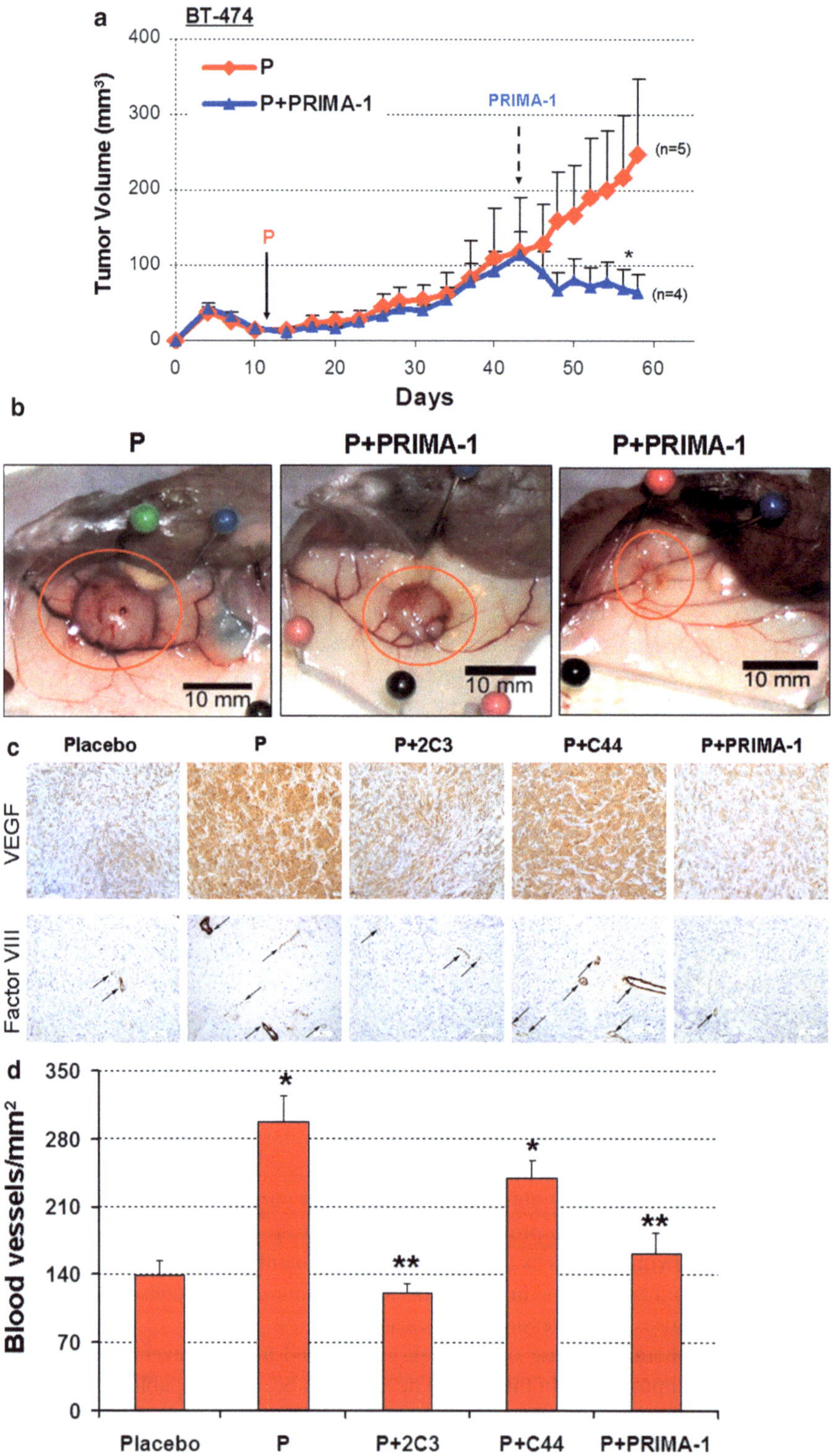
a
BT-474
P
P+PRIMA-1
PRIMA-1
P
(n=5)
(n=4)
*
Tumor Volume (mm³)
400
300
200
100
0
0
10
20
30
40
50
60
Days
b
P
P+PRIMA-1
P+PRIMA-1
10 mm
10 mm
10 mm
c
Placebo
P
P+2C3
P+C44
P+PRIMA-1
VEGF
Factor VIII
d
Blood vessels/mm²
350
280
210
140
70
0
*
*
**
**
Placebo
P
P+2C3
P+C44
P+PRIMA-1

they develop (tumor multiplicity); however, following the commencement of PRIMA-1 treatment, no new tumors were observed. The ability of PRIMA-1 to prevent the development of new tumors could have ramifications for its use as a chemopreventive agent, though this remains to be tested.

7.3.3 *Other Synthetic Compounds with Anti-angiogenic Properties*

The testing of other anti-angiogenic compounds which could be used to arrest in vivo progestin-accelerated breast tumor progression is ongoing. Studies are at a preliminary stage, though HIF-1α inhibitors appear to be very effective therapeutically, most likely through their regulation of VEGF. Publication of these findings is pending (Carroll et al. in preparation).

7.4 Mono- and Combination Therapies Using Anti-VEGF and Anti-vascular Antibodies to Target Hormone-Dependent Breast Cancer

As well as exploring the ability of naturally occurring and synthetic small molecular weight compounds to treat and prevent hormone-dependent breast cancer, our laboratory has also conducted studies to test the effectiveness of antibodies that target

Fig. 7.4 Activator of mutant p53 blocks the progression of progestin-driven breast cancer by reducing angiogenesis. (**a**) BT-474 cells were injected and tumor growth was monitored. When tumors began to regress, progesterone pellets (10 mg) were implanted (*solid arrow*) which resuscitated tumor growth as previously described for MPA in Fig. 7.3. When mean average tumor volume reached 100 mm^3, animals were injected with 50 mg/kg PRIMA-1 twice daily for 2 weeks (*broken arrow*). PRIMA-1 treatment significantly decreased tumor volume ($P<0.05$, *t* test). (**b**) Images of representative tumors are shown. (**c**) Immunohistochemical analysis of VEGF and factor VIII in progesterone-dependent xenograft tumors exposed to 2C3 or PRIMA-1. *Upper panel*: Progesterone treatment significantly increases VEGF expression compared with placebo. Progesterone-induced VEGF expression is suppressed by exposure to either 2C3 or PRIMA-1. *Lower panel*: Factor VIII staining shows more highly developed blood vessels in progesterone-treated xenografts but less well-developed blood vessels following exposure to either 2C3 or PRIMA-1. In xenografts exposed to P+C44 (control antibody), blood vessels were similar to P alone. (**d**) Blood vessel density was determined by counting factor VIII-positive staining. PRIMA-1 and 2C3 exposed xenografts exhibited significantly lower numbers of blood vessels compared with xenografts exposed to either *P* alone or P+C44 (**$P<0.05$). *Significantly different from placebo group

human VEGF and tumor-specific vasculature. A brief summary of our findings is provided below. Experiments in which antibodies were used alone will be described, as well as combination studies involving the administration of antibodies and other synthetic compounds.

7.4.1 Anti-VEGF Antibody

Jain et al. [103] inhibited tumor vascularization by indirectly targeting released pro-angiogenic growth factors and subsequently blocking their effects when they bind to receptors on cancer or stromal cells. With this in mind we examined the ability of anti-VEGF antibodies to prevent the growth of progestin-dependent xenograft tumors and found that 2C3, a human VEGF-specific antibody which blocks the binding of VEGF to VEGF receptor 2, reduced angiogenesis and suppressed tumor growth by fourfold (Fig. 7.4) [21]. This provided further evidence that the pro-tumorigenic effects of progestins involve mechanisms that include increased expression of VEGF. Anti-VEGF drugs such as Avastin (bevacizumab) effectively suppress angiogenesis and tumor growth [104]; however, in order to achieve greater effectiveness and increased rates of patient survival, combination therapies involving the administration of anti-VEGF drugs, including anti-VEGF antibodies, with other chemotherapeutic drugs, are necessary. The prolonged consumption of anti-VEGF drugs as monotherapy is risky, since it is associated with an increased possibility of selection of hypoxia-resistant cells, leading to tumor relapse [105]. This might explain the failure of anti-VEGF therapy in breast cancer trials [106, 107]. To counter this, we propose continuous treatment with a combination of anti-VEGF antibodies and other agents such as HIF-1α inhibitors. Such agents might include naturally occurring apigenin and curcumin, both of which target HIF-1α transcription factor, which regulates a number of survival-specific genes within tumor cells [108].

7.4.2 Anti-vascular Antibodies

We designed a novel strategy by which to combat breast cancer, combining the administration of PRIMA-1 with 2aG4, an antibody which specifically targets tumor vasculature [101]. Although these studies were not carried out strictly for the treatment of progestin-dependent tumors, their outcome indicates that such an approach could be an effective way to treat most types of breast cancer. 2aG4 exclusively targets anionic phospholipids that are selectively exposed on tumor blood vessels [101]. We believe strongly that employing a two-pronged attack of 2aG4 and PRIMA-1, which selectively activates mtp53 to functional p53, will produce synergistic effects with regard to antitumor treatment. We confirmed this hypothesis using mtp53-expressing breast cancer cells; a combination of 2aG4 and PRIMA-1 acted additively to lower tumor incidence, and in some animals tumors were completely eradicated. Tumor blood vessel density and tumor perfusion were both reduced as a consequence of such treatment. This approach was, furthermore, nontoxic to experimental animals. We are extremely optimistic that an approach based on this type of therapy could prove to be an extremely useful and safe means by which to control the progression of mtp53-containing tumors.

7.5 Conclusions

Clinical data indicates that combined progestin/estrogen HRT increases the risk of breast cancer in postmenopausal women, a phenomenon that is supported by a number of in vivo experimental studies showing that progestins promote tumor development. In 1998 we proposed that progesterone and synthetic progestins promote angiogenesis by inducing VEGF, a powerful angiogenic growth factor [53]. Others have described different roles for progestins in promoting breast cancer, for example, progestin-dependent increases in stemlike cells [70] and induction of other growth factors such as EGF [65] and RANKL [63, 64]. Thus we now have molecular targets through which it may be possible to treat progestin-dependent breast cancer. We have continued to examine mechanisms involved in progestin-dependent angiogenesis, focusing in particular on VEGF, which also plays a role in tumor cell survival [109] and which we contend offers a realistic molecular target for both treating and preventing this form of breast cancer. We demonstrated that the small molecular weight compound PRIMA-1, which activates mtp53 into its wild-type functional form, is thereby able to suppress progestin-dependent VEGF and bring about tumor regression. Similarly we showed that the naturally occurring compounds curcumin and apigenin, which also blocked progestin-dependent release of VEGF, are potentially good candidates as drugs for treating and preventing progestin-dependent breast cancer. Finally, we showed that anti-vascular antibodies which target both VEGF and tumor blood vessels also possess considerable potential as drugs which could be used to control the progression of breast cancers which are dependent upon progestins. We provide rationale for using a two-pronged approach of anti-vascular antibodies in combination with PRIMA-1, since the two could act additively and/or synergistically to reduce tumor burden and have the advantage of being nontoxic. Compounds which target other pathways, such as cholesterol biosynthesis, are also good candidates as anticancer agents. As tumors grow, cholesterol synthesis is essential for the normal membrane structure of many organelles, as well as providing the building block from which steroid hormones are synthesized. These studies, while preliminary, are encouraging and suggest that by blocking cholesterol production by tumor cells we might open up new possibilities for arresting angiogenesis and disrupting tumor growth.

The role of angiogenesis in breast cancer is now well recognized. With the recent success of clinically administered inhibitors of angiogenesis (see reviews by [110, 111]), it is feasible that such compounds might be used for "angio-prevention." In such a scenario their use would arrest tumor growth by suppressing the production of hormone-dependent VEGF and triggering the angiogenic switch, which is essential for small tumors or lesions to progress toward frank tumors. A strategy such as this would be especially useful when employed in a "personalized prevention" mode for post-menopausal women with mutations in tumor suppressors such as p53 and Brca1 [112, 113], prior to their being considered for hormone treatment. We are optimistic that many of the approaches discussed in this article will help us make significant progress against progestin-dependent breast cancer and thereby improve the lives of millions of women worldwide.

Acknowledgments We are grateful to the talented individuals in the lab, both past and present, who have been instrumental in obtaining the exciting data reported herein. These include Drs. Yayun Liang, Indira Benakanakere, Candace Carroll, and Jianbo Wu. We are also grateful to various collaborators for their contribution, in particular Dr. George Stancel (Univ. of Texas, Houston), in whose laboratory SMH began his research on hormones and angiogenesis, and Drs. Rolf Brekken and Philip Thorpe (UT Southwestern, Dallas, TX) who provided the anti-vascular antibodies for use in these studies. We also wish to thank the lab staff at RADIL/IDDEX for help with the immunohistochemical analysis.

Funding The following sources are acknowledged for supporting our laboratory studies: NIH grants CA-86916 and R56-CA86916, the Susan G. Komen Breast Cancer Foundation, Dept of Defense breast cancer awards (W81XWH-05-1-0416, W81XWH-06-1-0646 and W81XWH-12-1-0191), and COR grants from the College of Veterinary Medicine at the University of Missouri, Columbia. SMH is the Zalk Missouri Professor of Tumor Angiogenesis. Mrs. Zalk's bequest has been instrumental in allowing us to make key discoveries related to the role of progestins and angiogenic growth factors in breast cancer. The authors declare that there is no conflict of interest that would prejudice the impartiality of this scientific work.

References

1. Crawford Y, Ferrara N. VEGF inhibition: insights from preclinical and clinical studies. Cell Tissue Res. 2009;335:261–9.
2. Rosen EM, Goldberg ID. Regulation of angiogenesis by scatter factor. EXS. 1997;79:193–208.
3. Folkman J. Angiogenesis in cancer, vascular, rheumatoid and other disease. Nat Med. 1995;1:27–31.
4. DeSantis C, Siegel R, Bandi P, Jemal A. Breast cancer statistics. CA Cancer J Clin. 2011;61:409–18.
5. Kaklamani VG, Gradishar WJ. Adjuvant therapy of breast cancer. Cancer Invest. 2005;23:548–60.
6. Navolanic PM, McCubrey JA. Pharmacological breast cancer therapy (review). Int J Oncol. 2005;27:1341–4.
7. Chung AS, Kowanetz M, Wu X, Zhuang G, Ngu H, Finkle D, Komuves L, Peale F, Ferrara N. Differential drug-class specific metastatic effects following treatment with a panel of angiogenesis inhibitors. J Pathol. 2012. doi:10.1002/path.4052.
8. Ferrara N, Kerbel RS. Angiogenesis as a therapeutic target. Nature. 2005;438:967–74.
9. Hyder SM. Sex-steroid regulation of vascular endothelial growth factor in breast cancer. Endocr Relat Cancer. 2006;13:667–87.
10. Schneider BP, Miller KD. Angiogenesis of breast cancer. J Clin Oncol. 2005;23:1782–90.

11. Hyder SM, Stancel GM. Regulation of angiogenic growth factors in the female reproductive tract by estrogens and progestins. Mol Endocrinol. 1999;13:806–11.
12. Hyder SM, Stancel GM. Regulation of VEGF in the reproductive tract by sex-steroid hormones. Histol Histopathol. 2000;15:325–34.
13. Koutras AK, Starakis I, Lymperatou D, Kalofonos HP. Angiogenesis as a therapeutic target in breast cancer. Mini Rev Med Chem. 2012;12:1230–8.
14. Chlebowski RT, Anderson GL, Gass M, et al. Writing group for Women's Health Initiative Investigators. Estrogen plus progestin and breast cancer incidence and mortality in postmenopausal women. JAMA. 2010;304:1719–20.
15. Ross RK, Paganini-Hill A, Wan PC, Pike MC. Effects of hormone replacement therapy on breast cancer risk: estrogen versus estrogen plus progestin. J Natl Cancer Inst. 2000;92:328–32.
16. Benakanakere I, Williford CB, Schnell J, Brandt S, Ellersieck MR, Molinolo A, Hyder SM. Natural and synthetic progestins accelerate 7, 12-Dimethylbenz[a] Anthracene-initiated mammary tumors and increase angiogenesis in Sprague-Dawley rats. Clin Cancer Res. 2006;12:4062–71.
17. Benakanakere I, Besch-Williford C, Ellersieck MR, Hyder SM. Regression of progestin-accelerated 7,12-dimethylbenz[a]anthracene-induced mammary tumors in Sprague-Dawley rats by p53 reactivation and induction of massive apoptosis: a pilot study. Endocr Relat Cancer. 2009;16:85–98.
18. Benakanakere I, Besch-Williford C, Carroll CE, Hyder SM. Synthetic progestins differentially promote or prevent 7,12-Dimethylbenz(a)anthracene-induced mammary tumors in Sprague-Dawley rats. Cancer Prev Res. 2010;3:1157–67.
19. Liang Y, Wu J, Stancel GM, Hyder SM. p53-dependent inhibition of progestin-induced VEGF expression in human breast cancer cells. J Steroid Biochem Mol Biol. 2005;93:173–82.
20. Liang Y, Brekken RA, Hyder SM. Vascular endothelial growth factor induces proliferation of breast cancer cells and inhibits the anti-proliferative activity of anti-hormones. Endocr Relat Cancer. 2006;13:905–19.
21. Liang Y, Besch-Williford C, Brekken RA, Hyder SM. Progestin-dependent progression of human breast tumor xenografts: a novel model for evaluating anti-tumor therapeutics. Cancer Res. 2007;67:9929–36.
22. Liang Y, Benakanakere I, Besch-Williford CB, Ellersieck MR, Hyder SM. Synthetic progestins induce growth and metastasis of BT-474 human breast cancer xenografts in nude mice. Menopause. 2010;17:1040–7.
23. Bareschino MA, Schettino C, Colantuoni G, Rossi E, Rossi A, Maione P, Ciardiello F, Gridelli C. The role of antiangiogenetic agents in the treatment of breast cancer. Curr Med Chem. 2011;18:5022–32.
24. Dabrosin C, Palmer K, Muller WJ, Gauldie J. Estradiol promotes growth and angiogenesis in polyoma middle T transgenic mouse mammary tumor explants. Breast Cancer Res Treat. 2003;78:1–6.
25. Khosravi SP, Soria LA, Perez MG. Tumoral angiogenesis and breast cancer. Clin Transl Oncol. 2009;11:138–42.
26. Lichtenbeld HC, Barendsz-Janson AF, van Essen H, Struijker Boudier H, Griffioen AW, Hillen HF. Angiogenic potential of malignant and non-malignant human breast tissues in an in vivo angiogenesis model. Int J Cancer. 1998;77:455–9.
27. Brem SS, Gullino PM, Medina D. Angiogenesis: a marker for neoplastic transformation of mammary papillary hyperplasia. Science. 1977;195:880–2.
28. Jensen HM, Chen I, De Vault MF, Lewis AE. Angiogenesis induced by "normal" human breast tissue: a probable marker for precancer. Science. 1982;218:293–5.
29. Sledge Jr GW, Miller KD. Exploiting the hallmarks of cancer: the future conquest of breast cancer. Eur J Cancer. 2003;39:1668–75.
30. Zhang HT, Craft P, Scott PA, Ziche M, Weich HA, Harris AL, Bicknell R. Enhancement of tumor growth and vascular density by transfection of vascular endothelial cell growth factor into MCF-7 human breast carcinoma cells. J Natl Cancer Inst. 1995;87:231–19.

31. Fox SB, Generali DG, Harris AL. Breast tumour angiogenesis. Breast Cancer Res. 2007;181:207–12.
32. Toffoli S, Roegiers A, Feron O, et al. Intermittent hypoxia is an angiogenic inducer for endothelial cells: role of HIF-1. Angiogenesis. 2009;12:47–67.
33. Bergers G, Benjamin LE. Tumorigenesis and the angiogenic switch. Nat Rev Cancer. 2003;3:401–10.
34. Naumov GN, Bender E, Zurakowski D, Kang SY, Sampson D, Flynn E, Watnick RS, Straume O, Akslen LA, Folkman J, Almog N. A model of human tumor dormancy: an angiogenic switch from the nonangiogenic phenotype. J Natl Cancer Inst. 2006;98:316–25.
35. Dvorak HF, Detmar M, Claffey KP, Nagy JA, van de Water L, Senger DR. Vascular permeability factor/vascular endothelial growth factor: an important mediator of angiogenesis in malignancy and inflammation. Int Arch Allergy Immunol. 1995;107:233–5.
36. Rosenthal RA, Megyesi JF, Henzel WJ, Ferrara N, Folkman J. Conditioned medium from mouse sarcoma 180 cells contains vascular endothelial growth factor. Growth Factors. 1990;4:53–9.
37. Noh JJ, Maskarinec G, Pagano I, Cheung LW, Stanczyk FZ. Mammographic densities and circulating hormones: a cross-sectional study in premenopausal women. Breast. 2006;15:20–8.
38. Thomas T, Rhodin J, Clark L, Garces A. Progestins initiate adverse events of menopausal estrogen therapy. Climacteric. 2003;6:293–301.
39. Liang Y, Hyder SM. Proliferation of endothelial and tumor epithelial cells by progestin-induced vascular endothelial growth factor from human breast cancer cells: paracrine and autocrine effects. Endocrinology. 2005;146:3632–41.
40. Price DJ, Miralem T, Jiang S, Steinberg R, Avraham H. Role of vascular endothelial growth factor in the stimulation of cellular invasion and signaling of breast cancer cells. Cell Growth Differ. 2001;12:129–35.
41. Buteau-Lozano H, Velasco G, Cristofari M, Balaguer P, Perrot-Applanat M. Xenoestrogens modulate vascular endothelial growth factor secretion in breast cancer cells through an estrogen receptor-dependent mechanism. J Endocrinol. 2008;196(2):399–412.
42. Dadiani M, Seger D, Kreizman T, Badikhi D, Margalit R, Eilam R, Degani H. Estrogen regulation of vascular endothelial growth factor in breast cancer in vitro and in vivo: the role of estrogen receptor alpha and c-Myc. Endocr Relat Cancer. 2009;16:819–34.
43. Hyder SM, Stancel GM, Chiappetta C, Murthy L, Boettger-Tong HL, Makela S. Uterine expression of vascular endothelial growth factor is increased by estradiol and tamoxifen. Cancer Res. 1996;56:3954–60.
44. Hyder SM, Chiappetta C, Stancel GM. Pharmacological and endogenous progestins induce vascular endothelial growth factor expression in human breast cancer cells. Int J Cancer. 2001;92:469–73.
45. Wu J, Brandt S, Hyder SM. Ligand- and cell-specific effects of signal transduction pathway inhibitors on progestin-induced vascular endothelial growth factor levels in human breast cancer cells. Mol Endocrinol. 2005;19:312–26.
46. Axlund SD, Sartorius CA. Progesterone regulation of stem and progenitor cells in normal and malignant breast. Mol Cell Endocrinol. 2012;357:71–9.
47. Lanari C, Molinolo AA. Progesterone receptors-Animal models and cell signaling in breast cancer. Diverse activation pathways for the progesterone receptor possible implications for breast biology and cancer. Breast Cancer Res. 2002;4:240–3.
48. Hankinson SE, Eliassen AH. Circulating sex steroids and breast cancer risk in premenopausal women. Horm Cancer. 2010;1:2–10.
49. Gompel A, Malet C, Spritzer P, Lalardrie JP, Kuttenn F, Mauvais-Jarvis P. Progestin effect on cell proliferation and 17 beta-hydroxysteroid dehydrogenase activity in normal human breast cells in culture. J Clin Endocrinol Metab. 1986;63:1174–80.
50. Rajkumar L, Kittrell FS, Guzman RC, Brown PH, Nandi S, Medina D. Hormone-induced protection of mammary tumorigenesis in genetically engineered mouse model. Breast Cancer Res. 2007;9:R12.
51. Sivaraman L, Medina D. Hormone-induced protection against breast. J Mammary Gland Biol Neoplasia. 2002;7:77–92.

52. Goepfert TM, McCarthy M, Kittrell FS, Stephens C, Ullrich RL, Brinkley BR, Medina D. Progesterone facilitates chromosome instability (aneuploidy) in p53 null normal mammary epithelial cells. FASEB J. 2000;14:2221–9.
53. Hyder SM, Murthry L, Stancel GM. Progestin regulation of vascular endothelial growth factor in human breast cancer cells. Cancer Res. 1998;58:392–5.
54. Mirkin S, Wong BC, Archer DF. Effect of 17 beta-estradiol, progesterone, synthetic progestins, tibolone, and tibolone metabolites on vascular endothelial growth factor mRNA in breast cancer cells. Fertil Steril. 2005;84:485–91.
55. Mirkin S, Wong BC, Archer DF. Effects of 17beta-estradiol, progesterone, synthetic progestins, tibolone, and raloxifene on vascular endothelial growth factor and Thrombospondin-1 messenger RNA in breast cancer cells. Int J Gynecol Cancer. 2006;16 Suppl 2:560–3.
56. Chlebowski RT, Hendrix SI, Langer RD, Stefanick MI, et al. Influence of estrogen plus progestin on breast cancer mammography in healthy postmenopausal women. The Women's Health Initiative Random Trial. JAMA. 2003;289:3243–53.
57. Chlebowski RT, Anderson GL. Changing concepts: menopausal hormone therapy and breast cancer. J Natl Cancer Inst. 2012;104:517–27.
58. Lange CA, Richer JK, Horwitz KB. Hypothesis: progesterone primes breast cancer cells for cross-talk with proliferative or antiproliferative signals. Mol Endocrinol. 1999;13:829–36.
59. De Lignières B. Effects of progestogens on the postmenopausal breast. Climacteric. 2002;5:229–35.
60. Raafat AM, Hofseth LJ, Haklan SZ. Proliferative effects of combination estrogen and progesterone replacement therapy on the normal postmenopausal gland in a murine model. Am J Obstet Gynecol. 2001;183:1–14.
61. Soderqvist G, Isaksson E, Von Schoultz B, Carlstrom K, Tani E, Skoog L. Proliferation of breast epithelial cells in healthy women during the menstrual cycle. Am J Obstet Gynecol. 1997;176:123–8.
62. Saether S, Bakken K, Lund E. The risk of breast cancer linked to menopausal hormone therapy. Tidsskr Nor Laegeforen. 2012;132:1330–4.
63. Gonzalez-Suarez E, Jacob AP, Jones J, et al. RANK ligand mediates progestin-induced mammary epithelial proliferation and carcinogenesis. Nature. 2010;468:103–7.
64. Schramek D, Leibbrandt A, Sigl V, et al. Osteoclast differentiation factor RANKL controls development of progestin-driven mammary cancer. Nature. 2010;468:98–102.
65. Kariagina A, Xie J, Leipprandt JR, Haslam SZ. Amphiregulin mediates estrogen, progesterone and EGFR signaling in the normal rat mammary gland and in hormone-dependent rat mammary cancers. Horm Cancer. 2010;1:229–44.
66. Carvajal A, Espinoza N, Kato S, Pinto M, Sadarangani A, Monso C, et al. Progesterone pretreatment potentiates EGF pathway signaling in the breast cancer cell line ZR-75. Breast Cancer Res Treat. 2005;94:171–83.
67. Aldaz CM, Liao QY, LaBate M, Johnston DA. Medroxyprogesterone acetate accelerates the development and increases the incidence of mouse mammary tumors induced by dimethylbenzanthracene. Carcinogenesis. 1996;17:2069–72.
68. Blank EW, Wong PY, Lakshmanaswamy R, Guzman R, Nandi S. Both ovarian hormones estrogen and progesterone are necessary for hormonal mammary carcinogenesis in ovariectomized ACI rats. Proc Natl Acad Sci USA. 2008;105:3527–32.
69. Lanari C, Luthy I, Lamb CA, Fabris V, et al. Five novel hormone-responsive cell lines derived from murine mammary ductal carcinomas: In vivo and in vitro effects of estrogens and progestins. Cancer Res. 2001;61:293–302.
70. Horwitz KB, Sartorius CA. Progestins in hormone replacement therapies re-activate cancer stem cells in women with preexisting breast cancers: a hypothesis. J Clin Endocrinol Metab. 2008;93:3295–8.
71. Groshong SD, Owen GI, Grimison B, Schauer IE, et al. Biphasic regulation of breast cancer cell growth by progesterone: role of the cyclin-dependent kinase inhibitors, p21 and p27 (kip1). Mol Endocrinol. 1997;11:1593–607.
72. Matsubara Y, Matsubara K. Estrogen and progesterone play pivotal roles in endothelial progenitor cell proliferation. Reprod Biol Endocrinol. 2012;10:2–11.

73. Perou CM, Sortlie T, Eisen MB, Van de Rijn M, et al. Molecular portraits of human breast tumors. Nature. 2000;406:747–52.
74. Krämer EA, Seeger H, Krämer B, Wallwiener D, Mueck AO. The effect of progesterone, testosterone and synthetic progestogens on growth factor- and estradiol-treated human cancerous and benign breast cells. Eur J Obstet Gynecol Reprod Biol. 2006;129:77–83.
75. Russo J, Gusterson BA, Rogers AE, Russo IH, Wellings SR, Van Zwieten MJ. Comparative study of human and rat mammary tumorigenesis. Lab Invest. 1990;62:244–78.
76. Taylor D, Pearce CL, Hovanessian-Larsen L, Downey S, Spicer DV, Bartow S, Pike MC, Wu AH, Hawes D. Progesterone and estrogen receptors in pregnant and premenopausal non-pregnant normal human breast. Breast Cancer Res Treat. 2009;118:161–8.
77. Sartorius CA, Harvell DM, Shen T, Horwitz KB. Progestins initiate a luminal to myoepithelial switch in estrogen-dependent human breast tumors without altering growth. Cancer Res. 2005;65:9779–88.
78. Check JH, Dix E, Wilson C, Check D. Progesterone receptor antagonist therapy has therapeutic potential even in cancer restricted to males as evidenced from murine testicular and prostate cancer studies. Anticancer Res. 2010;30:4921–3.
79. Boehm T, Folkman J, Browder T, O'Reilly MS. Antiangiogenic therapy of experimental cancer does not induce acquired drug resistance. Nature. 1997;390:404–7.
80. Carroll CE, Benakanakere I, Besch-Williford C, Ellersierck MR, Hyder SM. Curcumin delays development of medroxyprogesterone acetate-accelerated 7,12-dimethybenz(a) anthracene-induced mammary tumors. Menopause. 2010;17:178–84.
81. Shehzad A, Khan S, Sup Lee Y. Curcumin molecular targets in obesity and obesity-related cancers. Future Oncol. 2012;8:179–90.
82. Sintara K, Thong-Ngam D, Patumraj S, Klaikeaw N. Curcumin attenuates gastric cancer induced by N-methyl-N-nitrosourea and saturated sodium chloride in rats. J Biomed Biotechnol. 2012;2012:915380.
83. Binion DG, Otterson MF, Rafiee P. Curcumin inhibits VEGF-mediated angiogenesis in human intestinal microvascular endothelial cells through COX-2 and MAPK inhibition. Gut. 2008;57:1509–17.
84. Darvesh AS, Carroll RT, Bishayee A, Novotny NA, Geldenhuys WJ, Van der Schyf CJ. Curcumin and neurodegenerative diseases: a perspective. Expert Opin Investig Drugs. 2012;21:1123–40.
85. Zava DT, Dollbaum CM, Blen M. Estrogen and progestin bioactivity of foods, herbs, and spices. Proc Soc Exp Biol Med. 1998;217:369–78.
86. Carroll CE, Ellersieck MR, Hyder SM. Curcumin inhibits MPA-induced secretion of VEGF from T47-D human breast cancer cells. Menopause. 2008;15:570–4.
87. Simon JA. Introduction: an overview of progesterone and progestins. Suppl J Fam Pract. 2007; 3–5.
88. Patel D, Shukla S, Gupta S. Apigenin and cancer chemoprevention: progress, potential and promise. Int J Oncol. 2007;30:233–45.
89. Shukla S, Gupta S. Apigenin: a promising molecule for cancer prevention. Pharm Res. 2010;27:962–78.
90. Mafuvadze B, Benakanakere I, Hyder SM. Apigenin blocks induction of VEGF mRNA and protein in progestin-treated human breast cancer cells. Menopause. 2010;17:1055–63.
91. Fang J, Zhou Q, Liu L, et al. Apigenin inhibits tumor angiogenesis through decreasing HIF-1α and VEGF expression. Carcinogenesis. 2007;28:858–64.
92. Mafuvadze B, Benakanakere I, Lopez Perez FR, Besch-Williford C, Ellersieck MR, Hyder SM. Apigenin prevents development of medroxyprogesterone acetate-accelerated 7,12-Dimethylbenz(a)anthracene-induced mammary tumors in Sprague-Dawley rats. Cancer Prev Res. 2011;4:1316–24.
93. Gupta S, Afaq F, Mukhtar H. Selective growth-inhibitory, cell-cycle deregulatory and apoptotic response of apigenin in normal versus human prostate carcinoma cells. Biochem Biophys Res Commun. 2001;287:914–20.

94. Mafuvadze B, Liang Y, Besch-Williford C, Hyder SM. Apigenin induces apoptosis and blocks growth of medroxyprogesterone acetate-dependent BT-474 xenograft tumors. Horm Cancer. 2012;3:160–71.
95. Liang Y, Besch-Williford C, Hyder SM. PRIMA-1 inhibits growth of breast cancer cells by re-activating mutant p53. Int J Oncol. 2009;35:1015–23.
96. Bartek J, Bartkova J, Vojtesek B, Staskova Z, Rejthar A, Kovarik J, Lane DP. Patterns of expression of the p53 tumor suppressor in human breast tissues and tumors in situ and in vitro. Int J Cancer. 1990;46:839–44.
97. Bottini A, Berruti A, Bersiga A, Brizzi MP, Brunelli A, et al. p53 but not bcl-2 immunostaining is predictive of poor clinical complete response to primary chemotherapy in breast cancer patients. Clin Cancer Res. 2000;6:2751–8.
98. Nigro JM, Baker SJ, Preisinger AC, Jessup JM, Hostetter R, Cleary K, Bigner SH, Davidson N, Baylin S, Devilee P, et al. Mutations in the p53 gene occur in diverse human tumour types. Nature. 1989;342:705–8.
99. Levine AJ. p53, the cellular gatekeeper for growth and division. Cell. 1997;88:323–31.
100. Munagala R, Aqil F, Gupta RC. Promising molecular targeted therapies in breast cancer. Ind J Pharmacol. 2011;43:236–45.
101. Liang Y, Besch-Williford C, Benakanakere I, Thorpe PE, Hyder SM. Targeting mutant p53 protein and the tumor vasculature: an effective combination therapy for advanced breast tumors. Breast Cancer Res Treat. 2011;125:407–20.
102. McVean M, Xiao H, Isobe K, Pelling JC. Increase in wild-type p53 stability and transactivation activity by the chemoprevention agent apigenin in keratinocytes. Carcinogenesis. 2000;21:633–9.
103. Jain RK, Duda DG, Clark JW, Loeffler JS. Lessons from phase III clinical trials on anti-VEGF therapy for cancer. Nat Clin Pract Oncol. 2006;3(1):24–40.
104. Jain RK. Antiangiogenic therapy for cancer: current and emerging concepts. Oncology (Williston Park). 2005;19(4 Suppl 3):7–16.
105. Blagosklonny MV. Antiangiogenic therapy and tumor progression. Cancer Cell. 2004;5:13–7.
106. Cao Y. Antiangiogenic cancer therapy: why do mouse and human patients respond in a different way to the same drug? Int J Dev Biol. 2011;55:557–62.
107. Zuniga RM, Torcuator R, Jain R, Anderson J, Doyle T, Schultz L, Mikkelsen T. Rebound tumour progression after the cessation of bevacizumab therapy in patients with recurrent high-grade glioma. J Neurooncol. 2010;99:237–42.
108. Tie J, Desai J. Antiangiogenic therapies targeting the vascular endothelia growth factor signaling system. Crit Rev Oncog. 2012;17:51–67.
109. Baek JH, Jang JE, Kang CM, Chung HY, Kim ND, Kim KW. Hypoxia-induced VEGF enhances tumor survivability via suppression of serum deprivation-induced apoptosis. Oncogene. 2000;19:4621–31.
110. Shojaei F. Anti-angiogenesis therapy in cancer: current challenges and future perspectives. Cancer Lett. 2012;320:130–7.
111. Wu JM, Staton CA. Anti-angiogenic drug discovery: lessons from the past and thoughts for the future. Expert Opin Drug Discov. 2012;7:723–43.
112. Shigetomi H, Higashiura Y, Kajihara H, Kobayashi H. Targeted molecular therapies for ovarian cancer: an update and future perspectives (Review). Oncol Rep. 2012;28:395–408.
113. Wang Z, Sun Y. Targeting p53 for novel anticancer therapy. Transl Oncol. 2010;3:1–12.

Chapter 8
New Insights on Estrogen Receptor Actions in Hormone-Responsive Breast Cancer Cells by Interaction Proteomics

Concetta Ambrosino, Roberta Tarallo, Giovanni Nassa, Francesca Cirillo, and Alessandro Weisz

Abstract Estrogens are tumor promoters for the mammary gland, due to their ability to control multiple functions of target cells and to stimulate their proliferation. The mechanisms that underlie control of cell proliferation by estrogens are still not fully defined, despite the important causal relationships between this hormonal action, mammary gland carcinogenesis, and breast cancer (BC) progression. Estrogens exert their actions in target tissues via two intracellular receptors, ERalpha (ERα) and ERbeta (ERβ), that show specific, and often antagonist, roles and can be found co-expressed in BC where, however, ERα appears to prevail in mediating estrogen actions. ERs are ligand-dependent transcription factors of the nuclear receptor superfamily of intracellular regulators, and their activity is tightly controlled by hormonal and non-hormonal ligands. This notion led to the design of synthetic ER antagonist ligands, including steroidal and nonsteroidal antiestrogens, that are effective to inhibit BC cell proliferation and, for this reason, widely used for treatment of hormone-responsive tumors. These drugs, however, exhibit side effects that limit their efficacy and use. Studies based on application of genomics and proteomics are revealing new insights on estrogen signaling in BC cells, with the discovery of novel ER partner proteins promising as potential novel drug targets. We review here the new insights on ER signaling derived by systematic application of interaction proteomics to map and characterize the intracellular network of proteins binding to agonist- and/or antagonist-activated ERα in a hormone-responsive human BC cell model.

C. Ambrosino
Department of Science for Biology, Geology and Environment, University of Sannio, Via Port'Arsa 11, 82100 Benevento, Italy

IRGS, Biogem, Via Camporeale, 83031 Ariano Irpino, Avellino, Italy

R. Tarallo • G. Nassa • F. Cirillo • A. Weisz (✉)
Laboratory of Molecular Medicine and Genomics, Department of Medicine and Surgery, University of Salerno, Via S. Allende 1, 84081 Baronissi, Salerno, Italy
e-mail: aweisz@unisa.it

H. Schatten (ed.), *Cell and Molecular Biology of Breast Cancer*,
DOI 10.1007/978-1-62703-634-4_8,

8.1 Estrogen Receptors in Breast Cancer

Clinical and experimental evidences point to female sex steroid hormones, estrogen and progesterone, as important pathogenic factors for breast cancer (BC). Therapies aiming at interfering with the actions of estrogen are widely used and effective for the prevention and cure of these diseases. Despite the numerous uncertainties surrounding the origins of these tumors, there is substantial evidence indicating that BC risk relates to endocrine and reproductive factors: only 5–10 % of these tumors can be related to Mendelian inheritance, and their development strongly depends on the ovarian and endocrine factors.

8.1.1 Estrogen Receptors: An Overview of Structural and Biological Properties

Two mammalian ER subtypes are known as ERα [1] and ERβ [2], showing distinct but often overlapping cellular and tissue distribution patterns. 17β-estradiol (E2) is the natural ER ligand, able to induce a conformational change driving receptor dimerization, enhancing its DNA-binding activity and recruitment of co-activator and corepressor complexes to chromatin-bound receptor and, thereby, its effects on gene transcription [3, 4]. ERs are nuclear proteins with identical modular domain organization [5], comprising an N-terminal domain (A/B domain) encoding a ligand-independent transcriptional activation domain termed AF-1; a central domain comprising the DNA-binding domain (DBD) and dimerization regions; a hinge region (D-domain) with an additional nuclear localization signal; an E-domain, representing the ligand-binding domain (LBD); and an F-domain, at the C terminus, also believed to be involved in ER dimerization, nuclear translocation, and ligand-dependent activation of gene expression [3, 4] (Fig. 8.1).

The expression patterns of the two ER subtypes and their splicing variants are partially overlapping. ERα is mainly expressed in the uterus, vagina, and mammary glands, while ERβ has been detected in several tissues, including in the male reproductive tracts and the central nervous system [6]. PCR and sequencing analyses identified ERα mRNA splice variants in various cancer cell lines and in breast, endometrial, and ovarian cancers (Fig. 8.1). Few truncated ERα variant isoforms have been examined in tumor samples and correlated with the clinical outcome [7]. It has been observed that truncated ERα variant ERα-36 mediates initial effects of estrogen signaling and its expression in the presence of wild-type ERα is associated with tamoxifen resistance in BC [6]. ERα-Δ3, lacking exon 3, unlike other isoforms is lost in cancer [8], while truncated isoform ERα-46 has been found significantly reduced in tamoxifen-resistant BC cells [9]. Once expressed in cancer cell lines, ERα isoforms modify the transcriptional activity of the wild-type receptor and influence cell growth through genomic and non-genomic pathways [6].

ERβ was identified more recently, in early 1990, and described initially in rat prostate and ovary and subsequently in human testis [10]. Sequencing data [11]

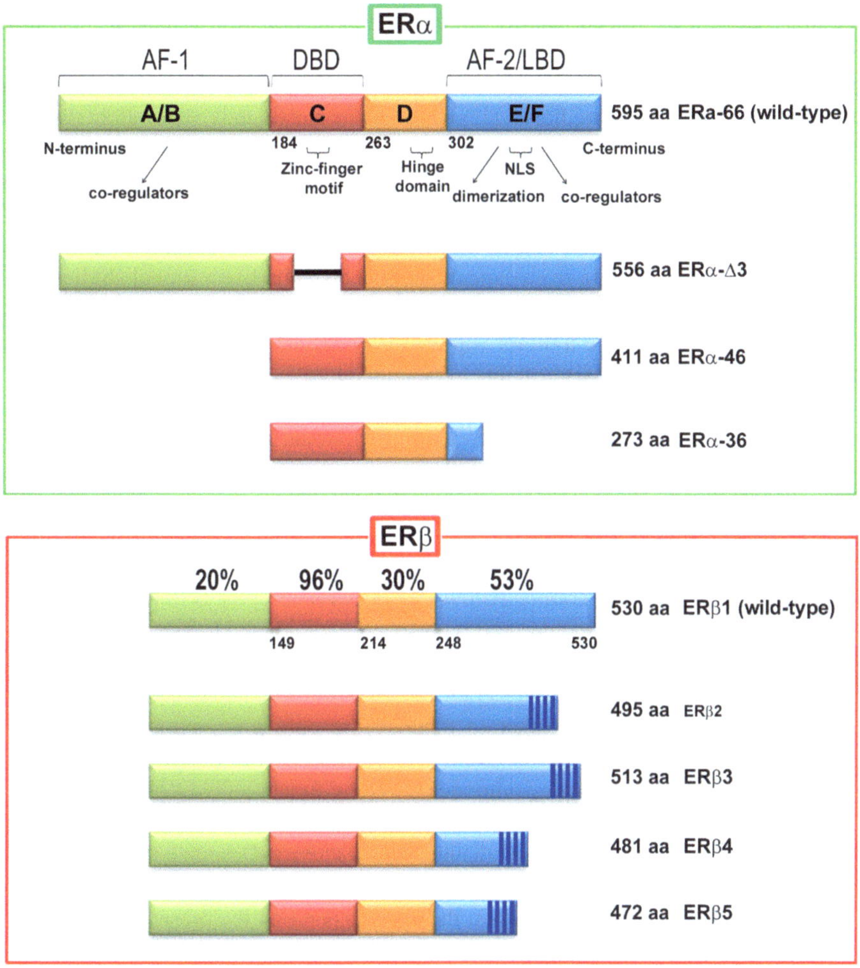

Fig. 8.1 ER**α** and ER**β** structural organization. *AF-1/2* activation function 1/2, *DBD* DNA-binding domain, *LBD* ligand-binding domain. Percentages reported above ER**β** represent domain homologies with respect to ER**α**. Below the structures of the wild-type receptors are reported receptor isoforms identified in BC and mentioned in the text

suggest the existence of several ERβ isoforms resulting from alternative splicing (Fig. 8.1). So far, four ERβ isoforms (ERβ2, ERβ3, ERβ4, and ERβ5) have been identified and characterized, with molecular weights of 59, 56, 54, and 53 kDa, respectively [12], and showing clear differences in the AF2 domain (C-terminus) and quite different biological activities, with ERβ1 and ERβ2 being able to form functional heterodimers with ERα and thereby inhibiting its signaling [12–14]. The major differences between ERβ1 and ERβ2 are that the first can counteract ERα signaling in two ways: by neutralizing ERα action via heterodimer formation and by directly triggering antiproliferative signals. In contrast, ERβ2 acts only according to the first mechanism. Recent studies suggest the existence of different ERβ cellular pools in the cytoplasm, in the mitochondria, and at the plasma membrane of BC cells. Interestingly, cytoplasmic localization of ERβ2, alone or in combination with the nuclear form, may predict a significantly worse overall survival in BC, as patients with only cytoplasmic ERβ2 suffer for a significantly poorer prognosis [15]. Moreover, nuclear ERβ2 was strongly predictive of a twofold greater response to endocrine therapy. At present, several lines of evidence support the presence of ERβ within the mitochondria and its association with mitochondrial proteins in BC cells [16–20]. Overall, data from cancer cell models and observational studies suggest that ERβ functions as a gatekeeper to inhibit tumor growth and progression. Although ERβ seems to be a tumor suppressor in numerous cell models, its role in human breast carcinogenesis remains to be elucidated.

ERα and ERβ share sequence homology within their DNA-binding and hormone-binding domains, but they have different transcriptional regulation properties, suggesting that each of them interacts with unique sets of nuclear factors and plays different roles in the control of gene expression [20, 21]. Estrogen-responsive cells can be endowed with one or both of these *trans*-acting transcription factors. In addition, estrogen can trigger rapid and transient cellular responses through mechanism(s) independent from this "genomic" pathway of steroid receptor action. Such "extra-genomic" estrogen effects include cell type-specific, rapid, and transient responses of signal transduction pathways, induction of intracellular calcium mobilization, and activation of membrane ion channels [21, 22]. The genomic and extra-genomic pathways do integrate with each other to mediate the mitogenic actions of estrogen, including activation of cell cycle controlling gene networks. Both molecular cascades are believed to involve multiple components that by functionally or physically interacting with ERs in specific cellular compartments (plasma membrane, cytoplasm, chromatin, etc.) modulate or/and mediate receptor activity. Despite extensive investigations, the mechanisms by which estrogens exert their growth regulatory actions are still not fully defined. It is well known that they depend on the presence of ERs and correlate with direct transcriptional regulation of cell cycle control genes, including proto-oncogenes and D-type cyclin genes [23, 24]. Activation of MAP kinase cascades by estrogens via an interaction of ERα with p*Src* has also been described and suggested as part of the mechanism for cell cycle regulation by these hormones [25]. Furthermore, ERα was shown to interact also with the regulatory subunit of phosphatidylinositol-3-OH-kinase, leading to activation of protein kinase B/Akt and other cellular effectors, while estrogens have been shown to activate membrane ion channels, such as the beta subunit of Maxi-K channel hSlo, and PKA in target cells [22, 24].

8.1.2 Estrogen Receptors Alpha and Beta in Breast Cancer

E2 is the ligand for both ERα and ERβ, but its interaction with each receptor results, however, in divergent transcriptional effects. The two ER subtypes are both detected in BC cells, but it is unclear how their combined activation by estrogen results in the overall cell response to the hormone [26]. Therefore, there is an increasing interest in understanding the biological significance of ERβ expression and whether this receptor subtype can be a valuable target for therapy of BC. Estrogen regulation of cell proliferation depends on the presence of ERα and -β and correlates with the transcriptional control of cell cycle genes [23, 24, 27]. The AF-1 and D domains represent the regions of highest divergence between ERα and ERβ [6]. Both receptors bind E2 with high affinity, but they differ in the affinities for various compounds and in the transcriptional response that will be elicited. For example, tamoxifen is a cell- and tissue-specific mixed agonist–antagonist for ERα and a pure antagonist for ERβ [6]. The two ERs can recruit the same co-activators or corepressors, but SRC-3 contributes more to the transcriptional activity of ERα than to that of ERβ [28]. Both ERs enhance transcription of a reporter plasmid containing an *E*strogen *R*esponse *E*lement (ERE); ERβ, however, is a weaker activator than ERα and the cellular context determines significant differences in its transcriptional activity [29].

These results suggest that the pattern of gene regulation by the two receptors is only partially overlapping and indeed, genes differentially regulated by ERα and -β in BC cells, such as cyclin D1 [20, 30] and fibulin-1C [20, 31], have been identified. ERα homodimers and ERα/β heterodimers show the same affinity for DNA that is instead lower for ERβ homodimers [32]. Many studies suggest that ERβ acts as a negative modulator of ERα transcriptional activity in BC cells by decreasing the cell sensibility to E2 [33] and acting on ERα-mediated regulation of estrogen target gene expression [20, 34], including those involved in DNA replication, cell cycle regulation, and proliferation [20, 35, 36]. Thus, despite ERα and ERβ being both key mediators of the estrogenic signal transduction cascade, they play different and/or antagonistic biological roles in BC. Overexpression of ERβ in ERα-positive BC cells inhibits cell proliferation in response to E2 by increasing the expression of antiproliferative genes and by decreasing the expression of proliferative and antiapoptotic ones. This growth-modulatory activity of ERβ can explain the better prognostic outcome of BC tumors expressing this receptor subtype [20, 37–39].

ERs are component of several multiprotein complexes [3, 4]. In human BC cells the presence of two distinct peaks of E2 binding has been demonstrated, detectable in low-salt extracts upon sucrose gradient centrifugation, containing ERα and ERβ, respectively. Differences among several breast samples in the amount of E2 bound and the ratio between the two peaks have been verified [40], suggesting that ERα is involved in larger protein complex than ERβ. In addition, it has been reported also that ERβ specifically interacts with MAP kinase-interacting kinase (Mnk2) [40] and that it is phosphorylated and activated by RSK2.

Overall, these observations suggest that different transduction pathways are involved in ERα and ERβ activity and that the overall cellular responses (gene expression and cell proliferation, for example) to hormone stimulation depend upon the presence and relative abundance in the cell of the two receptors and their respective functional partners.

8.2 Estrogen Receptors in Breast Cancer Biology

The etiology of BC is thought to involve a complex interplay of genetic, hormonal, and environmental factors that influence the physiological status of the host. In this respect, estrogens are important pathogenic factors in regulating differentiation and proliferation of normal as well as transformed breast epithelial cells.

8.2.1 Open Issues in Estrogen Receptor Actions in Breast Cancer Cells

Estrogens induce an array of cell type-specific responses in target tissues, ranging from the expression of specific differentiated cellular responses to cell proliferation. In BC, a subset of lesions shows a clear mitogenic response to estrogen, while others, otherwise indistinguishable, do not. This has long been explained merely by the presence or absence of ERs. A significant fraction of BCs are, at the time of diagnosis, hormone dependent and responsive to endocrine manipulations aiming at interfering with estrogen actions. When first discovered, this notion led to the introduction of the concept that certain biological indicators (in this case the presence of functional estrogen and progesterone receptors in cancer cells) could be used to predict the effectiveness of a therapeutic regimen, namely, endocrine therapy. This is still valid today, although extensive clinical evidence is indicating that steroid receptor status of BC cells is not a sufficient indicator of their hormone responsiveness. Endocrine therapy shows a response rate of 30 % in unselected patients, of about 50 % in ER-positive patients, and of 60–70 % if ER/PR is positive. There are, thus, about 30 % receptor-positive cancers that, despite all predictions, fail to respond to hormonal treatments, whereas 5–10 % of patients with receptor-negative tumors unexpectedly do respond. Tumor-specific estrogen receptor dysfunctions and mutations in their intracellular signaling are thought to contribute to this behavior. No substantial evidence in this sense is available to date, as many uncertainties still reside on the mechanisms that mediate the control of cell proliferation by estrogen in normal and transformed mammary gland cells.

More recently, it has become clear that hormone responsiveness is a more complex phenomenon, linked not only to receptor expression but also to other, still undefined, cellular factors. Such complexity can be observed daily in BC patients and is supported by strong biological evidences, the more convincing one being represented by the fact that hormone-independent BC cells cannot be transformed *in vitro* into hormone-responsive ones simply by the forced expression of exogenous ER. On the other hand, cell type-specific effects of estrogens are well known. In many target tissues, distinct and specific responses can be observed even in very similar cell types, all expressing ER to a similar extent.

Understanding the nature of the cellular factor(s) acting in concert with ERs to foster cell proliferation is a central issue for a better understanding of estrogen actions in BC, whose definition is required to elucidate the mechanisms underlying the hormone-responsive phenotype and, as a consequence, to devise new ways to exploit this phenotype to practical ends.

8.2.2 Application of "Omics" Technologies to Investigate Estrogen Receptor Signaling in Breast Cancer

ERs drive gene cascades comprising genes whose transcription is directly regulated by their physical interaction with regulatory sites (estrogen response element, ERE) in the genome or with signal transduction effectors (non-genomic pathway), as well as downstream genes whose expression depends on directly regulated ones. A significant effort has been made to identify in BC reliable predictors of tumor sensitivity to different drugs, markers of pharmacological resistance, and, finally, new therapeutic targets to overcome it. Owing to the advances made in genomic technologies, our understanding of breast tumorigenesis has significantly evolved in recent years [41, 42]. It is clear that, similarly to other cancer types, BC is a complex disease involving many mechanisms whose outcome is, among others, an alteration of transcriptional regulations. The analyses of the gene expression profiles conducted on BC specimen and cell lines have proven that the ERα-expressing breast tumors and cell lines share significant similarities in their transcriptomes. The expression profiles of estrogen-responsive gene sets identified *in vitro* have been found to be an intrinsic genetic signature of ERα-expressing breast tumors [43]. Gene expression profiling [44], proteome [45, 46], miRNome [47–49], and cistrome (defined as the combination of all target genes and binding sites of a regulatory factor in a given genome) [47, 50] analyses of ER and some of its cofactors shed light on the complexity of ER signaling in BC cells.

A number of laboratories focused their *omics* research on ERs in BC with the aim of characterizing fundamental processes engaged by ERs that will point to new molecular pathways that could be targeted for BC treatments. The fundamental aspect to be considered, however, is how to integrate these large sets of data in a useful and informative way. This is particularly true when considering that ERs activity in the cell involves multiple molecular components that control and/or mediate ER functions by functionally or physically interacting with these molecules in specific cellular compartments, such as the plasma membrane [51, 52], the cytoplasm, and chromatin [52–57]. Indeed, it is well known that most effects of estrogens are cell type specific, and this is achieved by differential expression not only of ERs but also of the functional partners of these receptors. These are believed to include transcriptional co-regulators, signaling effectors, molecular adapters, and other intracellular molecules, which participate in estrogen signal transduction within modular multiprotein complexes with different biological activities depending upon their absolute composition, stoichiometry, and conformation of their components [58–61]. Understanding the nature of the cellular proteins acting in concert with ERs to control cell functions is an open issue in BC biology [62–64]. To date, this issue has been addressed by the analysis of molecular profiles associated with hormone response and disease state in BC cells. Gene expression profiling [65–67] and quantitative proteomics analyses [68, 69] provided a blueprint of the effects of estrogen and other ER ligands in hormone-responsive cancer cells, revealing a complexity of ER-induced cellular responses that suggests the likelihood that ERs exist in the cell in multiple functional conformations. Indeed, the already-mentioned ability of ligand-activated ERs to form multiple complexes with key intracellular regulatory molecules represents a well-known mechanism to explain their multifaceted effects in key processes such as signal transduction and transcriptional regulation.

8.3 Analysis of the Estrogen Receptor Interactomes in Breast Cancer Cells

Proteins rarely act in isolation and their interactions are central to all biological functions. The interference with biomolecular networks often leads to disease. Protein–protein and protein–metabolite interactions have traditionally been studied case by case. Recently, powerful analytical technologies have been developed to enable large-scale investigations of protein–protein interaction networks, contributing to create comprehensive cartography of several pathways relevant to human diseases.

8.3.1 Interactome Analysis: Technical Aspects and Application of Tandem Affinity Purification to Investigate Estrogen Signaling

Discovering the molecular partners of a given protein results useful to understand its activity. Until few years ago the gold standard to reveal protein–protein interactions was considered yeast two-hybrid system [70], but it held several technical limitations due to the possibility to only detect binary interactions and to the high rate of false positives and false negatives [71]. Nevertheless, isolation of multiprotein complexes by immunoprecipitation of cell extracts with antibodies against one of the components of the complex is limited by the unavailability and poor quality of the antibodies. More recently, tandem affinity purification (TAP) followed by mass spectrometry has been introduced and used for high-throughput detection of multiprotein complexes. This approach allows ex vivo identification of most of the interacting partners of a given protein [72], where identification of the proteins in such samples is made possible by the availability of the genome sequence for many organisms. This method was firstly used in the analysis of protein–protein interaction in yeast [73] and lately employed for several organisms, including mammals [74–76].

The TAP-tagging procedure involves the fusion of the TAP tag to any target protein whose corresponding cDNA can be cloned and its introduction into the host cell. As shown in Fig. 8.2a, the original TAP tag consists of two IgG-binding

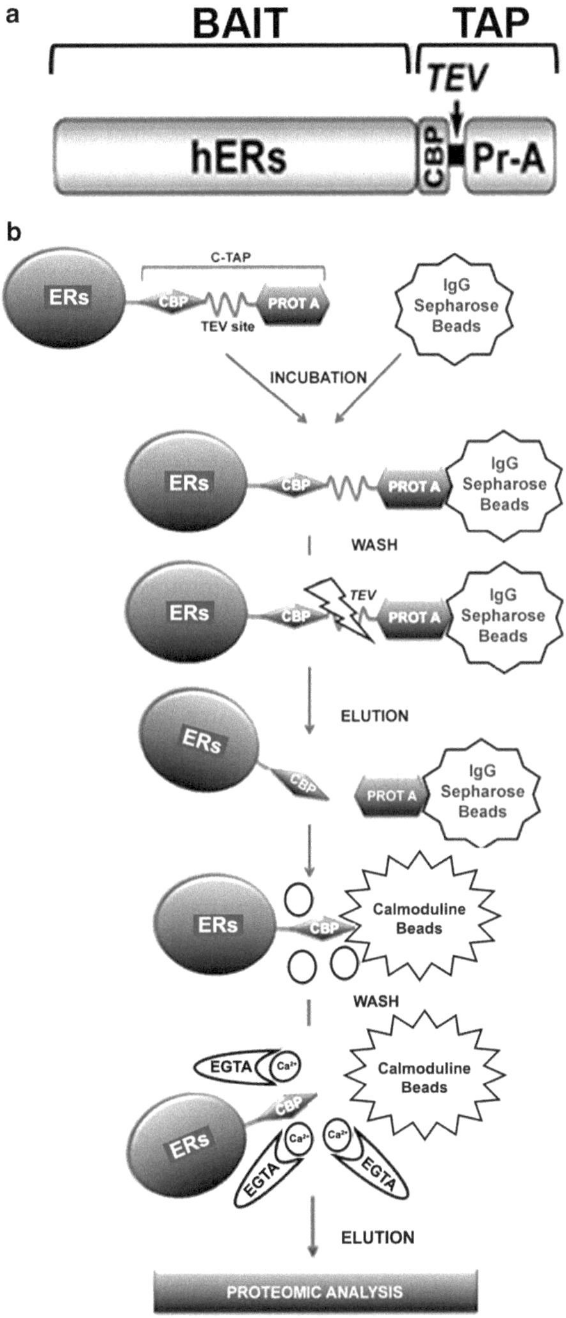

Fig. 8.2 ER interaction proteomics by tandem affinity purification. (**a**) TAP-ER fusion proteins used for ER interactome mapping. *hERs* human ERα or ERβ coding sequence, *CBP* calmodulin-binding peptide, *TEV* peptide comprising a TEV protease cleavage site, *Pr-A S. aureus* protein A. (**b**) Schematic representation of the different steps for ER-binding protein purification and identification

domains of *S. aureus* protein A (ProtA) and a calmodulin-binding peptide (CBP) separated by a TEV (Tobacco Etch Virus) protease cleavage site [77]. In order to reduce false associations, expression level of the recombinant protein should be maintained as close as possible to the natural one. Efficient protein complexes recovery is achieved under native conditions by two consecutive steps. Typically, cell protein extracts are prepared, and the fusion protein and its interactors are purified first by incubation with IgG sepharose beads (that bind with high affinity the ProtA domain of the tag); after washing ProtA is cleaved off with the TEV protease. The eluate deriving from the first step is then applied onto calmodulin-coated beads (that are recognized by the CBP peptide in the presence of Ca^{2+}), and following extensive washes, the bound material can be eluted by the addition of EGTA that removes Ca^{2+} destabilizing the calmodulin-CBP complexes (Fig. 8.2b). The second affinity step is useful as independent purification tool, and it removes also from samples of the TEV protease, thereby helping preserve the isolated complexes. The main advantages of this method are the gentle conditions used for complex purification, which maintain the native conformation of the proteins and thus their function, while showing a very low background and reproducible results. The procedure has some limitations, however, including the low yield in mammalian systems, requiring a huge amount of starting material, and the possible loss of transient protein interactions due to the long experimental times. For these reasons, more performing tags are being continuously designed [78]. To date several alternative dual-affinity tags have been developed, using different combinations to improve effectiveness in terms of protein recovery and flexibility in organisms other than yeast [79–81]. However, the use of the traditional TAP tagging still outnumbers the nontraditional ones. This approach guarantees reproducibility of the results, and transient interactions could be retained by performing single-step purification. Nevertheless, TAP was extensively used for mapping ERα and ERβ interactomes in BC cells [82–84], and the TAP protocol described above has been used to investigate systematically the multiprotein complexes in which ERs are involved, followed by the analysis of the role of their components in the regulation of BC cell functions by agonist- or antagonist-bound ERs. Efficient identification of proteins such as ERs and their molecular partners in BC cell extracts was achieved by analysis of purified complexes by nanoLC-MS/MS, a sensitive mass spectrometry technique allowing good discrimination among proteins also present at a low relative abundance. Briefly, ER-containing purified complexes were fractionated by SDS-PAGE, silver stained, gel excised, and subjected to tryptic digestion and mass spectrometry. Raw spectra obtained were then processed and submitted to MASCOT database for sequence searching.

8.3.2 Estrogen Receptor Alpha Interactomes

ERs, like all nuclear receptors, behave dynamically in the cell, and their kinetics allows them to rapidly interact with various co-regulatory proteins, chromatin, and DNA. Furthermore, ERs distribute to various cell compartments, where the local concentration of each partner protein is a key factor in determining the ability of the receptor to intervene in specific processes. With the aim of specifically investigating the multiprotein complexes in which ERs are involved, tandem affinity purification and mass spectrometry were applied to gain a comprehensive view of ER interactomes in different physiological conditions and cellular compartments of BC cells. The datasets obtained, combined with genomics and clinical data, provide a novel view of ER-driven molecular pathways involved in estrogen signaling in BC. MCF7 cells were used to this end, as they are the best-characterized model to investigate estrogen effect in hormone-responsive BC. The TAP tag described above was fused to the C-terminus of the coding region of human ERα (Fig. 8.1), stably transfected in MCF-7 cells to generate TAP-ERα-expressing cells [82] and used to identify E2-activated ERα nuclear proteins. Results [82, 83], independently confirmed with a different high-throughput approach [85], indicate that activated ERα is able to associate with a large number of protein components, playing roles in the definition of receptor-mediated cellular responses and in hormone-dependent ERα-mediated gene transcription in BC cells. Focusing on the identification of the nuclear and cytoplasmic partners of unliganded ERα, results confirmed the existence of a fine-tuned ER-dependent mechanism regulating signal transduction within BC cells, strongly supporting the hypothesis that key cellular events involving ERα are tightly dependent on an array of protein components that specifically associate with this receptor either in the absence or presence of the hormone (Fig. 8.3). Mass spectrometry analysis performed after TAP of cytosolic and nuclear protein extracts separately allowed the identification of 72 and 58 receptor partner proteins, respectively. The first striking evidence emerging in the dissection of unliganded ERα interactome of BC cells is that only one protein, myosin light chain kinase 2, is associated with the receptor in both compartments (Fig. 8.3a). This strengthens the assumption that several co-regulatory networks are differentially implicated in compartment- and condition-specific ERα activities. Although it has been often reported that in the absence of estrogen ERα is mainly sequestered within the cytoplasm in inactive complexes with molecular chaperones, other experimental evidences indicate that unliganded receptor partially localizes in the nucleus, thereby ensuring basal transcription of target genes [86, 87]. Nevertheless, under estrogen deprivation, both cytosolic and nuclear ERα seem to intervene in almost the same biological events, although in association with different partners. Functional annotation analysis highlights this concept, suggesting that in the absence of hormonal stimulus cytoplasmic ERα complexes are mainly involved in regulation of metabolic, biosynthetic, and ubiquitination processes (Fig. 8.3b). This could be exemplified by the presence of two factors: Serpin H1 (HSP47) and procollagen-lysine, 2-oxoglutarate 5-dioxygenase 3 (PLOD3), the former being involved as a chaperone in the biosynthetic pathway of collagen and the latter essential to confer stability to the intermolecular collagen cross-links. It has been reported that tumor cell lines derived from metastatic carcinomas synthesize higher levels of HSP47, suggesting that this protein may play an important role in tumor metastasis and could represent a prognostic marker in invasive ductal breast carcinoma [88]. Interestingly, the most representative nuclear processes involving ERα result to be metabolic, including RNA processing (Fig. 8.3c). These results suggest that in quiescent BC cells ERα is mainly involved in activities aiming at cell maintenance. Moreover, the metabolic role of

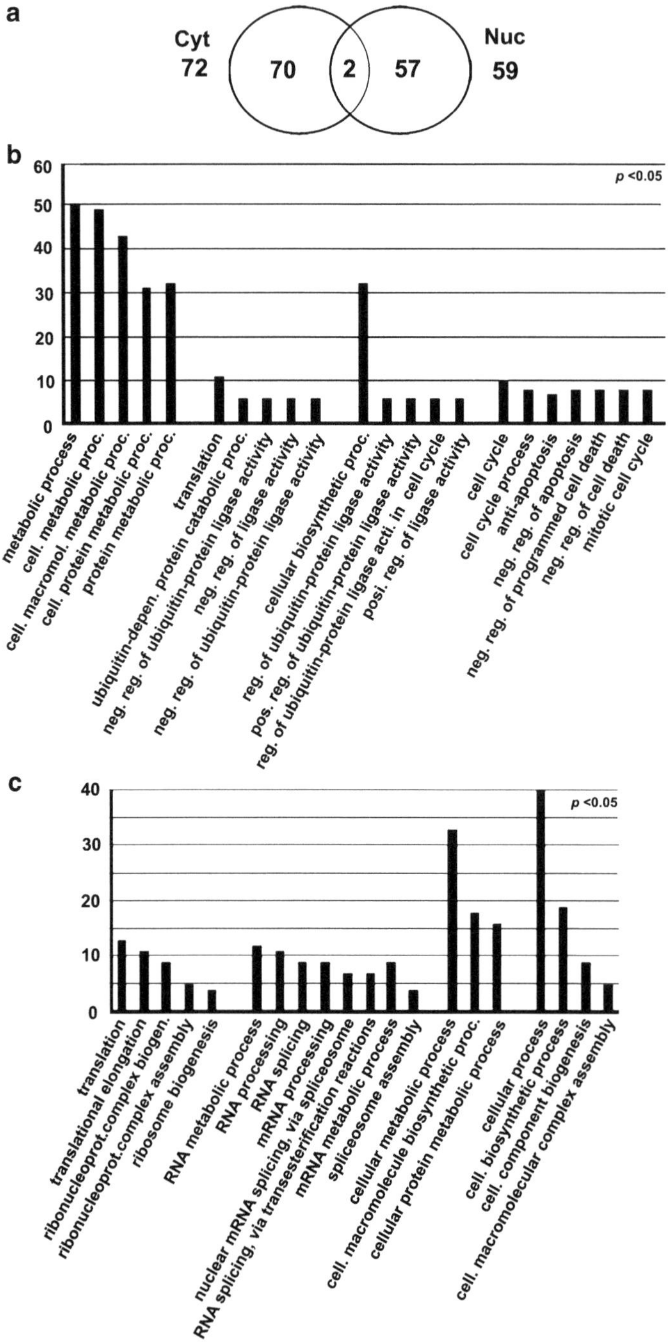

Fig. 8.3 Functional characterization of unliganded ERα interactomes. (**a**) Venn diagram summarizing the cytosolic (Cyt) and nuclear (Nuc) interactomes of ligand-free ERα mapped in BC cells. (**b**, **c**) Functional annotation tool according to Gene Ontology analysis of unliganded ERα cytosolic (**b**) and nuclear (**c**) proteins

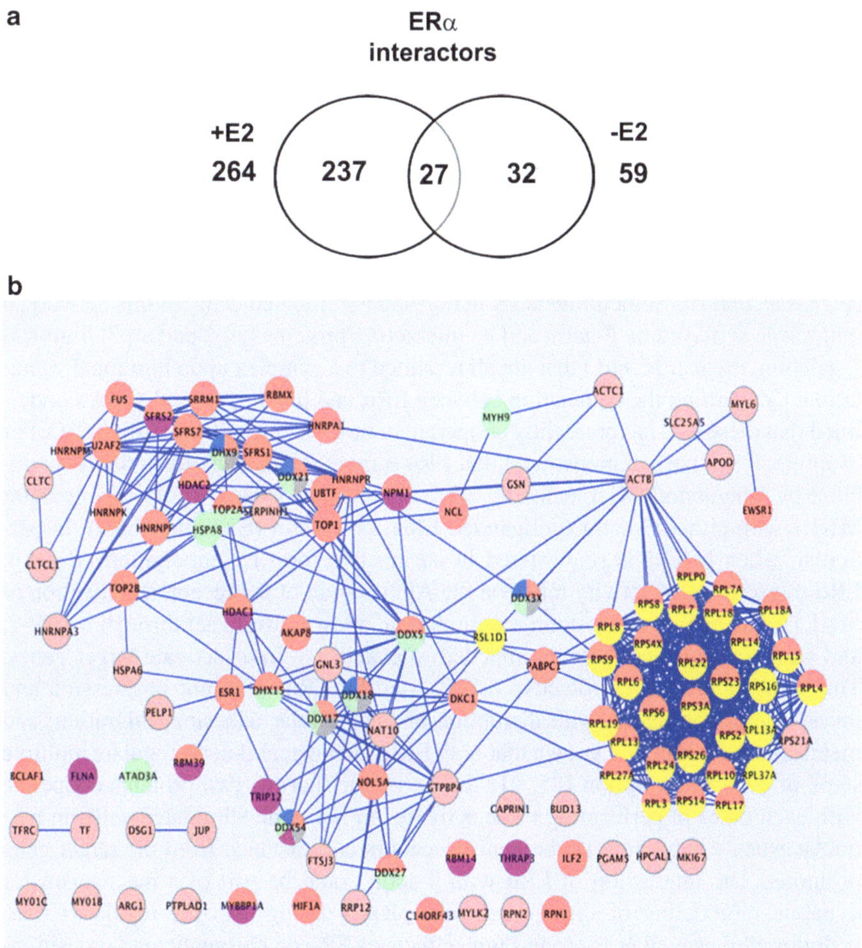

Fig. 8.4 Liganded and unliganded ERα interactomes in MCF-7 cell nuclei. (**a**) Venn diagram summarizing the liganded (+E2) and unliganded (–E2) ERα nuclear interactomes. (**b**) Network analysis of liganded ERα nuclear partners according to Molecular Function and Biological Process GO terms (analysis and visualization by Cytoscape). Statistically significant ($p<0.05$) Biological Processes represented cell organization and biogenesis (*blue*), actin binding (*light red*), RNA processing (*orange*), macromolecule metabolism (*yellow*), macromolecule biosynthesis (*violet*), DNA topological change (*gray*), estrogen receptor signaling pathway (*light blue*), and RNA processing (*green*)

ERα has been shown in other cell types to exert a role in insulin and glucose metabolism, acting on insulin sensitivity, lipogenesis and lipolysis, insulin secretion, gluconeogenesis, and energy homeostasis [89].

On the other hand, comparing nuclear ERα interactomes in the absence vs. the presence of estrogen, only 27 proteins were found associated with the receptor in both conditions (Fig. 8.4a), including predominantly splicing factors and ribosomal proteins. Additional 237 other proteins are specifically involved in estrogen-dependent signal transduction, RNA processing, and DNA topological changes (Fig. 8.4b), further confirming the role of ERα in the control of RNA biosynthesis and maturation and in β-actin nuclear network [83].

These results indicate the existence of common functions of ligand-free and E2-activated ERα, among which a central role appears to be played by ribosome biogenesis as this pathway, already involving the unliganded receptor, is significantly reinforced upon estrogen activation and sees ERα acting on this machinery via a pre-constituted, E2-dependent scaffold. Finally, ligand-activated ERα appears to act as a bridging factor, linking transcription with RNA splicing and processing, cell cycle, and ribosome biogenesis in hormone-stimulated cells. In this network, a major role is played by β-actin and its interacting proteins (arp2 and arp3, flightless I, gelsolin, myosin 1c, etc.) that are all recruited to a complex upon hormonal stimulation. Concerning the connection between ERα and β-actin, several studies underlined that these two factors tightly cooperate in modulating cell activities in BC. For example, it has been demonstrated that ERα transcriptional activity is finely modulated by Rho/actin/MKL1 pathway, via a specific association of the co-regulator MKL1 with either G-actin (unliganded ERα) or F-actin (E2-bound ERα). In particular, when MKL1 is sequestered in an inactive form by unpolymerized actin, ERα transcriptional activity relies on the AF-1 region of the receptor. Activation of MKL1, causing polymeric actin accumulation, allows ERα to act through its AF-2, and as a consequence, ERα is impeded to efficiently *trans*-activate target genes. This may suppress the protective role exerted by ERα on tumor progression and invasiveness, correlating with disruption of intercellular adhesion, migration, and metastasis [90]. It is also known that both ERα and nuclear β-actin regulate multiple steps of gene transcription [85, 91]. It is possible that the two proteins cooperate with each other in performing these activities in hormone-stimulated cells, in particular when ER binding to the genome occurs at a distance from the target gene promoter. The interaction of ERα with β-actin could be part of a mechanism for dynamic remodeling of multiprotein complexes during estrogen-regulated transcription [86] as well as for long-range effects of ERα on chromatin and repositioning of ER-responsive genes within the nucleus [53]. It has been also proposed that β-actin and actin-related proteins are required for the activity of SWI/SNF and for a stable association of remodeling complexes with chromatin. Considering the broadness of SWI/SNF complex, multiple protein–protein interactions, including actin-related proteins such as flightless, might be involved in its association with the transcription initiation complex assembled by ERα [92]. On the other hand, the interaction between ERα, β-actin, nucleophosmin, and ribosomal proteins indicates a direct molecular link between the mitogenic action of estrogen and ribosome biogenesis in BC cells. This possibility is suggested also by the known role of β-actin in ribosome biogenesis and maturation [93]. In conclusion, the β-actin network represents a bridging factor between ERα and the factors involved in splicing, ribonucleoparticles formation, and ribosome assembly and may play a key role in estrogen-dependent signaling in BC cells.

8.3.3 *Effects of Estrogen Receptor Beta Expression on Estrogen Receptor Alpha Interactome*

As described above, ERα and ERβ can elicit divergent transcriptional responses due to differences in their mechanism of action. Several experimental evidences point to the ability of the two receptor subtypes to regulate a variety of common and different genes in BC cells and suggest that the divergent transcriptional responses to these ligand-dependent transcription factors can be due to multiple cellular factors that can affect ER-mediated signal transduction pathway. Among these factors, a key role can be assigned to the property of ERs to be involved in modular multiprotein complexes that include specific transcriptional co-regulators, signaling effectors, molecular adapters, and other intracellular molecules that participate in estrogen signal transduction by physically interacting with the ERs and, thereby, convey them onto different biological activities, depending upon their absolute composition, stoichiometry, and conformation of their components. ERs are able to regulate the activity of each other by forming heterodimers both *in vitro* and *in vivo*, thereby influencing receptor-DNA interactions and, consequently, their effects on the genome. Indeed, ERβ appears to act as a dominant-negative regulator of estrogen signaling when co-expressed with ERα, opposing ERα-mediated transcription. This evidence suggested to investigate the molecular mechanisms sustaining ERβ interference upon ERα activity in BC cells by interactome analysis, based on the observation that E2 stimulation of ERβ-expressing MCF-7 cells yielded mainly ERα/β heterodimers, that as expected represent the predominant form of ERα under these conditions.

Using tandem affinity purification coupled to mass spectrometry, ERβ interactome was mapped and characterized in estrogen-responsive BC cells [84] and then compared to ERα interactome identified in ERβ cells [92]. In-depth computational analysis by protein–protein interaction topology and dissection of the two datasets through subcluster classification [94] showed that among the 264 proteins associated with ligand-activated ERα and 303 bound to ERα/β heterodimers in MCF-7 cells, only 70 were in common. Most of the interactors (194 and 234 specifically associated with ERα and ERβ, respectively) are selectively co-purified under the two conditions investigated, confirming that interaction proteomics is a useful approach to investigate ERβ on ERα molecular functions in the same cellular background. The results obtained by considering overrepresented Gene Ontology terms on the three lists of interactors identified (ERα specific, ERβ specific, and common to both receptors) show several differences (summarized in Fig. 8.5), pointing to biological processes converging on either ERα (ERα only), ERβ (ERβ only), or both receptors (ERα and ERβ). In particular, considering the Gene Ontology Biological Process (BP) terms found enriched (Fig. 8.5), ERα molecular partners are involved in RNA processing/splicing, chromosome organization and biogenesis, chromatin modification, and in regulation of actin polymerization and depolymerization and capping. Moreover, terms associated with transcriptional regulation, including mRNA transcription initiation and tRNA and rRNA transcription, and transcription factor

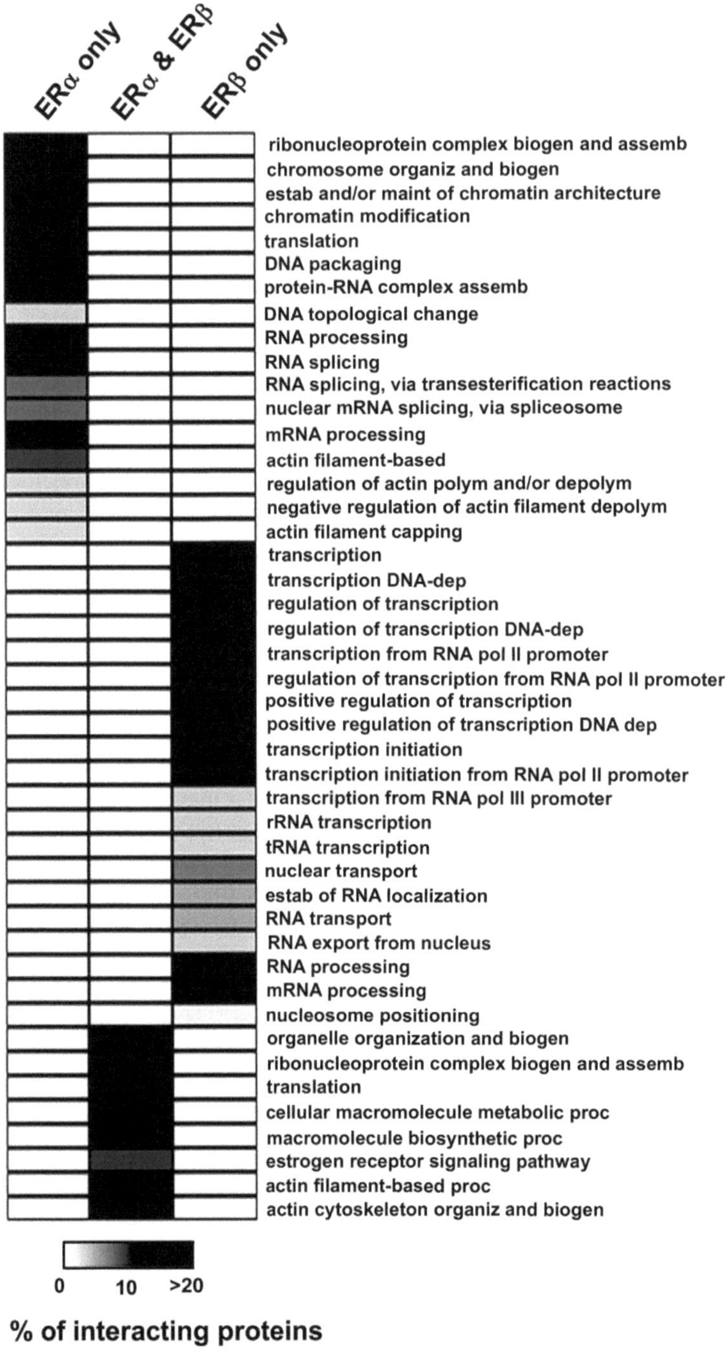

Fig. 8.5 Functional differences between ERα, ERβ, and ERα+β interactomes of estrogen-stimulated MCF-7 cells. HeatMap showing significant differences ($p<0.05$) in Biological Process Gene Ontology terms associated to ERα-specific, ERβ-specific, and ERα/ERβ complex-specific interacting proteins reporting, for each dataset, the percentage of interactors belonging to the indicated functional category. Enrichment was calculated in all cases with respect to all genes expressed in MCF-7 cells, detected by oligonucleotide microarray hybridization

Table 8.1 Recruitment of β-actin and actin-interacting proteins to ERα or ERβ following BC cell stimulation with 17β-estradiol (E2)

Swiss prot protein ID	Protein name	Gene name	Detection in purified ER samples ERα	ERβ
P03372	Estrogen receptor alpha	ESR1	+	+
P60709	Actin, cytoplasmic 1	ACTB	+	+
O00159	Myosin-Ic	MYO1C	+	+
P11142	Heat shock cognate 71 kDa protein	HSPA8	+	+
P06396	Gelsolin	GSN	+	+
P17844	Probable ATP-dependent RNA helicase DDX5	DDX5	+	−
P06748	Nucleophosmin	NPM1	+	−
P61160	Actin-related protein 2	ACTR2	+	−
P61158	Actin-related protein 3	ACTR3	+	+

(+) present, (−) absent

binding are found significantly represented among ERβ interactors. On the other hand, the two interactomes share similar functional annotations, in particular those referred to actin and ribosome organization and biogenesis and the generic category represented by estrogen signaling pathway. Considering instead the differences in composition of the two interaction networks, the one relative to ERβ comprises several known transcriptional co-regulators, such as SRC3 [95], TRIM24 [96], and MED1 [97], as well as proteins directly involved in the regulation of cell growth and apoptosis, like GTPBP4 and PES, that are known to be upregulated in BC compared to normal mammary epithelium [98, 99]. The association of ERβ with these factors could affect their ability to promote p53 downregulation and/or cyclin D1 upregulation, causing significant effects on cell survival and proliferation, in agreement with the established role of ERβ on these processes. GNL2, GNL3, mitochondrial pro-apoptotic protein MRPS29 [100], and BCLAF1, a transcriptional repressor localized to the nuclear envelope and able to induce cellular death [101], were also found associated to ERβ. Among the proteins found in common between the two networks, it is worth mentioning again β-actin (ACTB) for its role in liganded estrogen nuclear signaling [53, 82, 102], regulation of target gene activity, chromatin remodeling, and ribosome biogenesis. However, as shown in Table 8.1, co-expression of both ERs in the cell appears to interfere with recruitment of some ERα/β-actin interacting proteins such as DDX5, NPM1, and ACTR2. Concerning DDX5, a well-known ERα-interacting partner, recent evidences suggested that loss of this protein implies reorganization of actin cytoskeleton and reduction of cellular proliferation [103], results that relate to inhibition by ERβ of estrogen target genes activity mediated by ERα in BC cells, as shown, for example, in the case of the pS2 promoter [20, 104] and for nuclear processing of pri-miR-23b, -27b, and -24-1 [104].

8.3.4 Effects of Antiestrogens on Estrogen Receptor Alpha Interactome

Estrogens' role in BC led to the development of endocrine therapies against these tumors, based on antiestrogenic compounds able to compete with the estrogen and impinge their biological activities. These are synthetic compounds that antagonize hormone-induced proliferation and ERα-target gene expression in mammary tumor cells. Depending upon their functional effects, it is possible to distinguish two major classes of antiestrogens. The "Selective Estrogen Receptor Modulators" (SERMs) are compounds able to act both as ERα agonists and antagonists (partial antagonists), depending on the cellular and promoter context as well as on the targeted ER subtype. The "Selective Estrogen Receptor Downregulators" (SERDs), instead, completely block the activity of estradiol and are thus considered *pure* antiestrogens (full antagonists). SERDs increase receptor turnover and interfere with its nuclear localization, causing a significant reduction of ER concentration in treated cells both in vitro and in vivo.

The first antiestrogen introduced in the clinical practice is Tamoxifen that is, thus, the SERM prototype [105, 106]. Tamoxifen is a nonsteroidal antiestrogen that antagonizes the action of estrogen and is effective in both BC treatment [107, 108] and prevention [109]. Concerns have been raised, however, regarding the potential estrogenic effects of this drug on normal tissues, as Tamoxifen acts as estrogen agonist on bone, blood lipids, and endometrium [110], increasing the risk of endometrial cancer and thrombotic events [109, 111].

Raloxifene, a second-generation SERM, is a nonsteroidal antiestrogen produced by altering the triphenylethylene ring structure of tamoxifen to get a benzothiophene "fixed ring" structure. Originally it was not developed as an antiestrogen for BC [112] but to provide a new hormone replacement therapy to prevent osteoporosis and, as a beneficial side effect, to decrease the incidence of endometrial and mammary cancer in the general population [113]. Raloxifene is a potent antiestrogen and exhibits estrogen-like effects in bone cells, preserving the bone mineral density, but not in uterine cells. Furthermore it appears to cause a decrease in circulating cholesterol [114].

Fulvestrant, also known as ICI 182,780, the SERD prototype, is a steroidal molecule devoid of estrogen-like activity. It was synthesized in order to treat patients with hormone-sensitive breast tumors that, after first-line therapy with Tamoxifen, developed resistance to the drug and the negative side effects of the SERM in the gynecological tract [115]. Indeed, proliferation of Tamoxifen-resistant BC cell lines can be inhibited by Fulvestrant [116, 117], and clinical evidence suggests that the absence of an agonist activity of this SERD may indeed lead to overcoming the resistance which may develop following long-term therapy with Tamoxifen.

After binding to ERα, antiestrogens induce conformational changes of the receptor that are different from those induced by the endogenous agonist E2 and peculiar of each antiestrogen compound. More precisely, the different antagonists induce

Table 8.2 Differences in association of β-actin and its interacting proteins with ERα upon treatment of BC cells with a receptor agonist (17β-estradiol; E2) or antagonist (tamoxifen, TAM)

Swiss prot protein ID	Protein name	Gene name	Relative abundance	
			E2	TAM
P03372	Estrogen receptor alpha	ESR1	+++	+++
P60709	Actin, cytoplasmic 1	ACTB	+++	+
B2RTY4	Myosin isoforms	MYO	+++	+
P06748	Nucleophosmin	NPM1	++	−
P17844	Probable ATP-dependent RNA helicase DDX5	DDX5	++	−

(+++) high, (++) medium, (+) low, (−) undetected

diverse conformational changes of ERα and, consequently, differences in the external conformation of the receptor, in particular on the domains that interact with its molecular partners and the spatial reorganization of the receptor structure that occurs dependent thus strictly on the nature of the ligand [118]. This remodeling drives, or prevents, binding of specific co-regulatory proteins to ERα. The complex of all proteins that interact with the receptor, either directly or indirectly, makes up the receptor interactome [119] that, in turn, influences the nature of the biological response triggered by the ligand. Furthermore, the relative balance in a given co-activator and corepressor proteins also determines the response to each particular ligand. As stated above, the mitogenic effects of estrogen on the mammary gland are known to be mainly mediated by ERα, and this liaison led initially to the development of antiestrogen-based therapies that aims at inhibiting estrogen signaling via ERα. However, clinical experience has highlighted that approximately 30 % of receptor-positive neoplasms are not responsive to these therapies. Taken together, these facts indicate that a likely explanation for the lack of responsiveness of certain ERα-positive BCs could depend upon the lack of, or mutation, one or more components of the receptor interactome. As a consequence, knowing the nature and composition of each antiestrogen-dependent interactomes of BC cells is a logical approach to be able to decipher the molecular mechanisms of natural and acquired BC cell resistance to antihormone-based therapies. A recent study on MCF-7 cells treated with each of the three antiestrogen compounds describe above indeed showed very significant quantitative and qualitative differences of ERα nuclear interactome that relate to the nature of the ligand, i.e., E2, Tamoxifen, Raloxifene, or ICI 182,780 [120]. As exemplified in Table 8.2 for some actin-associated proteins known to be functionally important interactors of ERα in the presence of E2 [82–84], cell stimulation with Tamoxifen results in the recruitment of different complexes on the receptor and, therefore, on a different functional output of the complex.

In conclusion, the interactomics results reviewed here provide a first comprehensive view of the regulatory networks of hormone-responsive BC cells involving directly ERα and ERβ via protein–protein interactions occurring both in the nuclear and extranuclear compartments of the cell. These data represent a tool that can be exploited to elucidate estrogen signaling and its dysfunctions in BC.

Acknowledgments Supported by Italian Ministry for Education, University and Research (Grant PRIN 2010LC747T_002 and FIRB RBFR12W5V5_003), Fondazione con il Sud (Grant 2009-PdP-22), and Italian Association for Cancer Research (Grant IG-13176). FC is a student of the PhD Program in "Molecular Pathology and Physiopathology" of the University of Napoli "Federico II", GN is supported by a 'Mario and Valeria Rindi' fellowship of the Italian Foundation for Cancer Research.

References

1. Green S, Walter P, Kumar V, Krust A, Bornert JM, Argos P, Chambon P. Human oestrogen receptor cDNA: sequence, expression and homology to v-erb-A. Nature. 1986;320:134–9.
2. Kuiper GG, Enmark E, Pelto-Huikko M, Nilsson S, Gustafsson JA. Cloning of a novel receptor expressed in rat prostate and ovary. Proc Natl Acad Sci USA. 1996;93:5925–30.
3. Leclercq G. Molecular forms of the estrogen receptor in breast cancer. J Steroid Biochem Mol Biol. 2002;80:259–72.
4. Pettersson K, Gustafsson JA. Role of estrogen receptor beta in estrogen action. Annu Rev Physiol. 2001;63:165–92.
5. Kumar R, Thompson EB. The structure of the nuclear hormone receptors. Steroids. 1999;64:310–29.
6. Thomas C, Gustafsson JÅ. The different roles of ER subtypes in cancer biology and therapy. Nat Rev Cancer. 2011;11:597–608.
7. Townson SM, O'Connell P. Identification of estrogen receptor alpha variants in breast tumors: implications for predicting response to hormonal therapies. J Surg Oncol. 2006;94:271–3.
8. Erenburg I, Schachter B, Mira y Lopez R, Ossowski L. Loss of an estrogen receptor isoform (ER alpha delta 3) in breast cancer and the consequences of its reexpression: interference with estrogen-stimulated properties of malignant transformation. Mol Endocrinol. 1997;11: 2004–15.
9. Klinge CM, Riggs KA, Wickramasinghe NS, Emberts CG, McConda DB, Barry PN, Magnusen JE. Estrogen receptor alpha 46 is reduced in tamoxifen resistant breast cancer cells and re-expression inhibits cell proliferation and estrogen receptor alpha 66-regulated target gene transcription. Mol Cell Endocrinol. 2010;323:268–76.
10. Mosselman S, Polman J, Dijkema R. ER beta: identification and characterization of a novel human estrogen receptor. FEBS Lett. 1996;392:49–53.
11. Moore JT, McKee DD, Slentz-Kesler K, Moore LB, Jones SA, Horne EL, Su JL, Kliewer SA, Lehmann JM, Willson TM. Cloning and characterization of human estrogen receptor beta isoforms. Biochem Biophys Res Commun. 1998;247:75–8.
12. Leung YK, Mak P, Hassan S, Ho SM. Estrogen receptor (ER)-beta isoforms: a key to understanding ER-beta signaling. Proc Natl Acad Sci USA. 2006;103:13162–7.
13. Hall JM, McDonnell DP. The estrogen receptor beta-isoform (ERbeta) of the human estrogen receptor modulates ERalpha transcriptional activity and is a key regulator of the cellular response to estrogens and antiestrogens. Endocrinology. 1999;140:5566–78.
14. Peng B, Lu B, Leygue E, Murphy LC. Putative functional characteristics of human estrogen receptor-beta isoforms. J Mol Endocrinol. 2003;30:13–29.
15. Shaaban AM, Green AR, Karthik S, Alizadeh Y, Hughes TA, Harkins L, Ellis IO, Robertson JF, Paish EC, Saunders PT, Groome NP, Speirs V. Nuclear and cytoplasmic expression of ERbetal, ERbeta2, and ERbeta5 identifies distinct prognostic outcome for breast cancer patients. Clin Cancer Res. 2008;14:5228–35.
16. Chen JQ, Yager JD. Estrogen's effects on mitochondrial gene expression: mechanisms and potential contributions to estrogen carcinogenesis. Ann N Y Acad Sci. 2004;1028:258–72.

17. Chen JQ, Eshete M, Alworth WL, Yager JD. Binding of MCF-7 cell mitochondrial proteins and recombinant human estrogen receptors alpha and beta to human mitochondrial DNA estrogen response elements. J Cell Biochem. 2004;93:358–73.
18. Chen JQ, Delannoy M, Cooke C, Yager JD. Mitochondrial localization of ERalpha and ERbeta in human MCF7 cells. Am J Physiol Endocrinol Metab. 2004;286:E1011–22.
19. Clayton DA. Replication and transcription of vertebrate mitochondrial DNA. Annu Rev Cell Biol. 1991;7:453–78.
20. Grober OM, Mutarelli M, Giurato G, Ravo M, Cicatiello L, De Filippo MR, Ferraro L, Nassa G, Papa MF, Paris O, Tarallo R, Luo S, Schroth GP, Benes V, Weisz A. Global analysis of estrogen receptor beta binding to breast cancer cell genome reveals an extensive interplay with estrogen receptor alpha for target gene regulation. BMC Genomics. 2011;12:36.
21. Barkhem T, Carlsson B, Nilsson Y, Enmark E, Gustafsson J, Nilsson S. Differential response of estrogen receptor alpha and estrogen receptor beta to partial estrogen agonists/antagonists. Mol Pharmacol. 1998;54:105–12.
22. McInerney EM, Rose DW, Flynn SE, Westin S, Mullen TM, Krones A, Inostroza J, Torchia J, Nolte RT, Assa-Munt N, Milburn MV, Glass CK, Rosenfeld MG. Determinants of coactivator LXXLL motif specificity in nuclear receptor transcriptional activation. Genes Dev. 1998;12:3357–68.
23. Altucci L, Addeo R, Cicatiello L, Dauvois S, Parker MG, Truss M, Beato M, Sica V, Bresciani F, Weisz A. 17beta-Estradiol induces cyclin D1 gene transcription, p36D1-p34cdk4 complex activation and p105Rb phosphorylation during mitogenic stimulation of G(1)-arrested human breast cancer cells. Oncogene. 1996;12:2315–24.
24. Cicatiello L, Addeo R, Sasso A, Altucci L, Petrizzi VB, Borgo R, Cancemi M, Caporali S, Caristi S, Scafoglio C, Teti D, Bresciani F, Perillo B, Weisz A. Estrogens and progesterone promote persistent CCND1 gene activation during G1 by inducing transcriptional derepression via c-Jun/c-Fos/estrogen receptor (progesterone receptor) complex assembly to a distal regulatory element and recruitment of cyclin D1 to its own gene promoter. Mol Cell Biol. 2004;24:7260–74.
25. Mayer EL, Krop IE. Advances in targeting SRC in the treatment of breast cancer and other solid malignancies. Clin Cancer Res. 2010;16:3526–32.
26. Palmieri C, Cheng GJ, Saji S, Zelada-Hedman M, Wärri A, Weihua Z, Van Noorden S, Wahlstrom T, Coombes RC, Warner M, Gustafsson JA. Estrogen receptor beta in breast cancer. Endocr Relat Cancer. 2002;9:1–13.
27. Huang Y, Li X, Muyan M. Estrogen receptors similarly mediate the effects of 17β-estradiol on cellular responses but differ in their potencies. Endocrine. 2011;39:48–61.
28. Watanabe T, Inoue S, Ogawa S, Ishii Y, Hiroi H, Ikeda K, Orimo A, Muramatsu M. Agonistic effect of tamoxifen is dependent on cell type, ERE-promoter context, and estrogen receptor subtype: functional difference between estrogen receptors alpha and beta. Biochem Biophys Res Commun. 1997;236:140–5.
29. McInerney EM, Tsai MJ, O'Malley BW, Katzenellenbogen BS. Analysis of estrogen receptor transcriptional enhancement by a nuclear hormone receptor coactivator. Proc Natl Acad Sci USA. 1996;93:10069–73.
30. Webb P, Nguyen P, Valentine C, Lopez GN, Kwok GR, McInerney E, Katzenellenbogen BS, Enmark E, Gustafsson JA, Nilsson S, Kushner PJ. The estrogen receptor enhances AP-1 activity by two distinct mechanisms with different requirements for receptor transactivation functions. Mol Endocrinol. 1999;13:1672–85.
31. Liu MM, Albanese C, Anderson CM, Hilty K, Webb P, Uht RM, Price Jr RH, Pestell RG, Kushner PJ. Opposing action of estrogen receptors alpha and beta on cyclin D1 gene expression. J Biol Chem. 2002;277:24353–60.
32. Moll F, Katsaros D, Lazennec G, Hellio N, Roger P, Giacalone PL, Chalbos D, Maudelonde T, Rochefort H, Pujol P. Estrogen induction and overexpression of fibulin-1C mRNA in ovarian cancer cells. Oncogene. 2002;21:1097–107.

33. Cowley SM, Hoare S, Mosselman S, Parker MG. Estrogen receptors alpha and beta form heterodimers on DNA. J Biol Chem. 1997;272:19858–62.
34. Pettersson K, Delaunay F, Gustafsson JA. Estrogen receptor beta acts as a dominant regulator of estrogen signaling. Oncogene. 2000;19:4970–8.
35. Matthews J, Wihlén B, Tujague M, Wan J, Ström A, Gustafsson JA. Estrogen receptor (ER) beta modulates ERalpha-mediated transcriptional activation by altering the recruitment of c-Fos and c-Jun to estrogen-responsive promoters. Mol Endocrinol. 2006;20:534–43.
36. Williams C, Edvardsson K, Lewandowski SA, Ström A, Gustafsson JA. A genome-wide study of the repressive effects of estrogen receptor beta on estrogen receptor alpha signaling in breast cancer cells. Oncogene. 2008;27:1019–32.
37. Paruthiyil S, Parmar H, Kerekatte V, Cunha GR, Firestone GL, Leitman DC. Estrogen receptor beta inhibits human breast cancer cell proliferation and tumor formation by causing a G2 cell cycle arrest. Cancer Res. 2004;64:423–8.
38. Lin CY, Ström A, Li Kong S, Kietz S, Thomsen JS, Tee JB, Vega VB, Miller LD, Smeds J, Bergh J, Gustafsson JA, Liu ET. Inhibitory effects of estrogen receptor beta on specific hormone-responsive gene expression and association with disease outcome in primary breast cancer. Breast Cancer Res. 2007;9:R25.
39. Chen JQ, Russo PA, Cooke C, Russo IH, Russo J. ERbeta shifts from mitochondria to nucleus during estrogen-induced neoplastic transformation of human breast epithelial cells and is involved in estrogen-induced synthesis of mitochondrial respiratory chain proteins. Biochim Biophys Acta. 2007;1773:1732–46.
40. Knoblauch R, Garabedian MJ. Role for Hsp90-associated cochaperone p23 in estrogen receptor signal transduction. Mol Cell Biol. 1999;19:3748–59.
41. Slentz-Kesler K, Moore JT, Lombard M, Zhang J, Hollingsworth R, Weiner MP. Identification of the human Mnk2 gene (MKNK2) through protein interaction with estrogen receptor beta. Genomics. 2000;69:63–71.
42. Abba MC, Hu Y, Sun H, Drake JA, Gaddis S, Baggerly K, Sahin A, Aldaz CM. Gene expression signature of estrogen receptor alpha status in breast cancer. BMC Genomics. 2005;6:37.
43. Weisz A, Basile W, Scafoglio C, Altucci L, Bresciani F, Facchiano A, Sismondi P, Cicatiello L, De Bortoli M. Molecular identification of ERalpha-positive breast cancer cells by the expression profile of an intrinsic set of estrogen regulated genes. J Cell Physiol. 2004;200:440–50.
44. Hah N, Danko CG, Core L, Waterfall JJ, Siepel A, Lis JT, Kraus WL. A rapid, extensive, and transient transcriptional response to estrogen signaling in breast cancer cells. Cell. 2011;145:622–34.
45. Zhao J, Zhu K, Lubman DM, Miller FR, Shekhar MP, Gerard B, Barder TJ. Proteomic analysis of estrogen response of premalignant human breast cells using a 2-D liquid separation/mass mapping technique. Proteomics. 2006;6:3847–61.
46. Malorni L, Cacace G, Cuccurullo M, Pocsfalvi G, Chambery A, Farina A, Di Maro A, Parente A, Malorni A. Proteomic analysis of MCF-7 breast cancer cell line exposed to mitogenic concentration of 17beta-estradiol. Proteomics. 2006;6:5973–82.
47. Cicatiello L, Mutarelli M, Grober OM, Paris O, Ferraro L, Ravo M, Tarallo R, Luo S, Schroth GP, Seifert M, Zinser C, Chiusano ML, Traini A, De Bortoli M, Weisz A. Estrogen receptor alpha controls a gene network in luminal-like breast cancer cells comprising multiple transcription factors and microRNAs. Am J Pathol. 2010;176:2113–30.
48. Bhat-Nakshatri P, Wang G, Collins NR, Thomson MJ, Geistlinger TR, Carroll JS, Brown M, Hammond S, Srour EF, Liu Y, Nakshatri H. Estradiol-regulated microRNAs control estradiol response in breast cancer cells. Nucleic Acids Res. 2009;37:4850–61.
49. Ferraro L, Ravo M, Nassa G, Tarallo R, De Filippo MR, Giurato G, Cirillo F, Stellato C, Silvestro S, Cantarella C, Rizzo F, Cimino D, Friard O, Biglia N, De Bortoli M, Cicatiello L, Nola E, Weisz A. Effects of oestrogen on microRNA expression in hormone-responsive breast cancer cells. Horm Cancer. 2012;3(3):65–78.

50. Lupien M, Eeckhoute J, Meyer CA, Krum SA, Rhodes DR, Liu XS, Brown M. Coactivator function defines the active estrogen receptor alpha cistrome. Mol Cell Biol. 2009;29: 3413–23.
51. Watson CS, Alyea RA, Jeng YJ, Kochukov MY. Nongenomic actions of low concentration estrogens and xenoestrogens on multiple tissues. Mol Cell Endocrinol. 2007;274:1–7.
52. Prossnitz ER, Maggiolini M. Mechanisms of estrogen signaling and gene expression via GPR30. Mol Cell Endocrinol. 2009;308:32–8.
53. Hu Q, Kwon YS, Nunez E, Cardamone MD, Hutt KR, Ohgi KA, Garcia-Bassets I, Rose DW, Glass CK, Rosenfeld MG, Fu XD. Enhancing nuclear receptor-induced transcription requires nuclear motor and LSD1-dependent gene networking in interchromatin granules. Proc Natl Acad Sci USA. 2008;105:19199–204.
54. Kuiper GG, Carlsson B, Grandien K, Enmark E, Häggblad J, Nilsson S, Gustafsson JA. Comparison of the ligand binding specificity and transcript tissue distribution of estrogen receptors alpha and beta. Endocrinology. 1997;138:863–70.
55. Spiegelman BM, Heinrich R. Biological control through regulated transcriptional coactivators. Cell. 2004;119:157–67.
56. Métivier R, Penot G, Carmouche RP, Hübner MR, Reid G, Denger S, Manu D, Brand H, Kos M, Benes V, Gannon F. Transcriptional complexes engaged by apo-estrogen receptor-alpha isoforms have divergent outcomes. EMBO J. 2004;23:3653–66.
57. O'Malley BW. Coregulators: from whence came these "master genes". Mol Endocrinol. 2007;21:1009–13.
58. McKenna NJ, Cooney AJ, DeMayo FJ, Downes M, Glass CK, Lanz RB, Lazar MA, Mangelsdorf DJ, Moore DD, Qin J, Steffen DL, Tsai MJ, Tsai SY, Yu R, Margolis RN, Evans RM, O'Malley BW. Evolution of NURSA, the Nuclear Receptor Signaling Atlas. Mol Endocrinol. 2009;23:740–6.
59. Voss TC, Demarco IA, Booker CF, Day RN. Corepressor subnuclear organization is regulated by estrogen receptor via a mechanism that requires the DNA-binding domain. Mol Cell Endocrinol. 2005;231:33–47.
60. Lahusen T, Henke RT, Kagan BL, Wellstein A, Riegel AT. The role and regulation of the nuclear receptor co-activator AIB1 in breast cancer. Breast Cancer Res Treat. 2009;116:225–37.
61. Spears M, Bartlett J. The potential role of estrogen receptors and the SRC family as targets for the treatment of breast cancer. Expert Opin Ther Targets. 2009;13:665–74.
62. Ellmann S, Sticht H, Thiel F, Beckmann MW, Strick R, Strissel PL. Estrogen and progesterone receptors: from molecular structures to clinical targets. Cell Mol Life Sci. 2009;66:2405–26.
63. Ma CX, Sanchez CG, Ellis MJ. Predicting endocrine therapy responsiveness in breast cancer. Oncology. 2009;23:133–42.
64. Wu YL, Yang X, Ren Z, McDonnell DP, Norris JD, Willson TM, Greene GL. Structural basis for an unexpected mode of SERM-mediated ER antagonism. Mol Cell. 2005;18:413–24.
65. Cimino D, Fuso L, Sfiligoi C, Biglia N, Ponzone R, Maggiorotto F, Russo G, Cicatiello L, Weisz A, Taverna D, Sismondi P, De Bortoli M. Identification of new genes associated with breast cancer progression by gene expression analysis of predefined sets of neoplastic tissues. Int J Cancer. 2008;123:1327–38.
66. Mutarelli M, Cicatiello L, Ferraro L, Grober OM, Ravo M, Facchiano AM, Angelini C, Weisz A. Time-course analysis of genome-wide gene expression data from hormone-responsive human breast cancer cells. BMC Bioinformatics. 2008;9 Suppl 2:S12.
67. Scafoglio C, Ambrosino C, Cicatiello L, Altucci L, Ardovino M, Bontempo P, Medici N, Molinari AM, Nebbioso A, Facchiano A, Calogero RA, Elkon R, Menini N, Ponzone R, Biglia N, Sismondi P, De Bortoli M, Weisz A. Comparative gene expression profiling reveals partially overlapping but distinct genomic actions of different antiestrogens in human breast cancer cells. J Cell Biochem. 2006;98:1163–84.
68. Ou K, Kesuma D, Ganesan K, Yu K, Soon SY, Lee SY, Goh XP, Hooi M, Chen W, Jikuya H, Ichikawa T, Kuyama H, Matsuo E, Nishimura O, Tan P. Quantitative profiling of drug-

associated proteomic alterations by combined 2-nitrobenzenesulfenyl chloride (NBS) isotope labeling and 2DE/MS identification. J Proteome Res. 2006;5:2194–206.
69. Ou K, Yu K, Kesuma D, Hooi M, Huang N, Chen W, Lee SY, Goh XP, Tan LK, Liu J, Soon SY, Bin Abdul Rashid S, Putti TC, Jikuya H, Ichikawa T, Nishimura O, Salto-Tellez M, Tan P. Novel breast cancer biomarkers identified by integrative proteomic and gene expression mapping. J Proteome Res. 2008;7:1518–28.
70. Fields S, Sternglanz R. The two-hybrid system: an assay for protein-protein interactions. Trends Genet. 1994;8:286–92.
71. Legrain P, Wojcik J, Gauthier JM. Protein–protein interaction maps: a lead towards cellular functions. Trends Genet. 2001;17:346–52.
72. Drewes G, Bouwmeester T. Global approaches to protein-protein interactions. Curr Opin Cell Biol. 2003;15:199–205.
73. Puig O, Caspary F, Rigaut G, Rutz B, Bouveret E, Bragado-Nilsson E, Wilm M, Séraphin B. The tandem affinity purification (TAP) method: a general procedure of protein complex purification. Methods. 2001;3:218–29.
74. Cox DM, Du M, Guo X, Siu KW, McDermott JC. Tandem affinity purification of protein complexes from mammalian cells. Biotechniques. 2002;2:267–70.
75. Bürckstümmer T, Bennett KL, Preradovic A, Schütze G, Hantschel O, Superti-Furga G, Bauch A. An efficient tandem affinity purification procedure for interaction proteomics in mammalian cells. Nat Methods. 2006;12:1013–9.
76. Ewing RM, Chu P, Elisma F, Li H, Taylor P, Climie S, McBroom-Cerajewski L, Robinson MD, O'Connor L, Li M, Taylor R, Dharsee M, Ho Y, Heilbut A, Moore L, Zhang S, Ornatsky O, Bukhman YV, Ethier M, Sheng Y, Vasilescu J, Abu-Farha M, Lambert JP, Duewel HS, Stewart II, Kuehl B, Hogue K, Colwill K, Gladwish K, Muskat B, Kinach R, Adams SL, Moran MF, Morin GB, Topaloglou T, Figeys D. Large-scale mapping of human protein-protein interactions by mass spectrometry. Mol Syst Biol. 2007;3:89.
77. Rigaut G, Shevchenko A, Rutz B, Wilm M, Mann M, Séraphin B. A generic protein purification method for protein complex characterization and proteome exploration. Nat Biotechnol. 1999;17:1030–2.
78. Li Y. The tandem affinity purification technology: an overview. Biotechnol Lett. 2011;8:1487–99.
79. Collins MO, Choudhary JS. Mapping multiprotein complexes by affinity purification and mass spectrometry. Curr Opin Biotechnol. 2008;4:324–430.
80. Xu X, Song Y, Li Y, Chang J, Zhang H, An L. The tandem affinity purification method: an efficient system for protein complex purification and protein interaction identification. Protein Expr Purif. 2010;2:149–56.
81. Li Y. Commonly used tag combinations for tandem affinity purification. Biotechnol Appl Biochem. 2010;55:73–83.
82. Ambrosino C, Tarallo R, Bamundo A, Cuomo D, Franci G, Nassa G, Paris O, Ravo M, Giovane A, Zambrano N, Lepikhova T, Jänne OA, Baumann M, Nyman TA, Cicatiello L, Weisz A. Identification of a hormone-regulated dynamic nuclear actin network associated with estrogen receptor alpha in human breast cancer cell nuclei. Mol Cell Proteomics. 2010;9:1352–67.
83. Tarallo R, Bamundo A, Nassa G, Nola E, Paris O, Ambrosino C, Facchiano A, Baumann M, Nyman TA, Weisz A. Identification of proteins associated with ligand-activated estrogen receptor α in human breast cancer cell nuclei by tandem affinity purification and nano LC-MS/MS. Proteomics. 2011;11:172–9.
84. Nassa G, Tarallo R, Ambrosino C, Bamundo A, Ferraro L, Paris O, Ravo M, Guzzi PH, Cannataro M, Baumann M, Nyman TA, Nola E, Weisz A. A large set of estrogen receptor β-interacting proteins identified by tandem affinity purification in hormone-responsive human breast cancer cell nuclei. Proteomics. 2011;11:159–65.
85. Cheng PC, Chang HK, Chen SH. Quantitative nanoproteomics for protein complexes (QNanoPX) related to estrogen transcriptional action. Mol Cell Proteomics. 2010;9:209–24.

86. Métivier R, Penot G, Hübner MR, Reid G, Brand H, Kos M, Gannon F. Estrogen receptor-alpha directs ordered, cyclical, and combinatorial recruitment of cofactors on a natural target promoter. Cell. 2003;115:751–63.
87. Cardamone MD, Bardella C, Gutierrez A, Di Croce L, Rosenfeld MG, Di Renzo MF, De Bortoli M. ERalpha as ligand-independent activator of CDH-1 regulates determination and maintenance of epithelial morphology in breast cancer cells. Proc Natl Acad Sci USA. 2009;18:7420–5.
88. Nese N, Kandiloglu AR, Simsek G, Lekili M, Ozdamar A, Catalkaya A, Coskun T. Comparison of the desmoplastic reaction and invading ability in invasive ductal carcinoma of the breast and prostatic adenocarcinoma based on the expression of heat shock protein 47 and fascin. Anal Quant Cytol Histol. 2010;32:90–101.
89. Foryst-Ludwig A, Kintscher U. Metabolic impact of estrogen signalling through ERalpha and ERbeta. J Steroid Biochem Mol Biol. 2010;1–3:74–81.
90. Huet G, Mérot Y, Percevault F, Tiffoche C, Arnal JF, Boujrad N, Pakdel F, Métivier R, Flouriot G. Repression of the estrogen receptor-alpha transcriptional activity by the Rho/megakaryoblastic leukemia 1 signaling pathway. J Biol Chem. 2009;49:33729–39.
91. Zheng B, Han M, Bernier M, Wen JK. Nuclear actin and actin-binding proteins in the regulation of transcription and gene expression. FEBS J. 2009;276:2669–85.
92. Jeong KW, Lee YH, Stallcup MR. Recruitment of the SWI/SNF chromatin remodeling complex to steroid hormone-regulated promoters by nuclear receptor coactivator flightless-I. J Biol Chem. 2009;43:29298–309.
93. Percipalle P. The long journey of actin and actin-associated proteins from genes to polysomes. Cell Mol Life Sci. 2009;66:2151–65.
94. Nassa G, Tarallo R, Guzzi PH, Ferraro L, Cirillo F, Ravo M, Nola E, Baumann M, Nyman TA, Cannataro M, Ambrosino C, Weisz A. Comparative analysis of nuclear estrogen receptor alpha and beta interactomes in breast cancer cells. Mol Biosyst. 2011;7:667–76.
95. Chen L, Qiu J, Yang C, Yang X, Chen X, Jiang J, Luo X. Identification of a novel estrogen receptor beta1 binding partner, inhibitor of differentiation-1, and role of ERbeta1 in human breast cancer cells. Cancer Lett. 2009;2:210–9.
96. Thenot S, Bonnet S, Boulahtouf A, Margeat E, Royer CA, Borgna JL, Cavaillès V. Effect of ligand and DNA binding on the interaction between human transcription intermediary factor 1alpha and estrogen receptors. Mol Endocrinol. 1999;12:2137–50.
97. Warnmark A, Almlof T, Leers J, Gustafsson JA, Treuter E. Differential recruitment of the mammalian mediator subunit TRAP220 by estrogen receptors ERalpha and ERbeta. J Biol Chem. 2001;26:23397–404.
98. Li J, Yu L, Zhang H, Wu J, Yuan J, Li X, Li M. Down-regulation of pescadillo inhibits proliferation and tumorigenicity of breast cancer cells. Cancer Sci. 2009;100:2255–60.
99. Mannini L, Menga S, Musio A. The expanding universe of cohesin functions: a new genome stability caretaker involved in human disease and cancer. Hum Mutat. 2010;31:623–30.
100. Kim HR, Chae HJ, Thomas M, Miyazaki T, Monosov A, Monosov E, Krajewska M, Krajewski S, Reed JC. Mammalian dap3 is an essential gene required for mitochondrial homeostasis in vivo and contributing to the extrinsic pathway for 1178 apoptosis. FASEB J. 2007;21(1):188–96.
101. Kasof GM, Goyal L, White E. Btf, a novel death-promoting transcriptional 1180 repressor that interacts with Bcl-2-related proteins. Mol Cell Biol. 1999;19(6):4390–404.
102. Chakravarty D, Nair SS, Santhamma B, Nair BC, Wang L, Bandyopadhyay A, Agyin JK, Brann D, Sun LZ, Yeh IT, Lee FY, Tekmal RR, Kumar R, Vadlamudi RK. Extranuclear functions of ER impact invasive migration and metastasis by breast cancer cells. Cancer Res. 2010;70:4092–101.
103. Wang D, Huang J, Hu Z. RNA helicase DDX5 regulates microRNA expression and contributes to cytoskeletal reorganization in basal breast cancer cells. Mol Cell Proteomics. 2012;11(2):M111.011932.
104. Paris O, Ferraro L, Grober OM, Ravo M, De Filippo MR, Giurato G, Nassa G, Tarallo R, Cantarella C, Rizzo F, Di Benedetto A, Mottolese M, Benes V, Ambrosino C, Nola E, Weisz A. Direct regulation of microRNA biogenesis and expression by estrogen receptor beta in hormone-responsive breast cancer. Oncogene. 2012;31(38):4196–206.

105. Ward HW. Anti-oestrogen therapy for breast cancer: a trial of tamoxifen at two dose levels. Br Med J. 1973;1:13–4.
106. Fisher B, Costantino J, Redmond C, Poisson R, Bowman D, Couture J, Dimitrov NV, Wolmark N, Wickerham DL, Fisher ER, et al. A randomized clinical trial evaluating tamoxifen in the treatment of patients with node-negative breast cancer who have estrogen-receptor-positive tumors. N Engl J Med. 1989;320:479–84.
107. Cole MP, Jones CT, Todd ID. A new anti-oestrogenic agent in late breast cancer. An early clinical appraisal of ICI46474. Br J Cancer. 1971;25:270–5.
108. Early Breast Cancer Trialists' Collaborative Group. Tamoxifen for early breast cancer. Cochrane Database Syst Rev. 2001; CD000486.
109. Fisher B, Costantino JP, Wickerham DL, Redmond CK, Kavanah M, Cronin WM, Vogel V, Robidoux A, Dimitrov N, Atkins J, Daly M, Wieand S, Tan-Chiu E, Ford L, Wolmark N. Tamoxifen for prevention of breast cancer: report of the National Surgical Adjuvant Breast and Bowel Project P-1 Study. J Natl Cancer Inst. 1998;90:1371–88.
110. Jordan VC. Molecular mechanisms of antiestrogen action in breast cancer. Breast Cancer Res Treat. 1994;31:41–52.
111. Fisher B, Costantino JP, Redmond CK, Fisher ER, Wickerham DL, Cronin WM. Endometrial cancer in tamoxifen-treated breast cancer patients: findings from the National Surgical Adjuvant Breast and Bowel Project (NSABP) B-14. J Natl Cancer Inst. 1994;86:527–37.
112. Johnston SR. Systemic treatment of metastatic breast cancer. Hosp Med. 2001;62:289–95.
113. Jordan VC, Phelps E, Lindgren JU. Effects of anti-estrogens on bone in castrated and intact female rats. Breast Cancer Res Treat. 1987;10:31–5.
114. Draper MW, Flowers DE, Huster WJ, Neild JA, Harper KD, Arnaud C. A controlled trial of raloxifene (LY139481) HCl: impact on bone turnover and serum lipid profile in healthy postmenopausal women. J Bone Miner Res. 1996;11:835–42.
115. Howell A. New endocrine treatment strategies. Eur J Cancer Care. 1996;5(3 Suppl):2.
116. Carlson RW. The history and mechanism of action of fulvestrant. Clin Breast Cancer. 2005;6 Suppl 1:S5–8.
117. Gradishar W. Fulvestrant in the treatment of postmenopausal women with advanced breast cancer. Expert Rev Anticancer Ther. 2005;5:445–53.
118. Allegretto EA. Ligand-induced conformational changes in estrogen receptors-α and -β. In: Manni A, Verderame M, editors. Selective estrogen receptor modulators. Totowa, NJ: Humana Press; 2002.
119. Nalvarte I, Schwend T, Gustafsson JA. Proteomics analysis of the estrogen receptor alpha receptosome. Mol Cell Proteomics. 2010;7:1411–22.
120. Cirillo F, Nassa G, Tarallo R, Stellato C, De Filippo MR, Ambrosino C, Baumann M, Nyman TA, Weisz A. Molecular mechanisms of selective estrogen receptor modulator activity in human breast cancer cells: identification of novel nuclear cofactors of antiestrogen-ERα complexes by interaction proteomics. J Proteome Res. 2013;1:421–31.

Chapter 9
Reprogramming Breast Cancer Cells with Embryonic Microenvironments: Insights from Nodal Signaling

Gina Kirsammer and Mary J.C. Hendrix

Abstract Normal developmental pathways that determine cell fate, migration, and proliferative potential become reactivated in cancer to promote the most devastating aspect of the disease, namely, metastasis to new sites. However, unlike their function in the delimited process of normal growth and differentiation, developmental pathways in the context of metastatic cancer support an aberrant and unlimited morphogenic program. Understanding how specific morphogenic pathways function in normal development and how they become deregulated in cancer may provide insight into new therapeutic opportunities to limit cancer spread. Adding complexity, however, such developmental pathways do not function solely through linear, cell autonomous programs but rather as dynamic, iterative processes between cells and their microenvironment. Therefore, comprehensive strategies to treat cancer and limit recurrence and metastasis must consider the ever-changing, reciprocal developmental relationship of cancer cells with their microenvironment. One important developmental pathway that shapes the interdependent evolution of breast cancer cells and their microenvironment is signaling by the embryonic morphogen Nodal, a member of the TGF-β family and a promising, new therapeutic target. Herein we review the significance of bidirectional signaling with the microenvironment in tumor progression and the distorted recapitulation of normal developmental programs that promote tumor aggression. Further, this chapter examines the reemergence of Nodal signaling during breast cancer growth and, finally, the therapeutic potential of targeting cancer cell–microenvironment interactions in general, and particularly Nodal signaling, to reprogram these relationships and promote a more benign developmental course in malignant breast cancer.

G. Kirsammer • M.J.C. Hendrix (✉)
Children's Hospital of Chicago Research Center, Chicago Illinois and Northwestern University, Chicago, IL, USA
e-mail: g-kirsammer@northwestern.edu; m-hendrix@northwestern.edu

H. Schatten (ed.), *Cell and Molecular Biology of Breast Cancer*,
DOI 10.1007/978-1-62703-634-4_9,

Researchers have long noted that certain aspects of cancer resemble a distorted recapitulation of embryogenesis. Virchow famously first proposed that cancer might arise from persistent embryonic-like cells in the body. He presciently went on to describe the interconnectedness of biological processes and the importance of context in shaping cellular outcomes:

> "*Every animal presents itself as the sum of vital unities*, every one of which manifests all the characteristics of life.... Hence it follows that the structural composition of a body of considerable size, a so-called individual, always represents a kind of social arrangement of parts, an arrangement of a social kind, in which a number of individual existences are mutually dependent, but in such a way, that every element has its own special action, and, even though it derives its stimulus to activity from other parts, yet alone effects the actual performance of its duties." [1]

The "vital unities" he refers to are cells, and the "social arrangement" of cells foreshadows our current concept of the microenvironment. It was unclear in the 1850s, however, what factors might upset this social arrangement and allow cancer to develop.

Since that time, specific genetic mutations have been shown to perturb cellular homeostasis and cause cancer. Initially these discoveries led to the hope that targeting mutant proteins therapeutically could eradicate the disease and cure cancer patients; however, in most cases, and for breast cancer patients in particular, cancer cells become resistant to therapy and tumors recur, often with more aggressive characteristics. The ability of cancer cells to evolve and escape targeted treatments has refocused attention on the broader developmental process of cancer and the microenvironmental changes that might cooperate with genetic mutations to facilitate tumor progression and recurrence. Recent studies and advances in stem cell biology have revealed the importance of microenvironmental and epigenetic factors in shaping cancer development and suggest a need to address the whole developmental process of cancer in order to provide better outcomes for patients [2].

Similar to a normal developmental process, breast cancer growth may be hierarchically organized, with a small number of stem-like cells able to self-renew and fuel the growth of a large, complex, heterogeneous population of bulk tumor cells [3]. This hierarchy of tumor cells depends upon an equally complex and heterogeneous microenvironment for signals to maintain potency and viability and to differentiate and to respond to changes in the tumor context [4]. Far from a model where genetic mutations can be considered the solitary force of malignant transformation, current studies suggest that a reciprocal developmental relationship grows between cancer cells and their environment, each driving changes in the other and culminating in tumor progression as a developmental process but, importantly, one that seems to lack a regulated developmental endpoint.

The evolution of the tumor–microenvironment relationship ultimately rests on the remarkable plasticity of both cell types in adapting to changing stimuli. This extreme plasticity underlies the notorious ability of cancer cells to metastasize and evade treatment but may also hint at therapeutic opportunities to harness tumor cell developmental plasticity and "reprogram" tumor cells to a more limited, benign phenotype. Intriguingly, and reminiscent of Virchow's prediction, the plastic, aggressive phenotype of cancer may derive, in part, from the reemergence of embryonic signaling pathways in tumor cells. Moreover, studies suggest that the embryonic microenvironment, which restrains and directs the potency of embryonic stem (ES) cells, may hold clues to limiting or redirecting plasticity in cancer cells as well.

9.1 Breast Tumor Heterogeneity and the Microenvironment

Developing breast tumors are extraordinarily complex, harboring multiple evolving clones and genotypes and thereby complicating the development of targeted therapies. Additionally, the cells within a given clone are apparently not equal in their ability to instigate and maintain tumor growth. The cancer stem cell hypothesis suggests that a small population of the so-called cancer stem cells (CSCs) or tumor-initiating cells may harbor these functions and might also be responsible for both tumor metastasis and recurrence after treatment [5]. Whether or not CSCs can be considered true stem cells is currently debated, but regardless of how they may be named, targeting the malignant cells that are able to persist and repopulate cancerous tumors after therapy is obviously an important priority for patients. Unfortunately, identifying these cells has proven challenging. Adding complexity are reports that CSCs may not be phenotypically stable in some tumor types and that CSC behavior appears to be influenced by microenvironmental context, including methods used to isolate and culture them [6]. Therefore, a study of CSCs and their development requires improved understanding of the environmental factors that affect their phenotype and function.

Perhaps as complex as the tumor itself, the developing microenvironment of breast tumors supports tumor growth with a variety of essential functions and cell types. Importantly, the microenvironment provides a niche for maintaining breast cancer stem cells (BCSCs) and directing their functions. As BCSCs are thought to be responsible for the most devastating aspects of cancer progression, how these cells take instructions from their environment to self-renew, divide, adapt, and metastasize is under intense investigation. What is clear, however, is that bidirectional codevelopment of cancer cells and their microenvironment is as complex as any other developmental process, requiring sensitivity to environmental cues as well as developmental plasticity to respond to such stimuli and activate downstream phenotypic changes.

9.2 Breast Cancer Stem Cells

Following advances in stem cell biology and the isolation of stem cells in the hematopoietic system, Bonnet and Dick first demonstrated that cancer growth, like hematopoiesis, might occur as a stem cell-driven hierarchy of cell phenotypes [7]. In these studies, human leukemia cells were sorted into pools by the expression of the CD34 and CD38 cell surface antigens and transplanted into immunocompromised mice. It was shown that a fraction of human acute myeloid leukemia cells, defined by their CD34+/CD38− cell surface phenotype, harbor increased stem cell-like behavior and concentrated ability to establish and maintain leukemia in mice. Leukemia cells in the other pools were greatly diminished in their tumorigenicity, suggesting that all cancer cells may not be equal in aggressive potential and that tumor "stem cells" might be a high-priority therapeutic target.

In the subsequent search for solid tumor stem cells, Al-Hajj and colleagues demonstrated that human breast cancer may also be propagated through a small proportion of tumor cells, which are uniquely enriched in tumor-forming ability and defined by the $CD24^{low}/CD44^{high}$ cell surface profile [5]. Other studies, using markers such as EpCAM and ALDH1, have also demonstrated enrichment for tumor-forming ability in a subset of cancer cells [8, 9]. Importantly, while BCSCs are identified by their cell surface profile, the BCSC phenotype is sensitive to growth conditions and must, therefore, be functionally defined by the ability to form specific colonies in culture and to generate tumors in immunocompromised mice, either by mammary fat pad xenograft or intravenous injection.

Indeed, the microenvironment has a strong effect on CSC phenotype, but how this phenotype arises and in what original cell type are unresolved questions. The hypothesis that CSCs derive from normal tissue stem cells is attractive for several reasons. First, stem cells already harbor extensive self-renewal capacity and the ability to produce diverse progeny, two definitive characteristics of tumor stem cells. Tissue stem cells are the most plastic cells in a tissue, with the lowest level of epigenetic programming and, arguably, the lowest threshold to activation by internal events or extracellular stimuli. Additionally, normal stem cells occupy the stem cell niche, where they receive environmental cues for stem cell function and potency. A cell undergoing transformation at this site would be well poised to maintain a stem cell phenotype. Alternatively, it remains possible that more differentiated cells may acquire stem-like character through genetic or epigenetic changes to give rise to cancer. In either case, identification of the cell of origin in breast cancer has become more achievable with a better understanding of the features of mammary stem cells and how they develop stem cell features.

Subsequent to the identification of BCSCs, normal human mammary stem cells have successfully been enriched based on the $CD49+/Epcam^{lo}$ fraction and are defined by their ability to form nonadherent mammospheres in culture and mammary ductal structures in a cleared mouse mammary fat pat [10]. Identification of normal mammary stem cells and progenitors has allowed for breast cancer cell of origin studies in the mouse, which suggest that indeed mammary stem and progenitor cells likely sustain the changes that give rise to malignancy [11]. Furthermore, while transformed stem and progenitor cells may retain some of their original character, they demonstrate phenotypic plasticity and may not resemble their cell of origin. For example, several studies suggest that the mammary luminal progenitor population may give rise to both luminal- and basal-like breast carcinomas [12, 13].

Recent evidence suggests that transformed cells in the mammary gland may achieve phenotypic plasticity and gain stem cell character by accessing the programs used by fetal mammary stem cells. These embryonic counterparts to adult mammary stem cells remain bipotential for basal and luminal lineages until shortly after birth and harbor a transcriptional program that is distinct from that of adult mammary stem cells. After E15.5, fetal mammary stem cells have been shown to undergo dramatic expansion, concurrent with changes in the microenvironmental context, as the fetal mammary epithelium undergoes EMT and begins to invade the surrounding mesenchyme and ingress into the adjacent fat pad [14, 15]. Interestingly, the process of EMT itself has been shown to produce stem cell characteristics in mammary epithelium and breast cancer cell lines, and it has been suggested that the embryonic mammary stromal cues which produce EMT and invasion may simultaneously generate robust expansion of the fetal mammary stem cell compartment [15–17]. Determining the molecular mechanisms used to generate fetal mammary stem cells may provide insight as to whether these same pathways are accessed to generate BCSCs. Transcriptional profiling of fetal mammary stem cells and the surrounding stroma has revealed overlap with different subtypes of breast cancer which is distinct from the adult mammary stem cell profile, including coexpression of luminal and myoepithelial markers with vimentin, which is not typically observed in the adult mammary gland, but does accompany the transition to aggressive undifferentiated cancers [15]. Indeed, forced coexpression of keratins 8 and 18 with vimentin has been shown to cause increased proliferation and motility of breast cancer cell lines in vitro and lends support to the hypothesis that reactivation of the EMT program in breast cancer cells may resurrect embryonic pathways used to confer stem cell character and plasticity [18].

Interestingly, transcriptional profiles of breast cancer subtypes also reveal shared gene expression patterns with the fetal mammary stroma [15]. The dependence of stem cell behavior on contextual cues from the stroma is well demonstrated, and this finding suggests that breast cancer cells may evolve autocrine sources of signals that are typically delivered to the embryonic mammary epithelium through a stromal paracrine source. An important area of research, the cell of origin for breast carcinoma, and its development and phenotypic fate most likely depend on a constellation of genetic mutations and changing environmental influences. Increasing evidence suggests that reactivation of autocrine and paracrine embryonic signals may contribute to the development of plastic, proliferative stem cell character in breast cancer cells.

9.3 The Breast Cancer Microenvironment

The concept of a stem cell niche first emerged in the study of hematopoietic stem cells to describe a particular anatomical location that integrates diverse and dynamic extracellular stimuli to modulate stem cell function and tissue homeostasis [19]. Numerous studies in invertebrate and mammalian models have demonstrated that stem cells depend on interaction with the niche to retain proliferative potential and potency [20–24]. During development, the niche has been shown to both shape the development of stem cells and restrict inappropriate cell growth and behavior. The relationship between stem cells and their niche then is a delicate balance of permissive and restrictive cues supported by bidirectional communication and plasticity that evolve together during development. As it must respond to dynamic environmental signals, the niche itself retains a high level of plasticity. This plasticity can unfortunately be appropriated by cancer stem cells in order to silence restrictive cues and promote a pro-tumor context that appears to evolve alongside tumor cells, culminating in malignancy. Understanding factors that change the balance of signals in the CSC niche is essential to understanding the roots of tumorigenesis.

The acquisition and manifestations of plasticity by BCSCs are accompanied by the parallel development of the BCSC niche. Like its normal counterpart, the BCSC niche includes all the factors available to tumor cells to guide and support their development, including mesenchymal and immune cells, fibroblasts, ECM, blood supply, soluble and hormonal factors, and cell types recruited by the tumor itself. The cross talk between BCSCs and the niche is well demonstrated in activating diverse tumor processes at the local and systemic level. Tumor-niche communication is involved in creating reactive stroma, initiating invasion, recruiting bone marrow mesenchymal cells, and preparing distant sites for metastasis [25, 26]. Likewise, interaction with the microenvironment provides signals to breast cancer cells to support EMT and stem cell character and to initiate processes such as metastasis and vasculogenic mimicry, whereby tumor cells can form perfusive channels to nourish a growing tumor [4, 27]. In order to support breast cancer development, the normal regulatory function of the niche that supports homeostasis is subverted, and a new developmental program emerges, driven by communication with tumor stem cells. Indeed, a number of embryonic signaling pathways appear to be part of the cancer morphogenic process, including TGFβ family members, Shh, Notch, and Wnt pathways, which can induce EMT, maintain stem cell potency, initiate migration, and regulate growth, self-renewal, and differentiation processes in development and cancer [28].

The extent to which tumor–stroma interactions can be compared to normal development is, of course, limited. Cancer development is an abnormal process, complicated by genetic mutations and clonal selection. As mentioned above, breast cancer cells themselves have been shown to secrete factors normally provided by stroma, demonstrating that the activation of developmental pathways in cancer does not directly copy a developmental paradigm. However, embryonic developmental pathways do seem to support some processes in the context of cancer that are similar to those supported by the same pathways during embryogenesis, suggesting that improved knowledge of how these pathways function and how they can be regulated in embryogenesis and cancer may provide clues as to targeting developmental plasticity in breast cancer.

9.4 The Embryonic Microenvironment

Lewis Wolpert's famous quote "It is not birth, marriage, or death, but *gastrulation*, which is truly the most important time in your life" captures the significance of the establishment and differentiation of the embryonic microenvironment for subsequent development. During successful embryonic development, the infinite potential of ES cells is carefully winnowed and guided by interaction with the microenvironment to produce a specific outcome with remarkable reliability. If the interaction between embryonic cell domains is not properly established, development will be abnormal. Since the original observation that tumor development resembles an embryonic process, studies have shown that the interdependent development of tumor and stromal cells does indeed reproduce aspects of embryonic development, where plasticity and environmental context determine cell fate and eventual outcomes. In tumorigenesis, however, it appears that the process of clonal selection allows tumor cells to overcome the dominant influence of the normal microenvironment over cell fate, resulting in a chaotic and deleterious developmental program. Recent lines of evidence suggest that the embryonic microenvironment, capable of restraining embryonic signaling pathways and shaping highly plastic cells into differentiated states, may hold clues as to suppressing the reemergent embryonic pathways known to promote aggressive tumor behavior.

The relationship between embryonic and tumor developmental states has been investigated through a series of now classic experiments. First, unmodified ES cells transplanted into a blastocyst have been shown to give rise to normal adult tissues and germ cells in a mouse model, and yet the same cells transplanted subcutaneously can produce malignant teratocarcinomas [29, 30]. This experiment demonstrates the cancerous potential of normal ES cell properties in an alien context and the essential, dominant function of the environment in restraining this potential. Mintz and Ilmensee sought to test the implications of these conclusions for tumor biology in a set of experiments with embryonal carcinoma cells. After subcutaneous injection, this cancerous cell line formed robust tumors in the host; however, the same cells were able to generate normal, noncancerous tissues of a chimeric mouse when transferred into the blastocyst, demonstrating that cancerous cells with genetic alterations can adopt new phenotypes in response to embryonic environmental cues [31]. Bissell and colleagues have shown further that chicken embryonic cells infected with RSV, which expresses the potent oncogene pp60src, do not develop tumors, yet cells taken from these embryos and placed in culture immediately exhibited tumorigenic behavior [32]. Finally, Hochedlinger and colleagues were able to show that nuclei from malignant melanoma cells could be transplanted into oocytes and give rise to normal, albeit tumor-prone, tissues, demonstrating the interaction between genetic and epigenetic determinants at the cellular level [33]. Together these studies demonstrate the natural tumorigenic potential of embryogenesis and the importance of the embryonic microenvironment in restraining this potential, imposing cell fate and directing embryonic processes in a finite developmental course. For cancerous cells, while genetic mutations almost certainly instigate malignant transformation, the reprogramming of tumor cells by the embryonic environment suggests that the environment and cellular phenotype can modulate the effects of cancer-causing genotype in determining cell behavior. Further, the embryonic microenvironment, with its remarkable plasticity and instructional potency, may hold promise for developing cancer therapeutics. The molecular mechanisms underlying embryonic morphogenesis have been studied for some time; however, the dysfunction of embryonic-like signaling between tumor cells and their microenvironment is poorly understood.

In an effort to investigate mechanisms of bidirectional communication between cancer cells and the embryonic microenvironment, the embryonic zebrafish model has been used effectively as a biosensor for embryonically potent signaling by malignant melanoma cells. In these experiments, malignant melanoma cells were transplanted into developing zebrafish embryos at either the animal pole or the yolk margin and caused cranial outgrowth and axis duplication, respectively. Alternatively, if GFP-labeled melanoma cells were injected centrally in the embryo, they persisted in the animal until maturity, without giving rise to tumors or duplicate axes [34]. These experiments demonstrate that cancer cells are able to participate in specific embryonic processes in ways that change developmental outcomes and, moreover, that the phenotypes induced in receptive embryonic fields might provide insight into the morphogenic signaling pathways active in tumors.

The above experiments in zebrafish led to the discovery that melanoma cells secrete Nodal, an embryonic morphogen of the TGFβ family with demonstrated roles in plasticity and fate determination [34]. Expressed in early development, Nodal is essential to support embryonic stem cells in their undifferentiated state and to promote specific types of differentiation, including EMT, body axis specification, and left–right asymmetry (for review, see [35]). As a soluble factor released into the microenvironment, Nodal partners with a co-receptor, Cripto-1, and binds type I and II ALK family receptors to activate signaling. The active receptor complex phosphorylates signaling molecules Smad2 and Smad3, which are then able to bind Smad4, translocate to the nucleus and complex with transcription factors, such as FoxH1 and Mixer, to activate transcription of target genes. In normal development, transcriptional targets of Nodal include potent fate determinants such as Goosecoid and Nodal itself. Importantly, during embryogenesis, the activity of Nodal is limited temporally and spatially by expression of a Nodal inhibitor, termed Lefty. A soluble factor in the microenvironment, Lefty binds Nodal ligand, prevents its signaling through cell surface receptors, and thereby breaks the positive feedback cycle of Nodal signaling to attenuate Nodal expression. Cancer cells, however, appear to methylate the Lefty gene promoter, silencing its expression and leaving Nodal expression and the Nodal-driven morphogenic program unchecked. Reintroducing Lefty to the environment of cancer cells, either with ES cell-conditioned media or ES cell-derived Lefty, is sufficient to suppress both Nodal expression and tumorigenicity of these cells [36].

Subsequent to experiments in melanoma, Nodal signaling has been shown to support tumorigenicity in a number of other cancers, highlighting a potentially broad relevance for this signaling pathway in the development of prostate, pancreatic, and breast cancers [37–39].

9.5 Nodal Signaling in Mammary Development and Breast Cancer

Although Nodal had been thought to function uniquely in early embryogenesis, recent studies suggest that Nodal signaling may contribute to normal mammary development as well. In the developing mouse, Nodal is expressed in the growing mammary ducts along with Cripto-1. Disruption of Nodal signaling in the developing mammary gland leads to defects in ductal architecture and ductal branching and causes abnormal epithelial organization, suggesting that Nodal may exert its potent morphogenic effects on breast stem and progenitor cells during mammary development [40]. To further investigate the function of Nodal in mammary development, Kenney and colleagues implanted beads-secreting recombinant Nodal into the developing mouse mammary gland and demonstrated increased disorganized side branching of mammary ducts in the region of increased Nodal signaling. Moreover, ductal branching proceeded away from the source of recombinant Nodal, indicating that Nodal may provide positional and morphological cues to developing mammary tissue [40]. Similar to Nodal function in other contexts, Nodal signaling is attenuated as development proceeds, and normal mature human breast tissues do not express Nodal protein. As breast cancers develop, however, it appears that the Nodal signaling pathway is reactivated. Strizzi and colleagues have shown that human breast tumors display increased Nodal expression in correlation with tumor grade, and, similar to melanoma, these tumors do not express Lefty [39].

Intriguingly, exposure of malignant MDA-MB-231 breast cancer cells to an embryonic microenvironment is sufficient both to suppress Nodal expression and reduce malignant behavior. Alternatively, exposure of MDA-MB-231 cells to ES cell-derived Lefty suppresses Nodal expression in these cells along with the tumorigenic phenotype [36]. In order to investigate the dependency of breast carcinoma cells on Nodal signaling, short hairpin-mediated knockdown of Nodal expression in breast has shown a dramatic phenotype in multiple cell lines. In vitro, Nodal knockdown causes reduced growth, proliferation, invasion, and polyploidization culminating in increased apoptosis. Furthermore, in a mouse xenograft model, tumors derived from Nodal knockdown cells displayed significantly diminished growth and reduced tumor engraftment when compared with controls (our unpublished data). Ongoing studies will determine the mechanisms by which Nodal signaling can promote a tumorigenic program in breast cancer cells and whether Nodal participates in reactivating embryonic processes in breast cancer progression.

9.6 Reprogramming Breast Cancer Cells

The plasticity of BCSCs and their ability to adapt to changing stimuli are the notorious qualities that make the disease so devastating, and yet these qualities also represent a potential Achilles' heel of tumor aggression. First, as CSCs are receptive to certain environmental cues, providing appropriate cues may represent an opportunity to therapeutically reprogram breast cancer cells to a more benign phenotype. Second, while tumorigenesis begins with mutations in the DNA sequence of targeted cells, reversible epigenetic factors appear to play a large role in progression of the plastic, stem-like phenotype. Therapeutically targeting these epigenetic processes may deprive tumor cells of essential contextual cues and lead to a loss of plasticity and aggressive character. Furthermore, although it may one day be possible to "mend" DNA lesions in cancer cells, the epigenetic evolution of cancerous clones and their associated microenvironment may be far removed from the original genetic changes, representing an independent process that must be targeted in order to eradicate sources of disease recurrence. In recent years, better understanding of microenvironmental determinants and epigenetic programming have pushed these factors forward as promising new avenues to target the most devastating aspects of breast cancer.

Considered all heritable changes in DNA structure outside the DNA sequence itself, epigenetics is a broad term for many complex and interconnected processes that define how the genome can be used by a given cell. Epigenetic "programming" also determines how genetic factors will interact with the microenvironment to produce cell behavior or phenotype. While cancer cells and their environment can shape the evolution of each other's epigenetic state, it is the epigenetic state of the cell that determines how the cell will behave under various conditions. Therefore, only cells with an epigenetic stem or progenitor cell program can respond to cues to self-renew, proliferate, and differentiate into multiple lineages, while more differentiated cell types do not have access to the transcriptional programs necessary to effect plasticity processes. In breast cancer, epigenetics also help define the breast cancer stem cell phenotype, but the evolution of epigenetic state is most likely somewhat different than in normal development or homeostasis. Normal stem cells can be considered among the most "unprogrammed" cells, with the greatest plasticity and potential for future programming and also with the greatest need for tight regulation by the environment to guide these functions toward serving the needs of the whole individual. As discussed, cancer cells also display great plasticity; however, clonal selection among tumor and environmental cells promotes conditions that favor growth and favor the evolution of increasingly plastic epigenetic states. In cancerous cells, tumor suppressor genes, which restrain growth, may become silenced by both DNA and histone methylation, acetylation, ubiquitination, and other modifications, while oncogenes may retain unscheduled activity. DNA- and histone methyltransferases and other epigenetic modifiers themselves are often targeted for mutation and epigenetic misregulation, leading to global epigenetic changes, depending on tumor context [41–43]. Importantly, unlike genetic mutations, epigenetic changes are reversible and may therefore be promising targets for therapy.

Methods to epigenetically reprogram tumor cells as a therapeutic method can be thought of along the lines of two basic, overlapping approaches, both of which are tailored to tumor-type and molecular and epigenetic features (Fig. 9.1). The first can be thought of as "deprogramming," wherein drugs remove or prevent the accumulated epigenetic marks that promote malignancy. This process is similar to the global demethylation and deprogramming of oocytes upon fertilization that "resets" them for redevelopment. Clinically, multiple drugs designed to prevent hypoacetylation or hypermethylation have been developed and are employed in combination with traditional chemotherapy [42]. Emerging studies suggest that the tumor-targeting effects of epigenetic drugs are pleiotropic, but one important effect is the reactivation of tumor suppressor genes that have been inappropriately silenced. Reactivation of tumor suppressors leads to growth suppression and apoptosis and increased sensitivity to chemotherapy drugs. Furthermore, in experimental studies, epigenetic modifiers seem to target and diminish the BCSC population [44].

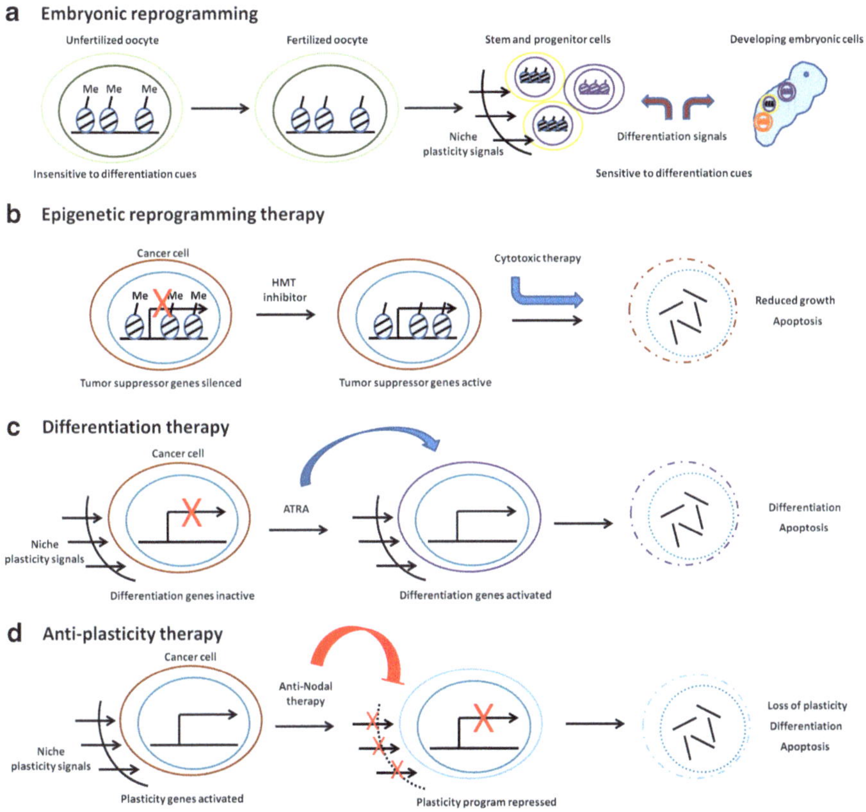

Fig. 9.1 Methods of nuclear reprogramming. (**a**) Upon fertilization, the oocyte genome undergoes rapid demethylation, effectively reprogramming the nucleus. The reprogrammed cell and its progeny can receive microenvironmental cues and participate in a developmental program. (**b**) Histone methyl transferase (HMT) inhibitors prevent gene silencing associated with histone methylations. Cancer therapy with HMT inhibitors leads to re-expression of tumor suppressor genes, which slow tumor growth and promote apoptosis, particularly in combination with cytotoxic drugs. (**c**) The aim of differentiation therapy is to provide differentiation cues to plastic tumor cells, thus pushing them toward a less plastic, less malignant fate, and making cells more sensitive to apoptotic stimuli. (**d**) Anti-plasticity therapy deprives cancer cells of the signals required to remain plastic, thereby allowing differentiation and enabling apoptosis

While these studies are promising and despite the success of epigenetic modifiers in treating leukemia patients, the same drugs have not been as effective in treating breast cancer patients.

It remains unclear why breast cancers are less responsive to epigenetic modifiers than hoped, despite promising in vitro experimentation, but several plausible explanations have been put forward. First, although such therapies appear to target CSCs, the model for the cancer developmental hierarchy may be shifting. Recent reports suggest that tumor cells may exhibit phenotypic plasticity, allowing non-CSCs in the tumor to assume a CSC fate in response to unknown cues and replenish the CSC population after treatment [45]. A related explanation is that epigenetic changes produced by therapy are reversible and that after "reprogramming," the remaining tumor cells may continue to receive instructive cues from the microenvironment to revert to an aggressive state. Both cases are compelling in the need for better understanding of the communication between tumor cells and their environment, the signaling pathways that are used, the origins and maintenance of cancer stem cells, and the methods that may be used to target epigenetic modifications in combination with microenvironmental factors and conventional antitumor therapy.

A second approach to epigenetic therapy, termed "differentiation therapy," is less well developed clinically and more specific to tumor type and characteristics. Differentiation therapy seeks to harness the plasticity of tumor cells to push them further down a developmental pathway, making them less plastic, less aggressive, and potentially more sensitive to cytotoxic drugs (Fig. 9.1). Theoretically, this approach includes both providing differentiation cues to the tumor cells themselves and modifying the tumor environment to limit stem cell potential and support differentiation. The first successful administration of differentiation therapy has been the treatment of acute promyelocytic leukemia (APL) patients harboring the PML-RARα translocation with all-trans retinoic acid (ATRA). Unlike the RARα protein, which activates transcription of a differentiation program, the PML-RARα fusion protein binds and represses RARα target genes, blocks differentiation, and causes leukemia. Administration of ATRA restores activation of these genes, promotes both granulocytic differentiation and apoptosis of leukemic blasts, and evokes remission in an otherwise uniformly fatal malignancy [46]. The example of APL illustrates both the potential success of differentiation therapy and also the difficulty in developing this type of therapy, as it must be specific for molecular mechanisms active and relevant in a specific tumor and its microenvironmental context. Exploration of the pathways that maintain pluripotency will hopefully aid in identifying new opportunities for differentiation therapy that may be active in multiple kinds of tumors. As embryonic pathways emerge as critical supports for cancer aggression, the molecules active in embryonic plasticity and differentiation may provide attractive targets with broad applicability among cancer types.

A number of embryonic signaling pathways that are reactivated in cancer have been investigated as potential therapeutic targets to limit plasticity, including Notch, Hedgehog, mTOR, and the TGFβ family, with some success and some limits to application imposed by side effects to normal stem cells. Although less well developed, Nodal is an intriguing therapeutic target, as it is expressed by both tumor cells and stroma of diverse cancer types, but minimally expressed in normal adult tissues (although the function of Nodal in select normal adult stem cells needs to be demonstrated) [47, 48]. Inhibiting Nodal may then limit the plasticity of tumor cells and stroma without affecting normal stem cells, thereby suppressing tumor growth and adaptation processes such as vasculogenic mimicry and metastasis, and thus curtailing the ability of the tumor stroma to induce and support CSC characteristics (Fig. 9.1).

While therapeutically suppressing Nodal signaling in vitro has been successful, accomplishing this task in vivo has been challenging. In a mouse xenograft model of melanoma, intraperitoneal delivery of anti-Nodal antibodies was able to reduce

Nodal signaling, slow tumor growth, and promote apoptosis [49]. In studies with pancreatic cancer, however, the Nodal/Activin inhibitor SB431542 was able to diminish populations of phenotypic CSCs in vitro, but not in vivo, apparently because the induced differentiation was reversible and treatment withdrawal led to a resurgence of the CSC phenotype. Interestingly, inclusion of Gemcitabine in the therapy regimen with SB431542 led to complete and irreversible elimination of CSCs in pancreatic cell line xenografts. This effect was not carried forward in xenografts of human pancreatic tumors, potentially because of the large amount of associated stroma, which is itself a source of Nodal signaling, and the inability of the drug to penetrate the tumor. Interestingly, several groups have shown that in the case of pancreatic cancer, inhibiting Shh signaling in the stroma may reduce Activin/Nodal signaling there and may have benefit in combined therapy aimed at reducing Nodal signaling [37, 50, 51].

Indeed, while embryonic signaling pathways are an integral part of breast tumor development, there are multiple other processes that contribute to disease progression, including genetic mutations and genomic instability, inflammation and immune factors, activation of the stroma, and global epigenetic deregulation. Successful treatment will likely involve a combination of therapeutics directed at these and other aspects of breast tumorigenesis.

9.7 Conclusions

Just as embryonic cells establish a relationship of mutual development with other cells in the embryo, recent evidence shows that cancer cells induce a mutual evolution of phenotypes with their environment which may represent a distorted recapitulation of certain embryonic processes. The functions of reactivated embryonic signaling pathways in breast cancer stem cells and their microenvironment are likely as complex, heterogeneous, and complicated by cross talk as the embryonic processes themselves. However, studies suggest that the investigation of these pathways and improved understanding of their effect on plasticity, proliferation, differentiation, and bidirectional communication with the microenvironment are beginning to define new therapeutic opportunities. Specifically, studies of the embryonic microenvironment, as a niche specialized to restrain plasticity and direct differentiation, may provide new directions in suppressing breast cancer plasticity and treating breast cancer patients.

References

1. Virchow R. Cellular pathology as based upon physiological and pathological histology. Philadelphia, PA: Lippincott; 1863.
2. Castano Z, et al. The bed and the bugs: interactions between the tumor microenvironment and cancer stem cells. Semin Cancer Biol. 2012;22(5–6):462–70.

3. Al-Hajj M, Clarke MF. Self-renewal and solid tumor stem cells. Oncogene. 2004;23(43):7274–82.
4. Bissell MJ, Hines WC. Why don't we get more cancer? A proposed role of the microenvironment in restraining cancer progression. Nat Med. 2011;17(3):320–9.
5. Al-Hajj M, et al. Prospective identification of tumorigenic breast cancer cells. Proc Natl Acad Sci USA. 2003;100(7):3983–8.
6. Quintana E, et al. Efficient tumour formation by single human melanoma cells. Nature. 2008;456(7222):593–8.
7. Bonnet D, Dick JE. Human acute myeloid leukemia is organized as a hierarchy that originates from a primitive hematopoietic cell. Nat Med. 1997;3(7):730–7.
8. Fillmore CM, Kuperwasser C. Human breast cancer cell lines contain stem-like cells that self-renew, give rise to phenotypically diverse progeny and survive chemotherapy. Breast Cancer Res. 2008;10(2):R25.
9. Park SY, et al. Heterogeneity for stem cell-related markers according to tumor subtype and histologic stage in breast cancer. Clin Cancer Res. 2010;16(3):876–87.
10. Stingl J, et al. Deciphering the mammary epithelial cell hierarchy. Cell Cycle. 2006;5(14):1519–22.
11. Proia TA, et al. Genetic predisposition directs breast cancer phenotype by dictating progenitor cell fate. Cell Stem Cell. 2011;8(2):149–63.
12. Molyneux G, et al. BRCA1 basal-like breast cancers originate from luminal epithelial progenitors and not from basal stem cells. Cell Stem Cell. 2010;7(3):403–17.
13. Lim E, et al. Aberrant luminal progenitors as the candidate target population for basal tumor development in BRCA1 mutation carriers. Nat Med. 2009;15(8):907–13.
14. Veltmaat JM, et al. Mouse embryonic mammogenesis as a model for the molecular regulation of pattern formation. Differentiation. 2003;71(1):1–17.
15. Spike BT, et al. A mammary stem cell population identified and characterized in late embryogenesis reveals similarities to human breast cancer. Cell Stem Cell. 2012;10(2):183–97.
16. Polyak K, Weinberg RA. Transitions between epithelial and mesenchymal states: acquisition of malignant and stem cell traits. Nat Rev Cancer. 2009;9(4):265–73.
17. Mani SA, et al. The epithelial-mesenchymal transition generates cells with properties of stem cells. Cell. 2008;133(4):704–15.
18. Hendrix MJ, et al. Experimental co-expression of vimentin and keratin intermediate filaments in human breast cancer cells results in phenotypic interconversion and increased invasive behavior. Am J Pathol. 1997;150(2):483–95.
19. Schofield R. The relationship between the spleen colony-forming cell and the haemopoietic stem cell. Blood Cells. 1978;4(1–2):7–25.
20. Xie T, Spradling AC. A niche maintaining germ line stem cells in the Drosophila ovary. Science. 2000;290(5490):328–30.
21. Kiger AA, White-Cooper H, Fuller MT. Somatic support cells restrict germline stem cell self-renewal and promote differentiation. Nature. 2000;407(6805):750–4.
22. Zhang J, et al. Identification of the haematopoietic stem cell niche and control of the niche size. Nature. 2003;425(6960):836–41.
23. Ohlstein B, Spradling A. The adult Drosophila posterior midgut is maintained by pluripotent stem cells. Nature. 2006;439(7075):470–4.
24. Micchelli CA, Perrimon N. Evidence that stem cells reside in the adult Drosophila midgut epithelium. Nature. 2006;439(7075):475–9.
25. McAllister SS, Weinberg RA. Tumor–host interactions: a far-reaching relationship. J Clin Oncol. 2010;28(26):4022–8.
26. Karnoub AE, et al. Mesenchymal stem cells within tumour stroma promote breast cancer metastasis. Nature. 2007;449(7162):557–63.
27. Hendrix MJ, et al. Molecular biology of breast cancer metastasis. Molecular expression of vascular markers by aggressive breast cancer cells. Breast Cancer Res. 2000;2(6):417–22.
28. Takebe N, Ivy SP. Controversies in cancer stem cells: targeting embryonic signaling pathways. Clin Cancer Res. 2010;16(12):3106–12.

29. Bradley A, et al. Formation of germ-line chimaeras from embryo-derived teratocarcinoma cell lines. Nature. 1984;309(5965):255–6.
30. Nagy A, et al. Derivation of completely cell culture-derived mice from early-passage embryonic stem cells. Proc Natl Acad Sci USA. 1993;90(18):8424–8.
31. Mintz B, Illmensee K. Normal genetically mosaic mice produced from malignant teratocarcinoma cells. Proc Natl Acad Sci USA. 1975;72(9):3585–9.
32. Dolberg DS, Bissell MJ. Inability of Rous sarcoma virus to cause sarcomas in the avian embryo. Nature. 1984;309(5968):552–6.
33. Hochedlinger K, et al. Reprogramming of a melanoma genome by nuclear transplantation. Genes Dev. 2004;18(15):1875–85.
34. Topczewska JM, et al. Embryonic and tumorigenic pathways converge via Nodal signaling: role in melanoma aggressiveness. Nat Med. 2006;12(8):925–32.
35. Schier AF. Nodal morphogens. Cold Spring Harb Perspect Biol. 2009;1(5):a003459.
36. Postovit LM, et al. Human embryonic stem cell microenvironment suppresses the tumorigenic phenotype of aggressive cancer cells. Proc Natl Acad Sci USA. 2008;105(11):4329–34.
37. Lonardo E, et al. Nodal/Activin signaling drives self-renewal and tumorigenicity of pancreatic cancer stem cells and provides a target for combined drug therapy. Cell Stem Cell. 2011;9(5):433–46.
38. Lawrence MG, et al. Reactivation of embryonic nodal signaling is associated with tumor progression and promotes the growth of prostate cancer cells. Prostate. 2011;71(11):1198–209.
39. Strizzi L, et al. Potential for the embryonic morphogen Nodal as a prognostic and predictive biomarker in breast cancer. Breast Cancer Res. 2012;14(3):R75.
40. Kenney NJ, Adkins HB, Sanicola M. Nodal and Cripto-1: embryonic pattern formation genes involved in mammary gland development and tumorigenesis. J Mammary Gland Biol Neoplasia. 2004;9(2):133–44.
41. Esteller M. Epigenetics in cancer. N Engl J Med. 2008;358(11):1148–59.
42. Huang Y, et al. Epigenetics in breast cancer: what's new? Breast Cancer Res. 2011;13(6):225.
43. Jovanovic J, et al. The epigenetics of breast cancer. Mol Oncol. 2010;4(3):242–54.
44. Tsai HC, et al. Transient low doses of DNA-demethylating agents exert durable antitumor effects on hematological and epithelial tumor cells. Cancer Cell. 2012;21(3):430–46.
45. Li Y, Laterra J. Cancer stem cells: distinct entities or dynamically regulated phenotypes? Cancer Res. 2012;72(3):576–80.
46. Jing Y. The PML-RARalpha fusion protein and targeted therapy for acute promyelocytic leukemia. Leuk Lymphoma. 2004;45(4):639–48.
47. Strizzi L, et al. Nodal expression and detection in cancer: experience and challenges. Cancer Res. 2012;72(8):1915–20.
48. Hooijkaas AI, et al. Expression of the embryological morphogen Nodal in stage III/IV melanoma. Melanoma Res. 2011;21(6):491–501.
49. Strizzi L, et al. Nodal as a biomarker for melanoma progression and a new therapeutic target for clinical intervention. Expert Rev Dermatol. 2009;4(1):67–78.
50. Shinozaki S, et al. Indian hedgehog promotes the migration of rat activated pancreatic stellate cells by increasing membrane type-1 matrix metalloproteinase on the plasma membrane. J Cell Physiol. 2008;216(1):38–46.
51. Bailey JM, et al. Sonic hedgehog promotes desmoplasia in pancreatic cancer. Clin Cancer Res. 2008;14(19):5995–6004.

Chapter 10
Metastatic Determinants: Breast Tumour Cells in Circulation

Nisha Kanwar and Susan J. Done

Abstract More than 90 % of deaths from breast cancer are a result of metastases, rather than the primary tumour. In recent years it has become possible to study the occult steps of metastasis that have previously been clinically undetectable—single tumour cells that have disseminated early on from the primary tumour and are en route to distant sites. High-resolution genomic and gene analyses of these rare cells show that they carry their own unique sets of aberrations and are frequently quite different from the primary tumours they originated from. They appear to be heterogeneous and in a transitional state, expressing genes necessary to allow them passage into the circulation as well as genes required to see them through survival or dormancy in blood vessels and metastatic niches such as the bone marrow and lymph nodes. They possess gene signatures ranging from up-regulation of genes associated with invasiveness and dormancy to expression of favourable growth factor receptors that facilitate extravasation and survival at secondary sites. Circulating tumour cells (CTCs) in the blood and disseminated tumour cells (DTCs) in the bone marrow have been reported to have strong prognostic relevance by predicting survival and relapse in both early and late stages of breast cancer. They are emerging as promising biomarkers for monitoring the response to treatment, whereby a drop in cell numbers is suggestive of a positive response, but persisting cells indicate resistance and a poor prognosis. It is apparent that not just the primary and metastatic tumours need to be targeted, but also the intermediate cells in transition that do not necessarily reflect the genetics of the tumour they originated from or the metastasis they may eventually give rise to. As more disseminated cell markers are being consecutively added to a growing panel, the heterogeneous nature of breast cancer is becoming more evident, paving the way for a systemic approach to experimental design and treatment regimens. Molecular characterization of single disseminated cells in the bloodstream

N. Kanwar • S.J. Done (✉)
The Campbell Family Institute for Breast Cancer Research, 620 University Avenue, Suite 7-504, Toronto, ON, Canada, M5G 2M9
e-mail: nisha.kanwar@mail.utoronto.ca; Susan.Done@uhn.ca

H. Schatten (ed.), *Cell and Molecular Biology of Breast Cancer*,
DOI 10.1007/978-1-62703-634-4_10,

will help address many of the questions surrounding the development of breast cancer metastasis.

10.1 Breast Cancer Metastasis

Despite significant advances in early screening and targeted therapies to reduce deaths from breast cancer, metastasis remains the leading cause of mortality amongst cancer patients. In metastasis, a select few cells acquire the ability to invade tissues that surround the primary tumour, thus allowing them to break free and enter the circulation via intravasation of the blood or lymphatic systems. Once in circulation, this 'occult' process goes undetected; they travel to distant organs, and in some cases like 'seeds', displaying specific receptors that home them to their respective 'soil'; and they are able to survive, extravasate, reinitiate aberrant cell division and propagate secondary tumours at new organs (Fig. 10.1). It is not the primary tumours, but these secondary tumours with varying potential that are responsible for cancer deaths.

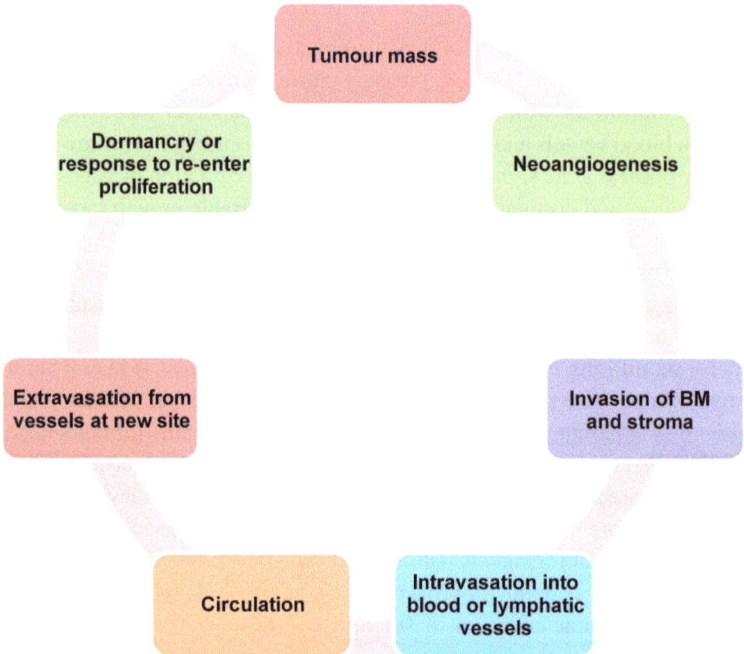

Fig. 10.1 The metastatic cascade: tumour cells within the primary tumour mass disseminate and invade through the basement membrane (BM) into the surrounding stromal tissue, where with the aid of stromal-released factors, they will intravasate into newly formed blood vessels or the lymphatic system and circulate to distant sites such as the bone marrow, lymph nodes or secondary organs. Here, they will either go through apoptosis or enter a period of dormancy until they receive favourable signals from the secondary site to reinitiate proliferation and form a new tumour. Once a secondary tumour has been established, cells from this tumour mass are potentially capable of going through the same cycle to seed new metastases or reseed the original tumour they were derived from

Although tumour cells are shed in a range of 10–1,000s early on in tumourigenesis, they must survive in vessels against shear stress forces, anoikis and unfavourable growth niches, and a majority of them will be in circulation for only a few days before undergoing apoptosis [1]. Metastasis is thus a highly inefficient process, where less than 0.01 % of cells that enter circulation succeed in arriving at a new site where they follow one of two fates: either they remain dormant for up to 20 or more years or they respond to favourable signals in their new microenvironment and enter cell division to form a new tumour [2]. Recently, in mouse models, it was shown that cells in circulation have the capability to essentially retrace their dissemination route back to the primary tumours, where they may multiply and subsequently re-disseminate [3]. It is this rare population of circulating cells with metastatic propensity that escapes chemotherapy and radiation, making them an aggressive subset that requires the attention of new targeted therapies to prevent their spread and stop the threat of metastasis. One of the hallmarks of curable cancer is early detection so that the tumour may be removed and treated locally by radiation before it spreads. If, however, cells are already in circulation at early time points, the approach needs to be a systemic one, where the cells in circulation must be targeted and eradicated before they receive the signals to propagate at new and distant sites.

10.2 CTCs and DTCs in Breast Cancer

The existence of tumour cells in the circulating blood of cancer patients was first reported in 1869 by a physician, Thomas R. Ashworth [4]. Over a century later, technology has provided us with refined detection tools that enable us to find disseminated tumour cells in the blood, bone marrow and other niche organs that sustain them. In the clinic however, these cells fall under the umbrella of minimal residual disease, go undetected by conventional radiological imaging techniques (CT, PET, MRI) and are technically impossible to resect or collectively remove. Tumour cells found in the circulating blood are referred to as circulating tumour cells or CTCs, while tumour cells found in the bone marrow are referred to as disseminated tumour cells or DTCs. It is still unclear as to which niche houses the earliest population of tumour cells that have left the primary site or what route they took to arrive there (haematogenous or lymphatic systems). Notably, although lymph node metastasis is an accurate prognostic indicator of distant metastases, there are about 20–30 % of patients whose cancer metastasizes without involvement of the lymph nodes, indicating haematogenous spread of tumour cells to the metastatic sites [5]. There is growing evidence that the secondary organ's microenvironment is a deciding factor in whether or not CTCs or DTCs will extravasate there and survive. For example, TGFβ expressed in the bone or lungs, TGFα in the liver and CXCR4/7 chemokine receptors expressed on tumour cells, will home to their ligands CXCL12 and CCL21 in the lung, liver and bone marrow, and thus influence the ability of metastatic cells to grow in the respective organs via the activation of various signalling pathways involved in migration, actin polymerization, proliferation and survival [1, 6]. Thus, it might not be an access point, but rather, pre-established signals from optimal sites that decide the route circulating cells will take.

Tumour cells in circulation are rare, with approximately 1 cell in 10^6–10^7 nucleated blood cells [7]. In order to study them, an enrichment step from the blood or bone marrow is first performed, followed by labelling for normal epithelial markers. Several methods for enrichment of CTCs have been described—immunomagnetic bead separation, density centrifugation, size-based exclusion, flow cytometric separation and more recently microfluidic devices such as the CTC chip and Herringbone chip which are coated with antibodies for tumour cell capture [1, 5, 8–10]. Each of these methods allows for either positive selection by targeting CTCs with epithelial markers (cytokeratin, EpCAM) or negative selection by targeting white blood cells with panleukocyte markers (CD45).

Currently it is impossible to differentiate between CTCs and DTCs of prognostic value or metastatic potential from those that will remain dormant, undergo apoptosis or be targeted by the immune system. CTCs and DTCs are identified with the same core set of markers—cytokeratins and EpCAM positive and CD45 or leukocyte negative. In 2008, the FDA approved the CellSearch™ System for the isolation and enumeration of tumour cells from the blood or bone marrow of metastatic breast cancer patients [11]. Since then there have been numerous publications validating its sensitivity in isolation of these rare tumour cells [12–16]. The system includes fixation and labelling of cells with markers for epithelial and white blood cells, followed by immunomagnetic separation of epithelial cells. An automated system then enumerates signals to provide an output of the number of epithelial cells per mL of blood or bone marrow analysed. The array of markers used to identify tumour cells is progressively increasing—HER2 was recently added to the breast cancer panel [8]. Cells in circulation that evade immune targeting or systemic chemotherapy treatments are hypothesized to be a more aggressive subpopulation of chemoresistant, stem cell-like (CD44+/CD24−) tumour cells, and as a result, stem cell and epithelial-to-mesenchymal transition (EMT) markers (CD44, CD24, ALDH1, Twist and AKT) are also included in some studies to determine if there are stem cell-like or tumour-initiating cells amongst the CTCs and DTCs, which are capable of survival and self-renewal [17, 18]. Ki67, EGFR, EMMPRIN and uPAR are amongst other frequently used markers for studying disseminated cells [5].

In addition to immunocytochemical methods, RT-PCR-based detection of tumour cells in circulation is by far the most sensitive assay in use. After enrichment, transcripts of tumour origin may be detected in a range of 1 per 20 million cells, although the downside is that the specificity is riddled with confounding factors such as false positives, pseudogenes, non-specific PCR products and contamination [1]. Previously, we carried out quantitative detection of mRNA transcripts for KRT19 after spiking healthy volunteer blood with breast cancer cells. We reported that RT-PCR assays had a reliable detection limit of 5 cells/mL in 1 mL of blood, and CTC load was accurately reflected by linear amplification of KRT19 mRNA [19]. Thus, RT-PCR could also be developed into an enumerative assay without absolute counting of CTCs. Another advantageous feature of RT-PCR-based assays is the ability to multiplex and measure larger panels of cancer- or tissue-specific markers, which overcomes the issue of specificity of the cells in question. Some common markers include KRT19, CEA, TERT, TWIST1, MUC1, mammaglobin and EGFR [1]. The caveat, however, in any enrichment method that involves labelling tumour cells is the heterogeneity of breast cancers where there will be variability in expression of these markers depending on molecular subtype in the case of EGFR or differentiation in the case of MGB2 or even EMT in the case of KRT19, EPCAM and TWIST.

10.3 Metastasis Is an Early Event

Recent studies have shown that primary tumours may have a gene expression signature that is predictive of metastasis [20–22]. Furthermore, primary tumour expression signatures also define the route of metastatic spread—haematogenous or lymphatic. Woelfle et al. compared primary tumours with and without DTCs or lymph node metastases and found distinct signatures were able to predict bone marrow versus lymphatic micrometastases, with minimal overlap of only nine genes [23]. Genes involved in the DTC-positive primary tumour signature included JAK/STAT and the HIF-1α pathways, implicated in tumour cell survival, invasion and angiogenesis. HIF-1α also activates other genes such as lysyl oxidase which activates focal adhesion kinases to enhance invasion, and CXCR4 involved in homing and survival of cancer cells at secondary sites [23]. There is also collective evidence to show that the dissemination of tumour cells is in fact an early event. The previous notion that larger tumours gave rise to metastases has been challenged by findings that suggest dissemination can even occur in the earliest invasive stages of cancer progression in both murine models and human breast cancer, and it occurs independently of tumour size [24]. Disseminated tumour cells have been found in niches such as the bone marrow and lymph nodes before the onset of overt metastases [1, 23, 24]. Furthermore, disseminated tumour cells were found in circulation in patients with small (less than 2 mm) tumours and also in patients with undetectable primary tumours (less than 7 % of cases of breast cancer) [21, 25]. These cells can remain dormant in niches such as the bone marrow or a future site of metastasis such as the lung, and upon induction by growth stimuli of the microenvironment, they can be reprogrammed to establish secondary tumours [26]. These observations provide evidence that the initial steps of metastasis occur early on within primary tumours and are programmed to allow dissemination via predetermined routes to potentially predetermined niches for either a dormant phase or an aggressive proliferative phase. This also raises an important caveat in the current methods to produce these molecular signatures—should we be analysing bulk tumours or should we be addressing heterogeneity by analysing tumours on a compartmentalized or single-cell basis? The latter method would indeed lead to the discovery of low-level signatures present in those selective cells destined for dissemination and lead to novel targets to prevent this metastatic step altogether. Schmidt et al. showed that in multifocal prostate cancer, CTCs originated from distinct foci, even if they were as small as 0.2 cm, again suggesting that the blueprint for dissemination is probably in the primary tumour, and is detectable in single cells [27]. With the advent of high-resolution next-generation sequencing technology, single cells have been sequenced from primary tumours and their matched metastases, which reveal heterogeneity and clonal evolution of one or few clones in a sequence of events [28]. It has been reported that significant heterogeneity can occur on a single-nucleotide mutation level. Using next-generation sequencing in breast cancer progression, 6 of 32 somatic mutations found in a metastasized tumour were also found in the primary tumour from 9 years earlier and were detected at low frequencies (1–13 %) [29]. Similarly, Ding et al. reported significant changes in allelic frequencies of mutations found in brain metastases, suggestive of low frequencies of these cells existing in the original breast primary tumour [30]. In the future, this type of analysis will provide the added benefit of using markers of these small fractions of cells for prognostic and monitoring purposes over time.

Schardt et al. used comparative genomic hybridization (CGH) to show that DTCs in patients without metastases had fewer aberrations and were more heterogeneous compared to those found in patients with metastatic late-stage cancer [22]. This suggests that DTCs have the capability to accumulate genetic abnormalities parallel to the primary tumour, with clonal selection of aggressive cells occurring in later stages. Hüsemann et al. showed in their murine model that pre-invasive lesions stained positively and more intensely for the EMT marker Twist compared to the invasive edge of tumours [24]. Also early DTCs and CTCs often express the poor prognostic aggressive disease marker HER2, although they originated from HER2-negative primary tumours [24]. Both of these observations are indicative of transitional changes occurring early in primary tumours allowing cells within them to disseminate. This also highlights a requirement for treating breast cancer in a systemic manner, based on markers activated in the transitional steps that lead to overt metastasis, rather than just the genotype of the primary tumour.

10.4 Molecular Characteristics of DTCs and CTCs

Two main models of metastasis have been proposed. The first is the linear progression model where a primary tumour's malignant genetic status is complete and decided first, and the disseminated tumour cells evolve from these founder cells [31]. The second is the parallel progression model, where tumour cells disseminate early on and evolve into their own genetically malignant entities, independently of the primary tumour, at distant sites [32]. Klein et al. showed through molecular characterization of primary tumours and their matched DTCs that once cells have left the primary tumour at early stages, they develop independently with a unique set of aberrations and are more heterogeneous compared to DTCs in patients with distant metastases at late stages [33]. Furthermore, genomic aberrations that are characteristic of the breast primary tumours (16q−, 13q−, 17p− and 8q+) were absent from the genomic profiles of DTCs analysed by CGH, although their malignant origin could be confirmed by microdeletions also found in primary tumours (16q22−, 8q11−) [22, 33]. The prevailing pattern observed was that the DTCs were distinct from primary tumours, which were more like their lymph node metastases than the DTCs [21, 34]. Such discrepancies were also reported in CTCs when compared to primary tumours for markers such as ER, PR, HER2 and EGFR mutations. These observations are in concordance with studies which showed discrepancies between molecular alterations of primary and metastasized secondary tumours. Flores et al. designed a study to establish the relationship between HER2 status of primary tumours, CTCs and metastatic lesions in 75 patients with breast cancer using fluorescence in situ hybridization (FISH) [8]. Interestingly, patients with HER2-positive primaries had HER2-positive CTCs 98 % of the time, compared to patients with HER2-negative primaries, where 33 % of patients showed discordance with HER2-positive CTCs [8]. The unexpected finding was that in these 33 % of patients, 90 % of the metastatic lesions matched the primary tumour [8]. Other studies with similar endpoints have reported HER2 discrepancies of up to 40 % between primary tumours and matched CTCs [35]. This observation brings forward the phenomenon of aggressive characteristics being acquired transitionally to progress select cells through individual steps in metastasis such as invasion and intravasation, which are lost at later stages not requiring these functions. Logically, it follows through that a disseminated tumour cell that is in a transitional state between a primary and a metastatic tumour should in fact have such a transitional genetic profile, subject to change as it encounters new environments and selective pressures. This also explains the difference in genomic profiles of DTCs isolated from lymph nodes versus bone marrow [36]. Furthermore, it is likely that a common set of genomic events exists to allow for the survival of cells in this state, for example, HER2 gain has been described as the most frequent region of gain in DTCs whether they disseminated via the blood or lymphatic routes, and it is not concordant with the primary tumours [32, 36]. Other efforts for single-cell gene expression analysis of DTCs from three individual patients showed heterogeneous expression of genes involved in cell cycle progression (STK12, CCNA), ECM degradation (cathepsins B, D, L, uPA, MMP7), invasion (RAC1, ROCK1, CDC42), replication and growth arrest (PRKDC, CDKN1A) and highly homogenous expression of an extracellular matrix metalloproteinase inducer EMMPRIN, expressed at gene and protein levels in 61–82 % of patients with lung, breast and prostate cancer [33].

Lu et al. addressed the question of EMT and the inadequacy of most of the methods used to enrich for CTCs or DTCs using epithelial cell markers such as cytokeratin19 and EpCAM. It has been proposed that cells in circulation are a subset of tumour progenitor cells fitting the phenotype of aggressiveness, low proliferation and resistance to therapy. In this study, CTCs were enriched based on invasive function rather than inconsistent expression of a marker. The collagen adhesion matrix (CAM) selects for cells that are able to invade, remove and ingest CAM fragments by formation of invadopodia [37]. They were able to conclude that the presence of invasive CTCs correlated to higher stage, lymph node positivity and poorer survival of patients with early breast cancer [37]. Furthermore, if propagated in culture, the gene expression signature of these cells showed that they had properties of EMT stem cells expressing TWIST1 and CD44 [37]. FACS analysis showed that this was not a property held by all CTCs, as they could be separated into three distinct populations highlighting the heterogeneous nature of these cells—one showing epithelial lineage as EpCAM+, the other showing progenitor cell lineage as CD44+ and the third showing the intersection of cells expressing both markers. Other tumour-specific markers expressed were TERT, MUC16, ER and PR [37]. Surprisingly HER2, VIM and other aggressiveness markers were not expressed in a specific population [37]. Gradilone et al. have further propelled the idea of CTCs or DTCs having a stem cell-like phenotype, whereby they investigated the expression of ATP-binding cassette family genes—or the multidrug resistance-related proteins (MRPs) [35]. They reported a significant correlation between the stem cell marker ALDH1 and the expression of a number of MRPs, in addition to the co-expression of HER2 and ERα in a discordant manner compared to primary tumours [35]. These results taken together paint a clearer picture of aggressive disseminated cells gaining a proliferative, invasive and survival advantage with the divergence of their expression profiles.

Schwarzenbach et al. attempted to determine the correlation of cell-free DNA, DTCs and primary tumours in 22 patients and found that there was no concordance between the LOH markers for cell-free DNA in serum and the primary tumours (only 3/22 patients showed concordance of LOH status) [38]. The authors of this study speculated that the origin of cell-free DNA could be CTCs, which have a short half life in the blood (less than 24 h), and if these cells had originated from multifocal heterogeneous areas of the primary tumour, the LOH discrepancies were well founded [38]. A significant region associated with the relapse of patients with high-grade breast cancer was 3p24.2–25, consisting of CDKN2 which is a negative regulator of the cell cycle, the loss of which contributes to increased proliferative capacity of tumour cells [38]. Cells that have left the primary tumour may be more similar to the metastases than the primary. Studies carried out on matched primary and metastatic tumours support divergence, for example, TP53, KRAS and EGFR mutations show 20–80 % discrepancy in colorectal and lung cancers [32, 36]. These phenomena again bring attention to the need to treat breast cancer in a systemic manner and not just based on the characteristics of the primary tumour. Cells in circulation from early time points, present in the absence of lymph node metastases or even once the primary tumour has been resected, are persisting threats with unique aberrations that lead to distant metastases and must be molecularly defined in order to target them.

Smirnov et al. were able to perform global gene expression profiling of CTCs from prostate, colorectal and breast cancers and elegantly showed that these profiles were indeed tissue specific, although they also showed some commonality. They found that genes such as KRT19 and AGR2 were expressed in CTCs from all samples and not expressed in normals [39]. The tissue-specific genes were S100A14/16 and CEACAM5 for breast and colorectal cancers; KLK2/3, MSMB, DDC, AR and HPN for prostate cancers; and SCGB2A1/2 and PIP for breast cancers alone [39]. Most of these genes function in cell proliferation, migration and oncogenesis. The combination of this gene signature was able to classify tumour and normal correctly with 79.3 % accuracy, which was comparable with the classification power of gene expression signatures obtained from primary tumours [39].

Fig. 10.2 Distribution of functionally relevant genes up-regulated in DTCs and down-regulated in matched bone metastases from the same patient. DTCs are heterogeneous and likely in a balance between proliferation and apoptosis signals that lead to quiescence and growth arrest. They also likely express receptors to attract interacting macrophages and platelets that aid in extravasation, evasion from the immune system and survival during dormant phases

In our own laboratory we have progressed with work aimed at gene expression profiling of DTCs and matched bone metastases in patients with metastatic breast cancer. We reported genes that were uniquely expressed in each tumour sample as well as a few that were shared across matched samples from the same patients. Amongst genes that were more highly expressed in DTCs compared to matched bone metastases were cell cycle and cell division-related genes CCNB2, CCNJ, CCNM2 and CDCA7L; genes involved in platelet aggregation such as C-type lectins and integrin ITGA2B; as well as the actin cytoskeletal gene TUBAL3 (Fig. 10.2) [40]. These were in contrast to genes down-regulated which included cadherins 2, 5, 11 and 15, claudin 5 and collagens COL10A1, 1A1, 1A2 and 8A1 all involved in cell–cell and cell–extracellular matrix interactions (Fig. 10.3) [40]. Interestingly, HER2 was shared between both DTCs and bone metastases, in addition to chemokine CCL20 and IGF-binding proteins, suggestive of a population of late stage DTCs with aggressive proclivity and receptive to signals for reactivation of growth [40].

In a separate study, using immunomagnetic sorting and laser capture microdissection, we isolated CTCs from the peripheral blood of breast cancer patients with and without distant metastases or lymph node metastases. We then amplified the whole genome of CTCs as well as matched normal leukocytes from the same patient ($n=17$). DNA was processed on the high-resolution Affymetrix Human Genome-Wide SNP 6.0 Array, where each tumour sample was normalized using its own unique reference sample (normal leukocyte fraction) as a baseline [41]. Genomic copy number gains and losses were identified by a paired sample approach using an HMM algorithm (Partek Genomics Suite, version 6.6). Our preliminary analysis

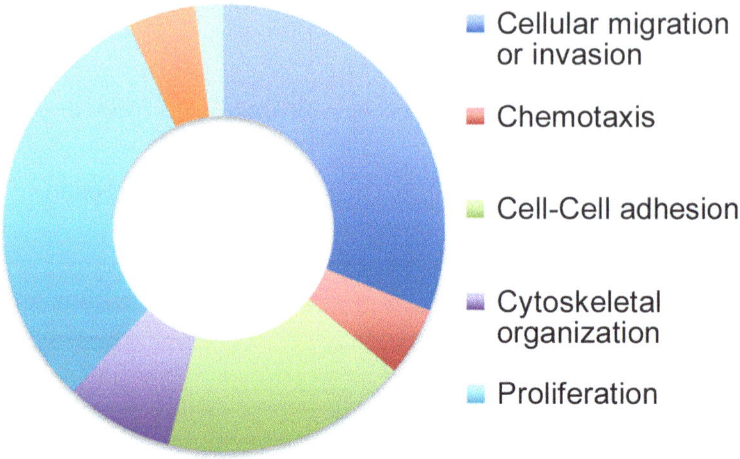

Fig. 10.3 Distribution of functionally relevant genes down-regulated in DTCs and up-regulated in matched bone metastases from the same patient. Tumour cells within bone metastases showed predominant expression of invasive, cell–cell adhesion and proliferation-related genes. These are likely functions down-regulated in DTCs that are in a single-cell-detached state, non-proliferative to evade chemotherapy and past the initial stage of metastasis where they invaded the BM to get into circulation

shows that there are a number of regions of altered DNA in CTCs. In general, there were significantly more regions with gains compared to losses. The most frequently altered regions involved gains of chromosomes 1p36.3, 1q21.3, 4q26, 6q16.3, 8q23, 9q34.1–q34.3, 11p15.4–15.5, 17q21–25, 20p13, 22q11.2–13.3 and many regions across the entire chromosome 19.

Our findings are concordant with other studies of genes involved in cancer progression and metastasis. A recent study performed global gene expression analysis of CTCs from patients with breast, prostate and colorectal metastatic cancers. KLK2 and KLK3 (19q13.4) were a few of the highly expressed genes in CTCs of metastatic prostate cancer patients [39]. CEACAM5 (19q13.3) and PIP genes (7q34) were highly expressed in CTCs from metastatic breast cancer [39]. In conjunction with expression patterns of a few other candidate genes, this panel was able to discriminate the CTCs by tissue of origin [39]. Most interestingly, genes belonging to the S100 family of calcium-binding proteins (1q21.3) were identified as the strongest candidates for CTC-specific expression markers, S100A13 for all types of metastatic cancer and more specifically S100A14/16 for metastatic breast cancer [39]. Survivin (17q25) has also been found to have elevated gene expression in CTCs from patients with breast cancer and was associated with advanced clinical pathological markers [42]. MUC1 (1q21.3) is a widely used marker for all cancers [1, 39]. TGFβ (19q13.3) induces expression of angiopoietin-like 4 protein and enhances metastatic capacity in oestrogen-receptor-negative breast cancers [43]. ANGPTL4 plays a critical role in the extravasation of tumour cells through endothelial cell–cell junctions in the lung [43]. In a meta-analysis, out of 3,131 cancers

(26 histological types), 158 regions of altered copy number were identified across cancer types [44]. On chromosome 19q13, this region contained CCNE1 ($p < 1 \times 10^{-30}$) [44]. The importance of intravasation to metastasis was emphasized by Zijlstra et al. who recently showed that a function-blocking antibody directed toward the integrin-associated protein CD151 (11p15.4) prevents intravasation and consequently inhibits metastasis by greater than 90 % [45]. Genomic gains that are characteristic of CTCs may define these cells as the more aggressive component of a tumour, capable of invasion, intravasation and survival, and could be developed into CTC-specific markers to identify and isolate these cells from blood.

10.5 CTCs and DTCs: Specific Functions and Related Genes

It is evident that the specialized abilities of single cells that persist independently from the primary tumour mass include invasion of the organ microenvironment (primary and secondary); intravasation into the blood vessels that route them to optimal growth or survival niches, survival and dormancy for extended periods of time; and aggressiveness whether it is through the successful accomplishment of the many steps of the metastatic cascade or their resistance to conventional therapies and novel attempts to stop their spread. Each one of these features most likely defines a cell at a specific phase during its molecular evolution. The acquisition of genomic alterations that confer specialized functions to these cells could be temporal and spatial depending on when and where they intersect a particular step of metastasis.

10.5.1 Invasion of the Organ Microenvironment

Epithelial organs are surrounded by a basement membrane (BM), beyond which is a microenvironment consisting of stromal cells (fibroblasts, endothelial cells and macrophages). In order for single cells to escape from epithelial primary tumours, they need to dissolve and break through the BM and act in concert with the stromal cells to eventually gain access to a circulatory blood or lymphatic vessel. Deregulation of proteolytic enzymes such as metalloproteinases MMP1 and MMP7 have been widely implicated in the degradation of the BM [46]. Various cytokines and growth factors such as IL6, CCL2, CCL5, CXCL12, TGFβ, EGF, HGF and FGF2 are also released which in turn aid in tumourigenesis via cell proliferation and survival [46]. The tumour cells themselves will go through a reprogramming in order to detach, invade and migrate through the degraded BM. This reprogramming is termed EMT (epithelial-mesenchymal-transition) and involves the loss of E-cadherin, tight junctions and cell polarity proteins and simultaneous up-regulation of N-cadherin, vimentin and various transcription factors: Slug, Snail, Twist, ZEB1 and 2 as well as miR-200 family members [46]. Adbelkarim et al. recently showed the in vitro selection of cells that were able to invade Matrigel, a BM, indeed displayed a separate gene expression profile compared to those cells that were less invasive. These cells were defined by loss of adhesion to the extracellular matrix (ECM), via down-regulation of genes involved in cell–cell adhesion such as OCLN, up-regulation of genes involved in EMT and mesenchymal morphology such as TGFβ2 and up-regulation of anti-apoptotic genes TIMP-1 and IRAK1 [47]. Additionally, there was down-regulation of cyclin D2 and caspase 4 genes which render the cells desensitized to certain types of drug-induced apoptosis [47]. Angiogenesis genes VEGF and NRP-1 were also expressed at a higher level in these invasive cells [47].

Once invasion has occurred, single disseminated tumour cells are in direct contact with stromal cells referred to as carcinoma-associated fibroblasts (CAFs). CAFs become activated by tumour progression and are involved in recruitment of inflammatory cells and pro-angiogenic factors, growth factors and cytokines such as EGFR signalling, IL-4, IL-6, TNFα, TGFβ, CCL2, etc. which form a tumour-sustaining microenvironment [48]. There have been a few studies that performed microarray profiling of tumour-associated stroma and revealed prognostic signatures in stromal cells for metastatic outcome. Finak et al. described a stromal-derived prognostic predictor consisting of up-regulated genes in stromal tissue that influenced the progression of primary breast tumours and was thus able to predict patient outcome. Stromal gene signatures were clustered into six distinct biological subtypes—matrix remodelling (MMPs, CXCL13, S100 family), immune Th1-type response (high interferon), activated fibroblasts, ER signalling (ESR1, FOXA1), angiogenic response and a hypoxic-type signature [49]. Of these, the matrix remodelling, angiogenic and activated fibroblast signatures predicted poor outcome [49].

10.5.2 Intravasation and Extravasation

Intravasation and extravasation involve the migration of a tumour cell into or out of the lumen of a lymphatic or blood vessel. Some studies have shown that the entry of tumour cells into vessels is influenced by a chemoattractant gradient involving tumour-associated macrophages (TAMs) [48]. EGF is produced by TAMs which interacts with the EGF receptor expressed on tumour cells, and conversely, CSF-1 expressed by tumour cells interacts with the CSF-1 receptor on macrophages leading to a synergistic movement of tumour cells enhanced by TAMs into the blood vessels [46, 48]. In addition, tumour cells express VEGF that stimulates the production of new blood vessels—neoangiogenesis. These vessels possess dual functions: to transport tumour cells into and out of the main circulatory system as well as feed a growing tumour with a dedicated blood supply. These new blood vessels tend to be not as structurally sophisticated as the body's main circulatory system; they are leaky and not well formed, which facilitates the movement of tumour cells through them [46].

One striking difference between the mechanisms of intravasation and extravasation is the microenvironment. Once CTCs or DTCs have arrived at their secondary site of arrest, they must extravasate from the circulatory system's vasculature into a new organ microenvironment, the only major difference being the absence of a pre-existing primary tumour. Essentially the blood vessels and BM in this case are lacking CAFs, TAMs and leaky neoangiogenic blood vessels. Surprisingly, it was shown that these barriers are overcome by the same distant primary tumour by secreting signalling molecules such as lysyl oxidase, which results in the migration of VEGF receptor expressing haematopoietic progenitor cells from the bone marrow to a predetermined metastatic site [46]. These cells then prime the new site with EREG, COX-2, MMP-1, MMP-2, ADAMTS1 and VEGF to aid in the disruption of endothelial cell-cell junctions for extravasation to occur [46]. An up-regulation of ANGPTL4 has been reported in the lungs specifically [43]. This organ-specific response is likely to be seen within the genetic architecture of the extravasating cell itself, probably early on, and is the focus of future studies which isolate tumour cells en route to distant sites of arrest.

10.5.3 Survival and Dormancy

It is still relatively unknown as to how long CTCs and DTCs are alive in circulation. They must overcome the first response—anoikis, which is apoptosis as a result of lack of cell–cell interaction once it has detached from the primary tumour. Next, these cells must evade shear stress forces flowing through the blood vessels that were designed for optimal flow of white blood cells with much smaller diameters (8 μM versus 20–30 μM) [6]. Lastly, these cells must go undetected by the immune responses. Tumour cells have been shown to attach themselves to platelets, forming microemboli which are able to accomplish evasion from immune detection as well as decrease shear forces of the blood vessels directly acting upon them [46, 48]. Platelet aggregation occurs via the expression of L/P selectins or tissue factors which bind coagulation factors of platelets [46]. Others have also suggested a macrophage-tumour cell hybrid which is formed transiently allowing survival until it reaches its secondary site [48].

Another attribute of CTCs and DTCs is to, in essence, reverse or silence their acquired invasiveness, uncontrolled proliferation and angiogenic properties and enter a state of dormancy at a distant site. In the case of breast cancers, metastatic secondary tumours may arise after dormant periods of 20 or more years. Some studies have suggested that tumour cells are in fact still dividing but also undergoing apoptosis, and it is this balanced proliferation and death that prevents the formation of macrometastases at secondary sites [26]. Nonetheless, dormant tumour cells may very well be the rare tumour-initiating cells or stem cells that are quiescent, in G_0–G_1 arrest, and resistant to chemotherapy. Activation of metastasis-suppressor genes such as NM23 and KISS1 has been shown to reduce metastatic output in mouse models, as has moderate activation of the stress pathway JNK/p38 which will not induce apoptosis but rather growth arrest [26]. This state continues until these cells receive positive signals from the secondary microenvironment to start proliferation, most likely growth factors, chemokines, or up-regulation of proliferative pathways via EGFR or MAPK, for example, high EGFR/p38 ratios permit proliferation, compared to the aforementioned high p38 stress response which would propagate dormancy [26].

10.6 Clinical Value of Cells in Circulation

The application of early screening and adjuvant therapies has certainly led to a decrease in mortality; however, it has in return created a population of larger numbers of patients who require beneficial targeted therapies. Overtreatment and undertreatment of patients remain a problem, and clinicians and scientists strive to better define markers of optimal prognostic and predictive value. The current standard of care that guides treatment options involves a comprehensive analysis of the primary tumour for clinicopathological parameters such as tumour size, grade, lymph node involvement, lymphovascular invasion and the status of hormone receptors ER/PR and the HER2 receptor [25]. Significant progress beyond these standards is yet to be made, although new markers such as Ki 67, cyclin D1, and cyclin E and various multigene assays such as MammaPrint and Oncotype DX are emerging as promising markers which are useful in conjunction with existing tools [25]. Another emerging field of biomarkers is microRNAs, where several miRNAs—10b, 21, 31, 126, 335 and 373 have been correlated with metastatic outcome [46]. The mere presence of CTCs and DTCs in patients with breast cancer is a negative prognostic indicator; however, the 2007 recommendations of the American Society of Clinical Oncology have not approved their use pending larger-scale studies with more definitive and consistent results [50]. Breast cancer is not a single disease but rather is comprised of various subtypes with characteristic gene expression profiles and prognostic markers [20]. Therefore, the use of multiple markers to assess metastatic propensity will be more beneficial.

There have been several studies that showed that the presence of DTCs in the bone marrow of patients with primary breast cancer has a negative prognostic impact. A multicenter retrospective study included 4,703 patients with metastatic breast cancer across Europe and the USA and concluded that over a 10-year follow-up period, patients with DTCs had significantly decreased overall and disease-free survival compared to patients who did not have DTCs in their bone marrow [51]. Presence of DTCs was also correlated with higher tumour grade, presence of lymph node metastases as well as overt metastases, and a poorer prognosis, independent of tumour size [51]. Since there is the issue of invasiveness and patient discomfort when it comes to obtaining repetitive bone marrow samples for the isolation of DTCs, as well as the difficulty in obtaining metastatic samples, CTCs measured in a blood sample provide an easy to collect and relatively non-invasive method of monitoring disease progression and its response to therapies in real time, providing prognostic information by probing for specific molecular markers. The prognostic relevance of CTCs in metastatic breast cancer has also been demonstrated by numerous groups, pioneered by a prospective study where 177 metastatic breast cancer patients were shown to have a reduced overall and disease-free survival if they had five or more CTCs per 7.5 mL of blood [52]. In a follow-up study with the same group of patients, it was reported that the number of CTCs was a better indicator of disease progression than traditional techniques such as imaging with PET, CT or MRI scans [53–56]. Ignatiadis et al. were one of the first groups to show prognostic importance of CTCs present in the blood of early breast cancer patients and have been succeeded by larger cohorts such as the German SUCCESS trial of 1,489 patients which recently showed that the presence of even one CTC had prognostic value in early breast cancer [57].

A study found that metastatic breast cancer patients treated with chemotherapy or endocrine therapy, with five or more CTCs, were 6.3 times more likely to have radiographic disease progression compared with patients who had less than five CTCs [56]. Using enumeration of CTCs as an endpoint, it has been shown that patients who respond well to systemic chemotherapy show a decrease in CTC number. Pachmann et al. and Xenidis et al. showed that patients treated with tamoxifen who had persisting CTCs were more likely to relapse and had shorter disease-free and overall survival [58, 59]. Similarly a study of 696 patients showed that the persistence of DTCs in patients after chemotherapy was an indicator of relapse and reduced survival [60]. Sequencing of EGFR mutations from primary tumours and CTCs of the same patients has enabled the identification of the evolution of tumour resistance during the course of treatment of small cell lung cancers with EGFR tyrosine kinase inhibitors [61]. Regions of aberrant DNA were observed in CTCs that were absent in corresponding primaries, and vice versa, providing novel targets to study mechanisms of resistance [61]. There are ongoing studies such as GeparQuattro and SUCCESS which will determine whether these decreases in CTC number are associated with better prognosis and improved survival of patients [1].

10.7 Conclusions and Future Perspectives

The presence of DTCs and CTCs is now an established phenomenon in early stage as well as metastatic breast cancer. Currently, researchers are narrowing in on arrays of markers which when combined are able to paint a much more accurate and informative picture of how early these cells disseminated (analysis of the extent of genomic instability and genetic aberrations), the degree of their metastatic potential (analysis of known markers of aggressiveness such as HER2, EGFR and EMT markers such as TWIST, VIM), potential response to a selected regimen of systemic therapy (evolution of HER2 and EGFR status discordant from the primary tumours, expression of stem cell markers CD44, ALDH1, low proliferative marker via Ki67), as well as which secondary soil they are destined to 'seed' (cytokines and growth factors specifically expressed in the lungs or bone). Clinically, the prognostic relevance of cells in circulation is becoming increasingly evident as multicenter prospective investigations including thousands of patients provide conclusive results of few numbers of disseminated cells being associated with poorer patient survival and response to treatments. Cutting-edge advancements such as next-generation sequencing and whole genome amplification of single cells for high-resolution genetic analysis will shed more light on the heterogeneity that underlies first- and second-generation expression profiles of primary tumours alone. A study on intratumoural heterogeneity using interphase FISH techniques showed that there are distinct chromosomal regions which are gained through aneuploidy that define a pro-metastatic type of cell [62]. Interestingly they also found that in a proportion of cases, a minor clone present in the primary tumour was the most highly represented clone in its metastases [62]. Another study attempted to classify primary ductal breast carcinomas as 'monogenomic' tumours with a more homogenous genome profile or as 'polygenomic' tumours with multiple subpopulations of genomic clones [63]. They analysed individual tumours as sectors using both aCGH and FISH techniques. Their results showed a significant proportion of both tumour types, where polygenomic tumours had up to three major tumour subpopulations with clear clonal evolution [63]. Such studies bring forward not only the collective profiles necessary for the progression of tumour cells through the steps of metastasis but also highlight the specific differences seen in this heterogeneous cell population to enable the formation of new hypotheses to be tested with regards to tumour-initiating aggressive subpopulations which must be targeted.

Exploration of this novel field of prognostic and diagnostic relevance warrants the recognition of important caveats—a standardized marker does not exist for the selection of single disseminated cells from the blood or bone marrow. Thus, we cannot avoid biased enrichment of subpopulations, which will differ across separation platforms, patient cohorts and lead to discrepant conclusions to the same research questions. As with gene expression profiles of primary tumours, it is also observed that gene expression profiles of single DTCs and CTCs are quite heterogeneous with minimal overlap across studies. It is likely that although the genes up- or down-regulated in the snapshot of malignant progression in samples may be different, we must make note if they are part of the same umbrella functional pathway and if they will lead to similar endpoints (relapse or survival), as well as if these endpoints are concordant with current prognostic assays—clinical markers, pathological immunohistochemical markers and array-based tests such as MammaPrint and the multiplex RT-PCR assay, Oncotype DX.

In conclusion, CTCs and DTCs might be the earliest detectable cells with metastatic abilities and are emerging as promising biomarkers for breast cancer progression. These cells may affect cancer prognosis years before the onset of overt metastasis and therefore improve risk assessment and help identify patients in need of additional treatment. The cells themselves may provide new targets for therapy to prevent their spread to distant sites. By profiling the whole genome of CTCs and DTCs, we can identify novel genomic alterations specific to cells in transition, which may be utilized in the development of specific markers for circulating cells in the blood and bone marrow. Altered genomes, transcriptomes and proteomes in circulating cells from early breast cancer patients hold valuable prognostic genetic information about the progression of early breast cancer to metastasis.

References

1. Pantel K, Brakenhoff RH, Brandt B. Detection, clinical relevance and specific biological properties of disseminating tumour cells. Nat Rev Cancer. 2008;8:329–40.
2. Chiang AC, Massagué J. Molecular basis of metastasis. N Engl J Med. 2008;359:2814–23.
3. Kim M-Y, et al. Tumor self-seeding by circulating cancer cells. Cell. 2009;139:1315–26.

4. Ashworth TR. A case of cancer in which cells similar to those in the tumours were seen in the blood after death. Aust Med J. 1869;14:146–7.
5. Pantel K, Brakenhoff RH. Dissecting the metastatic cascade. Nat Rev Cancer. 2004;4:448–56.
6. Chambers AF, Groom AC, MacDonald IC. Metastasis: dissemination and growth of cancer cells in metastatic sites. Nat Rev Cancer. 2002;2:563–72.
7. Krishnamurthy S, et al. Detection of minimal residual disease in blood and bone marrow in early stage breast cancer. Cancer. 2010;116:3330–7.
8. Flores LM, et al. Improving the yield of circulating tumour cells facilitates molecular characterisation and recognition of discordant HER2 amplification in breast cancer. Br J Cancer. 2010;102:1495–502.
9. Stott SL, et al. Isolation of circulating tumor cells using a microvortex-generating herringbone-chip. Proc Natl Acad Sci USA. 2010;107:18392–7.
10. Nagrath S, et al. Isolation of rare circulating tumour cells in cancer patients by microchip technology. Nature. 2007;450:1235–9.
11. Van der Auwera I, et al. Circulating tumour cell detection: a direct comparison between the cell search system, the AdnaTest and CK-19/mammaglobin RT-PCR in patients with metastatic breast cancer. Br J Cancer. 2010;102:276–84.
12. Giordano A, Cristofanilli M. CTCs in metastatic breast cancer. Recent Results Cancer Res. 2012;195:193–201.
13. Andreopoulou E, et al. Comparison of assay methods for detection of circulating tumor cells in metastatic breast cancer: AdnaGen AdnaTest BreastCancer Select/Detect™ versus Veridex Cell Search™ system. Int J Cancer. 2012;130:1590–7.
14. Lucci A, et al. Circulating tumour cells in non-metastatic breast cancer: a prospective study. Lancet Oncol. 2012;13:688–95.
15. Farace F, et al. A direct comparison of Cell Search and ISET for circulating tumour-cell detection in patients with metastatic carcinomas. Br J Cancer. 2011;105:847–53.
16. Kraan J, et al. External quality assurance of circulating tumor cell enumeration using the Cell Search® system: a feasibility study. Cytometry B Clin Cytom. 2011;80:112–8.
17. Aktas B, et al. Stem cell and epithelial-mesenchymal transition markers are frequently overexpressed in circulating tumor cells of metastatic breast cancer patients. Breast Cancer Res. 2009;11:R46.
18. Kasimir-Bauer S, Hoffmann O, Wallwiener D, Kimmig R, Fehm T. Expression of stem cell and epithelial-mesenchymal transition markers in primary breast cancer patients with circulating tumor cells. Breast Cancer Res. 2012;14:R15.
19. Iakovlev VV, Goswami RS, Vecchiarelli J, Arneson NCR, Done SJ. Quantitative detection of circulating epithelial cells by Q-RT-PCR. Breast Cancer Res Treat. 2007;107:145–54.
20. van de Vijver MJ, et al. A gene-expression signature as a predictor of survival in breast cancer. N Engl J Med. 2002;347:1999–2009.
21. Schmidt-Kittler O, et al. From latent disseminated cells to overt metastasis: genetic analysis of systemic breast cancer progression. Proc Natl Acad Sci USA. 2003;100:7737–42.
22. Schardt JA, et al. Genomic analysis of single cytokeratin-positive cells from bone marrow reveals early mutational events in breast cancer. Cancer Cell. 2005;8:227–39.
23. Woelfle U, et al. Molecular signature associated with bone marrow micrometastasis in human breast cancer. Cancer Res. 2003;63:5679–84.
24. Hüsemann Y, et al. Systemic spread is an early step in breast cancer. Cancer Cell. 2008;13:58–68.
25. Weigelt B, Peterse JL, van't Veer L. Breast cancer metastasis: markers and models. Nat Rev Cancer. 2005;5:591–602.
26. Wikman H, Vessella R, Pantel K. Cancer micrometastasis and tumour dormancy. APMIS. 2008;116:754–70.
27. Schmidt H. Asynchronous growth of prostate cancer is reflected by circulating tumor cells delivered from distinct, even small foci, harboring loss of heterozygosity of the PTEN gene. Cancer Res. 2006;66:8959–65.
28. Russnes HG, Navin N, Hicks J, Borresen-Dale A-L. Insight into the heterogeneity of breast cancer through next-generation sequencing. J Clin Invest. 2011;121:3810–8.

29. Shah SP, et al. Mutational evolution in a lobular breast tumour profiled at single nucleotide resolution. Nature. 2009;461:809–13.
30. Ding L, et al. Genome remodelling in a basal-like breast cancer metastasis and xenograft. Nature. 2010;464:999–1005.
31. Klein CA. Parallel progression of primary tumours and metastases. Nat Rev Cancer. 2009;9:302–12.
32. Stoecklein NH, Klein CA. Genetic disparity between primary tumours, disseminated tumour cells, and manifest metastasis. Int J Cancer. 2010;126:589–98.
33. Klein CA, et al. Genetic heterogeneity of single disseminated tumour cells in minimal residual cancer. Lancet. 2002;360:683–9.
34. Mathiesen RR, et al. High-resolution analyses of copy number changes in disseminated tumor cells of patients with breast cancer. Int J Cancer. 2012;131:E405–15.
35. Gradilone A, et al. Circulating tumor cells (CTCs) in metastatic breast cancer (MBC): prognosis, drug resistance and phenotypic characterization. Ann Oncol. 2011;22:86–92.
36. Klein CA, Stoecklein NH. Lessons from an aggressive cancer: evolutionary dynamics in esophageal carcinoma. Cancer Res. 2009;69:5285–8.
37. Lu J, et al. Isolation of circulating epithelial and tumor progenitor cells with an invasive phenotype from breast cancer patients. Int J Cancer. 2010;126:669–83.
38. Schwarzenbach H, et al. Comparative evaluation of cell-free tumor DNA in blood and disseminated tumor cells in bone marrow of patients with primary breast cancer. Breast Cancer Res. 2009;11:R71.
39. Smirnov DA, et al. Global gene expression profiling of circulating tumor cells. Cancer Res. 2005;65:4993–7.
40. Cawthorn TR, et al. Mechanisms and pathways of bone metastasis: challenges and pitfalls of performing molecular research on patient samples. Clin Exp Metastasis. 2009;26:935–43.
41. Kanwar N, Done S. Circulating tumour cells: implications and methods of detection. In: Done S, editor. Breast cancer—recent advances in biology, imaging and therapeutics. Rijeka: InTech Publishers; 2011. http://www.intechopen.com/books/breast-cancer-recent-advances-in-biology-imaging-and-therapeutics/circulating-tumour-cells-implications-and-methods-of-detection.
42. Yie S, et al. Detection of survivin-expressing circulating cancer cells (CCCs) in peripheral blood of patients with gastric and colorectal cancer reveals high risks of relapse. Ann Surg Oncol. 2008;15:3073–82.
43. Padua D, et al. TGFbeta primes breast tumors for lung metastasis seeding through angiopoietin-like 4. Cell. 2008;133:66–77.
44. Beroukhim R, et al. The landscape of somatic copy-number alteration across human cancers. Nature. 2010;463:899–905.
45. Zijlstra A, Lewis J, DeGryse B, Stuhlmann H, Quigley JP. The inhibition of tumor cell intravasation and subsequent metastasis via regulation of in vivo tumor cell motility by the tetraspanin CD151. Cancer Cell. 2008;13:221–34.
46. Valastyan S, Weinberg RA. Tumor metastasis: molecular insights and evolving paradigms. Cell. 2011;147:275–92.
47. Abdelkarim M, et al. Invading basement membrane matrix is sufficient for MDA-MB-231 breast cancer cells to develop a stable in vivo metastatic phenotype. PLoS One. 2011;6:e23334.
48. Khamis ZI, Sahab ZJ, Sang Q-XA. Active roles of tumor stroma in breast cancer metastasis. Int J Breast Cancer. 2012;2012:1–10.
49. Finak G, et al. Stromal gene expression predicts clinical outcome in breast cancer. Nat Med. 2008;14:518–27.
50. Harris L, et al. American Society of Clinical Oncology 2007 update of recommendations for the use of tumor markers in breast cancer. J Clin Oncol. 2007;25:5287–312.
51. Braun S, et al. A pooled analysis of bone marrow micrometastasis in breast cancer. N Engl J Med. 2005;353:793–802.
52. Cristofanilli M, et al. Circulating tumor cells, disease progression, and survival in metastatic breast cancer. N Engl J Med. 2004;351:781–91.

53. Cristofanilli M, et al. Circulating tumor cells in metastatic breast cancer: biologic staging beyond tumor burden. Clin Breast Cancer. 2007;7:471–9.
54. Bidard F-C, et al. Prognosis of women with stage IV breast cancer depends on detection of circulating tumor cells rather than disseminated tumor cells. Ann Oncol. 2008;19:496–500.
55. Nelson NJ. Circulating tumor cells: will they be clinically useful? J Natl Cancer Inst. 2010;102:146–8.
56. Liu MC, et al. Circulating tumor cells: a useful predictor of treatment efficacy in metastatic breast cancer. J Clin Oncol. 2009;27:5153–9.
57. Ignatiadis M, et al. Prognostic value of the molecular detection of circulating tumor cells using a multimarker reverse transcription-PCR assay for cytokeratin 19, mammaglobin A, and HER2 in early breast cancer. Clin Cancer Res. 2008;14:2593–600.
58. Pachmann K, et al. Monitoring the response of circulating epithelial tumor cells to adjuvant chemotherapy in breast cancer allows detection of patients at risk of early relapse. J Clin Oncol. 2008;26:1208–15.
59. Xenidis N, et al. Cytokeratin-19 mRNA-positive circulating tumor cells after adjuvant chemotherapy in patients with early breast cancer. J Clin Oncol. 2009;27:2177–84.
60. Janni W, et al. Persistence of disseminated tumor cells in the bone marrow of breast cancer patients predicts increased risk for relapse—a European pooled analysis. Clin Cancer Res. 2011;17:2967–76.
61. Maheswaran S, et al. Detection of mutations in EGFR in circulating lung-cancer cells. N Engl J Med. 2008;359:366–77.
62. Sayagués JM, et al. Intratumoural cytogenetic heterogeneity of sporadic colorectal carcinomas suggests several pathways to liver metastasis. J Pathol. 2010;221:308–19.
63. Navin N, et al. Inferring tumor progression from genomic heterogeneity. Genome Res. 2010;20:68–80.

Chapter 11
Breast Cancer Epigenetics: Biomarkers and Therapeutic Potential

Nancy H. Nabilsi, Carolina E. Pardo, Maria Zajac-Kaye, and Michael P. Kladde

Abstract In this chapter, we review the central roles that epigenetic mechanisms, including DNA methylation, histone posttranslational modifications, and nucleosome positioning, play in governing gene expression and how their dysregulation contributes to carcinogenesis. Dramatic improvements in high-throughput DNA sequencing, so-called next-generation technologies, have driven a veritable revolution in the large-scale assessment of cellular "epigenomes." We also review several reports that have profiled epigenomic differences between normal tissue and diverse types of breast tumors, emphasizing how they inform our understanding of disease complexity and the utility of epigenetic-based therapies for breast cancer treatment. Finally, we discuss a methodology that interrogates multiple epigenetic as well as

N.H. Nabilsi • C.E. Pardo
Department of Biochemistry and Molecular Biology, University of Florida College of Medicine, Gainesville, FL, USA

University of Florida Shands Cancer Center Program in Cancer Genetics, Epigenetics and Tumor Virology, University of Florida College of Medicine, Gainesville, FL, USA

M. Zajac-Kaye
University of Florida Shands Cancer Center Program in Cancer Genetics, Epigenetics and Tumor Virology, University of Florida College of Medicine, Gainesville, FL, USA

Department of Anatomy and Cell Biology, University of Florida College of Medicine, Gainesville, FL, USA

M.P. Kladde (✉)
Department of Biochemistry and Molecular Biology, University of Florida College of Medicine, Gainesville, FL, USA

University of Florida Shands Cancer Center Program in Cancer Genetics, Epigenetics and Tumor Virology, University of Florida College of Medicine, Gainesville, FL, USA

Department of Biochemistry and Molecular Biology, University of Florida Shands Cancer Center, University of Florida College of Medicine, 2033 Mowry Road, Box 103633, Gainesville, FL 32610-3633, USA
e-mail: kladde@ufl.edu

H. Schatten (ed.), *Cell and Molecular Biology of Breast Cancer*,
DOI 10.1007/978-1-62703-634-4_11, © Springer Science+Business Media New York 2013

genetic features at the single-molecule level and how it may be utilized to increase our understanding of intratumoral heterogeneity.

List of Abbreviations

5-aza	5-Azacytidine, DNA demethylating agent
ACTR	AIB1, RAC3, and TRAM1, transcriptional coactivator
ADD	ATRX-DNMT3A-DNMT3L, cysteine-rich domain in DNMT3L that interacts with H3 unmethylated at K4
APC	Adenomatous polyposis coli, tumor suppressor gene
ATP	Adenosine triphosphate
ATRX	Alpha-thalassemia X-linked mental retardation, ATP-dependent helicase chromatin remodeler
BGS	Bisulfite genome sequencing
BRCA1	Breast cancer 1 susceptibility protein
BRCA2	Breast cancer type 2 susceptibility protein
BRG1	Brahma-related gene 1, part of the large ATP-dependent chromatin remodeling complex SWI/SNF
Bromodomain	Protein domain that preferentially binds acetylated lysine
CDK	Cyclin-dependent kinase
CDKN1A	Cyclin-dependent kinase inhibitor 1A
Cfp1	CXXC finger protein 1, binds non-methylated CGs in a majority of CGIs and associates with the human Set1 complex
CG/CpG	Dinucleotide with cytosine followed by guanine
CGI	CG island
C/EBPα	CCAAT/enhancer binding protein alpha, binds responsive elements and recruits coactivators
C/EBPβ	CCAAT/enhancer binding protein beta, binds responsive elements and recruits coactivators
ChIP	Chromatin immunoprecipitation
ChIP-seq	ChIP followed by next-generation sequencing
Chromo	Chromatin organization modifier, domain in chromatin-interacting proteins that binds to methylated histones
CIMP	CGI methylator phenotype
CREB1	Cyclic AMP-responsive element binding protein 1, stimulates transcription
CSC	Cancer stem cell
CTCF	CCCTC-binding factor, site-specific DNA-binding factor with 11 zinc fingers that organizes chromatin
DNMT	DNA methyltransferase
Dnmt1	DNA methyltransferase 1; maintenance methyltransferase
Dnmt3A	DNA methyltransferase 3A2; de novo methyltransferase

Dnmt3B	DNA methyltransferase B; de novo methyltransferase
Dnmt3L	DNA methyltransferase 3 like; no enzymatic activity
DOT1L	DOT1 (disruptor of telomeric silencing) like, histone H3 methyltransferase
E2F1	E2F transcription factor 1
ER	Estrogen receptor
ERE	Estrogen-responsive element
ES cells	Embryonic stem cells
EZH2	Enhancer of zeste homolog 2; the methyltransferase component of PRC2 that trimethylates H3K27
FOXA1	Forkhead box protein A1, transcriptional activator
FOXP3	Forkhead box P3, transcription factor involved in immune response
GCN5	General control non-derepressible 5
H2A, H2B, H3, and H4	Core histones 2A, 2B, 3, and 4, respectively
H3K4	The fourth residue, K or lysine, on the N-terminus of histone H3
H3K9	The ninth residue, K or lysine, on the N-terminus of histone H3
H3K27	The 27th residue, K or lysine, on the N-terminus of histone H3
HAT	Histone acetyltransferase
HBO1	Human acetylase binding to ORC1, a HAT
HCC1954	Adherent, invasive ductal carcinoma cell line, ER and PR negative, HER2 overexpressed
HDAC	Histone deacetylase
HER2	Human epidermal growth factor receptor 2
HMEC	Human mammary epithelial cells
HP1	Heterochromatin protein 1, involved in gene repression
ICR	Imprinting control region
IGF2	Insulin-like growth factor 2, an imprinted gene
IL-6 and IL-8	Interleukin 6 and 8 respectively, cytokines
KAT	Lysine acetyltransferase
KMT	Lysine methyltransferase
LMR	Low-methylated regions
LOI	Loss of imprinting
LSD1 (KDM1)	Lysine-specific demethylase 1, demethylates H3K4me2 to H3K4me1 or H3K4me0
m^5C	Cytosine methylated on carbon 5
MAPit	Methyltransferase accessibility protocol for individual templates
MBD	Methyl-CG-binding domain
MBP	Methyl-CG-binding protein
MCF10A	Spontaneously immortalized, disease-free human mammary epithelial cell line

M.CviPI	First DNA methyltransferase from *Chlorella virus* P
MDA-MB-231	Invasive, adherent mammary adenocarcinoma cell line, ER negative, mutant p53
MDS	Myelodysplastic syndrome, hematological condition that often progresses to acute myelogenous leukemia
MeCP2	Methyl-CpG-binding protein 2, transcriptional regulator
MEK/ERK	Chain of extracellular signal-regulated kinases downstream of mitogen-activated Ras pathway
miRNA	MicroRNA, small noncoding RNA that regulates gene expression
MLH1	MutL homolog 1, TSG that encodes a mismatch repair protein
MLL	Myeloid/lymphoid, or mixed-lineage, leukemia protein, an HMT
MMTV	Mouse mammary tumor virus
MMTV-PyMT	Transgene encoding the MMTV long terminal repeat promoter upstream of the polyoma virus middle T antigen. Transgenic females develop palpable mammary tumors at 5–40 weeks of age that metastasize to the lung with 80–94 % incidence
NaB	Sodium butyrate, HDAC inhibitor
ncRNA	Noncoding RNA
NDR	Nucleosome-depleted region
NFR	Nucleosome-free region
NURF	Nucleosome remodeling factor, ISWI-containing chromatin remodeling complex that catalyzes ATP-dependent nucleosome sliding
ORC1	Origin recognition complex subunit 1
OSN	Osteopontin, small integrin-binding ligand *N*-linked glycoprotein
$p15^{INK4B}$	Cyclin-dependent kinase inhibitor protein, inhibits cell cycle progression
p300	E1A binding protein p300, transcriptional coactivator with HAT activity
p53	Protein 53 kDa, a tumor suppressor
PcG	Polycomb group
PCNA	Proliferating cell nuclear antigen
PHD	Plant homeodomain, protein domain that binds to various forms of methyl-lysine
PR	Progesterone receptor
PRC2	Polycomb repressive complex 2; the complex trimethylates H3K27, a mark associated with inactive transcription
RAD51	Radiation sensitive 51, RecA homolog involved in homologous recombination at DNA double-strand breaks
RAS	Rat sarcoma, family of small GTPases

RB	Retinoblastoma protein
RNAi	RNA interference
SAHA	Suberoylanilide hydroxamic acid, HDAC inhibitor
SANT	Swi3, Ada2, N-CoR, and TFIIIB
SATB1	Special AT-rich sequence-binding protein 1, global chromatin organizer
SET	Drosophila Su(var)3-9, Enhancer of zeste, and Trithorax
SIM2	Single-minded 2 gene, putative TSG in breast
SNF5	Sucrose non-fermenting 5, core subunit of the SWI/SNF chromatin remodeling complex
SUV39H1	Suppressor of variegation 3-9 H1, H3K9 KMT
SWI/SNF	Switch/sucrose non-fermenting; ATP-dependent chromatin remodeling complex
TBP	TATA binding protein
TET1–3	Chromosome ten-eleven translocation 1–3, three separate genes encoding methylcytosine dioxygenases
TFIID	Transcription factor IID
TGF-β	Transforming growth factor beta, cytokine
TNF	Tumor necrosis factor
TRAIL	TNF-related apoptosis-inducing ligand, cytokine
TRIM24	Tripartite motif-containing 24, an E3 ubiquitin ligase
TSA	Trichostatin A, HDAC inhibitor
TSG	Tumor suppressor gene
TSS	Transcription start site
UHRF1	Ubiquitin-like plant homeodomain and RING finger domain-containing protein 1, binds hemimethylated DNA and recruits HDACs
VPA	Valproic acid, HDAC inhibitor

11.1 Introduction

Breast cancer remains a worldwide clinical problem largely due to substantial molecular heterogeneity that complicates disease diagnosis and treatment decisions. Importantly, the frequency and degree of disease heterogeneity cannot be explained solely by genetic determinants. The discovery of and crucial control exerted over gene expression by epigenetic mechanisms that do not alter DNA sequence have placed them at the front line of cancer research. Breakdown in the tightly interwoven, but dynamic, regulation of these various epigenetic layers facilitates abnormal patterns of gene expression that, in turn, disrupts genome homeostasis and proper cell behavior. Epigenetic perturbations have been found at early stages of breast carcinogenesis and additionally in every cancer type studied to date and thus qualify as major determinants of cancer initiation and progression. Unlike diseases that involve genetic mutation, those driven by epigenetic alterations are potentially reversible and therefore constitute attractive therapeutic targets.

11.1.1 DNA Methylation

In differentiated mammalian cells, DNA methylation takes place largely, if not exclusively, at carbon 5 of the cytosine ring (m^5C) within CpG dinucleotides (hereafter, CG). Mammalian genomes are depleted for CG due to spontaneous hydrolytic deamination of methylated CG to TG over evolutionary time [1, 2]. By contrast, in conspicuously GC-rich regions known as CG islands (CGIs), CG frequency approaches the expected value, after normalizing for overall genomic G+C composition [3]. Approximately 60–70 % of human gene promoters contain one or more CGI within −2 kb to +1 kb of their transcription start site (TSS) [4–6]. The majority of housekeeping gene promoters possess at least one CGI, whereas only 40 % of promoters of tissue-specific genes do [7–9]. While most CG sites distributed throughout the genome are methylated (75 % of total CGs), CGs associated with promoter CGIs are maintained in an unmethylated state. Unmethylated CGIs are characterized by transcriptionally permissive chromatin that allows efficient gene expression in response to appropriate transcription factors [8–10]. Only 6 % of tissue-specific CGIs become methylated during early development or in differentiated tissues, rendering the associated genes transcriptionally silent [11, 12]. In general, promoter CGI hypermethylation has been strongly associated with stable transcriptional silencing. Not surprisingly, then, CGIs are crucial epigenetic regulatory sites in the genome (reviewed in [13–17]). Transcriptional silencing by promoter CGI methylation is thought to be dependent on the density of methylated CGs; however, detailed analysis of specific CG sites within a promoter CGI has provided evidence for regulation by a subset of CpG sites with crucial regulatory roles [18, 19]. Likewise, methylation of recently described CGI "shores," areas low in CG density located as far as 2 kb away from CGIs, has been shown to correlate with changes in gene expression [20, 21]. This regulation occurs irrespective of the methylation status of the CGI proximal to the promoter. CGI shores can also be methylated in a tissue-dependent manner, suggesting a role in alternative TSS usage.

A significant number of CGIs reside within intragenic regions and at the 3′ end of genes rather than within gene promoters. While such CGIs tend to be methylated, their role in transcription is not well understood; however, several lines of evidence point towards transcriptional silencing. Therefore, hypermethylated CGIs largely correlate with transcriptional downregulation, regardless of a genic *versus* intergenic location. A subset of intragenic CGIs may be associated with TSSs of uncharacterized genes with highly regulated expression patterns. For example, it has been shown that rare transcripts expressed during specific developmental stages originate from TSSs localized within intragenic CGIs [22–25]. Alternatively, intragenic CGIs can be associated with TSSs of regulatory noncoding RNAs (ncRNAs) or antisense transcripts, which can negatively regulate transcription of the primary gene with which they are associated. Prominent examples include *HOTAIR* [26], *AIR* [27], and *XIST* [28], which participate in epigenetic silencing of the *HOXD* gene cluster during mammalian axial development, the paternally imprinted *Igf2r*/*Slc22a2*/*Slc22a3* gene cluster, and one of the two X chromosomes in female mammals, respectively.

As mentioned above, at locations outside of known promoters, CGIs are prone to methylation that represses expression of these highly regulated transcripts or ncRNAs [29, 30]. In the same manner, and consistent with the high degree of CG methylation found at non-CGI locations, gene bodies of ubiquitously expressed genes are characterized by high levels of CG methylation. Gene body methylation has been proposed to protect against spurious transcription and to promote efficient transcriptional elongation [30–32].

Although largely correlated with promoter silencing, there are instances in which CG methylation promotes gene activation. As an archetype of this type of regulation, paternal methylation of the *H19*/*Igf2* imprinting center in the mouse relieves DNA-binding and enhancer-blocking activity of the insulator protein CTCF, allowing activation of *Igf2* expression [33–35]. Likewise, loss of CTCF binding due to DNA methylation of a CGI in the first intron of the human *BCL6* gene upregulated its transcription in lymphoma cells [36]. As opposed to the classical paradigm whereby DNA methylation blocks protein–DNA interactions, recent work of Vinson and colleagues has shown that DNA methylation enhances binding of C/EBPα and C/EBPβ to half cAMP-responsive elements [37, 38]. C/EBPα associates with methylated promoters with low CG content (i.e., non-CGI) to drive active transcription of several hundred differentiation-specific genes.

DNA methylation in vertebrates is catalyzed by the C-5 DNA methyltransferase (DNMT) family of enzymes, which enlist the universal methyl donor *S*-adenosyl-L-methionine. At least three DNMTs, namely, DNMT1, DNMT3A, and DNMT3B, are responsible for establishing and/or maintaining DNA methylation patterns in mammals [39, 40]. DNMT1 was the first mammalian DNMT described [41]. The enzyme is essential for embryonic development and is required to maintain bulk genomic methylation as shown by a mouse genetic knockout [42]. These studies and a 30- to 60-fold preference for hemimethylated DNA [10, 43–45] have led to the view that DNMT1 is the maintenance enzyme, despite its prominent de novo DNMT activity in biochemical studies [46]. Complete knockout of *DNMT1* in differentiated cells results in p53-mediated cell death [47]. Interestingly, knockout of *DNMT1* in human colorectal cancer cells also results in cell death, suggesting that cancer cell survival is dependent on DNMT1 activity as well [48].

DNMT1 is highly and ubiquitously expressed in proliferating cells, whereas low levels of the protein are present in nondividing cells. The transcript encoding DNMT1 also exhibits cell cycle-dependent regulation, achieving maximal levels during S phase [43, 47, 49, 50]. Additionally, DNMT1 interacts with the DNA polymerase-associated proliferating cell nuclear antigen (PCNA), which localizes to replication forks during S phase [51]. Association of DNMT1 with PCNA is essential for post-replicative conversion of hemi- to dimethylated DNA and thus maintenance of genomic patterns of DNA methylation [52–54]. Recently, DNMT1 has also been shown to interact with the ubiquitin-like plant homeodomain and RING finger domain-containing protein 1 (UHRF1), which binds to hemimethylated DNA [55–57]. This provides an additional mechanism for recruiting DNMT1 to hemimethylated DNA, possibly in stages of the cell cycle other than S phase.

DNMT3A and DNMT3B are referred to as de novo DNMTs and are thought to be responsible for establishing DNA methylation patterns, especially during early embryogenesis and germ cell development. Consistent with this view, the de novo DNMTs are highly expressed in embryonic tissues and undifferentiated embryonic stem (ES) cells and are downregulated in differentiated cells [16, 48, 58]. In contrast to DNMT1, DNMT3A and DNMT3B have no activity preference for hemimethylated over unmethylated DNA [45, 58]. A third member of the DNMT3 family, DNA methyltransferase 3-like (DNMT3L), is a catalytically inactive regulatory factor that directly interacts with and stimulates enzymatic activity of DNMT3A and DNMT3B [59, 60]. Targeted deletions of *Dnmt3a* and *Dnmt3b* in mouse models have demonstrated that the enzymes are essential for normal development [42, 58]. Interestingly, DNMT3A and DNMT3B can also catalyze non-CG methylation at CA and CT dinucleotide sites [61]. Non-CG methylation is prevalent in ES cells where it constitutes as much as 25 % of the DNA methylation content [62–64]. The biological function(s) of non-CG methylation remains unknown, although the modification is crucial for hematopoiesis, pluripotent stem cell differentiation, epigenetic reprogramming of zygotes, and leukemogenesis [65, 66].

The classification of DNMTs as either maintenance or de novo enzymes is an oversimplification, as several observations suggest that mechanisms by which DNA methylation is established and maintained are more complex than previously appreciated. For example, members of the DNMT3 family are required for maintenance of DNA methylation patterns of certain sequences, and DNMT1 cannot maintain established patterns of DNA methylation in ES cells in the absence of DNMT3 members [67, 68]. In other cases, DNA methylation patterns are maintained in the absence of DNMT1 [58, 67, 68].

DNA hypermethylation strongly correlates with repression or silencing of transcription in both normal and pathological regulatory environments, such as silencing of tumor suppressor gene (TSG) expression in cancer. While the precise mechanisms by which DNA methylation affects transcriptional activity remain elusive, two main views have emerged. The presence of methyl groups on the C-5 atom in the major groove can directly interfere with transcription factors binding to DNA elements, such as CREB1 binding to its cAMP-responsive element [37, 69]. Alternatively, transcription of methylated sequences can be silenced by methyl-CG-binding proteins (MBPs) [70–76]. A subset of MBPs contain domains that bind to 5-methyl-C, that is, methyl binding domains (MBDs), in various sequence contexts [77]. MBPs, such as the founding member methyl-CG-binding protein 2 (MeCP2), have been shown to be components of stable biochemical complexes that contain histone-modifying enzymes, such as histone deacetylases (HDACs) [70, 71, 78], which generally exert a repressive effect on transcription. Further evidence for the coordinated action of MBPs and HDACs comes from the observation that HDACs interact with chromatin remodeling complexes that also repress transcription [73, 79]. These early studies seeded the idea that transcriptional gene silencing mediated by DNA hypermethylation involves coordination between histone modifications and nucleosome occupancy/positioning, which regulates the access of *trans*-acting factors to DNA. It should be noted that, using domains distinct from its MBD,

MeCP2 is able to bind to and mediate higher-order folding of chromatin fibers containing unmethylated DNA [80–82], which may contribute to repression of transcription independent of histone deacetylation.

11.1.2 Nucleosome Positioning

Epigenetic regulation by DNA methylation within the eukaryotic nucleus occurs in the context of chromatin, the full complement of DNA associated with histone and nonhistone proteins. Chromatin can be subdivided into two broad classes: transcriptionally active euchromatin and inactive heterochromatin. These chromatin states profoundly influence diverse biological activities, including replication, repair, recombination, transcription, and mitotic chromatin condensation. The fundamental repeating unit of eukaryotic chromatin is the nucleosome, which is organized into arrays along chromosomes. Each nucleosome consists of a core particle, the histone octamer (two copies each of histones H2A, H2B, H3, and H4) wrapped by 1.65 turns (~147 bp) of DNA, plus a variable length of histone-free linker DNA [83].

The location of nucleosomes relative to each other and to a given point in DNA is termed translational positioning. Generally, access to DNA in nucleosome cores is impaired, especially at sites closer to the pseudodyad center. By contrast, sequences within linkers between nucleosome cores are relatively accessible to DNA-binding factors. Thus, nucleosome positioning is a key mechanism for regulating accessibility of *trans*-acting factors to their cognate *cis*-binding sites and hence the biological activity of DNA (reviewed in [84]).

The advent of innovative nucleosome mapping techniques that use next-generation sequencing has led to a proliferation of genome-wide analyses of nucleosome positioning (reviewed in [85]). A consistent theme emerging from these studies is the conserved, nonrandom positioning of nucleosomes across budding yeast, *Drosophila*, *C. elegans*, and human genomes. In the yeast genome, which has been extensively studied due to its small size, the propensity of a given DNA sequence to position nucleosomes is determined in part by its ability to bend around the histone octamer surface [86, 87]. Sequence-dependent flexibility of DNA explains the positioning of ~50 % of nucleosomes in the yeast genome, suggesting that other cellular events must govern nucleosome positioning in vivo (reviewed in [88]).

Thus, another nonrandom, highly conserved feature of genomic chromatin is the organization of open regions of chromatin, termed nucleosome-depleted regions (NDRs; also called nucleosome-free regions, NFRs), at regulatory regions, such as promoters and enhancers (reviewed in [84, 89–91]). In particular, at the 5′ end of genes, NDRs are located just upstream of or encompass the TSSs of transcriptionally permissive promoters. In humans, most promoters that contain CGIs and are transcriptionally active maintain an organization with an ~200 bp NDR localized upstream or near the TSS [92]. The open chromatin of NDRs at many promoters is thought to play a crucial role in transcription by presenting a platform for assembly of the basal transcriptional machinery. Consistent with this, in colon cancer cells, we and others have shown loss of a promoter NDR that correlates with CGI hypermethylation and transcriptional silencing [93, 94].

Importantly, even though the presence of an NDR is compatible with and supports transcriptional activation, it is not sufficient for transcription to occur. Many genes that are transcriptionally quiescent, but poised, share this stable and common open chromatin conformation [95–97]. Conversely, some constitutively repressed tissue-specific genes and some highly inducible genes lack NDRs and have a positioned nucleosome immediately upstream of their TSS [98, 99].

Promoter NDRs are flanked by arrays of positioned nucleosomes, usually discernible for two upstream (–2 and –1) and as many as six downstream (+1 to +6) nucleosomes. Each of these nucleosomes becomes progressively less well positioned with increasing distance from the TSS [98, 100–102], suggesting that the NDR is the "organizing center." An elegant study by Zhang et al. [103] showed that genome-wide formation of NDRs and their flanking positioned nucleosomes could be recapitulated in in vitro-assembled chromatin by addition of a whole cell extract that supports ATP-dependent chromatin remodeling. This highly dynamic repositioning of nucleosomes is likely to control transcription by facilitating or impeding access of transcription factors to their corresponding regulatory DNA sequences in gene promoters [99, 104]. The importance of chromatin remodeling to the proper orchestration of gene regulation is evidenced by frequent mutation of genes for various chromatin remodelers in cancer [105].

11.1.3 Histone Modifications

The core histones are subject to a myriad of posttranslational covalent modifications or marks, including acetylation, methylation, phosphorylation, ADP-ribosylation, ubiquitination, and sumoylation (reviewed in [106–110]). Many of these sites of modification reside in the N-terminal tails of the core histones, whereas others map to the globular domain of the nucleosome core (reviewed in [110, 111]). These modifications are also reversibly targeted to their respective lysine, arginine, threonine, and serine residues by an ever-increasing number of histone modifiers [110]. In this vein, steady-state levels of histone acetylation are governed by the opposing activities of histone or lysine acetyltransferases (HATs or KATs) [112] and histone deacetylases (HDACs). In addition, accumulated levels of histone methylation are modulated by the coordinated activity of histone or lysine methyltransferases (HMTs or KMTs) and histone or lysine demethylases (KDMs) [113]. Histone-modifying enzymes vary in their levels of specificity. For example, KATs and HDACs generally target multiple residues in core histones as well as other protein substrates. In contrast, KMT, KDM, and kinases apparently exhibit high residue specificity [110, 114]. Therefore, histone marks constitute a dynamic and complex regulatory network that influences genome functions.

Posttranslational modifications of histones are key players in the global transcription landscape of both euchromatin and heterochromatin [108, 110, 115, 116].

High levels of histone acetylation correlate with transcriptional activation, while histone hypoacetylation is associated with transcriptional repression, both globally and at the level of specific genes. Consistent with this, histone acetylation is enriched at promoters and enhancers and is maintained at low levels in gene bodies to prevent spurious transcription [108, 110, 117].

Histone modifications, either singly or in combination, can drive specific signaling outputs in chromatin (reviewed in [118]). Structurally, the hypoacetylated N-termini of the core histones mediate interactions between adjacent nucleosomes, facilitating higher-order compaction [119–125]. Acetylation of histone H4K16 alone is sufficient to inhibit compaction of 30 nm-like fibers [126]. Posttranslational histone modifications also recruit biochemical activities to nucleosomes by serving as docking sites on nucleosomes for distinct protein domains contained within various nuclear proteins [85, 110, 127]. These domains include the bromodomain that binds to acetyllysine residues in histone N-termini as well as the chromodomain and plant homeodomain (PHD) that interact with various states of methyl-lysine, i.e., un-, mono-, di-, and trimethylated (reviewed in [128, 129]). Many such domains are found within subunits of various chromatin remodeling complexes and in histone-modifying enzymes themselves, thereby creating positive feedback that reinforces specific chromatin states (reviewed in [107, 110, 114, 130]). For example, several KATs, e.g., GCN5 and P300, contain bromodomains that tether them to their acetyl-lysine chromatin product, creating a positive feedback loop of acetylation [131–134]. Histone modifications can also effect nucleosome repositioning (also called nucleosome sliding) and/or disassembly through interactions with ATP-dependent chromatin remodeling complexes (reviewed in [127]). For example, Hassan et al. [132, 135] showed that acetylated nucleosomes stabilize bromodomain-dependent recruitment of SWI/SNF.

Methylation of specific histone residues also elicits distinct outcomes with respect to transcription. Trimethylation of histone 3 lysine 4 (H3K4me3) preferentially localizes to the TSSs of recently transcribed promoters [136–139]. H3K4me3 has been shown to mediate gene activation through interaction with nucleosomes of the PHD motifs in subunits of chromatin-associated complexes, such as the general transcription factor TFIID, chromatin remodeler NURF, and HBO1 KAT complex [140–145]. Alternatively, H3K4me3 can lead to active gene repression through PHD-mediated recruitment of HDAC-containing complexes [146, 147]. Histone 3 lysine 36 trimethylation (H3K36me3), although strongly associated with actively transcribed gene bodies (reviewed in [148]), recruits a complex with HDAC activity via two subunits, one of which contains a chromodomain and the other a PHD motif [114, 149, 150]. This HDAC recruitment is thought to suppress inappropriately initiated or cryptic transcription at sites besides TSSs by aiding refolding of the chromatin fiber in the wake of advancing RNA polymerase II [116, 151]. On the other hand, trimethylation of histone H3 lysine 9 (H3K9me3) and histone H3 lysine 27 (H3K27me3) are strongly associated with transcriptional silencing. H3K9me3 and H3K27me3 are catalyzed, respectively, by a family of SET (*Drosophila* Su(var)3-9, Enhancer of zeste, and Trithorax)-domain-containing KMTs (e.g., G9a and Suvar39) and the SET-domain protein EZH2, which is a component of the PRC2 Polycomb complex.

Regulation of histone posttranslational modifications is often redundant, as placement or removal of one mark can be catalyzed by more than one enzyme. In the same manner, different signaling pathways can be used to achieve the same transcriptional output. For example, H3K9me3 and H3K27me3 are both thought to influence chromatin accessibility and induce transcriptional silencing. Comparative genome-wide studies have shown that most genes are silenced by either H3K9me3 or H3K27me3 [152, 153]. Genes silenced in the presence of either of these marks exhibit histone deacetylation (specifically at H3K9 and H3K14) and decreased H3K4me3. However, H3K27me3-mediated silencing can be achieved independent of DNA methylation, while the proposed model for H3K9me3-mediated silencing requires DNA methylation [154–156].

11.1.4 Interplay of Epigenetic Mechanisms

The coordinated action of DNA methylation, histone modifications, and nucleosome positioning is crucial in governing gene expression in diverse cellular processes. These interactions can produce linear signaling outputs or can employ redundant mechanisms that ultimately reinforce each other (reviewed in [157–161]). Despite intense study, our knowledge of the complex signaling networks between the various layers of epigenetic regulation remains limited.

Nevertheless, several lines of evidence support cross talk between various levels of epigenetic regulation. For example, TSG promoter silencing has been linked to co-localization of DNA hypermethylation and repressive histone marks. Heterochromatin protein 1 (HP1), which contains a chromodomain that binds to repressive H3K9me3, is present in a biochemical complex with the SUV39H1 H3K9 KMT, creating positive reinforcement between heterochromatin posttranslational marks and DNA methylation [162]. The association of DNA hypermethylation with nuclease-resistant chromatin, characteristic of regions with dense arrays of nucleosomes, connects DNA methylation and nucleosome occupancy [159, 163]. MBD-containing proteins are also found in complexes with HDAC and H3K9 KMT activities [70, 71, 73, 79, 162], which may increase nucleosome occupancy by inhibiting recruitment of chromatin remodelers that disassemble nucleosomes.

More recently, an interesting connection was described between H3K4me demethylation, the LSD1 (KDM1) histone demethylase, and DNMT1, whereby conditional deletion of the *Lsd1* gene in mouse ES cells led to progressive loss of global DNA methylation [164]. Interestingly, this effect was traced to methylation-dependent destabilization of the DNMT1 enzyme. While loss of LSD1 in this system did not change total levels of K3K4me3, in pluripotent human ES cells, K3K4me3 shows a strong inverse correlation with DNA methylation [165]. Another possible explanation for this inverse relationship is provided by the demonstration that the cysteine-rich ADD (ATRX-DNMT3A-DNMT3L) domain in the C-terminus of DNMT3L interacts preferentially with histone H3 that is unmethylated at Lys 4 [166]. Consistent with these results, the protein Cfp1 selectively binds non-methylated CGs in a majority of CGIs and associates with the human Set1 complex, an H3K4 KMT [167–170].

Elucidating the complex interplay between distinct epigenetic layers is a key challenge facing the field. For example, it is generally accepted that CG methylation influences nucleosome positioning/occupancy. However, recent data also suggest the reverse: that nucleosome positioning can dictate accessibility of DNMTs to DNA and hence DNA methylation patterns, confounding attribution of causality [127, 171]. An important step in unraveling the interconnectivity of epigenetic regulation is the development of integrative epigenetic methods that can directly correlate different epigenetic features, not only in a single experiment but on the same DNA molecule. Our lab has developed a technique that, combined with bisulfite sequencing, allows for simultaneous single-molecule level detection of DNA methylation and chromatin accessibility. The DNA methyltransferase accessibility protocol for individual templates (MAPit) localizes protein–DNA interactions by probing with cytosine-modifying DNA methyltransferases that differ in specificity from endogenous DNMTs. Sequencing individually cloned DNA products after amplification of bisulfite-converted sequences permits assignment of the methylation status of every enzyme target site along a single DNA strand. Use of the GC-methylating enzyme M.CviPI [172] allows simultaneous mapping of chromatin accessibility and endogenous CG methylation [173, 174]. In addition, three groups have recently integrated chromatin immunoprecipitation (ChIP) with bisulfite genomic sequencing to directly link specific histone marks with DNA methylation on single molecules [175–177]. Such integrative methods should prove valuable in illuminating the mechanisms by which distinct epigenetic events work together to establish and maintain transcriptional programming in both normal and neoplastically transformed cells.

11.1.5 Epigenetics in Tumorigenesis

Cancer cells are characterized by misregulated epigenetic environments compared to their nonmalignant counterparts. Alterations in each layer of epigenetic control have been described in essentially every cancer type. Cancer cells exhibit global DNA hypomethylation (around 20–60 % overall reduction in m^5C) when compared to normal cells. The exact role of global DNA hypomethylation in cancer initiation and progression remains poorly understood [178–180]. It has been proposed that global loss of m^5C in cancer cells compromises gene repression of loci that are normally transcriptional inactive or silenced, like repetitive sequences, retrotransposons, and centromeres [181]. Hence, the main contribution of hypomethylation to cancer etiology is thought to be through the reactivation of retrotransposons, endoparasitic elements, and increased incidence of chromosomal rearrangements at repetitive sequences [182–185]. Thus, epigenetic changes in DNA methylation have deleterious consequences on the integrity of the cancer cell genome. Gene-specific hypomethylation also commonly contributes to cancer progression through the reactivation of proto-oncogenes, e.g., N-*myc* [186]. Loss of imprinting (LOI) of methylated alleles of *IGF2* in breast, colon, lung, and liver cancers has also been reported [187–189]. Reduced DNA methylation in gene bodies may also contribute to aberrant cryptic transcription and gene expression, but more research in this area is needed [31].

A well-characterized epigenetic disruption in cancer cells is DNA methylation-dependent silencing of specific tumor suppressor genes (TSG) [15]. Gene-specific hypermethylation can occur at CGIs in promoters or shores, resulting in transcriptional silencing of the associated gene. Of note, CGI shore DNA methylation can affect TSS selection and contribute to cancer [17, 21, 190]. TSG silencing has been shown to affect important cellular pathways, including DNA repair, cell cycle control, Ras signaling, apoptosis, metastasis, detoxification, and responses to hormones and vitamins. Inactivation of these pathways enhances cancer cell proliferative capacity and genetic instability (reviewed in [160]). The correlation between promoter CGI hypermethylation and TSG silencing in cancer has been firmly established. As a result, silenced TSGs have proven useful as biomarkers for diagnostic and prognosis as well as to inform treatment of several cancer types (reviewed in [191]). Nevertheless, the mechanism(s) by which aberrant DNA methylation is established at these previously transcriptionally active chromatin regions and how it contributes to stable aberrant gene silencing remain ill defined (reviewed in [30, 161, 190]).

Histone modification patterns are also severely altered in human tumors. Genome-wide studies of histone H4 modifications in normal and tumor-derived cell lines show that transformed cells exhibit global reduction of H4K16ac1 and H4K20me3. Global loss of H3K4me3 as well as H3 and H4 acetylation has also been reported in cancer cells. These losses can lead to disruption of heterochromatic regions associated with repetitive sequences and parasitic elements, thereby generating genomic instability and predisposing cells to cancer development [163, 192, 193]. Histone modifications also contribute to gene-specific TSG silencing in cancer cells. Aberrant hypermethylation of TSG promoter sequences is accompanied by a reduction of active histone marks (e.g., histone acetylation and H3K4me3) and accumulation of repressive marks (e.g., H3K9me3 and/or H3K27me3).

Little data regarding nucleosome positioning in cancer are available. We and others have shown that, in cancer, methylated TSG promoters undergo chromatin reorganization that involves loss of nucleosome depletion at TSSs [93, 94]. Nucleosome positioning and chromatin organization are profoundly affected by deregulation of chromatin remodeling complexes, which mobilize nucleosomes in an ATP-dependent manner [194–196]. For example, mutations in genes coding for several subunits of the highly conserved SWI/SNF chromatin remodeler have been associated with oncogenesis [104]. For instance, mutant forms of BRG1, an ATPase subunit of the SWI/SNF remodeler, have been reported in colon, breast, pancreatic, and lung cancers. Mice heterozygous for *SNF5*, encoding an essential core subunit of SWI/SNF, exhibit a high predisposition to aggressive and metastatic cancers [197, 198]. Further studies are needed to determine if the oncogenic effect of SNF5 and BRG1 is dependent solely on the remodeling activity of SWI/SNF. Further investigation is also needed to resolve current controversies concerning the temporal sequence of molecular events accompanying epigenetic gene silencing and to shed light on how epigenetic events contribute to different stages of cancer progression.

11.2 Breast Cancer Epigenetics

11.2.1 DNA Methylation in Breast Cancer

Global DNA hypomethylation is a hallmark of cancer progression and is a frequent occurrence in breast tumors [199, 200]. This aberrant hypomethylation is mainly observed at repetitive DNA sequences and satellite DNA that are heavily methylated in normal cells. It is proposed that hypomethylation induces transcriptional reactivation of transposable repetitive elements, thus leading to insertional mutagenesis and contributing to genomic instability. The cause(s) of global hypomethylation remains elusive. Loss of DNMT expression is rarely observed in breast cancer; rather increased expression, especially of DNMT3B, is frequent. It has been suggested that aberrant expression of DNMT splice variants may contribute to the methylation changes (reviewed in [201]). An alternate possibility is that active DNA demethylation may be stimulated. The chromosome ten-eleven translocation (TET1-3) proteins have recently been implicated as DNA demethylases (reviewed in [202]). These proteins are proposed to oxidize m^5C to an intermediate metabolite (either 5-hydroxymethyl-C, 5-formyl-C, or 5-carboxyl-C) which is then recognized and excised by DNA glycosylases and resolved through base excision repair [65, 203, 204]. It has been reported that the tumor suppressor adenomatous polyposis coli (APC) regulates expression of proteins involved in active demethylation and that loss of APC function results in aberrant DNA hypomethylation [205]. This suggests that active demethylation may play a role in breast tumorigenesis. Interestingly, a recent study by Thillainadesan et al. [206] found that stimulation of TGF-β signaling caused rapid demethylation at the $p15^{INK4B}$ tumor suppressor locus as well as global decreases in m^5C levels in breast cancer cells. Note that TGF-β is initially tumor suppressive [207, 208] and often dysregulated and associated with malignancy in breast cancer cells (reviewed in [209]).Thus, oncogene overexpression and/or tumor suppressor gene silencing may facilitate epigenetic reprogramming through the aberrant regulation of the DNA demethylase machinery.

Global hypomethylation is often accompanied by local promoter hypermethylation and silencing of tumor suppressor genes. Though not clearly understood, it appears that the process of aberrant de novo methylation is not random. Rather, certain genes are consistently hypermethylated and silenced among diverse cancer types, while some hypermethylation events are restricted to particular tumor types or to a particular subtype or stage during tumor progression. These differentially methylated regions could potentially be exploited for prognostic or diagnostic purposes.

Gene silencing by DNA hypermethylation is a common event in a variety of neoplasms. In colorectal cancer, a distinct subset of tumors was identified as having a set of consistent, concordantly methylated genes. This phenotype, referred to as the "CGI methylator phenotype" (CIMP), has been observed in additional tumor types and often informs disease progression, treatment decisions, and patient survival outcomes (reviewed in [210, 211]). Several studies examining global DNA methylation profiles in breast cancer cell lines and in tumor tissues have been conducted.

Table 11.1 Hypermethylated gene biomarkers for breast cancer

Subtype specific [212]	
Luminal A	*MMP7, PEG10, RASSF1*
Luminal B	*GSTP1, CHI3L2, KIT, LYN, MMP7, MYBL2, RASSF1*
HER2⁺	*GSTP1*
Basal-like	*ARGHDIB, CHCL9, GRB7*
Receptor associated [213]	
ER⁺	*EVI1, ETS1, IRF7, LYN, PDXK, PTGS2, RUNX3, VIM, ACADL, ADAMTSL1, ARFGAP3, B3GAT1, CDCA7, FAM78A, FAM89A, FLJ31951 (RNF145), FLJ34922 (SLFN11), GAS6, HAAO, HEY2, HOXB9, ITGA11, NETO, PROX1, PSAT1, RECK, SMOC1, SND1, TNFSF9*
ER⁻	*DAB2IP, HSD17B4, PER1, ADHFE1, DYNLRB2, HSD17B8, PISD, PDXK, WNK4*
CIMP associated [214]	
CIMP⁺	*ALX4, ARHGEF7, SOX8, FBN1, FOXL2, RASGRF2*
Tumor suppressor genes (reviewed in Yang et al. 2001 [216])	
	CDKN2A, HME1, ESR1, BRCA1, RARB2, CDH1, PGR, TIMP3

These studies (summarized below) have provided important insights into different mechanisms that may drive subtype-specific gene expression and tumor characteristics. DNA methylation biomarkers identified in breast cancer are listed in Table 11.1.

In Holm et al. [212], DNA methylation was profiled at 801 cancer-specific genes in 189 breast tumors and 4 normal breast samples. They observed that 25 % of molecular classifiers based on gene expression in breast cancer were subject to subtype-specific DNA methylation. This indicates an important role for DNA methylation in regulating gene expression in specific subtypes of breast cancer. DNA methylation profiles also differ between tumors mutated in *BRCA1* and *BRCA2*, indicating that epigenetic features can vary in hereditary as well as sporadic disease. It should be noted that BRCA1 has recently been reported to directly increase DNMT1 activity which could in part explain the observed methylation differences [216]. Subtype-specific DNA methylation signatures discriminated between basal-like, luminal A, and luminal B tumors, whereas normal-like and *HER2*-amplified tumors did not present a specific signature. Overall, the DNA methylation levels among the identified biomarkers were highest in luminal B tumors, lower in luminal A tumors, and lowest in basal-like tumors. Interestingly, basal-like tumors expressed higher levels than luminal tumors of the histone methyltransferase EZH2 and its chromatin mark H3K27me3. PRC2-mediated gene silencing through H3K27me3 is commonly utilized by stem and progenitor cells to maintain pluripotency [217, 218]. In addition, previous studies have suggested that basal-like tumors develop from transformed mammary progenitor cells, whereas luminal tumors develop from transformation of more differentiated luminal epithelial cells [219]. The DNA methylation results are consistent with this cell-of-origin distinction.

Subsequent global studies supported and expanded on the Holm et al. [212] results. In Fackler et al. [213], CGs at over 14,000 human genes (mostly in promoters) were queried in 103 breast tumor and 21 normal samples. A DNA methylation signature was obtained that could stratify tumors based on estrogen receptor (ER) status with increased DNA methylation observed in ER^+ tumors. A "recurrence set" was also identified, whereby the DNA methylation status at 100 loci could predict disease recurrence with greater sensitivity than gene expression profiling. Notably, 20 % of the 100 recurrence loci were from homeobox-containing genes, the products of which are likely involved in developmental regulation.

Though the above mentioned studies hinted at a potential CIMP in breast cancer, strong evidence for global CIMP in breast cancer was recently reported by Fang et al. [214]. By profiling 39 tumors at over 14,000 genes (mostly promoters), they observed that 44 % of the tumors tested were, by their definition, $CIMP^+$. These tumors were almost exclusively ER^+ and PR^+ (94 %) and associated with a lower propensity for metastasis and a better clinical outcome than $CIMP^-$ tumors. The $CIMP^-$ tumors comprised some ER^+ and PR^+ (45 %), but mostly basal-like (55 %), tumors. By testing the DNA methylation status at their three most informative loci in a validation set of 132 additional tumors, $CIMP^+$ tumors demonstrated a significantly lower risk for metastatic relapse and death independently of other known prognostic factors. Notably, many of the hypermethylated genes occurred at Polycomb group (PcG) targets, in agreement with the Holm et al. [212] findings. It was recently determined that, in addition to PcG targets in adult stem/progenitor cells, loci identified as hypermethylated in breast cancer are commonly occupied by bivalent chromatin marks in ES cells and are largely directed towards stable silencing of developmental regulators [220]. The breast cancer $CIMP^+$ signature was compared to that of glioma and colon cancer, and while some overlap occurred, many genes informing $CIMP^+$ tend to be tissue-type specific and associated with cancer-type-specific clinical outcomes. Whether breast cancer CIMP will emerge to be as important for prognosis or personalized treatment decisions as colon cancer CIMP remains to be determined.

It should be noted that the DNA methylation information reported in the above mentioned studies was obtained using approaches based on early versions of array hybridization. Although this strategy offers the advantage of profiling DNA methylation at thousands of genes in hundreds of samples, a disadvantage is that the CpG probes tend to be biased towards gene promoters and that the methylation status of a particular gene is classified based on the average methylation status at only 1 or 2 CG sites. This precludes interrogation of DNA methylation in distal non-promoter regulatory sites, e.g., CG shores, which can often exert transcriptional control over adjacent genes [21]. Also, some studies have shown that disease-relevant perturbations exist in regions that are not completely hypermethylated but rather exhibit partial methylation [221]. The suggestion is that these regions are more "plastic" and can be more readily modified in response to environmental cues over the course of tumorigenesis than regions that are hypermethylated and tightly silenced.

Shotgun bisulfite sequencing of HCC1954 breast cancer and normal human mammary epithelial cell (HMEC) DNA has provided a detailed view of global DNA methylation patterns and how they differ between normal and cancer cells [222]. This unbiased view of global DNA methylation found that the $HER2^+$ cell line was epigenetically distinct from normal cells. The cancer cells exhibited extensive hypomethylation of coding and intergenic regions that correlated directly with gene silencing and were enriched for H3K9me3 or H3K27me3. These blocks of repressive chromatin were also often peppered with promoter hypermethylation. Their data suggested that, in addition to gene silencing through local promoter hypermethylation, global hypomethylation in breast cancer cells may lead to tumor suppressor gene silencing through the formation and expansion of repressive chromatin domains. This mechanism of long blocks of repressive chromatin effecting selective gene silencing is also characteristic of ES cells [62]. Thus, more extensive characterization of breast cancer methylomes and chromatin features may further strengthen the sensitivity of epigenetic-based molecular subtyping and may expand this strategy to include subtypes that were indistinguishable when limiting the analysis to DNA methylation at gene promoters.

It is now clear that the DNA methylomes of breast cancer cells are markedly different than those of normal cells. A pervading question is, what drives or contributes to the re-patterning of DNA methylation during tumorigenesis? Though often evaluated separately, increasing evidence indicates that global and local DNA methylation patterns are associated with specific active or repressive histone marks [223]. The concept of chromatin-directed DNA methylation is supported by several studies. For example, it has been shown that DNA in nucleosomes is protected from modification by DNMT3A, which preferentially methylates linkers [224–226]. Also, certain structural domains on proteins bind to specific histone marks that can either exclude or direct DNMTs to the neighboring DNA [156, 227]. Studies supporting the opposing view of DNA methylation-directed chromatin remodeling are also prevalent. The presence of DNA methylation can prevent the binding of certain proteins to DNA. For example, it has been shown that DNA methylation prevents binding of the PRC2 complex to chromatin [228–230]. Further, it has been observed that inhibition of DNA methylation by 5-azacytidine in human embryonic kidney cells results in global increases of H3K9me3 and H3K27me3 [231]. These data also support the observation that DNA methylation and H3K27me3 are mutually exclusive repressive marks [222], though this mutual exclusivity was recently challenged based on data acquired using a newer, more integrative chromatin technology that queries histones marks and DNA methylation on the same DNA strands [176, 177]. Though the initiating mechanism(s) remains elusive, aberrant DNA methylation and chromatin structure are both important variables in understanding breast cancer biology.

11.2.2 Chromatin Modifiers in Breast Cancer

Chromatin-modifying enzymes exert control over multiple target regions in the genome; thus, the altered expression of thousands of tumor-promoting genes could be attributed to aberrant expression of a single chromatin regulator. SATB1 is a nuclear protein that functions as a "genome organizer." Essential for proper T-cell development, SATB1 mediates a functional nuclear architecture that has a "cage-like" protein structure surrounding heterochromatin [232]. SATB1 regulates gene expression by recruiting and tethering chromatin remodeling and modifying enzymes and transcription factors to DNA. On T-helper cell activation, SATB1 becomes expressed and folds the 200 kb locus encoding T-helper 2 cytokine on mouse chromosome 11I into dense loops for rapid induction of multiple cytokine genes [233]. In breast cancer, SATB1 becomes expressed during malignancy and markedly induces gene expression to promote an aggressive phenotype that supports both tumor growth and metastasis. Positive nuclear staining for SATB1 also served as an independent marker of poor prognosis [234]. While the mechanism of SATB1 overexpression in aggressive breast cancer is unclear, a recent study found that SATB1 transcript and protein expression could be repressed by FOXP3 and its downstream miRNA targets [235]. FOXP3 is a member of the forkhead/winged helix family of transcription factors, first identified in T-regulatory cells as essential for cellular identity and function [236]. FOXP3 is an X-linked gene and has been described as having tumor-suppressive functions [237]. Furthermore, loss of FOXP3 function has been reported in primary breast cancers, and it has been suggested that loss of FOXP3 function may support SATB1 overexpression. As the X-inactive allele of FOXP3 is likely functional (i.e., genetically wild type), epigenetic rescue of FOXP3 expression may be a viable strategy for repressing SATB1 expression in malignant breast cancer.

Defects in histone modifiers including KAT and HDAC enzymes and aberrant expression of variant histones have been observed in various stages of breast cancer progression [238, 239]. As histone-modifying enzymes are considered highly "druggable" targets, this constitutes an exciting area for pharmacologic development (discussed in Sect. 11.3). Differences in the expression of chromatin-modifying enzymes have been observed in breast cancer. Genes encoding EZH2 and other PcG proteins responsible for H3K27me3 are frequently overexpressed in breast cancers [240, 241], and point mutations that lead to loss or gain of function have also been reported in these genes [242–244]. As described in the previous section, gene silencing in $HER2^{+}$ breast cancer cells seems to be characterized by long blocks of repressive histone marks that could be a consequence of aberrant PcG protein expression. Additional histone modifiers that have been reported to be aberrantly expressed in breast cancer are listed in Table 11.2.

11.2.3 Chromatin Landscapes in Breast Cancer

In addition to aberrant expression of chromatin-modifying or-organizing proteins, large-scale redistribution of transcription factor binding also contributes to global changes in chromatin structure in cancer cells. Differences in breast cancer-associated ER chromatin occupancy have recently been reported. Ross-Innes et al. [264] mapped ERα occupancy using ChIP followed by deep sequencing (ChIP-seq) of primary breast tumors and metastases. Differential binding analysis revealed two sets of ERα-bound genomic regions: the set from the good prognosis tumor group mostly contained estrogen-responsive elements (EREs), whereas the set associated with the poor prognosis tumor group contained EREs and FOXA1 binding motifs

Table 11.2 Histone modifiers aberrantly expressed in breast tumors

Enzyme	Type	Substrate	Alteration	Reference
EZH2	KMT	H3K27	Mutation, amplification	[240, 245–247]
PRDM14	KMT	Unknown	Amplified	[248–250]
SMYD3	KMT	H3K4	Increased expression	[251]
WHSC1L1	KMT	H3K36	Amplified	[252, 253]
PRDM5	KMT	Unknown	Reduced expression	[254]
CARM1	RMT	H3R17	Increased expression	[255, 256]
KDM1 (LSD1)	KDM	H3K4me, H3K9me	Increased expression	[257]
KDM4A (JMJD2A)	KDM	H3K9me2/3, H3K36me2/3	Increased expression	[239]
KDM6A (UTX)	KDM	H3K27me2/3	Increased expression	[239]
KDM5B (PLU-1)	KDM	H3K4	Increased expression	[258]
HBO1	HAT	H4K5, H4K12	Increased expression	[259]
MOF	HAT	H4K16	Decreased expression	[260]
HDAC1	HDAC	Multiple	Increased expression	[239]
HDAC2	HDAC	Multiple	Decreased expression	[261]
HDAC5	HDAC	Multiple	Increased expression	[239]
HDAC3	HDAC	Multiple	Increased expression	[262]
HDAC6	HDAC	Multiple	Decreased expression	[263]
CHD5	CHD	Multiple	Decreased expression	[263]

Abbreviations: *KMT* lysine methyltransferase, *RMT* arginine methyltransferase, *KDM* lysine demethylase, *HAT* histone acetyltransferase, *HDAC* histone deacetylase, *CHD* chromodomain/helicase/DNA-binding protein

[264]. FOXA1 is a pioneer factor that has been shown to direct ER chromatin binding [265]. Together with additional experiments in breast cancer cell lines, the data from both papers indicates that FOXA1 reprograms ER target gene selection in treatment-resistant cancer cells from EREs to FOXA1-bound loci. Though they did not assess the effects of global ER redistribution on overall chromatin structure or DNA methylation, one can envision a scenario whereby loss of ER binding to EREs, recruitment of ER to FOXA1-bound chromatin, or both could facilitate nucleosome repositioning to accommodate loss or gain of factor binding. This scenario has been observed in ChIP-seq experiments conducted on androgen-stimulated prostate cancer cells, whereby androgen treatment resulted in nucleosome repositioning and depletion from androgen receptor binding sites [266, 267]. The repositioning of nucleosomes could then support further epigenetic reprogramming by redirecting DNA methylation accordingly. Recall that the breast cancer CIMP$^+$ phenotype was restricted to ER$^+$ tumors, further supporting a link between ER and DNA methylation. The redirection of ER binding from EREs to FOXA1-bound chromatin may also lead to increased DNA methylation at EREs. This is possible because transcription factor binding has recently been shown to create low-methylated regions (LMR) in mouse ES cells and loss of factor binding potentiated accumulation of DNA methylation [268].

The modifications applied to histone tails by histone-modifying enzymes must be "read" and "interpreted" by additional chromatin-interacting proteins. The E3-ubiquitin ligase, tripartite motif-containing 24 (TRIM24), is a chromatin regulator that has been reported to function as a reader of dual histone marks through tandem PHD and bromodomain motifs [269]. Structural and biochemical analyses suggest that the PHD-bromodomain region of TRIM24 serves as a functional unit for combined recognition of unmodified H3K4 and acetylated H3K23 within the same histone tail. TRIM24 chromatin binding overlaps with ER binding, especially upon estrogen stimulation, and augments activation of estrogen-dependent genes associated with cellular proliferation and tumor development. Global analysis of chromatin interactions showed estrogen-stimulated binding of TRIM24 and ERα at sites that exhibit loss of H3K4me2 and gain of histone acetylation. Interestingly, while estrogen-stimulated recruitment of TRIM24 to chromatin overlaps with a similar number of ER binding regions as FOXA1, very few regions are bound by both TRIM24 and FOXA1. This could be because TRIM24 preferentially binds to EREs depleted of H3K4me2, but enriched for H3 acetylation [269], whereas FOXA1 binding occurs primarily at distal enhancers enriched in H3K4me2 [270]. However, it should be noted that ERα recruits KATs [271] and LSD1 (demethylates H3K4me2/1) to chromatin [272]. Therefore, without high-resolution time course experiments, it is difficult to determine whether the chromatin differences existed before or after differential binding by each transcription factor. Regardless, it is clear that estrogen stimulation regulates different classes of target genes depending on specific chromatin features, such as H3K4 methylation status. Importantly, aberrant overexpression of TRIM24 was frequently observed in breast cancer and directly correlated with poor survival in patients with either ER^+ or ER^- tumors [269]. These results support further evaluation of TRIM24 as a therapeutic target for breast cancer.

11.3 Targeting Epigenetics in Cancer Treatment

11.3.1 Demethylation Agents

The first epigenetic modulator in clinical use was the demethylating agent, 5-azacytidine (5-aza). In a Phase 1 clinical trial reported in 1972, 30 patients with advanced solid tumors were treated with 5-azacytidine [273]. Responses were seen in 7 of 11 patients with breast cancer, 2 of 5 patients with melanoma, and 2 of 6 patients with colon cancer. However, significant toxicities dampened interest in the drug. In 1993, the results of a single-arm Phase 1/2 trial in patients with myelodysplastic syndrome (MDS) treated with 5-aza indicated that longer assessments were necessary to see a significant effect on outcome [274]. Following a Phase 3 trial, 5-aza was approved for use in patients with MDS in 2004.

To date, the most common DNA demethylating agents in clinical use as anticancer drugs are 5-aza and 5-aza-2′-deoxycytidine (decitabine). Though originally developed as a nucleoside antimetabolite, it was later determined that 5-aza treatment results in loss of DNA methylation. The proposed mechanism is that 5-aza incorporates into RNA and DNA and forms a covalent bond with DNMTs, thus "trapping" them to DNA and effecting DNMT depletion. Decitabine is a structurally similar compound that selectively incorporates into DNA and was therefore expected to be a less toxic and more specific alternative to 5-aza [275–278]. Decitabine has been extensively tested in the clinic and approved by the U.S. Food and Drug Administration (FDA) for treatment of acute myelogenous leukemia (AML). However, both 5-aza and decitabine are toxic and highly unstable in aqueous solutions (reviewed in [279]). This makes both drugs difficult to use in clinical settings, especially in treating solid tumors. Thus, development of DNMT inhibitors with more favorable properties remains of high interest.

A new DNMT inhibitor with promising translational potential is zebularine (reviewed in [280]). Zebularine is also a cytidine analog, but it is more stable and less toxic than 5-aza and decitabine. Zebularine was originally developed as a cytidine deaminase inhibitor but was later shown to potently inhibit DNMT activity [281] and cancer cell growth [282–284]. The mechanism of action is largely the same as 5-aza and decitabine; however, lower toxicity and longer biological half-life make it an attractive candidate for preclinical testing. It has been shown that zebularine can be administered orally due to its longer half-life than other DNMT inhibitors and can inhibit tumor growth and induce expression of tumor suppressor genes [133, 134, 285]. Subsequent xenograft studies showed that short-term treatment with zebularine [286] or even a single injection [287] can inhibit tumor growth. Studies in genetically engineered mice have also been conducted to evaluate long-term therapy with an oral formulation of zebularine for intestinal adenomas [288] and mammary tumors [289]. In our study, high-dose zebularine treatment delayed breast tumor growth and reduced tumor burden in an MMTV-PyMT mouse model. In the Yoo et al. [288] study, continuous treatment of Min transgenic mice with low-dose zebularine prevented intestinal polyp formation in the majority of treated animals, whereas controls all developed polyps. These studies suggest that zebularine is an attractive target for future clinical trials.

11.3.2 Histone Deacetylase Inhibitors

Histone deacetylase (HDAC) inhibitors were discovered in the 1970s, when it was shown that treatment of cells with the short-chain fatty acid sodium butyrate (NaB) led to hyperacetylation of histones [290]. In the following two decades, several more promising antitumor agents that inhibit HDACs were discovered. These include the hydroxamic acid-derived compounds such as Trichostatin A (TSA), suberoylanilide hydroxamic acid (SAHA, trade name Vorinostat), and Scriptaid; aliphatic acids such as valproic acid (VPA); the benzamide derivative MS-275 (Entinostat); and cyclic tetrapeptides such as depsipeptide (FK228 or FR901228, trade name Romidepsin). Each of these inhibits HDACs by binding to various positions on the zinc-containing catalytic domain (reviewed in [291, 292]). It should be noted that inhibition of deacetylation of nonhistone proteins as well as histones is observed. Many cancer-relevant transcription factors exhibit increased acetylation in the presence of HDAC inhibitors, including p53, RB, E2F1, and ACTR, among others (reviewed in [293]).

The three HDAC inhibitors in most widespread clinical use today are VPA, SAHA, and depsipeptide. VPA has the longest clinical history as it has been used for treatment of epilepsy since the 1960s. It inhibits proliferation of cultured cancer cells at millimolar concentrations, shows synergistic effects in combination with decitabine or hydralazine, and is currently in clinical trials for treatment of multiple types of cancer. SAHA also induces growth inhibition, differentiation, or apoptosis in cultured cancer cells [294, 295] and inhibits cancer cell growth synergistically with decitabine or zebularine at micromolar concentrations; however, it has a short biological half-life [296]. Depsipeptide, discovered in 1994 [297], preferentially targets class I HDACs [298]. It shows synergistic effects in combination with decitabine, zebularine, Trichostatin A, and 5-aza and inhibits cancer cell proliferation at sub-micromolar concentrations. Currently, SAHA and depsipeptide have been approved by the FDA for treatment of cutaneous T-cell lymphoma, while VPA clinical trials are ongoing.

11.3.3 Combination Therapies with DNMT and HDAC Inhibitors

Thus far, DNMT and HDAC inhibitor monotherapy has shown a less impressive clinical track record in solid tumors compared to hematological malignancies. One problem is decreased drug efficacy in solid tumors. This could be due to the presence of tumor cells that replicate slowly or not at all. Decreased efficacy could also be attributed to limited drug exposure given the instability of the compound. There are several compelling clinical reasons to use epigenetic modalities in combination. Given the synergistic biological activities of DNMTs and HDACs, combining a DNA demethylating agent and an HDAC inhibitor might be expected to improve efficacy. In addition, the potential synergistic combination could decrease toxicity by allowing administration of lower doses of either drug. In general, treatment with demethylating agents or HDAC inhibitors alone can cause significant toxicity. Recently, however, the combination of decitabine and depsipeptide was demonstrated to synergistically inhibit growth of lung and breast cancer cells, remarkably, at 1,000-fold lower doses of depsipeptide than current clinical usage [133, 134]. Several ongoing Phase 1 and 2 clinical trials are using DNA methylation or HDAC inhibitors in combination in breast cancer patients (reviewed in [299]). Epigenetic agents can also resensitize refractory tumors to traditional therapies. Preclinical experiments using cancer cell lines and murine models may help inform optimal therapeutic combinations that would yield better clinical outcomes.

Inhibitors of DNMTs and HDACs in combination show synergistic growth inhibition in cancer cell lines and in animal models of cancer. These inhibitors synergistically affect chromatin state and lead to more pronounced re-expression of epigenetically silenced genes for tumor suppressors and cell cycle regulators [154]. They can also synergize with radiotherapy [300, 301], CDK inhibitors [302], TRAIL cytokine [303], and conventional chemotherapy agents, such as cisplatin [304], paclitaxel, doxorubicin, and 5-fluorouracil [305]. Furthermore, they can induce a response in tumors that are resistant to conventional chemotherapeutic agents [306]. The effects of epigenetic drugs on tumor growth in xenograft and in genetically engineered mouse models are similar to those observed in cancer cell lines. Treatment of mice with DNMT1 and HDAC inhibitors causes induction of genes coding for tumor suppressors and pro-apoptotic proteins, leading to inhibition of tumor growth and apoptosis. The effects of epigenetic agents were similar using several distinct drugs and tumor models—e.g., p21 protein expression was induced with five different drugs (SAHA, MS-275, PXD101, LBH-589, and decitabine) in xenograft models of breast, lung, and ovarian cancer. CDKN1A/p21 expression was also induced by zebularine in a transgenic mouse model of breast cancer [289].

11.3.4 Histone Methyltransferase Inhibitors

Though much of the focus for epigenetic drug development has been on DNMT and HDAC inhibitors, histone methyltransferases (HMTs) have recently emerged as viable drug targets for cancer. Mutations in or aberrant expression of HMTs has been documented in a variety of human tumors, including breast cancer. Importantly, though known HMTs comprise a 96-member family of enzymes, they exhibit a high degree of structural and biochemical diversity, allowing for biological specificity and pharmacologic selectivity. As such, they are considered highly favorable "druggable" targets, and many companies have developed small-molecule HMT inhibitors that are in various phases of preclinical development. Most of the small molecules bind within the binding pocket for the *S*-adenosyl-L-methionine methyl donor of the targeted HMT and thus competitively inhibit HMT function.

Mixed-lineage leukemia (MLL) patients are likely to be the initial participants in HMT inhibitor trials. A hallmark of MLL disease is a large number of chromosomal translocations involving the *MLL* gene that create chimeric proteins [307]. The *MLL* gene encodes for an HMT that contains a SET domain and catalyzes H3K4me at specific gene loci [308, 309]. In the MLL translocations, the SET domain is lost and the remaining MLL protein is fused to a variety of partners that interact directly or indirectly with the HMT DOT1L [310–315]. DOT1L catalyzes methylation of H3K79 and leads to enhanced expression of leukemogenic genes [316–318]. DOT1L-specific inhibitors have been developed and preclinical evaluation of one compound, EPZ004777, shows that it is highly specific to DOT1L, causes depletion of H3K79 methylation (but not other histone methylation marks), selectively inhibits proliferation and induces apoptosis in cells with MLL translocations, and increases survival in a murine MLL model [319]. If successfully translated, the use of DOT1L inhibitors in MLL patients will serve as a proof of principle for epigenetic therapy of cancer with a genetic defect.

Lead optimization for pharmacological inhibition of several disease-associated HMTs is underway. In addition to DOT1L, EZH2 inhibitors are in preclinical development. EZH2 is a SET-domain-containing HMT in the PRC2 complex that catalyzes the repressive mark H3K27me3. EZH2 is aberrantly expressed in several tumors; it is commonly mutated in non-Hodgkin's lymphoma and amplified in breast, prostate, colon, gastric, bladder, liver, and skin cancers (reviewed in [320]). As described in Sect. 11.2.2, widespread H3K27me3 repressive chromatin domains associated with decreased gene expression are observed in aggressive breast cancers. Selective EZH2 knockdown using RNA interference (RNAi) leads to decreased breast cancer cell proliferation in vitro and decreased tumor growth in MDA-MB-231 xenografts in vivo [321]. These studies indicate that EZH2 inhibition may be an attractive treatment strategy for EZH2-overexpressing breast cancers. However, current selective EZH2 inhibitors are directed towards specific mutant forms of EZH2 [243, 322]. At least in non-Hodgkin's lymphoma, cells expressing such EZH2 mutants are more sensitive to these compounds than cells harboring wild-type EZH2. Therefore, translating EZH2 inhibition to targeted breast cancer therapy will require preclinical testing of existing compounds in breast cancer models and/or lead optimization of additional compounds.

11.3.5 Next-Generation Compounds

The clinical efficacy of DNMT or HDAC inhibitors as single agents has been disappointing, especially towards solid tumors. This could be due to the apparent interplay between DNMTs and histone modifiers in regulating gene expression, which suggests these agents would be more effective when combined. Alternatively, efficacy could be increased if delivery of these agents is improved. Decitabine and 5-aza are both highly unstable, which may limit effective treatment in the clinic. HDAC inhibitors are more stable than DNMT inhibitors, and although they exhibit cancer cell-selective anti-proliferative effects, their maximum therapeutic window is adversely affected by their systemic delivery to all cells rather than just tumor cells. Re-tailoring existing compounds to increase bioavailability or tumor cell selectivity may also increase their clinical efficacy.

The compound SGI-110 (formerly S-110) is a dinucleotide of decitabine followed by 2′-deoxyguanosine. This formulation allows for increased patient exposure to drug by inhibiting in vivo deamination of decitabine by cytidine deaminase [323]. SGI-110 has shown demethylation and antitumor activities in xenograft models and in primates [324–326]. The compound is also currently being evaluated in a randomized Phase 1-2 first-in-human clinical trial for treatment of relapsed/refractory MDS and AML. Interim results from this trial were recently released and indicated that SGI-110 was more effective than decitabine when delivered intravenously. Major responses were seen for 3 of 7 refractory AML patients, accompanied by hypomethylation of DNA. Importantly, the optimal biologically effective dose was reached before the maximum-tolerated dose, allowing for expansion of the study to enroll previously untreated MDS patients and elderly AML patients [327].

Though HDAC inhibitors are being used with some success in clinical trials, a recent study by Pazolli et al. [328] suggests that effects of HDAC inhibitors on cells in the tumor microenvironment may limit their efficacy. More specifically, normal fibroblasts treated with HDAC inhibitors potently increase expression of senescence-associated secretory factors, including interleukins 6 and 8 (IL-6 and IL-8) as well as osteopontin (OSN). These factors are implicated in promoting tumor growth and pretreatment of fibroblasts with NaB or SAHA enhanced growth of keratinocytes compared to vehicle-treated fibroblasts that were co-injected into immunocompromised mice. In addition, primary breast fibroblasts also exhibited increased expression of IL-6, IL-8, and OSN upon treatment with either NaB or SAHA. Though preliminary, these data suggest that it will be important to assess molecular changes in the breast stroma following HDAC inhibitor therapy and possibly other treatment modalities. If tumor-promoting cytokine expression is elevated in patients following treatment with these agents, then future HDAC inhibitor development could benefit from identifying compounds that inhibit tumor growth without stimulating production of detrimental paracrine factors from cells in the microenvironment.

Chemical-based strategies have also been employed to improve the efficacy of HDAC inhibitors. The compound CHR-2845 is a fusion of a novel HDAC inhibitor with a cleavable esterase-sensitive motif (ESM) moiety. Presumably, the HDAC inhibitor will remain inactive until delivered to a cell that expresses an appropriate esterase, which recognizes and hydrolyzes the compound to release an acid that inhibits HDACs. For this compound, the cells expressing the relevant esterase are macrophages, and thus, this strategy exploits the biological accumulation of macrophages in tumors. Preclinical benefits observed for ESM-based HDAC inhibitors include improved tolerability, increased potency, and increased duration of action. CHR-2845 is currently in Phase 1 trials for hematological disease and lymphoid malignancy. A related compound, CHR-4487, is a SAHA-ESM fusion currently in preclinical development and has shown improved outcomes compared to SAHA treatment in murine arthritis models [329].

Though these strategies using next-generation compounds are promising, future testing is required to determine if maximizing bioavailability and cell-type-restricted drug delivery will increase the clinical efficacy of current DNMT and HDAC inhibitors.

11.3.6 Targeted Epigenetic Therapy in Breast Cancer

In addition to amplification of EZH2, other features of aggressive breast cancer can be exploited to maximize the benefits of using epigenetic agents in combination or in addition to traditional therapies. For example, ER expression is lost in a subset of breast tumors rendering these tumors refractory to endocrine therapy. In many cases, loss of ER expression has been attributed to epigenetic mechanisms in breast cancer cell lines. Treatment with demethylating agents, HDAC inhibitors, or both leads to re-expression of ER and, furthermore, sensitizes such cells to endocrine therapies, e.g., tamoxifen [330–333]. Not surprisingly, clinical trials combining epigenetic and endocrine therapy have been conducted. In a recent Phase 2 trial, the combination of the HDAC inhibitor Vorinostat (SAHA) and tamoxifen was tested in patients with hormone therapy-resistant breast cancer [334]. The clinical benefit rate (response or stable disease >24 weeks) was 40 %. Histone hyperacetylation and higher baseline HDAC2 levels correlated with response. This study suggests that the combination of Vorinostat and tamoxifen is well tolerated and exhibits encouraging activity in reversing hormone resistance. However, it should be noted that a randomized trial is required to determine the effects of the combination over the potential efficacy of tamoxifen alone. Also, it was determined that the absence of bulk histone acetylation could be used as an early negative predictor for patients who are not likely to benefit. This biomarker could be used to identify patients likely to benefit in future clinical trials.

Epigenetic agents are more effective when key tumor suppressor genes, such as *CDKN2A* (*P16*) and *P18*, are epigenetically silenced [335], suggesting that their derepression may be critical for therapeutic effect. For example, efficacy of the DNMT inhibitors decitabine or zebularine combined with the HDAC inhibitor depsipeptide in lung and breast tumor cells with defined *CDKN2A* status was described recently [133, 134]. This study showed that non-small cell lung cancer cells with methylated *CDKN2A* were significantly more sensitive to methylation inhibitors than cell lines with deleted *CDKN2A* [133, 134]. In addition, the combination of zebularine and depsipeptide resulted in a synergistic effect on cell growth inhibition that was also linked with the presence of epigenetically silenced *CDKN2A*. These data strongly support the importance of prospective preselection of patients in future clinical trials and suggest *CDKN2A* status as a key biomarker for DNMT/HDAC inhibition studies.

11.4 Breast Tumor Heterogeneity and Therapy

11.4.1 Intratumoral Heterogeneity

One major barrier to successful therapeutic intervention of any type of cancer is tumor heterogeneity. In addition to the histological, molecular, and epidemiological interindividual heterogeneity that impacts treatment decisions, intratumoral heterogeneity has emerged as a crucial factor affecting long-term treatment efficacy. Increasing evidence suggests that many human cancers, including breast cancer, are initiated and promoted by a small subpopulation of cells within the bulk tumor. These cells often display characteristics associated with adult tissue stem cells, including similar cell surface antigen expression, drug efflux capacity, and slow replication rates [336]. Accordingly, these cells are often classified as tumor-initiating, tumor-promoting, or cancer stem cells (CSCs), which are presumed to undergo metastasis and underlie treatment resistance and disease recurrence. It should be noted that the classification of CSC does not imply that these cells originate from normal adult tissue stem cells that have become neoplastically transformed. Rather, CSC denotes specific functional characteristics, including the ability to initiate tumors, capacity for self-renewal, and capacity to generate non-tumorigenic progeny. Several studies have provided evidence that CSCs are more tolerant to both chemotherapy and radiotherapy [337–341]. This tolerance of CSCs in solid tumors has been implicated as a possible mechanism for the high degree of persistent disease observed following current treatment modalities.

Though it is highly likely that CSCs are epigenetically distinct from bulk tumor cells, differences between the epigenomes of CSCs and their non-tumorigenic counterparts have not been evaluated. Interestingly, a study by Sharma et al. [342] examining multiple human cancer cell lines consistently identified a subpopulation (0.3–5 %) of drug-tolerant cells that were enriched for CSC markers. Characterization of these cells indicated that the drug-tolerance phenotype was reversible and dependent on an altered chromatin state that could be pharmacologically reversed using HDAC inhibitors. Epigenetic characterization of breast CSCs offers an attractive opportunity towards identifying additional avenues for CSC targeting. For example, in Chang et al. [241], it was reported that breast CSCs isolated from primary tumors were enriched for EZH2 expression. Increased EZH2 expression correlated with decreased expression of the DNA repair protein RAD51. Ectopic expression of EZH2 increased the CSC subpopulation and culminated in amplification of RAF1 expression and increased MEK/ERK signaling. Importantly, treatment with a MEK/ERK inhibitor significantly suppressed the number of cells in the CSC subpopulation. Thus, targeting of the breast CSC population is feasible, and further characterization of these cells may unveil novel therapeutic strategies.

11.4.2 Considerations for Molecular Profiling of Tumor Heterogeneity

Several technical challenges to characterizing the global gene expression and epigenome of relevant CSCs exist. For one, because CSCs comprise such a small fraction of cells in a tumor, traditional population-averaged techniques, such as microarrays or hybridization-based epigenetic assays, lack the sensitivity to detect CSC contributions to bulk tumor gene expression or epigenomic features. Sorting cells based on CSC characteristics prior to such analyses is helpful, but technically difficult and not comprehensive. Many different molecular breast CSC markers have been reported; however, most have been challenged. Furthermore, antigen signatures identified in cells taken from a single tumor, but isolated based on different CSC markers, do not overlap [343]. One can use functional characteristics, such as separating slow- from fast-dividing cells or resistance to a particular drug, but this requires cell culturing rather than direct interrogation of primary tumor specimens. It would also be unclear what characteristics were intrinsic to the cells and which were acquired due to drug treatment or cell culture conditions. With the increased accessibility and affordability of massively parallel sequencing technologies, future studies providing single-molecule-based information may allow for interrogation of epigenetic features with sufficient coverage to allow for detection of unique epigenetic subpopulations.

While DNA methylation is the most often queried epigenetic feature, the results from Sharma et al. [342] highlight the potential for the importance of differential chromatin structure features as well. As previously mentioned, our lab has

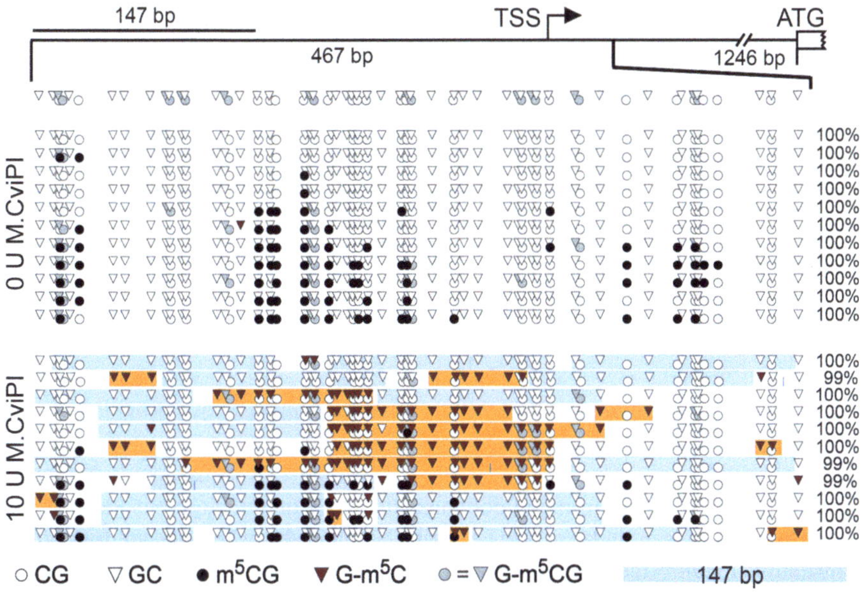

Fig. 11.1 MAPit analysis of the TSS region of human *SIM2* in MCF10A cells. Nuclei (10^6) were probed with 10 U of wild-type M.CviPI, a GC DNMT, for 30 min at 37 °C. *SIM2* is expressed in MCF-10A cells. Each *horizontal line* represents 524 bp of chromatin from a single cell. *Circles* represent CG sites and *triangles* represent GC sites. *Black-filled circles* and *red-filled triangles* represent m^{5}CG and G-m^{5}C, respectively. GCG sites are represented by both *gray triangle* and *circles*. GCG site methylation cannot rigorously be discriminated as being placed by endogenous or exogenous DNMT, but this can often be inferred from context. *Blue highlighted areas* represent 147 bp of contiguous M.CviPI DNA footprint. Note that about half of the alleles have relatively high levels of endogenous methylation (*black-filled circles*). Based on molecules from cells not treated with M.CviPI, it can be inferred that *gray* GCG sites in these densely methylated MCF-10A alleles were likely methylated by endogenous DNMTs. The other half of the molecules is almost free of endogenous methylation but shows an accessible, nucleosome-length region high in M. CviPI methylation (*red triangles*) highlighted in *red*. No other technique can determine this bipartite pattern of chromosome structure. The high accessibility to M.CviPI is probably due to histone depletion near the TSS. In contrast, this putative histone-free region is flanked by protected spans of median length around 150 bp. *Numbers at the right* of each molecule depiction indicate the percentage of C conversion to T in non-CG and non-GC sequences. Nucleotides that failed to convert or reverted to a C during PCR amplification are indicated by *vertical blue tick marks*. Reprinted with permission from John Wiley & Sons, Inc

developed a technique that, combined with bisulfite sequencing, allows for simultaneous single-molecule level detection of DNA methylation and chromatin accessibility (MAPit) [173, 174]. MAPit is currently, to our knowledge, the only footprinting method that can detect subpopulations of molecules with distinct patterns of protein binding or chromatin architecture and correlate them directly with the occurrence of endogenous methylation. Using this technology in MCF10A cells, we were able to identify two subpopulations with epigenetically distinct features at *SIM2* (Fig. 11.1, reprinted from [174]), a putative tumor suppressor gene in the breast [344–346]. Notably, we could associate a particular nucleosome occupancy profile with endogenous DNA methylation within a heterogeneous cell population. Coupling MAPit with next-generation sequencing offers an opportunity for large-scale interrogation of chromatin structure in complex samples made up of heterogeneous cell populations [347]. DNA target enrichment strategies that provide sufficient enrichment of particular gene targets should allow for detection of chromatin states among rare tumor cell subpopulations as well as provide a measure for the level of epigenetic heterogeneity observed within a given biological sample.

11.5 Outlook

Whole genome assessments of breast cancer cell epigenomes have provided details suggesting that cancer-relevant epigenetic perturbations are not restricted to hypermethylation of tumor suppressor gene TSSs. Global chromatin rearrangements, expansion of repressive chromatin domains, and DNA methylation outside of TSSs or CGIs are all evident in breast cancer cells. Whether these features are functional contributors to tumorigenesis or passive by-products of an underlying dysregulation will require further study, especially since the mechanisms for the normal physiologic functions of these features remain unresolved.

Tumor heterogeneity has long been a barrier to successful therapeutic intervention. Further development of assays that permit epigenetic study at the level of single cells and/or molecules coupled with decreased costs of next-generation sequencing should allow for an unprecedented, in-depth view of epigenetic heterogeneity in human tumors. Characterization of epigenetically distinct subpopulations could provide important information regarding which pathways could be uniquely targeted in these cells and may provide a unique platform by which to combat metastatic tumor cells.

References

1. Bird AP. DNA methylation and the frequency of CpG in animal DNA. Nucleic Acids Res. 1980;8:1499–504.
2. Deaton AM, Bird A. CpG islands and the regulation of transcription. Genes Dev. 2011;25:1010–22. doi:10.1101/gad.2037511.
3. Bird AP. CpG-rich islands and the function of DNA methylation. Nature. 1986;321:209–13. doi:10.1038/321209a0.
4. Gardiner-Garden M, Frommer M. CpG islands in vertebrate genomes. J Mol Biol. 1987;196:261–82.
5. Takai D, Jones PA. Comprehensive analysis of CpG islands in human chromosomes 21 and 22. Proc Natl Acad Sci USA. 2002;99:3740–5. doi:10.1073/pnas.052410099.
6. Wang Y, Leung FCC. An evaluation of new criteria for CpG islands in the human genome as gene markers. Bioinformatics. 2004;20:1170–7. doi:10.1093/bioinformatics/bth059.
7. Saxonov S, Berg P, Brutlag DL. A genome-wide analysis of CpG dinucleotides in the human genome distinguishes two distinct classes of promoters. Proc Natl Acad Sci USA. 2006;103:1412–7. doi:10.1073/pnas.0510310103.

8. Weber M, Hellmann I, Stadler MB, Ramos L, Paabo S, Rebhan M, Schübeler D. Distribution, silencing potential and evolutionary impact of promoter DNA methylation in the human genome. Nat Genet. 2007;39:457–66. doi:10.1038/ng1990.
9. Zhu J, He FH, Hu SN, Yu J. On the nature of human housekeeping genes. Trends Genet. 2008;24:481–4. doi:10.1016/j.tig.2008.08.004.
10. Jurkowska RZ, Jurkowski TP, Jeltsch A. Structure and function of mammalian DNA methyltransferases. Chembiochem. 2011;12:206–22. doi:10.1002/cbic.201000195.
11. Straussman R, Nejman D, Roberts D, Steinfeld I, Blum B, Benvenisty N, Simon I, Yakhini Z, Cedar H. Developmental programming of CpG island methylation profiles in the human genome. Nat Struct Mol Biol. 2009;16:564–71. doi:10.1038/nsmb.1594.
12. Isagawa T, Nagae G, Shiraki N, Fujita T, Sato N, Ishikawa S, Kume S, Aburatani H. DNA methylation profiling of embryonic stem cell differentiation into the three germ layers. PLoS One. 2011;6:e26052. doi:10.1371/journal.pone.0026052.
13. Bird AP, Wolffe AP. Methylation-induced repression–belts, braces, and chromatin. Cell. 1999;99:451–4.
14. Esteller M. CpG island hypermethylation and tumor suppressor genes: a booming present, a brighter future. Oncogene. 2002;21:5427–40. doi:10.1038/sj.onc.1205600.
15. Herman JG, Baylin SB. Mechanisms of disease: gene silencing in cancer in association with promoter hypermethylation. N Engl J Med. 2003;349:2042–54.
16. Esteller M. Epigenetic gene silencing in cancer: the DNA hypermethylome. Hum Mol Genet. 2007;16:R50–9. doi:10.1093/hmg/ddm018.
17. Sinčić N, Herceg Z. DNA methylation and cancer: ghosts and angels above the genes. Curr Opin Oncol. 2011;23:69–76. doi:10.1097/CCO.0b013e3283412eb4.
18. Irvine RA, Lin IG, Hsieh CL. DNA methylation has a local effect on transcription and histone acetylation. Mol Cell Biol. 2002;22:6689–96. doi:10.1128/mcb.22.19.6689-6696.2002.
19. van Vlodrop IJH, Niessen HEC, Derks S, Baldewijns M, van Criekinge W, Herman JG, van Engeland M. Analysis of promoter CpG island hypermethylation in cancer: location, location, location! Clin Cancer Res. 2011;17:4225–31. doi:10.1158/1078-0432.ccr-10-3394.
20. Rauch TA, Zhong XY, Wu XW, Wang M, Kernstine KH, Wang ZD, Riggs AD, Pfeifer GP. High-resolution mapping of DNA hypermethylation and hypomethylation in lung cancer. Proc Natl Acad Sci USA. 2008;105:252–7. doi:10.1073/pnas.0710735105.
21. Irizarry RA, Ladd-Acosta C, Wen B, Wu Z, Montano C, Onyango P, Cui H, Gabo K, Rongione M, Webster M, Ji H, Potash JB, Sabunciyan S, Feinberg AP. The human colon cancer methylome shows similar hypo- and hypermethylation at conserved tissue-specific CpG island shores. Nat Genet. 2009;41:178–86. doi:10.1038/ng.298.
22. Gardiner-Garden M, Frommer M. Transcripts and CpG islands associated with the proopiomelanocortin gene and other neurally expressed genes. J Mol Endocrinol. 1994;12: 365–82.
23. Macleod D, Ali RR, Bird A. An alternative promoter in the mouse major histocompatibility complex class II I-Aβ gene: implications for the origin of CpG islands. Mol Cell Biol. 1998;18:4433–43.
24. Kleinjan DA, Seawright A, Childs AJ, van Heyningen V. Conserved elements in *Pax6* intron 7 involved in (auto)regulation and alternative transcription. Dev Biol. 2004;265:462–77. doi:10.1016/j.ydbio.2003.09.011.
25. Rauch TA, Wu XW, Zhong X, Riggs AD, Pfeifer GP. A human B cell methylome at 100-base pair resolution. Proc Natl Acad Sci USA. 2009;106:671–8. doi:10.1073/pnas.0812399106.
26. Rinn JL, Kertesz M, Wang JK, Squazzo SL, Xu X, Brugmann SA, Goodnough LH, Helms JA, Farnham PJ, Segal E, Chang HY. Functional demarcation of active and silent chromatin domains in human *HOX* loci by noncoding RNAs. Cell. 2007;129:1311–23. doi:10.1016/j.cell.2007.05.022.
27. Sleutels F, Zwart R, Barlow DP. The non-coding Air RNA is required for silencing autosomal imprinted genes. Nature. 2002;415:810–3.
28. Panning B, Jaenisch R. DNA hypomethylation can activate *Xist* expression and silence X-linked genes. Genes Dev. 1996;10:1991–2002. doi:10.1101/gad.10.16.1991.

29. Nguyen C, Liang GM, Nguyen TT, Tsao-Wei D, Groshen S, Lubbert M, Zhou JH, Benedict WF, Jones PA. Susceptibility of nonpromoter CpG islands to de novo methylation in normal and neoplastic cells. J Natl Cancer Inst. 2001;93:1465–72.
30. Illingworth RS, Bird AP. CpG islands – 'a rough guide'. FEBS Lett. 2009;583:1713–20. doi:10.1016/j.febslet.2009.04.012.
31. Hellman A, Chess A. Gene body-specific methylation on the active X chromosome. Science. 2007;315:1141–3. doi:10.1126/science.1136352.
32. Ball MP, Li JB, Gao Y, Lee JH, LeProust EM, Park IH, Xie B, Daley GQ, Church GM. Targeted and genome-scale strategies reveal gene-body methylation signatures in human cells. Nat Biotechnol. 2009;27:361–8. doi:10.1038/nbt.1533.
33. Bell AC, Felsenfeld G. Methylation of a CTCF-dependent boundary controls imprinted expression of the *Igf2* gene. Nature. 2000;405:482–5. doi:10.1038/35013100.
34. Hark AT, Schoenherr CJ, Katz DJ, Ingram RS, Levorse JM, Tilghman SM. CTCF mediates methylation-sensitive enhancer-blocking activity at the *H19/Igf2* locus. Nature. 2000;405:486–9. doi:10.1038/35013106.
35. Kanduri C, Pant V, Loukinov D, Pugacheva E, Qi CF, Wolffe A, Ohlsson R, Lobanenkov VV. Functional association of CTCF with the insulator upstream of the *H19* gene is parent of origin-specific and methylation-sensitive. Curr Biol. 2000;10:853–6.
36. Lai AY, Fatemi M, Dhasarathy A, Malone C, Sobol SE, Geigerman C, Jaye DL, Mav D, Shah R, Li L, Wade PA. DNA methylation prevents CTCF-mediated silencing of the oncogene *BCL6* in B cell lymphomas. J Exp Med. 2010;207:1939–50. doi:10.1084/jem.20100204.
37. Rishi V, Bhattacharya P, Chatterjee R, Rozenberg J, Zhao J, Glass K, Fitzgerald P, Vinson C. CpG methylation of half-CRE sequences creates C/EBPα binding sites that activate some tissue-specific genes. Proc Natl Acad Sci USA. 2010;107:20311–6. doi:10.1073/pnas.1008688107.
38. Chatterjee R, Vinson C. CpG methylation recruits sequence specific transcription factors essential for tissue specific gene expression. Biochim Biophys Acta. 2012;1819:763–70. doi:10.1016/j.bbagrm.2012.02.014.
39. Bestor TH. The DNA methyltransferases of mammals. Hum Mol Genet. 2000;9:2395–402.
40. Goll MG, Bestor TH. Eukaryotic cytosine methyltransferases. Annu Rev Biochem. 2005;74:481–514.
41. Bestor T, Laudano A, Mattaliano R, Ingram V. Cloning and sequencing of a cDNA-encoding DNA methyltransferase of mouse cells: the carboxyl-terminal domain of the mammalian enzyme is related to bacterial restriction methyltransferases. J Mol Biol. 1988;203:971–83.
42. Li E, Bestor TH, Jaenisch R. Targeted mutation of the DNA methyltransferase gene results in embryonic lethality. Cell. 1992;69:915–26. doi:10.1016/0092-8674(92)90611-f.
43. Goyal R, Reinhardt R, Jeltsch A. Accuracy of DNA methylation pattern preservation by the Dnmt1 methyltransferase. Nucleic Acids Res. 2006;34:1182–8. doi:10.1093/nar/gkl002.
44. Jeltsch A. On the enzymatic properties of Dnmt1 specificity, processivity, mechanism of linear diffusion and allosteric regulation of the enzyme. Epigenetics. 2006;1:63–6.
45. Hashimoto H, Vertino PM, Cheng XD. Molecular coupling of DNA methylation and histone methylation. Epigenomics. 2010;2:657–69.
46. Pradhan S, Bacolla A, Wells RD, Roberts RJ. Recombinant human DNA (cytosine-5) methyltransferase. I. Expression, purification, and comparison of de novo and maintenance methylation. J Biol Chem. 1999;274:33002–10.
47. Robertson KD, Keyomarsi K, Gonzales FA, Velicescu M, Jones PA. Differential mRNA expression of the human DNA methyltransferases (DNMTs) 1, 3a and 3b during the G_0/G_1 to S phase transition in normal and tumor cells. Nucleic Acids Res. 2000;28:2108–13. doi:10.1093/nar/28.10.2108.
48. Chen TP, Hevi S, Gay F, Tsujimoto N, He T, Zhang BL, Ueda Y, Li E. Complete inactivation of *DNMT1* leads to mitotic catastrophe in human cancer cells. Nat Genet. 2007;39:391–6. doi:10.1038/ng1982.
49. Kimura F, Seifert HH, Florl AR, Santourlidis S, Steinhoff C, Swiatkowski S, Mahotka C, Gerharz CD, Schulz WA. Decrease of DNA methyltransferase 1 expression relative to cell proliferation in transitional cell carcinoma. Int J Cancer. 2003;104:568–78. doi:10.1002/ijc.10988.

50. Hermann A, Goyal R, Jeltsch A. The Dnmt1 DNA-(cytosine-C5)-methyltransferase methylates DNA processively with high preference for hemimethylated target sites. J Biol Chem. 2004;279:48350–9. doi:10.1074/jbc.M403427200.
51. Leonhardt H, Page AW, Weier HU, Bestor TH. A targeting sequence directs DNA methyltransferase to sites of DNA replication in mammalian nuclei. Cell. 1992;71: 865–73.
52. Chuang LSH, Ian HI, Koh TW, Ng HH, Xu GL, Li BFL. Human DNA (cytosine-5) methyltransferase PCNA complex as a target for p21^{WAF1}. Science. 1997;277:1996–2000. doi:10.1126/science.277.5334.1996.
53. Schermelleh L, Haernmer A, Spada F, Rosing N, Meilinger D, Rothbauer U, Cardoso MC, Leonhardt H. Dynamics of Dnmt1 interaction with the replication machinery and its role in postreplicative maintenance of DNA methylation. Nucleic Acids Res. 2007;35:4301–12. doi:10.1093/nar/gkm432.
54. Spada F, Haemmer A, Kuch D, Rothbauer U, Schermelleh L, Kremmer E, Carell T, Längst G, Leonhardt H. DNMT1 but not its interaction with the replication machinery is required for maintenance of DNA methylation in human cells. J Cell Biol. 2007;176:565–71. doi:10.1083/jcb.200610062.
55. Bostick M, Kim JK, Esteve PO, Clark A, Pradhan S, Jacobsen SE. UHRF1 plays a role in maintaining DNA methylation in mammalian cells. Science. 2007;317:1760–4. doi:10.1126/science.1147939.
56. Sharif J, Muto M, Takebayashi S, Suetake I, Iwamatsu A, Endo TA, Shinga J, Mizutani-Koseki Y, Toyoda T, Okamura K, Tajima S, Mitsuya K, Okano M, Koseki H. The SRA protein Np95 mediates epigenetic inheritance by recruiting Dnmt1 to methylated DNA. Nature. 2007;450:908–12. doi:10.1038/nature06397.
57. Arita K, Ariyoshi M, Tochio H, Nakamura Y, Shirakawa M. Recognition of hemi-methylated DNA by the SRA protein UHRF1 by a base-flipping mechanism. Nature. 2008;455:818–21. doi:10.1038/nature07249.
58. Okano M, Bell DW, Haber DA, Li E. DNA methyltransferases Dnmt3a and Dnmt3b are essential for de novo methylation and mammalian development. Cell. 1999;99:247–57.
59. Chen ZX, Mann JR, Hsieh CL, Riggs AD, Chedin F. Physical and functional interactions between the human DNMT3L protein and members of the de novo methyltransferase family. J Cell Biochem. 2005;95:902–17. doi:10.1002/jcb.20447.
60. Holz-Schietinger C, Reich NO. The inherent processivity of the human de novo methyltransferase 3A (DNMT3A) is enhanced by DNMT3L. J Biol Chem. 2010;285:29091–100. doi:10.1074/jbc.M110.142513.
61. Ramsahoye BH, Biniszkiewicz D, Lyko F, Clark V, Bird AP, Jaenisch R. Non-CpG methylation is prevalent in embryonic stem cells and may be mediated by DNA methyltransferase 3a. Proc Natl Acad Sci USA. 2000;97:5237–42. doi:10.1073/pnas.97.10.5237.
62. Lister R, Pelizzola M, Dowen RH, Hawkins RD, Hon G, Tonti-Filippini J, Nery JR, Lee L, Ye Z, Ngo QM, Edsall L, Antosiewicz-Bourget J, Stewart R, Ruotti V, Millar AH, Thomson JA, Ren B, Ecker JR. Human DNA methylomes at base resolution show widespread epigenomic differences. Nature. 2009;462:315–22. doi:10.1038/nature08514.
63. Harris RA, Wang T, Coarfa C, Nagarajan RP, Hong CB, Downey SL, Johnson BE, Fouse SD, Delaney A, Zhao YJ, Olshen A, Ballinger T, Zhou X, Forsberg KJ, Gu JC, Echipare L, O'Geen H, Lister R, Pelizzola M, Xi YX, Epstein CB, Bernstein BE, Hawkins RD, Ren B, Chung WY, Gu HC, Bock C, Gnirke A, Zhang MQ, Haussler D, Ecker JR, Li W, Farnham PJ, Waterland RA, Meissner A, Marra MA, Hirst M, Milosavljevic A, Costello JF. Comparison of sequencing-based methods to profile DNA methylation and identification of monoallelic epigenetic modifications. Nat Biotechnol. 2010;28:1097–105. doi:10.1038/nbt.1682.
64. Xie W, Barr CL, Kim A, Yue F, Lee AY, Eubanks J, Dempster EL, Ren B. Base-resolution analyses of sequence and parent-of-origin dependent DNA methylation in the mouse genome. Cell. 2012;148:816–31. doi:10.1016/j.cell.2011.12.035.
65. Wu H, Zhang Y. Mechanisms and functions of Tet protein-mediated 5-methylcytosine oxidation. Genes Dev. 2011;25:2436–52. doi:10.1101/gad.179184.111.

66. Ziller MJ, Muller F, Liao J, Zhang YY, Gu HC, Bock C, Boyle P, Epstein CB, Bernstein BE, Lengauer T, Gnirke A, Meissner A. Genomic distribution and inter-sample variation of non-CpG methylation across human cell types. PLoS Genet. 2011;7:e1002389. doi:10.1371/journal.pgen.1002389.
67. Rhee I, Jair KW, Yen RWC, Lengauer C, Herman JG, Kinzler KW, Vogelstein B, Baylin SB, Schuebel KE. CpG methylation is maintained in human cancer cells lacking DNMT1. Nature. 2000;404:1003–7.
68. Liang GG, Chan MF, Tomigahara Y, Tsai YC, Gonzales FA, Li E, Laird PW, Jones PA. Cooperativity between DNA methyltransferases in the maintenance methylation of repetitive elements. Mol Cell Biol. 2002;22:480–91. doi:10.1128/mcb.22.2.480-491.2002.
69. Iguchi-Ariga SM, Schaffner W. CpG methylation of the cAMP-responsive enhancer/promoter sequence TGACGTCA abolishes specific factor binding as well as transcriptional activation. Genes Dev. 1989;3:612–9.
70. Jones PL, Veenstra GJC, Wade PA, Vermaak D, Kass SU, Landsberger N, Strouboulis J, Wolffe AP. Methylated DNA and MeCP2 recruit histone deacetylase to repress transcription. Nat Genet. 1998;19:187–91.
71. Nan XS, Ng HH, Johnson CA, Laherty CD, Turner BM, Eisenman RN, Bird A. Transcriptional repression by the methyl-CpG-binding protein MeCP2 involves a histone deacetylase complex. Nature. 1998;393:386–9.
72. Knoepfler PS, Bergstrom DA, Uetsuki T, Dac-Korytko I, Sun YH, Wright WE, Tapscott SJ, Kamps MP. A conserved motif N-terminal to the DNA-binding domains of myogenic bHLH transcription factors mediates cooperative DNA binding with pbx-Meis1/Prep1. Nucleic Acids Res. 1999;27:3752–61. doi:10.1093/nar/27.18.3752.
73. Zhang Y, Ng HH, Erdjument-Bromage H, Tempst P, Bird A, Reinberg D. Analysis of the NuRD subunits reveals a histone deacetylase core complex and a connection with DNA methylation. Genes Dev. 1999;13:1924–35.
74. Klose RJ, Bird AP. Genomic DNA methylation: the mark and its mediators. Trends Biochem Sci. 2006;31:89–97. doi:10.1016/j.tibs.2005.12.008.
75. Choy MK, Movassagh M, Goh HG, Bennett MR, Down TA, Foo RS. Genome-wide conserved consensus transcription factor binding motifs are hyper-methylated. BMC Genomics. 2010;11:519. doi:10.1186/1471-2164-11-519.
76. Palacios D, Summerbell D, Rigby PWJ, Boyes J. Interplay between DNA methylation and transcription factor availability: Implications for developmental activation of the mouse *Myogenin* gene. Mol Cell Biol. 2010;30:3805–15. doi:10.1128/mcb.00050-10.
77. Defossez PA, Stancheva I. Biological functions of methyl-CpG-binding proteins. Prog Mol Biol Transl Sci. 2011;101:377–98. doi:10.1016/B978-0-12-387685-0.00012-3.
78. Qui Y, Shabashvili D, Li Y, Gopalan PK, Chen M, Zajac-Kaye M. DNA methylation and histone deacetylation: interplay and combined therapy in cancer. In: Tatarinova T, Kerton O, editors. DNA methylation from genomics to technology. INTECH, 2012; p. 227–88.
79. Wade PA, Gegonne A, Jones PL, Ballestar E, Aubry F, Wolffe AP. Mi-2 complex couples DNA methylation to chromatin remodelling and histone deacetylation. Nat Genet. 1999;23:62–6.
80. Georgel PT, Horowitz-Scherer RA, Adkins N, Woodcock CL, Wade PA, Hansen JC. Chromatin compaction by human MeCP2: assembly of novel secondary chromatin structures in the absence of DNA methylation. J Biol Chem. 2003;278:32181–8. doi:10.1074/jbc.M305308200.
81. McBryant SJ, Adams VH, Hansen JC. Chromatin architectural proteins. Chromosome Res. 2006;14:39–51. doi:10.1007/s10577-006-1025-x.
82. Nikitina T, Ghosh RP, Horowitz-Scherer RA, Hansen JC, Grigoryev SA, Woodcock CL. MeCP2-chromatin interactions include the formation of chromatosome-like structures and are altered in mutations causing Rett syndrome. J Biol Chem. 2007;282:28237–45. doi:10.1074/jbc.M704304200.
83. Luger K, Mader AW, Richmond RK, Sargent DF, Richmond TJ. Crystal structure of the nucleosome core particle at 2.8 Å resolution. Nature. 1997;389:251–60.
84. Jiang CZ, Pugh BF. Nucleosome positioning and gene regulation: advances through genomics. Nat Rev Genet. 2009;10:161–72. doi:10.1038/nrg2522.

85. Zhang Z, Pugh BF. High-resolution genome-wide mapping of the primary structure of chromatin. Cell. 2011;144:175–86. doi:10.1016/j.cell.2011.01.003.
86. Widom J. Role of DNA sequence in nucleosome stability and dynamics. Q Rev Biophys. 2001;34:269–324.
87. Kaplan N, Moore IK, Fondufe-Mittendorf Y, Gossett AJ, Tillo D, Field Y, LeProust EM, Hughes TR, Lieb JD, Widom J, Segal E. The DNA-encoded nucleosome organization of a eukaryotic genome. Nature. 2009;458:362–6. doi:10.1038/nature07667.
88. Struhl K, Segal E. Determinants of nucleosome positioning. Nat Struct Mol Biol. 2013;20:267–73. doi:10.1038/nsmb.2506.
89. Lieb JD, Clarke ND. Control of transcription through intragenic patterns of nucleosome composition. Cell. 2005;123:1187–90. doi:10.1016/j.cell.2005.12.010.
90. Schones DE, Zhao K. Genome-wide approaches to studying chromatin modifications. Nat Rev Genet. 2008;9:179–91. doi:10.1038/nrg2270.
91. Travers AA, Vaillant C, Arneodo A, Muskhelishvili G. DNA structure, nucleosome placement and chromatin remodelling: a perspective. Biochem Soc Trans. 2012;40:335–40. doi:10.1042/BST20110757.
92. Heintzman ND, Stuart RK, Hon G, Fu Y, Ching CW, Hawkins RD, Barrera LO, Van Calcar S, Qu C, Ching KA, Wang W, Weng Z, Green RD, Crawford GE, Ren B. Distinct and predictive chromatin signatures of transcriptional promoters and enhancers in the human genome. Nat Genet. 2007;39:311–8.
93. Lin JC, Jeong S, Liang G, Takai D, Fatemi M, Tsai YC, Egger G, Gal-Yam EN, Jones PA. Role of nucleosomal occupancy in the epigenetic silencing of the *MLH1* CpG island. Cancer Cell. 2007;12:432–44.
94. Pardo CE, Carr IM, Hoffman CJ, Darst RP, Markham AF, Bonthron DT, Kladde MP. MethylViewer: computational analysis and editing for bisulfite sequencing and methyltransferase accessibility protocol for individual templates (MAPit) projects. Nucleic Acids Res. 2010;39:e5. doi:10.1093/nar/gkq716.
95. Albert I, Mavrich TN, Tomsho LP, Qi J, Zanton SJ, Schuster SC, Pugh BF. Translational and rotational settings of H2A.Z nucleosomes across the *Saccharomyces cerevisiae* genome. Nature. 2007;446:572–6. doi:10.1038/nature05632.
96. Mavrich TN, Jiang CZ, Ioshikhes IP, Li XY, Venters BJ, Zanton SJ, Tomsho LP, Qi J, Glaser RL, Schuster SC, Gilmour DS, Albert I, Pugh BF. Nucleosome organization in the *Drosophila* genome. Nature. 2008;453:358–62. doi:10.1038/nature06929.
97. Shivaswamy S, Bhinge A, Zhao YJ, Jones S, Hirst M, Iyer VR. Dynamic remodeling of individual nucleosomes across a eukaryotic genome in response to transcriptional perturbation. PLoS Biol. 2008;6:e65. doi:10.1371/journal.pbio.0060065.
98. Schones DE, Cui KR, Cuddapah S, Roh TY, Barski A, Wang ZB, Wei G, Zhao KJ. Dynamic regulation of nucleosome positioning in the human genome. Cell. 2008;132:887–98. doi:10.1016/j.cell.2008.02.022.
99. Cairns BR. The logic of chromatin architecture and remodelling at promoters. Nature. 2009;461:193–8. doi:10.1038/nature08450.
100. Ozsolak F, Song JS, Liu XS, Fisher DE. High-throughput mapping of the chromatin structure of human promoters. Nat Biotechnol. 2007;25:244–8. doi:10.1038/nbt1279.
101. Boyle AP, Davis S, Shulha HP, Meltzer P, Margulies EH, Weng Z, Furey TS, Crawford GE. High-resolution mapping and characterization of open chromatin across the genome. Cell. 2008;132:311–22. doi:10.1016/j.cell.2007.12.014.
102. Tillo D, Kaplan N, Moore IK, Fondufe-Mittendorf Y, Gossett AJ, Field Y, Lieb JD, Widom J, Segal E, Hughes TR. High nucleosome occupancy is encoded at human regulatory sequences. PLoS One. 2010;5:e9129. doi:10.1371/journal.pone.0009129.
103. Zhang Z, Wippo CJ, Wal M, Ward E, Korber P, Pugh BF. A packing mechanism for nucleosome organization reconstituted across a eukaryotic genome. Science. 2011;332:977–80. doi:10.1126/science.1200508.
104. Radman-Livaja M, Rando OJ. Nucleosome positioning: how is it established, and why does it matter? Dev Biol. 2010;339:258–66. doi:10.1016/j.ydbio.2009.06.012.

105. DeCristofaro MF, Betz BL, Rorie CJ, Reisman DN, Wang W, Weissman BE. Characterization of SWI/SNF protein expression in human breast cancer cell lines and other malignancies. J Cell Physiol. 2001;186:136–45.
106. Jenuwein T, Allis CD. Translating the histone code. Science. 2001;293:1074–80.
107. Zhang Y, Reinberg D. Transcription regulation by histone methylation: interplay between different covalent modifications of the core histone tails. Genes Dev. 2001;15:2343–60.
108. Kouzarides T. Histone methylation in transcriptional control. Curr Opin Genet Dev. 2002;12:198–209.
109. Stancheva I. Caught in conspiracy: cooperation between DNA methylation and histone H3K9 methylation in the establishment and maintenance of heterochromatin. Biochem Cell Biol. 2005;83:385–95. doi:10.1139/005-043.
110. Kouzarides T. Chromatin modifications and their function. Cell. 2007;128:693–705. doi:10.1016/j.cell.2007.02.005.
111. Cosgrove MS, Wolberger C. How does the histone code work? Biochem Cell Biol. 2005;83:468–76. doi:10.1139/o05-137.
112. Allis CD, Berger SL, Côté J, Dent S, Jenuwien T, Kouzarides T, Pillus L, Reinberg D, Shi Y, Shiekhattar R, Shilatifard A, Workman J, Zhang Y. New nomenclature for chromatin-modifying enzymes. Cell. 2007;131:633–6. doi:10.1016/j.cell.2007.10.039.
113. Vermeulen M, Timmers HT. Grasping trimethylation of histone H3 at lysine 4. Epigenomics. 2010;2:395–406. doi:10.2217/epi.10.11.
114. Li B, Carey M, Workman JL. The role of chromatin during transcription. Cell. 2007;128:707–19. doi:10.1016/j.cell.2007.01.015.
115. Bernstein BE, Liu CL, Humphrey EL, Perlstein EO, Schreiber SL. Global nucleosome occupancy in yeast. Genome Biol. 2004;5:11. doi:10.1016/j.cell.2005.01.001.
116. Carrozza MJ, Li B, Florens L, Suganuma T, Swanson SK, Lee KK, Shia WJ, Anderson S, Yates J, Washburn MP, Workman JL. Histone H3 methylation by Set2 directs deacetylation of coding regions by Rpd3S to suppress spurious intragenic transcription. Cell. 2005;123:581–92.
117. Vogelauer M, Wu J, Suka N, Grunstein M. Global histone acetylation and deacetylation in yeast. Nature. 2000;408:495–8.
118. Rando OJ. Combinatorial complexity in chromatin structure and function: revisiting the histone code. Curr Opin Genet Dev. 2012;22:148–55. doi:10.1016/j.gde.2012.02.013.
119. Hansen JC. Conformational dynamics of the chromatin fiber in solution: determinants, mechanisms, and functions. Annu Rev Biophys Biomol Struct. 2002;31:361–92. doi:10.1146/annurev.biophys.31.101101.140858.
120. Dorigo B, Schalch T, Bystricky K, Richmond TJ. Chromatin fiber folding: requirement for the histone H4 N-terminal tail. J Mol Biol. 2003;327:85–96.
121. Dorigo B, Schalch T, Kulangara A, Duda S, Schroeder RR, Richmond TJ. Nucleosome arrays reveal the two-start organization of the chromatin fiber. Science. 2004;306:1571–3. doi:10.1126/science.1103124.
122. Zheng C, Lu X, Hansen JC, Hayes JJ. Salt-dependent intra- and internucleosomal interactions of the H3 tail domain in a model oligonucleosomal array. J Biol Chem. 2005;280:33552–7. doi:10.1074/jbc.M507241200.
123. Kan PY, Hayes JJ. Detection of interactions between nucleosome arrays mediated by specific core histone tail domains. Methods. 2007;41:278–85. doi:10.1016/j.ymeth.2006.08.012.
124. Kan PY, Lu X, Hansen JC, Hayes JJ. The H3 tail domain participates in multiple interactions during folding and self-association of nucleosome arrays. Mol Cell Biol. 2007;27:2084–91. doi:10.1128/MCB.02181-06.
125. Kan PY, Caterino TL, Hayes JJ. The H4 tail domain participates in intra- and internucleosome interactions with protein and DNA during folding and oligomerization of nucleosome arrays. Mol Cell Biol. 2009;29:538–46. doi:10.1128/MCB.01343-08.
126. Shogren-Knaak M, Ishii H, Sun JM, Pazin MJ, Davie JR, Peterson CL. Histone H4-K16 acetylation controls chromatin structure and protein interactions. Science. 2006;311:844–7. doi:10.1126/science.1124000.

127. Bai L, Morozov AV. Gene regulation by nucleosome positioning. Trends Genet. 2010;26: 476–83. doi:10.1016/j.tig.2010.08.003.
128. Ruthenburg AJ, Li H, Patel DJ, Allis CD. Multivalent engagement of chromatin modifications by linked binding modules. Nat Rev Mol Cell Biol. 2007;8:983–94. doi:10.1038/nrm2298.
129. Taverna SD, Li H, Ruthenburg AJ, Allis CD, Patel DJ. How chromatin-binding modules interpret histone modifications: lessons from professional pocket pickers. Nat Struct Mol Biol. 2007;14:1025–40. doi:10.1038/nsmb1338.
130. Klose RJ, Yi Z. Regulation of histone methylation by demethylimination and demethylation. Nat Rev Mol Cell Biol. 2007;8:307–18. doi:10.1038/nrm2143.
131. Dhalluin C, Carlson JE, Zeng L, He C, Aggarwal AK, Zhou MM. Structure and ligand of a histone acetyltransferase bromodomain. Nature. 1999;399:491–6. doi:10.1038/20974.
132. Hassan AH, Prochasson P, Neely KE, Galasinski SC, Chandy M, Carrozza MJ, Workman JL. Function and selectivity of bromodomains in anchoring chromatin-modifying complexes to promoter nucleosomes. Cell. 2002;111:369–79.
133. Chen J, Ghazawi FM, Li Q. Interplay of bromodomain and histone acetylation in the regulation of p300-dependent genes. Epigenetics. 2010;5:509–15.
134. Chen M, Voeller D, Marquez VE, Kaye FJ, Steeg PS, Giaccone G, Zajac-Kaye M. Enhanced growth inhibition by combined DNA methylation/HDAC inhibitors in lung tumor cells with silenced *CDKN2A*. Int J Oncol. 2010;37:963–71.
135. Hassan AH, Neely KE, Workman JL. Histone acetyltransferase complexes stabilize SWI/SNF binding to promoter nucleosomes. Cell. 2001;104:817–27.
136. Santos-Rosa H, Schneider R, Bannister AJ, Sherriff J, Bernstein BE, Emre NC, Schreiber SL, Mellor J, Kouzarides T. Active genes are tri-methylated at K4 of histone H3. Nature. 2002;419:407–11. doi:10.1038/nature01080.
137. Ng HH, Robert F, Young RA, Struhl K. Targeted recruitment of Set1 histone methylase by elongating Pol II provides a localized mark and memory of recent transcriptional activity. Mol Cell. 2003;11:709–19.
138. Bernstein BE, Kamal M, Lindblad-Toh K, Bekiranov S, Bailey DK, Huebert DJ, McMahon S, Karlsson EK, Kulbokas EJ, Gingeras TR, Schreiber SL, Lander ES. Genomic maps and comparative analysis of histone modifications in human and mouse. Cell. 2005;120:169–81. doi:10.1016/j.cell.2005.01.001.
139. Koch CM, Andrews RM, Flicek P, Dillon SC, Karaoz U, Clelland GK, Wilcox S, Beare DM, Fowler JC, Couttet P, James KD, Lefebvre GC, Bruce AW, Dovey OM, Ellis PD, Dhami P, Langford CF, Weng Z, Birney E, Carter NP, Vetrie D, Dunham I. The landscape of histone modifications across 1% of the human genome in five human cell lines. Genome Res. 2007;17:691–707. doi:10.1101/gr.5704207.
140. Li H, Ilin S, Wang W, Duncan EM, Wysocka J, Allis CD, Patel DJ. Molecular basis for site-specific read-out of histone H3K4me3 by the BPTF PHD finger of NURF. Nature. 2006;442:91–5. doi:10.1038/nature04802.
141. Peña PV, Davrazou F, Shi X, Walter KL, Verkhusha VV, Gozani O, Zhao R, Kutateladze TG. Molecular mechanism of histone H3K4me3 recognition by plant homeodomain of ING2. Nature. 2006;442:100–3. doi:10.1038/nature04814.
142. Wysocka J, Swigut T, Xiao H, Milne TA, Kwon SY, Landry J, Kauer M, Tackett AJ, Chait BT, Badenhorst P, Wu C, Allis CD. A PHD finger of NURF couples histone H3 lysine 4 trimethylation with chromatin remodelling. Nature. 2006;442:86–90. doi:10.1038/nature04815.
143. Vermeulen M, Mulder KW, Denissov S, Pijnappel WW, van Schaik FM, Varier RA, Baltissen MP, Stunnenberg HG, Mann M, Timmers HT. Selective anchoring of TFIID to nucleosomes by trimethylation of histone H3 lysine 4. Cell. 2007;131:58–69. doi:10.1016/j.cell.2007.08.016.
144. Hung T, Binda O, Champagne KS, Kuo AJ, Johnson K, Chang HY, Simon MD, Kutateladze TG, Gozani O. ING4 mediates crosstalk between histone H3 K4 trimethylation and H3 acetylation to attenuate cellular transformation. Mol Cell. 2009;33:248–56. doi:10.1016/j.molcel.2008.12.016.

145. Saksouk N, Avvakumov N, Champagne KS, Hung T, Doyon Y, Cayrou C, Paquet E, Ullah M, Landry AJ, Côté V, Yang XJ, Gozani O, Kutateladze TG, Côté J. HBO1 HAT complexes target chromatin throughout gene coding regions via multiple PHD finger interactions with histone H3 tail. Mol Cell. 2009;33:257–65. doi:10.1016/j.molcel.2009.01.007.
146. Shi X, Hong T, Walter KL, Ewalt M, Michishita E, Hung T, Carney D, Pena P, Lan F, Kaadige MR, Lacoste N, Cayrou C, Davrazou F, Saha A, Cairns BR, Ayer DE, Kutateladze TG, Shi Y, Côté J, Chua KF, Gozani O. ING2 PHD domain links histone H3 lysine 4 methylation to active gene repression. Nature. 2006;442:96–9. doi:10.1038/nature04835.
147. Cai C, He HH, Chen S, Coleman I, Wang H, Fang Z, Nelson PS, Liu XS, Brown M, Balk SP. Androgen receptor gene expression in prostate cancer is directly suppressed by the androgen receptor through recruitment of lysine-specific demethylase 1. Cancer Cell. 2011;20:457–71. doi:10.1016/j.ccr.2011.09.001.
148. Wagner EJ, Carpenter PB. Understanding the language of Lys36 methylation at histone H3. Nat Rev Mol Cell Biol. 2012;13:115–26. doi:10.1038/nrm3274.
149. Joshi AA, Struhl K. Eaf3 chromodomain interaction with methylated H3-K36 links histone deacetylation to Pol II elongation. Mol Cell. 2005;20:971–8. doi:10.1016/j.molcel.2005.11.021.
150. Keogh MC, Kurdistani SK, Morris SA, Ahn SH, Podolny V, Collins SR, Schuldiner M, Chin K, Punna T, Thompson NJ, Boone C, Emili A, Weissman JS, Hughes TR, Strahl BD, Grunstein M, Greenblatt JF, Buratowski S, Krogan NJ. Cotranscriptional Set2 methylation of histone H3 lysine 36 recruits a repressive Rpd3 complex. Cell. 2005;123:593–605. doi:10.1016/j.cell.2005.10.025.
151. Kaplan CD, Laprade L, Winston F. Transcription elongation factors repress transcription initiation from cryptic sites. Science. 2003;301:1096–9. doi:10.1126/science.1087374.
152. Barski A, Cuddapah S, Cui KR, Roh TY, Schones DE, Wang ZB, Wei G, Chepelev I, Zhao KJ. High-resolution profiling of histone methylations in the human genome. Cell. 2007;129:823–37. doi:10.1016/j.cell.2007.05.009.
153. Komashko VM, Acevedo LG, Squazzo SL, Iyengar SS, Rabinovich A, O'Geen H, Green R, Farnham PJ. Using ChIP-chip technology to reveal common principles of transcriptional repression in normal and cancer cells. Genome Res. 2008;18:521–32. doi:10.1101/gr.074609.107.
154. Cameron EE, Bachman KE, Myohanen S, Herman JG, Baylin SB. Synergy of demethylation and histone deacetylase inhibition in the re-expression of genes silenced in cancer. Nat Genet. 1999;21:103–7.
155. Fahrner JA, Eguchi S, Herman JG, Baylin SB. Dependence of histone modifications and gene expression on DNA hypermethylation in cancer. Cancer Res. 2002;62:7213–8.
156. Lehnertz B, Ueda Y, Derijck AA, Braunschweig U, Perez-Burgos L, Kubicek S, Chen T, Li E, Jenuwein T, Peters AH. Suv39h-mediated histone H3 lysine 9 methylation directs DNA methylation to major satellite repeats at pericentric heterochromatin. Curr Biol. 2003;13:1192–200.
157. Schreiber SL, Bernstein BE. Signaling network model of chromatin. Cell. 2002;111:771–8.
158. Fuks F. DNA methylation and histone modifications: teaming up to silence genes. Curr Opin Genet Dev. 2005;15:490–5. doi:10.1016/j.gde.2005.08.002.
159. Ting AH, McGarvey KM, Baylin SB. The cancer epigenome – components and functional correlates. Genes Dev. 2006;20:3215–31. doi:10.1101/gad.1464906.
160. Esteller M. Cancer epigenomics: DNA methylomes and histone-modification maps. Nat Rev Genet. 2007;8:286–98. doi:10.1038/nrg2005.
161. McCabe MT, Brandes JC, Vertino PM. Cancer DNA methylation: molecular mechanisms and clinical implications. Clin Cancer Res. 2009;15:3927–37. doi:10.1158/1078-0432.ccr-08-2784.
162. Fuks F, Hurd PJ, Deplus R, Kouzarides T. The DNA methyltransferases associate with HP1 and the SUV39H1 histone methyltransferase. Nucleic Acids Res. 2003;31:2305–12. doi:10.1093/nar/gkg332.
163. Fraga MF, Esteller M. Towards the human cancer epigenome: a first draft of histone modifications. Cell Cycle. 2005;4:1377–81.

164. Wang J, Hevi S, Kurash JK, Lei H, Gay F, Bajko J, Su H, Sun W, Chang H, Xu G, Gaudet F, Li E, Chen T. The lysine demethylase LSD1 (KDM1) is required for maintenance of global DNA methylation. Nat Genet. 2009;41:125–9. doi:10.1038/ng.268.
165. Hawkins RD, Hon GC, Lee LK, Ngo Q, Lister R, Pelizzola M, Edsall LE, Kuan S, Luu Y, Klugman S, Antosiewicz-Bourget J, Ye Z, Espinoza C, Agarwahl S, Shen L, Ruotti V, Wang W, Stewart R, Thomson JA, Ecker JR, Ren B. Distinct epigenomic landscapes of pluripotent and lineage-committed human cells. Cell Stem Cell. 2010;6:479–91. doi:10.1016/j.stem.2010.03.018.
166. Ooi SKT, Qiu C, Bernstein E, Li KQ, Jia D, Yang Z, Erdjument-Bromage H, Tempst P, Lin SP, Allis CD, Cheng XD, Bestor TH. DNMT3L connects unmethylated lysine 4 of histone H3 to *de novo* methylation of DNA. Nature. 2007;448:714–7. doi:10.1038/nature05987.
167. Lee JH, Skalnik DG. CpG-binding protein (CXXC finger protein 1) is a component of the mammalian Set1 histone H3-Lys4 methyltransferase complex, the analogue of the yeast Set1/COMPASS complex. J Biol Chem. 2005;280:41725–31. doi:10.1074/jbc.M508312200.
168. Lee JH, Tate CM, You JS, Skalnik DG. Identification and characterization of the human Set1B histone H3-Lys4 methyltransferase complex. J Biol Chem. 2007;282:13419–28. doi:10.1074/jbc.M609809200.
169. Lee W, Tillo D, Bray N, Morse RH, Davis RW, Hughes TR, Nislow C. A high-resolution atlas of nucleosome occupancy in yeast. Nat Genet. 2007;39:1235–44. doi:10.1038/ng2117.
170. Thomson JP, Skene PJ, Selfridge J, Clouaire T, Guy J, Webb S, Kerr AR, Deaton A, Andrews R, James KD, Turner DJ, Illingworth R, Bird A. CpG islands influence chromatin structure via the CpG-binding protein Cfp1. Nature. 2010;464:1082–6. doi:10.1038/nature08924.
171. Hinshelwood RA, Melki JR, Huschtscha LI, Paul C, Song JZ, Stirzaker C, Reddel RR, Clark SJ. Aberrant *de novo* methylation of the $p16^{INK4A}$ CpG island is initiated post gene silencing in association with chromatin remodelling and mimics nucleosome positioning. Hum Mol Genet. 2009;18:3098–109. doi:10.1093/hmg/ddp251.
172. Xu M, Kladde MP, Van Etten JL, Simpson RT. Cloning, characterization and expression of the gene coding for a cytosine-5-DNA methyltransferase recognizing GpC. Nucleic Acids Res. 1998;26:3961–6.
173. Kilgore JA, Hoose SA, Gustafson TL, Porter W, Kladde MP. Single-molecule and population probing of chromatin structure using DNA methyltransferases. Methods. 2007;41:320–32. doi:10.1016/j.ymeth.2006.08.008.
174. Pardo CE, Darst RP, Nabilsi NH, Delmas AL, Kladde MP. Simultaneous single-molecule mapping of protein-DNA interactions and DNA methylation by MAPit. Curr Protoc Mol Biol. 2011;Chapter 21:Unit 21.22.1–21.22.18. doi:10.1002/0471142727.mb2122s95.
175. Kagey JD, Kapoor-Vazirani P, McCabe MT, Powell DR, Vertino PM. Long-term stability of demethylation after transient exposure to 5-aza-2′-deoxycytidine correlates with sustained RNA polymerase II occupancy. Mol Cancer Res. 2010;8:1048–59. doi:10.1158/1541-7786.MCR-10-0189.
176. Brinkman AB, Gu H, Bartels SJ, Zhang Y, Matarese F, Simmer F, Marks H, Bock C, Gnirke A, Meissner A, Stunnenberg HG. Sequential ChIP-bisulfite sequencing enables direct genome-scale investigation of chromatin and DNA methylation cross-talk. Genome Res. 2012;22:1128–38. doi:10.1101/gr.133728.111.
177. Statham AL, Robinson MD, Song JZ, Coolen MW, Stirzaker C, Clark SJ. Bisulphite-sequencing of chromatin immunoprecipitated DNA (BisChIP-seq) directly informs methylation status of histone-modified DNA. Genome Res. 2012. doi:10.1101/gr.132076.111.
178. Diala ES, Hoffman RM. DNA methylation levels in normal and chemically-transformed mouse 3T3 cells. Biochem Biophys Res Commun. 1982;104:1489–94.
179. Feinberg AP, Vogelstein B. Hypomethylation distinguishes genes of some human cancers form their normal counterparts. Nature. 1983;301:89–92.
180. Goelz SE, Vogelstein B, Hamilton SR, Feinberg AP. Hypomethylation of DNA from benign and malignant human-colon neoplasms. Science. 1985;228:187–90.
181. Suzuki MM, Bird A. DNA methylation landscapes: provocative insights from epigenomics. Nat Rev Genet. 2008;9:465–76. doi:10.1038/nrg2341.

182. Eden A, Gaudet F, Waghmare A, Jaenisch R. Chromosomal instability and tumors promoted by DNA hypomethylation. Science. 2003;300:455.
183. Gaudet F, Hodgson JG, Eden A, Jackson-Grusby L, Dausman J, Gray JW, Leonhardt H, Jaenisch R. Induction of tumors in mice by genomic hypomethylation. Science. 2003;300:489–92.
184. Howard G, Eiges R, Gaudet F, Jaenisch R, Eden A. Activation and transposition of endogenous retroviral elements in hypomethylation induced tumors in mice. Oncogene. 2008;27:404–8. doi:10.1038/sj.onc.1210631.
185. Iskow RC, McCabe MT, Mills RE, Torene S, Pittard WS, Neuwald AF, Van Meir EG, Vertino PM, Devine SE. Natural mutagenesis of human genomes by endogenous retrotransposons. Cell. 2010;141:1253–61. doi:10.1016/j.cell.2010.05.020.
186. Strieder V, Lutz W. Regulation of N-*myc* expression in development and disease. Cancer Lett. 2002;180:107–19.
187. Watt PM, Kumar R, Kees UR. Promoter demethylation accompanies reactivation of the *HOX11* proto-oncogene in leukemia. Genes Chromosomes Cancer. 2000;29:371–7.
188. Cui HM. Loss of imprinting of *IGF2* as an epigenetic marker for the risk of human cancer. Dis Markers. 2007;23:105–12.
189. Cheung HH, Lee TL, Rennert OM, Chan WY. DNA methylation of cancer genome. Birth Defects Res Part C Embryo Today. 2009;87:335–50. doi:10.1002/bdrc.20163.
190. Portela A, Esteller M. Epigenetic modifications and human disease. Nat Biotechnol. 2010;28:1057–68. doi:10.1038/nbt.1685.
191. Lima SC, Hernandez-Vargas H, Herceg Z. Epigenetic signatures in cancer: implications for the control of cancer in the clinic. Curr Opin Mol Ther. 2010;12:316–24.
192. Fraga MF, Ballestar E, Villar-Garea A, Boix-Chornet M, Espada J, Schotta G, Bonaldi T, Haydon C, Ropero S, Petrie K, Iyer NG, Perez-Rosado A, Calvo E, Lopez JA, Cano A, Calasanz MJ, Colomer D, Piris MA, Ahn N, Imhof A, Caldas C, Jenuwein T, Esteller M. Loss of acetylation at Lys16 and trimethylation at Lys20 of histone H4 is a common hallmark of human cancer. Nat Genet. 2005;37:391–400. doi:10.1038/ng1531.
193. Kanwal R, Gupta S. Epigenetic modifications in cancer. Clin Genet. 2012;81:303–11. doi:10.1111/j.1399-0004.2011.01809.x.
194. Whitehouse I, Flaus A, Cairns BR, White MF, Workman JL, Owen-Hughes T. Nucleosome mobilization catalysed by the yeast SWI/SNF complex. Nature. 1999;400:784–7. doi:10.1038/23506.
195. Saha A, Wittmeyer J, Cairns BR. Chromatin remodeling through directional DNA translocation from an internal nucleosomal site. Nat Struct Mol Biol. 2005;12:747–55.
196. Clapier CR, Cairns BR. The biology of chromatin remodeling complexes. Annu Rev Biochem. 2009;78:273–304. doi:10.1146/annurev.biochem.77.062706.153223.
197. Roberts CWM, Leroux MM, Fleming MD, Orkin SH. Highly penetrant, rapid tumorigenesis through conditional inversion of the tumor suppressor gene *Snf5*. Cancer Cell. 2002;2: 415–25. doi:10.1016/s1535-6108(02)00185-x.
198. Roberts CWM, Orkin SH. The SWI/SNF complex – chromatin and cancer. Nat Rev Cancer. 2004;4:133–42. doi:10.1038/nrc1273.
199. Bernardino J, Roux C, Almeida A, Vogt N, Gibaud A, Gerbault-Seureau M, Magdelenat H, Bourgeois CA, Malfoy B, Dutrillaux B. DNA hypomethylation in breast cancer: an independent parameter of tumor progression. Cancer Genet Cytogenet. 1997;97:83–9.
200. Soares J, Pinto AE, Cunha CV, André S, Barão I, Sousa JM, Cravo M. Global DNA hypomethylation in breast carcinoma: correlation with prognostic factors and tumor progression. Cancer. 1999;85:112–8.
201. Veeck J, Esteller M. Breast cancer epigenetics: from DNA methylation to microRNAs. J Mammary Gland Biol Neoplasia. 2010;15:5–17. doi:10.1007/s10911-010-9165-1.
202. Kinney SR, Pradhan S. Ten eleven translocation enzymes and 5-hydroxymethylation in mammalian development and cancer. Adv Exp Med Biol. 2013;754:57–79. doi:10.1007/978-1-4419-9967-2_3.
203. Cortellino S, Xu J, Sannai M, Moore R, Caretti E, Cigliano A, Le Coz M, Devarajan K, Wessels A, Soprano D, Abramowitz LK, Bartolomei MS, Rambow F, Bassi MR, Bruno T,

Fanciulli M, Renner C, Klein-Szanto AJ, Matsumoto Y, Kobi D, Davidson I, Alberti C, Larue L, Bellacosa A. Thymine DNA glycosylase is essential for active DNA demethylation by linked deamination-base excision repair. Cell. 2011;146:67–79. doi:10.1016/j.cell.2011.06.020.
204. He YF, Li BZ, Li Z, Liu P, Wang Y, Tang Q, Ding J, Jia Y, Chen Z, Li L, Sun Y, Li X, Dai Q, Song CX, Zhang K, He C, Xu GL. Tet-mediated formation of 5-carboxylcytosine and its excision by TDG in mammalian DNA. Science. 2011;333:1303–7. doi:10.1126/science.1210944.
205. Rai K, Sarkar S, Broadbent TJ, Voas M, Grossmann KF, Nadauld LD, Dehghanizadeh S, Hagos FT, Li Y, Toth RK, Chidester S, Bahr TM, Johnson WE, Sklow B, Burt R, Cairns BR, Jones DA. DNA demethylase activity maintains intestinal cells in an undifferentiated state following loss of APC. Cell. 2010;142:930–42. doi:10.1016/j.cell.2010.08.030.
206. Thillainadesan G, Chitilian JM, Isovic M, Ablack JN, Mymryk JS, Tini M, Torchia J. TGF-β-dependent active demethylation and expression of the *p15*^{ink4b} tumor suppressor are impaired by the ZNF217/CoREST Complex. Mol Cell. 2012;46:636–49. doi:10.1016/j.molcel.2012.03.027.
207. Zugmaier G, Ennis BW, Deschauer B, Katz D, Knabbe C, Wilding G, Daly P, Lippman ME, Dickson RB. Transforming growth factors type beta 1 and beta 2 are equipotent growth inhibitors of human breast cancer cell lines. 1989;141:353–61.
208. Basolo F, Fiore L, Ciardiello F, Calvo S, Fontanini G, Conaldi PG, Toniolo A. Response of normal and oncogene-transformed human mammary epithelial cells to transforming growth factor beta 1 (TGF-β 1): lack of growth-inhibitory effect on cells expressing the simian virus 40 large-T antigen. 1994;56:736–42.
209. Barcellos-Hoff MH, Akhurst RJ. Transforming growth factor-β in breast cancer: too much, too late. 2009;11:202.
210. Issa JP, Shen L, Toyota M. CIMP, at last. Gastroenterology. 2005;129:1121–4. doi:10.1053/j.gastro.2005.07.040.
211. Issa JP. Colon cancer: it's CIN or CIMP. Clin Cancer Res. 2008;14:5939–40. doi:10.1158/1078-0432.CCR-08-1596.
212. Holm K, Hegardt C, Staaf J, Vallon-Christersson J, Jönsson G, Olsson H, Borg A, Ringnér M. Molecular subtypes of breast cancer are associated with characteristic DNA methylation patterns. Breast Cancer Res. 2010;12:R36. doi:10.1186/bcr2590.
213. Fackler MJ, Umbricht CB, Williams D, Argani P, Cruz LA, Merino VF, Teo WW, Zhang Z, Huang P, Visvananthan K, Marks J, Ethier S, Gray JW, Wolff AC, Cope LM, Sukumar S. Genome-wide methylation analysis identifies genes specific to breast cancer hormone receptor status and risk of recurrence. Cancer Res. 2011;71:6195–207. doi:10.1158/0008-5472.CAN-11-1630.
214. Fang F, Turcan S, Rimner A, Kaufman A, Giri D, Morris LG, Shen R, Seshan V, Mo Q, Heguy A, Baylin SB, Ahuja N, Viale A, Massague J, Norton L, Vahdat LT, Moynahan ME, Chan TA. Breast cancer methylomes establish an epigenomic foundation for metastasis. Sci Transl Med. 2011;3:75ra25. doi:10.1126/scitranslmed.3001875.
215. Shukla V, Coumoul X, Lahusen T, Wang RH, Xu X, Vassilopoulos A, Xiao C, Lee MH, Man YG, Ouchi M, Ouchi T, Deng CX. BRCA1 affects global DNA methylation through regulation of DNMT1. Cell Res. 2010;20:1201–15. doi:10.1038/cr.2010.128.
216. Yang X, Yan L, Davidson NE. DNA methylation in breast cancer. 2001;8:115–27.
217. Boyer LA, Plath K, Zeitlinger J, Brambrink T, Medeiros LA, Lee TI, Levine SS, Wernig M, Tajonar A, Ray MK, Bell GW, Otte AP, Vidal M, Gifford DK, Young RA, Jaenisch R. Polycomb complexes repress developmental regulators in murine embryonic stem cells. Nature. 2006;441:349–53. doi:10.1038/nature04733.
218. Lee TI, Jenner RG, Boyer LA, Guenther MG, Levine SS, Kumar RM, Chevalier B, Johnstone SE, Cole MF, Isono K, Koseki H, Fuchikami T, Abe K, Murray HL, Zucker JP, Yuan B, Bell GW, Herbolsheimer E, Hannett NM, Sun K, Odom DT, Otte AP, Volkert TL, Bartel DP, Melton DA, Gifford DK, Jaenisch R, Young RA. Control of developmental regulators by Polycomb in human embryonic stem cells. Cell. 2006;125:301–13. doi:10.1016/j.cell.2006.02.043.
219. Lim E, Vaillant F, Wu D, Forrest NC, Pal B, Hart AH, Asselin-Labat ML, Gyorki DE, Ward T, Partanen A, Feleppa F, Huschtscha LI, Thorne HJ, Fox SB, Yan M, French JD, Brown MA,

Smyth GK, Visvader JE, Lindeman GJ. Aberrant luminal progenitors as the candidate target population for basal tumor development in *BRCA1* mutation carriers. Nat Med. 2009;15:907–13. doi:10.1038/nm.2000.
220. Easwaran H, Johnstone SE, Van Neste L, Ohm J, Mosbruger T, Wang Q, Aryee MJ, Joyce P, Ahuja N, Weisenberger D, Collisson E, Zhu J, Yegnasubramanian S, Matsui W, Baylin SB. A DNA hypermethylation module for the stem/progenitor cell signature of cancer. Genome Res. 2012. doi:10.1101/gr.131169.111.
221. Hansen KD, Timp W, Bravo HC, Sabunciyan S, Langmead B, McDonald OG, Wen B, Wu H, Liu Y, Diep D, Briem E, Zhang K, Irizarry RA, Feinberg AP. Increased methylation variation in epigenetic domains across cancer types. Nat Genet. 2011;43:768–75. doi:10.1038/ng.865.
222. Hon GC, Hawkins RD, Caballero OL, Lo C, Lister R, Pelizzola M, Valsesia A, Ye Z, Kuan S, Edsall LE, Camargo AA, Stevenson BJ, Ecker JR, Bafna V, Strausberg RL, Simpson AJ, Ren B. Global DNA hypomethylation coupled to repressive chromatin domain formation and gene silencing in breast cancer. Genome Res. 2012;22:246–58. doi:10.1101/gr.125872.111.
223. Hawkins RD, Hon GC, Yang C, Antosiewicz-Bourget JE, Lee LK, Ngo QM, Klugman S, Ching KA, Edsall LE, Ye Z, Kuan S, Yu P, Liu H, Zhang X, Green RD, Lobanenkov VV, Stewart R, Thomson JA, Ren B. Dynamic chromatin states in human ES cells reveal potential regulatory sequences and genes involved in pluripotency. Cell Res. 2011;21:1393–409. doi:10.1038/cr.2011.146.
224. Okuwaki M, Verreault A. Maintenance DNA methylation of nucleosome core particles. J Biol Chem. 2004;279:2904–12. doi:10.1074/jbc.M310111200.
225. Takeshima H, Suetake I, Tajima S. Mouse Dnmt3a preferentially methylates linker DNA and is inhibited by histone H1. J Mol Biol. 2008;383:810–21. doi:10.1016/j.jmb.2008.03.001.
226. Zhang Y, Jurkowska R, Soeroes S, Rajavelu A, Dhayalan A, Bock I, Rathert P, Brandt O, Reinhardt R, Fischle W, Jeltsch A. Chromatin methylation activity of Dnmt3a and Dnmt3a/3L is guided by interaction of the ADD domain with the histone H3 tail. Nucleic Acids Res. 2010;38:4246–53. doi:10.1093/nar/gkq147.
227. Viré E, Brenner C, Deplus R, Blanchon L, Fraga M, Didelot C, Morey L, Van Eynde A, Bernard D, Vanderwinden JM, Bollen M, Esteller M, Di Croce L, de Launoit Y, Fuks F. The Polycomb group protein EZH2 directly controls DNA methylation. Nature. 2006;439:871–4. doi:10.1038/nature04431.
228. Lindroth AM, Park YJ, McLean CM, Dokshin GA, Persson JM, Herman H, Pasini D, Miro X, Donohoe ME, Lee JT, Helin K, Soloway PD. Antagonism between DNA and H3K27 methylation at the imprinted *Rasgrf1* locus. PLoS Genet. 2008;4:e1000145. doi:10.1371/journal.pgen.1000145.
229. Bartke T, Vermeulen M, Xhemalce B, Robson SC, Mann M, Kouzarides T. Nucleosome-interacting proteins regulated by DNA and histone methylation. Cell. 2010;143:470–84. doi:10.1016/j.cell.2010.10.012.
230. Wu H, Coskun V, Tao J, Xie W, Ge W, Yoshikawa K, Li E, Zhang Y, Sun YE. Dnmt3a-dependent nonpromoter DNA methylation facilitates transcription of neurogenic genes. Science. 2010;329:444–8. doi:10.1126/science.1190485.
231. Komashko VM, Farnham PJ. 5-azacytidine treatment reorganizes genomic histone modification patterns. Epigenetics. 2010;5:229–40.
232. Alvarez JD, Yasui DH, Niida H, Joh T, Loh DY, Kohwi-Shigematsu T. The MAR-binding protein SATB1 orchestrates temporal and spatial expression of multiple genes during T-cell development. Genes Dev. 2000;14:521–35.
233. Cai S, Lee CC, Kohwi-Shigematsu T. SATB1 packages densely looped, transcriptionally active chromatin for coordinated expression of cytokine genes. Nat Genet. 2006;38:1278–88. doi:10.1038/ng1913.
234. Han HJ, Russo J, Kohwi Y, Kohwi-Shigematsu T. SATB1 reprogrammes gene expression to promote breast tumour growth and metastasis. Nature. 2008;452:187–93. doi:10.1038/nature06781.
235. McInnes N, Sadlon TJ, Brown CY, Pederson S, Beyer M, Schultze JL, McColl S, Goodall GJ, Barry SC. FOXP3 and FOXP3-regulated microRNAs suppress SATB1 in breast cancer cells. Oncogene. 2012;31:1045–54. doi:10.1038/onc.2011.293.

236. Fontenot JD, Gavin MA, Rudensky AY. Foxp3 programs the development and function of $CD4^{+}CD25^{+}$ regulatory T cells. Nat Immunol. 2003;4:330–6. doi:10.1038/ni904.
237. Ladoire S, Arnould L, Mignot G, Coudert B, Rebe C, Chalmin F, Vincent J, Bruchard M, Chauffert B, Martin F, Fumoleau P, Ghiringhelli F. Presence of Foxp3 expression in tumor cells predicts better survival in HER2-overexpressing breast cancer patients treated with neoadjuvant chemotherapy. Breast Cancer Res Treat. 2011;125:65–72. doi:10.1007/s10549-010-0831-1.
238. Dalvai M, Bystricky K. The role of histone modifications and variants in regulating gene expression in breast cancer. J Mammary Gland Biol Neoplasia. 2010;15:19–33. doi:10.1007/s10911-010-9167-z.
239. Patani N, Jiang WG, Newbold RF, Mokbel K. Histone-modifier gene expression profiles are associated with pathological and clinical outcomes in human breast cancer. Anticancer Res. 2011;31:4115–25.
240. Kleer CG, Cao Q, Varambally S, Shen R, Ota I, Tomlins SA, Ghosh D, Sewalt RG, Otte AP, Hayes DF, Sabel MS, Livant D, Weiss SJ, Rubin MA, Chinnaiyan AM. EZH2 is a marker of aggressive breast cancer and promotes neoplastic transformation of breast epithelial cells. Proc Natl Acad Sci USA. 2003;100:11606–11. doi:10.1073/pnas.1933744100.
241. Chang CJ, Yang JY, Xia W, Chen CT, Xie X, Chao CH, Woodward WA, Hsu JM, Hortobagyi GN, Hung MC. EZH2 promotes expansion of breast tumor initiating cells through activation of RAF1-β-catenin signaling. Cancer Cell. 2011;19:86–100. doi:10.1016/j.ccr.2010.10.035.
242. Dalgliesh GL, Furge K, Greenman C, Chen L, Bignell G, Butler A, Davies H, Edkins S, Hardy C, Latimer C, Teague J, Andrews J, Barthorpe S, Beare D, Buck G, Campbell PJ, Forbes S, Jia M, Jones D, Knott H, Kok CY, Lau KW, Leroy C, Lin ML, McBride DJ, Maddison M, Maguire S, McLay K, Menzies A, Mironenko T, Mulderrig L, Mudie L, O'Meara S, Pleasance E, Rajasingham A, Shepherd R, Smith R, Stebbings L, Stephens P, Tang G, Tarpey PS, Turrell K, Dykema KJ, Khoo SK, Petillo D, Wondergem B, Anema J, Kahnoski RJ, Teh BT, Stratton MR, Futreal PA. Systematic sequencing of renal carcinoma reveals inactivation of histone modifying genes. Nature. 2010;463:360–3. doi:10.1038/nature08672.
243. Morin RD, Johnson NA, Severson TM, Mungall AJ, An J, Goya R, Paul JE, Boyle M, Woolcock BW, Kuchenbauer F, Yap D, Humphries RK, Griffith OL, Shah S, Zhu H, Kimbara M, Shashkin P, Charlot JF, Tcherpakov M, Corbett R, Tam A, Varhol R, Smailus D, Moksa M, Zhao Y, Delaney A, Qian H, Birol I, Schein J, Moore R, Holt R, Horsman DE, Connors JM, Jones S, Aparicio S, Hirst M, Gascoyne RD, Marra MA. Somatic mutations altering EZH2 (Tyr641) in follicular and diffuse large B-cell lymphomas of germinal-center origin. Nat Genet. 2010;42:181–5. doi:10.1038/ng.518.
244. Yap DB, Chu J, Berg T, Schapira M, Cheng SW, Moradian A, Morin RD, Mungall AJ, Meissner B, Boyle M, Marquez VE, Marra MA, Gascoyne RD, Humphries RK, Arrowsmith CH, Morin GB, Aparicio SA. Somatic mutations at EZH2 Y641 act dominantly through a mechanism of selectively altered PRC2 catalytic activity, to increase H3K27 trimethylation. Blood. 2011;117:2451–9. doi:10.1182/blood-2010-11-321208.
245. Bracken AP, Pasini D, Capra M, Prosperini E, Colli E, Helin K. EZH2 is downstream of the pRB-E2F pathway, essential for proliferation and amplified in cancer. EMBO J. 2003;22:5323–35. doi:10.1093/emboj/cdg542.
246. Gong Y, Huo L, Liu P, Sneige N, Sun X, Ueno NT, Lucci A, Buchholz TA, Valero V, Cristofanilli M. Polycomb group protein EZH2 is frequently expressed in inflammatory breast cancer and is predictive of worse clinical outcome. Cancer. 2011;117:5476–84. doi:10.1002/cncr.26179.
247. Kunju LP, Cookingham C, Toy KA, Chen W, Sabel MS, Kleer CG. EZH2 and ALDH-1 mark breast epithelium at risk for breast cancer development. Mod Pathol. 2011;24:786–93. doi:10.1038/modpathol.2011.8.
248. Nishikawa N, Toyota M, Suzuki H, Honma T, Fujikane T, Ohmura T, Nishidate T, Ohe-Toyota M, Maruyama R, Sonoda T, Sasaki Y, Urano T, Imai K, Hirata K, Tokino T. Gene amplification and overexpression of PRDM14 in breast cancers. Cancer Res. 2007;67:9649–57. doi:10.1158/0008-5472.CAN-06-4111.
249. Moelans CB, de Weger RA, Monsuur HN, Maes AH, van Diest PJ. Molecular differences between ductal carcinoma in situ and adjacent invasive breast carcinoma: a multiplex ligation-

dependent probe amplification study. Anal Cell Pathol (Amst). 2010;33:165–73. doi:10.3233/ACP-CLO-2010-0546.
250. Moelans CB, de Weger RA, Monsuur HN, Vijzelaar R, van Diest PJ. Molecular profiling of invasive breast cancer by multiplex ligation-dependent probe amplification-based copy number analysis of tumor suppressor and oncogenes. Mod Pathol. 2010;23:1029–39. doi:10.1038/modpathol.2010.84.
251. Hamamoto R, Silva FP, Tsuge M, Nishidate T, Katagiri T, Nakamura Y, Furukawa Y. Enhanced SMYD3 expression is essential for the growth of breast cancer cells. Cancer Sci. 2006;97:113–8. doi:10.1111/j.1349-7006.2006.00146.x.
252. Angrand PO, Apiou F, Stewart AF, Dutrillaux B, Losson R, Chambon P. Nsd3, a new set domain-containing gene, maps to 8p12 and is amplified in human breast cancer cell lines. Genomics. 2001;74:79–88. doi:10.1006/geno.2001.6524.
253. Garcia MJ, Pole JC, Chin SF, Teschendorff A, Naderi A, Ozdag H, Vias M, Kranjac T, Subkhankulova T, Paish C, Ellis I, Brenton JD, Edwards PA, Caldas C. A 1 Mb minimal amplicon at 8p11-12 in breast cancer identifies new candidate oncogenes. Oncogene. 2005;24:5235–45. doi:10.1038/sj.onc.1208741.
254. Deng Q, Huang S. PRDM5 is silenced in human cancers and has growth suppressive activities. Oncogene. 2004;23:4903–10. doi:10.1038/sj.onc.1207615.
255. Frietze S, Lupien M, Silver PA, Brown M. CARM1 regulates estrogen-stimulated breast cancer growth through up-regulation of E2F1. Cancer Res. 2008;68:301–6. doi:10.1158/0008-5472.CAN-07-1983.
256. Al-Dhaheri M, Wu J, Skliris GP, Li J, Higashimato K, Wang Y, White KP, Lambert P, Zhu Y, Murphy L, Xu W. CARM1 is an important determinant of ERα-dependent breast cancer cell differentiation and proliferation in breast cancer cells. Cancer Res. 2011;71:2118–28. doi:10.1158/0008-5472.CAN-10-2426.
257. Lim S, Janzer A, Becker A, Zimmer A, Schule R, Buettner R, Kirfel J. Lysine-specific demethylase 1 (LSD1) is highly expressed in ER-negative breast cancers and a biomarker predicting aggressive biology. Carcinogenesis. 2010;31:512–20. doi:10.1093/carcin/bgp324.
258. Lu PJ, Sundquist K, Baeckstrom D, Poulsom R, Hanby A, Meier-Ewert S, Jones T, Mitchell M, Pitha-Rowe P, Freemont P, Taylor-Papadimitriou J. A novel gene (*PLU-1*) containing highly conserved putative DNA/chromatin binding motifs is specifically up-regulated in breast cancer. J Biol Chem. 1999;274:15633–45.
259. Iizuka M, Takahashi Y, Mizzen CA, Cook RG, Fujita M, Allis CD, Frierson Jr HF, Fukusato T, Smith MM. Histone acetyltransferase Hbo1: catalytic activity, cellular abundance, and links to primary cancers. Gene. 2009;436:108–14. doi:10.1016/j.gene.2009.01.020.
260. Pfister S, Rea S, Taipale M, Mendrzyk F, Straub B, Ittrich C, Thuerigen O, Sinn HP, Akhtar A, Lichter P. The histone acetyltransferase hMOF is frequently downregulated in primary breast carcinoma and medulloblastoma and constitutes a biomarker for clinical outcome in medulloblastoma. Int J Cancer. 2008;122:1207–13. doi:10.1002/ijc.23283.
261. Suzuki J, Chen YY, Scott GK, Devries S, Chin K, Benz CC, Waldman FM, Hwang ES. Protein acetylation and histone deacetylase expression associated with malignant breast cancer progression. Clin Cancer Res. 2009;15:3163–71. doi:10.1158/1078-0432.CCR-08-2319.
262. Feng Y, Wang X, Xu L, Pan H, Zhu S, Liang Q, Huang B, Lu J. The transcription factor ZBP-89 suppresses p16 expression through a histone modification mechanism to affect cell senescence. FEBS J. 2009;276:4197–206. doi:10.1111/j.1742-4658.2009.07128.x.
263. Wu X, Zhu Z, Li W, Fu X, Su D, Fu L, Zhang Z, Luo A, Sun X, Dong JT. Chromodomain helicase DNA binding protein 5 plays a tumor suppressor role in human breast cancer. Breast Cancer Res. 2012;14:R73. doi:10.1186/bcr3182.
264. Ross-Innes CS, Stark R, Teschendorff AE, Holmes KA, Ali HR, Dunning MJ, Brown GD, Gojis O, Ellis IO, Green AR, Ali S, Chin SF, Palmieri C, Caldas C, Carroll JS. Differential oestrogen receptor binding is associated with clinical outcome in breast cancer. Nature. 2012;481:389–93. doi:10.1038/nature10730.
265. Hurtado A, Holmes KA, Ross-Innes CS, Schmidt D, Carroll JS. FOXA1 is a key determinant of estrogen receptor function and endocrine response. Nat Genet. 2011;43:27–33. doi:10.1038/ng.730.

266. He HH, Meyer CA, Shin H, Bailey ST, Wei G, Wang Q, Zhang Y, Xu K, Ni M, Lupien M, Mieczkowski P, Lieb JD, Zhao K, Brown M, Liu XS. Nucleosome dynamics define transcriptional enhancers. Nat Genet. 2010;42:343–7. doi:10.1038/ng.545.
267. Andreu-Vieyra C, Lai J, Berman BP, Frenkel B, Jia L, Jones PA, Coetzee GA. Dynamic nucleosome-depleted regions at androgen receptor enhancers in the absence of ligand in prostate cancer cells. Mol Cell Biol. 2011;31:4648–62. doi:10.1128/MCB.05934-11.
268. Stadler MB, Murr R, Burger L, Ivanek R, Lienert F, Schöler A, van Nimwegen E, Wirbelauer C, Oakeley EJ, Gaidatzis D, Tiwari VK, Schübeler D. DNA-binding factors shape the mouse methylome at distal regulatory regions. Nature. 2011;480:490–5. doi:10.1038/nature10716.
269. Tsai WW, Wang Z, Yiu TT, Akdemir KC, Xia W, Winter S, Tsai CY, Shi X, Schwarzer D, Plunkett W, Aronow B, Gozani O, Fischle W, Hung MC, Patel DJ, Barton MC. TRIM24 links a non-canonical histone signature to breast cancer. Nature. 2010;468:927–32. doi:10.1038/nature09542.
270. Lupien M, Eeckhoute J, Meyer CA, Wang Q, Zhang Y, Li W, Carroll JS, Liu XS, Brown M. FoxA1 translates epigenetic signatures into enhancer-driven lineage-specific transcription. Cell. 2008;132:958–70. doi:10.1016/j.cell.2008.01.018.
271. Yanagisawa J, Kitagawa H, Yanagida M, Wada O, Ogawa S, Nakagomi M, Oishi H, Yamamoto Y, Nagasawa H, McMahon SB, Cole MD, Tora L, Takahashi N, Kato S. Nuclear receptor function requires a TFTC-type histone acetyl transferase complex. Mol Cell. 2002;9:553–62.
272. Fullwood MJ, Liu MH, Pan YF, Liu J, Xu H, Mohamed YB, Orlov YL, Velkov S, Ho A, Mei PH, Chew EG, Huang PY, Welboren WJ, Han Y, Ooi HS, Ariyaratne PN, Vega VB, Luo Y, Tan PY, Choy PY, Wansa KD, Zhao B, Lim KS, Leow SC, Yow JS, Joseph R, Li H, Desai KV, Thomsen JS, Lee YK, Karuturi RK, Herve T, Bourque G, Stunnenberg HG, Ruan X, Cacheux-Rataboul V, Sung WK, Liu ET, Wei CL, Cheung E, Ruan Y. An oestrogen-receptor-α-bound human chromatin interactome. Nature. 2009;462:58–64. doi:10.1038/nature08497.
273. Weiss AJ, Stambaugh JE, Mastrangelo MJ, Laucius JF, Bellet RE. Phase I study of 5-azacytidine (NSC-102816). Cancer Chemother Rep. 1972;56:413–9.
274. Silverman LR, Holland JF, Weinberg RS, Alter BP, Davis RB, Ellison RR, Demakos EP, Cornell Jr CJ, Carey RW, Schiffer C, et al. Effects of treatment with 5-azacytidine on the in vivo and in vitro hematopoiesis in patients with myelodysplastic syndromes. Leukemia. 1993;7 Suppl 1:21–9.
275. Vesely J, Cihak A. Incorporation of a potent antileukemic agent, 5-aza-2′-deoxycytidine, into DNA of cells from leukemic mice. Cancer Res. 1977;37:3684–9.
276. Bouchard J, Momparler RL. Incorporation of 5-aza-2′-deoxycytidine-5′-triphosphate into DNA: interactions with mammalian DNA polymerase α and DNA methylase. Mol Pharmacol. 1983;24:109–14.
277. Christman JK. 5-Azacytidine and 5-aza-2′-deoxycytidine as inhibitors of DNA methylation: mechanistic studies and their implications for cancer therapy. Oncogene. 2002;21:5483–95. doi:10.1038/sj.onc.1205699.
278. Lyko F, Brown R. DNA methyltransferase inhibitors and the development of epigenetic cancer therapies. J Natl Cancer Inst. 2005;97:1498–506. doi:10.1093/jnci/dji311.
279. Stresemann C, Lyko F. Modes of action of the DNA methyltransferase inhibitors azacytidine and decitabine. Int J Cancer. 2008;123:8–13. doi:10.1002/ijc.23607.
280. Brueckner B, Kuck D, Lyko F. DNA methyltransferase inhibitors for cancer therapy. Cancer J. 2007;13:17–22. doi:10.1097/PPO.0b013e31803c7245.
281. Zhou L, Cheng X, Connolly BA, Dickman MJ, Hurd PJ, Hornby DP. Zebularine: a novel DNA methylation inhibitor that forms a covalent complex with DNA methyltransferases. J Mol Biol. 2002;321:591–9.
282. Cheng JC, Weisenberger DJ, Gonzales FA, Liang G, Xu GL, Hu YG, Marquez VE, Jones PA. Continuous zebularine treatment effectively sustains demethylation in human bladder cancer cells. Mol Cell Biol. 2004;24:1270–8.
283. Yoo CB, Cheng JC, Jones PA. Zebularine: a new drug for epigenetic therapy. Biochem Soc Trans. 2004;32:910–2. doi:10.1042/BST0320910.

284. Balch C, Yan P, Craft T, Young S, Skalnik DG, Huang TH, Nephew KP. Antimitogenic and chemosensitizing effects of the methylation inhibitor zebularine in ovarian cancer. Mol Cancer Ther. 2005;4:1505–14. doi:10.1158/1535-7163.MCT-05-0216.
285. Cheng JC, Matsen CB, Gonzales FA, Ye W, Greer S, Marquez VE, Jones PA, Selker EU. Inhibition of DNA methylation and reactivation of silenced genes by zebularine. J Natl Cancer Inst. 2003;95:399–409.
286. Dote H, Cerna D, Burgan WE, Carter DJ, Cerra MA, Hollingshead MG, Camphausen K, Tofilon PJ. Enhancement of in vitro and in vivo tumor cell radiosensitivity by the DNA methylation inhibitor zebularine. Clin Cancer Res. 2005;11:4571–9. doi:10.1158/1078-0432.CCR-05-0050.
287. Neureiter D, Zopf S, Leu T, Dietze O, Hauser-Kronberger C, Hahn EG, Herold C, Ocker M. Apoptosis, proliferation and differentiation patterns are influenced by Zebularine and SAHA in pancreatic cancer models. Scand J Gastroenterol. 2007;42:103–16. doi:10.1080/00365520600874198.
288. Yoo CB, Chuang JC, Byun HM, Egger G, Yang AS, Dubeau L, Long T, Laird PW, Marquez VE, Jones PA. Long-term epigenetic therapy with oral zebularine has minimal side effects and prevents intestinal tumors in mice. Cancer Prev Res (Phila). 2008;1:233–40. doi:10.1158/1940-6207.CAPR-07-0008.
289. Chen M, Shabashvili D, Nawab A, Yang SX, Dyer LM, Brown KD, Hollingshead M, Hunter KW, Kaye FJ, Hochwald SN, Marquez VE, Steeg P, Zajac-Kaye M. DNA methyltransferase inhibitor, zebularine, delays tumor growth and induces apoptosis in a genetically engineered mouse model of breast cancer. Mol Cancer Ther. 2012;11:370–82. doi:10.1158/1535-7163.MCT-11-0458.
290. Candido EP, Reeves R, Davie JR. Sodium butyrate inhibits histone deacetylation in cultured cells. Cell. 1978;14:105–13.
291. Marks PA, Dokmanovic M. Histone deacetylase inhibitors: discovery and development as anticancer agents. Expert Opin Investig Drugs. 2005;14:1497–511. doi:10.1517/13543784.14.12.1497.
292. Martinet N, Bertrand P. Interpreting clinical assays for histone deacetylase inhibitors. Cancer Manag Res. 2011;3:117–41. doi:10.2147/CMR.S9661.
293. Drummond DC, Noble CO, Kirpotin DB, Guo Z, Scott GK, Benz CC. Clinical development of histone deacetylase inhibitors as anticancer agents. Annu Rev Pharmacol Toxicol. 2005;45:495–528. doi:10.1146/annurev.pharmtox.45.120403.095825.
294. Richon VM, Webb Y, Merger R, Sheppard T, Jursic B, Ngo L, Civoli F, Breslow R, Rifkind RA, Marks PA. Second generation hybrid polar compounds are potent inducers of transformed cell differentiation. Proc Natl Acad Sci USA. 1996;93:5705–8.
295. Butler LM, Agus DB, Scher HI, Higgins B, Rose A, Cordon-Cardo C, Thaler HT, Rifkind RA, Marks PA, Richon VM. Suberoylanilide hydroxamic acid, an inhibitor of histone deacetylase, suppresses the growth of prostate cancer cells in vitro and in vivo. Cancer Res. 2000;60:5165–70.
296. Kelly WK, O'Connor OA, Krug LM, Chiao JH, Heaney M, Curley T, MacGregore-Cortelli B, Tong W, Secrist JP, Schwartz L, Richardson S, Chu E, Olgac S, Marks PA, Scher H, Richon VM. Phase I study of an oral histone deacetylase inhibitor, suberoylanilide hydroxamic acid, in patients with advanced cancer. J Clin Oncol. 2005;23:3923–31. doi:10.1200/JCO.2005.14.167.
297. Ueda H, Nakajima H, Hori Y, Goto T, Okuhara M. Action of FR901228, a novel antitumor bicyclic depsipeptide produced by *Chromobacterium violaceum* no. 968, on Ha-*ras* transformed NIH3T3 cells. Biosci Biotechnol Biochem. 1994;58:1579–83.
298. Furumai R, Matsuyama A, Kobashi N, Lee KH, Nishiyama M, Nakajima H, Tanaka A, Komatsu Y, Nishino N, Yoshida M, Horinouchi S. FK228 (depsipeptide) as a natural prodrug that inhibits class I histone deacetylases. Cancer Res. 2002;62:4916–21.
299. Pathiraja TN, Stearns V, Oesterreich S. Epigenetic regulation in estrogen receptor positive breast cancer–role in treatment response. J Mammary Gland Biol Neoplasia. 2010;15:35–47. doi:10.1007/s10911-010-9166-0.

300. Chinnaiyan P, Vallabhaneni G, Armstrong E, Huang SM, Harari PM. Modulation of radiation response by histone deacetylase inhibition. Int J Radiat Oncol Biol Phys. 2005;62:223–9. doi:10.1016/j.ijrobp.2004.12.088.
301. Munshi A, Kurland JF, Nishikawa T, Tanaka T, Hobbs ML, Tucker SL, Ismail S, Stevens C, Meyn RE. Histone deacetylase inhibitors radiosensitize human melanoma cells by suppressing DNA repair activity. Clin Cancer Res. 2005;11:4912–22. doi:10.1158/1078-0432.CCR-04-2088.
302. Almenara J, Rosato R, Grant S. Synergistic induction of mitochondrial damage and apoptosis in human leukemia cells by flavopiridol and the histone deacetylase inhibitor suberoylanilide hydroxamic acid (SAHA). Leukemia. 2002;16:1331–43. doi:10.1038/sj.leu.2402535.
303. Kaminskyy VO, Surova OV, Vaculova A, Zhivotovsky B. Combined inhibition of DNA methyltransferase and histone deacetylase restores caspase-8 expression and sensitizes SCLC cells to TRAIL. Carcinogenesis. 2011;32:1450–8. doi:10.1093/carcin/bgr135.
304. Shang D, Liu Y, Matsui Y, Ito N, Nishiyama H, Kamoto T, Ogawa O. Demethylating agent 5-aza-2′-deoxycytidine enhances susceptibility of bladder transitional cell carcinoma to cisplatin. Urology. 2008;71:1220–5. doi:10.1016/j.urology.2007.11.029.
305. Mirza S, Sharma G, Pandya P, Ralhan R. Demethylating agent 5-aza-2′-deoxycytidine enhances susceptibility of breast cancer cells to anticancer agents. Mol Cell Biochem. 2010;342:101–9. doi:10.1007/s11010-010-0473-y.
306. Plumb JA, Strathdee G, Sludden J, Kaye SB, Brown R. Reversal of drug resistance in human tumor xenografts by 2′-deoxy-5-azacytidine-induced demethylation of the *hMLH1* gene promoter. Cancer Res. 2000;60:6039–44.
307. Smith E, Lin C, Shilatifard A. The super elongation complex (SEC) and MLL in development and disease. Genes Dev. 2011;25:661–72. doi:10.1101/gad.2015411.
308. Milne TA, Briggs SD, Brock HW, Martin ME, Gibbs D, Allis CD, Hess JL. MLL targets SET domain methyltransferase activity to *Hox* gene promoters. Mol Cell. 2002;10:1107–17.
309. Nakamura T, Mori T, Tada S, Krajewski W, Rozovskaia T, Wassell R, Dubois G, Mazo A, Croce CM, Canaani E. ALL-1 is a histone methyltransferase that assembles a supercomplex of proteins involved in transcriptional regulation. Mol Cell. 2002;10:1119–28.
310. Okada Y, Feng Q, Lin Y, Jiang Q, Li Y, Coffield VM, Su L, Xu G, Zhang Y. hDOT1L links histone methylation to leukemogenesis. Cell. 2005;121:167–78. doi:10.1016/j.cell.2005.02.020.
311. Bitoun E, Oliver PL, Davies KE. The mixed-lineage leukemia fusion partner AF4 stimulates RNA polymerase II transcriptional elongation and mediates coordinated chromatin remodeling. Hum Mol Genet. 2007;16:92–106. doi:10.1093/hmg/ddl444.
312. Mueller D, Bach C, Zeisig D, Garcia-Cuellar MP, Monroe S, Sreekumar A, Zhou R, Nesvizhskii A, Chinnaiyan A, Hess JL, Slany RK. A role for the MLL fusion partner ENL in transcriptional elongation and chromatin modification. Blood. 2007;110:4445–54. doi:10.1182/blood-2007-05-090514.
313. Mohan M, Herz HM, Takahashi YH, Lin C, Lai KC, Zhang Y, Washburn MP, Florens L, Shilatifard A. Linking H3K79 trimethylation to Wnt signaling through a novel Dot1-containing complex (DotCom). Genes Dev. 2010;24:574–89. doi:10.1101/gad.1898410.
314. Park S, Osmers U, Raman G, Schwantes RH, Diaz MO, Bushweller JH. The PHD3 domain of MLL acts as a CYP33-regulated switch between MLL-mediated activation and repression. Biochemistry. 2010;49:6576–86. doi:10.1021/bi1009387.
315. Yokoyama A, Lin M, Naresh A, Kitabayashi I, Cleary ML. A higher-order complex containing AF4 and ENL family proteins with P-TEFb facilitates oncogenic and physiologic MLL-dependent transcription. Cancer Cell. 2010;17:198–212. doi:10.1016/j.ccr.2009.12.040.
316. Milne TA, Martin ME, Brock HW, Slany RK, Hess JL. Leukemogenic MLL fusion proteins bind across a broad region of the *Hox a9* locus, promoting transcription and multiple histone modifications. Cancer Res. 2005;65:11367–74. doi:10.1158/0008-5472.CAN-05-1041.
317. Guenther MG, Lawton LN, Rozovskaia T, Frampton GM, Levine SS, Volkert TL, Croce CM, Nakamura T, Canaani E, Young RA. Aberrant chromatin at genes encoding stem cell regulators in human mixed-lineage leukemia. Genes Dev. 2008;22:3403–8. doi:10.1101/gad.1741408.

318. Krivtsov AV, Feng Z, Lemieux ME, Faber J, Vempati S, Sinha AU, Xia X, Jesneck J, Bracken AP, Silverman LB, Kutok JL, Kung AL, Armstrong SA. H3K79 methylation profiles define murine and human MLL-AF4 leukemias. Cancer Cell. 2008;14:355–68. doi:10.1016/j.ccr.2008.10.001.
319. Daigle SR, Olhava EJ, Therkelsen CA, Majer CR, Sneeringer CJ, Song J, Johnston LD, Scott MP, Smith JJ, Xiao Y, Jin L, Kuntz KW, Chesworth R, Moyer MP, Bernt KM, Tseng JC, Kung AL, Armstrong SA, Copeland RA, Richon VM, Pollock RM. Selective killing of mixed lineage leukemia cells by a potent small-molecule DOT1L inhibitor. Cancer Cell. 2011;20: 53–65. doi:10.1016/j.ccr.2011.06.009.
320. Chang CJ, Hung MC. The role of EZH2 in tumour progression. Br J Cancer. 2012;106: 243–7. doi:10.1038/bjc.2011.551.
321. Gonzalez ME, Li X, Toy K, DuPrie M, Ventura AC, Banerjee M, Ljungman M, Merajver SD, Kleer CG. Downregulation of EZH2 decreases growth of estrogen receptor-negative invasive breast carcinoma and requires BRCA1. Oncogene. 2009;28:843–53. doi:10.1038/onc.2008.433.
322. Sneeringer CJ, Scott MP, Kuntz KW, Knutson SK, Pollock RM, Richon VM, Copeland RA. Coordinated activities of wild-type plus mutant EZH2 drive tumor-associated hypertrimethylation of lysine 27 on histone H3 (H3K27) in human B-cell lymphomas. Proc Natl Acad Sci USA. 2010;107:20980–5. doi:10.1073/pnas.1012525107.
323. Yoo CB, Jeong S, Egger G, Liang G, Phiasivongsa P, Tang C, Redkar S, Jones PA. Delivery of 5-aza-2′-deoxycytidine to cells using oligodeoxynucleotides. Cancer Res. 2007;67: 6400–8. doi:10.1158/0008-5472.CAN-07-0251.
324. Chuang JC, Warner SL, Vollmer D, Vankayalapati H, Redkar S, Bearss DJ, Qiu X, Yoo CB, Jones PA. S110, a 5-Aza-2′-deoxycytidine-containing dinucleotide, is an effective DNA methylation inhibitor in vivo and can reduce tumor growth. Mol Cancer Ther. 2010;9: 1443–50. doi:10.1158/1535-7163.MCT-09-1048.
325. Lavelle D, Saunthararajah Y, Vaitkus K, Singh M, Banzon V, Phiasivongsva P, Redkar S, Kanekal S, Bearss D, Shi C, Inloes R, DeSimone J. S110, a novel decitabine dinucleotide, increases fetal hemoglobin levels in baboons (*P. anubis*). J Transl Med. 2010;8:92.
326. Foulks JM, Parnell KM, Nix RN, Chau S, Swierczek K, Saunders M, Wright K, Hendrickson TF, Ho KK, McCullar MV, Kanner SB. Epigenetic drug discovery: targeting DNA methyltransferases. J Biomol Screen. 2012;17:2–17. doi:10.1177/1087057111421212.
327. Issa J. Interim results from a randomized Phase 1-2 first-in-human-(FIH) study of PK/PD guided escalating doses of SGI-110, a novel subcutaneous (SQ) second generation hypomethylating agent (HMA) in relapsed/refractory MDS and AML. American Association for Cancer Research (AACR) Annual Meeting; 2012; Chicago, IL.
328. Pazolli E, Alspach E, Milczarek A, Prior J, Piwnica-Worms D, Stewart SA. Chromatin remodeling underlies the senescence-associated secretory phenotype of tumor stromal fibroblasts that supports cancer progression. Cancer Res. 2012;72:2251–61. doi:10.1158/0008-5472.CAN-11-3386.
329. Needham LA, Davidson AH, Bawden LJ, Belfield A, Bone EA, Brotherton DH, Bryant S, Charlton MH, Clark VL, Davies SJ, Donald A, Day FA, Krige D, Legris V, McDermott J, McGovern Y, Owen J, Patel SR, Pintat S, Testar RJ, Wells GM, Moffat D, Drummond AH. Drug targeting to monocytes and macrophages using esterase-sensitive chemical motifs. J Pharmacol Exp Ther. 2011;339:132–42. doi:10.1124/jpet.111.183640.
330. Keen JC, Yan L, Mack KM, Pettit C, Smith D, Sharma D, Davidson NE. A novel histone deacetylase inhibitor, scriptaid, enhances expression of functional estrogen receptor alpha (ER) in ER negative human breast cancer cells in combination with 5-aza 2′-deoxycytidine. Breast Cancer Res Treat. 2003;81:177–86. doi:10.1023/A:1026146524737.
331. Sharma D, Saxena NK, Davidson NE, Vertino PM. Restoration of tamoxifen sensitivity in estrogen receptor-negative breast cancer cells: tamoxifen-bound reactivated ER recruits distinctive corepressor complexes. Cancer Res. 2006;66:6370–8. doi:10.1158/0008-5472.CAN-06-0402.
332. Fan J, Yin WJ, Lu JS, Wang L, Wu J, Wu FY, Di GH, Shen ZZ, Shao ZM. ERα negative breast cancer cells restore response to endocrine therapy by combination treatment with both

HDAC inhibitor and DNMT inhibitor. J Cancer Res Clin Oncol. 2008;134:883–90. doi:10.1007/s00432-008-0354-x.
333. Giacinti L, Giacinti C, Gabellini C, Rizzuto E, Lopez M, Giordano A. Scriptaid effects on breast cancer cell lines. J Cell Physiol. 2012;227:3426–33. doi:10.1002/jcp.24043.
334. Munster PN, Thurn KT, Thomas S, Raha P, Lacevic M, Miller A, Melisko M, Ismail-Khan R, Rugo H, Moasser M, Minton SE. A phase II study of the histone deacetylase inhibitor vorinostat combined with tamoxifen for the treatment of patients with hormone therapy-resistant breast cancer. Br J Cancer. 2011;104:1828–35. doi:10.1038/bjc.2011.156.
335. Gore SD, Baylin S, Sugar E, Carraway H, Miller CB, Carducci M, Grever M, Galm O, Dauses T, Karp JE, Rudek MA, Zhao M, Smith BD, Manning J, Jiemjit A, Dover G, Mays A, Zwiebel J, Murgo A, Weng LJ, Herman JG. Combined DNA methyltransferase and histone deacetylase inhibition in the treatment of myeloid neoplasms. Cancer Res. 2006;66:6361–9. doi:10.1158/0008-5472.CAN-06-0080.
336. Al-Hajj M, Wicha MS, Benito-Hernandez A, Morrison SJ, Clarke MF. Prospective identification of tumorigenic breast cancer cells. Proc Natl Acad Sci USA. 2003;100:3983–8. doi:10.1073/pnas.0530291100.
337. Bao S, Wu Q, McLendon RE, Hao Y, Shi Q, Hjelmeland AB, Dewhirst MW, Bigner DD, Rich JN. Glioma stem cells promote radioresistance by preferential activation of the DNA damage response. Nature. 2006;444:756–60. doi:10.1038/nature05236.
338. Phillips TM, McBride WH, Pajonk F. The response of $CD24^{-/low}/CD44^{+}$ breast cancer-initiating cells to radiation. J Natl Cancer Inst. 2006;98:1777–85. doi:10.1093/jnci/djj495.
339. Yu F, Yao H, Zhu P, Zhang X, Pan Q, Gong C, Huang Y, Hu X, Su F, Lieberman J, Song E. *let-7* regulates self renewal and tumorigenicity of breast cancer cells. Cell. 2007;131: 1109–23. doi:10.1016/j.cell.2007.10.054.
340. Li X, Lewis MT, Huang J, Gutierrez C, Osborne CK, Wu MF, Hilsenbeck SG, Pavlick A, Zhang X, Chamness GC, Wong H, Rosen J, Chang JC. Intrinsic resistance of tumorigenic breast cancer cells to chemotherapy. J Natl Cancer Inst. 2008;100:672–9. doi:10.1093/jnci/djn123.
341. Creighton CJ, Li X, Landis M, Dixon JM, Neumeister VM, Sjolund A, Rimm DL, Wong H, Rodriguez A, Herschkowitz JI, Fan C, Zhang X, He X, Pavlick A, Gutierrez MC, Renshaw L, Larionov AA, Faratian D, Hilsenbeck SG, Perou CM, Lewis MT, Rosen JM, Chang JC. Residual breast cancers after conventional therapy display mesenchymal as well as tumor-initiating features. Proc Natl Acad Sci USA. 2009;106:13820–5. doi:10.1073/pnas.0905718106.
342. Sharma SV, Lee DY, Li B, Quinlan MP, Takahashi F, Maheswaran S, McDermott U, Azizian N, Zou L, Fischbach MA, Wong KK, Brandstetter K, Wittner B, Ramaswamy S, Classon M, Settleman J. A chromatin-mediated reversible drug-tolerant state in cancer cell subpopulations. Cell. 2010;141:69–80. doi:10.1016/j.cell.2010.02.027.
343. Ginestier C, Hur MH, Charafe-Jauffret E, Monville F, Dutcher J, Brown M, Jacquemier J, Viens P, Kleer CG, Liu S, Schott A, Hayes D, Birnbaum D, Wicha MS, Dontu G. ALDH1 is a marker of normal and malignant human mammary stem cells and a predictor of poor clinical outcome. Cell Stem Cell. 2007;1:555–67. doi:10.1016/j.stem.2007.08.014.
344. Kwak HI, Gustafson T, Metz RP, Laffin B, Schedin P, Porter WW. Inhibition of breast cancer growth and invasion by single-minded 2s. Carcinogenesis. 2007;28:259–66. doi:10.1093/carcin/bgl122.
345. Laffin B, Wellberg E, Kwak HI, Burghardt RC, Metz RP, Gustafson T, Schedin P, Porter WW. Loss of singleminded-2s in the mouse mammary gland induces an epithelial-mesenchymal transition associated with up-regulation of SLUG and matrix metalloprotease 2. Mol Cell Biol. 2008;28:1936–46. doi:10.1128/mcb.01701-07.
346. Gustafson TL, Wellberg E, Laffin B, Schilling L, Metz RP, Zahnow CA, Porter WW. Ha-Ras transformation of MCF10A cells leads to repression of Singleminded-2s through NOTCH and C/EBPβ. Oncogene. 2009;28:1561–8. doi:10.1038/onc.2008.497.
347. Darst RP, Nabilsi NH, Pardo CE, Riva A, Kladde MP. DNA methyltransferase accessibility protocol for individual templates by deep sequencing. Methods Enzymol. 2012;513: 185–204. doi:10.1016/B978-0-12-391938-0.00008-2.

Chapter 12
The Impact of Centrosome Abnormalities on Breast Cancer Development and Progression with a Focus on Targeting Centrosomes for Breast Cancer Therapy

Heide Schatten

Abstract New research on the role of centrosomes in cancer cell proliferation has led to significant new insights into the multiple functions of this important organelle that serves not only as microtubule-organizing center (MTOC) in interphase and mitosis but also as important cellular communication center for signal transduction pathways and metabolic activities. Cancer cell centrosomes are distinguished from centrosomes in noncancer cells by specific abnormalities that include phosphorylation abnormalities, overexpression of centrosomal proteins, abnormalities in centriole and centrosome duplication, formation of multipolar spindles that play a role in aneuploidy and genomic instability, and several others that are highlighted in this chapter. Because of their critical role in cancer cell proliferation, several lines of research have started to target centrosomes for therapeutic intervention to inhibit abnormal cancer cell proliferation and control tumor progression. While many aberrant mechanisms leading to centrosome dysfunctions are common to all tumor types, there are specific abnormalities observed in breast cancer that will be reviewed in the present chapter in which breast cancer-specific therapies will also be discussed.

12.1 Introduction

The significant role of centrosomes in cancer cell proliferation had been remarkably well recognized by Theodor Boveri [1] (translated into English in [2]) who laid the foundation for modern research on cancer cell centrosomes that is now being pursued on multiple levels including genetics and cell and molecular biology (reviewed

H. Schatten (✉)
Department of Veterinary Pathobiology, University of Missouri,
1600 E Rollins Street, Columbia, MO, USA
e-mail: SchattenH@missouri.edu

H. Schatten (ed.), *Cell and Molecular Biology of Breast Cancer*,
DOI 10.1007/978-1-62703-634-4_12,

in [3, 4]). In mitosis, centrosomes form the bipolar mitotic spindle, but cancer cell centrosomes can form multipolar spindles resulting in aneuploidy and genomic instability. In recent years, numerous studies have highlighted the role of this important organelle not only as main microtubule-organizing center (MTOC) but also as important cellular communication hub for signal transduction pathways. The centrosome holds key roles in cell cycle regulation and in several other complex cellular functions that directly or indirectly affect cell cycle progression and cellular metabolism. It serves as docking station for enzyme-carrying vesicles that are translocated along microtubules, and through its cell cycle-specific microtubule organization capabilities, the centrosome controls and directs translocation of macromolecular complexes and cellular organelles such as mitochondria. Centrosome dysfunctions have been implicated in numerous diseases (reviewed in [5]), and centrosome abnormalities are strongly associated with cancer development and progression (reviewed in [6]). While cause and effect studies are still under investigation to determine when centrosomes become dysfunctional during the cascade of events leading to the observed abnormalities (reviewed in [6]), it has become clear that cancer cell centrosomes are significantly different from noncancer cell centrosomes in several aspects including their state of phosphorylation [7]. Unlike centrosomes in somatic noncancer cells, cancer cell centrosomes are phosphorylated at inappropriate times throughout the cell cycle, as had first been recognized when examining breast adenocarcinoma cells [7], while centrosomes in noncancer cells undergo precise cell cycle regulation and become phosphorylated only at the entry into mitosis during a process termed centrosome maturation that allows centrosomes to become division competent. The abnormal phosphorylation of centrosomes in cancer cells may indicate that cancer cell centrosomes remain division competent in all cell cycle stages and have lost phosphorylation control. Numerous kinases are involved in the transition from G2 to mitosis [8–11] that play a role in centrosome protein phosphorylation while dephosphorylation takes place when cells exit mitosis as will be discussed in specific sections below.

Increased abnormal phosphorylation of cancer cell centrosomes [7] can lead to increases in microtubule organization with consequences for aberrant segregation of chromosomes to the dividing daughter cells resulting in loss of tumor suppressor genes in some cells and increases in tumor promoter genes in others. Other factors involved in cancer cell centrosome dysregulation have recently been reviewed by several investigators and include disruption of centrosome duplication [12], DNA damage caused by radiation [13], protein degradation dysfunctions [8, 9, 14], and numerous others including effects by environmental factors [4, 15, 16].

While many aberrant mechanisms leading to centrosome dysfunctions are common to all tumor types, there are specific abnormalities observed in breast cancer that will be reviewed in the present chapter in which breast cancer-specific therapies will also be discussed. The present review will address (1) structure and function of centrosomes and abnormalities in breast cancer, (2) regulation of the centriole–centrosome complex, (3) the role of primary cilia in breast cancer, and (4) centrosomes as target for breast cancer therapy and prevention.

12.2 Structure and Function of Centrosomes and Abnormalities in Breast Cancer

In a typical mammalian somatic cell, the centrosome organelle consists of a centrally positioned perpendicularly oriented centriole pair that is embedded in a centrosomal matrix (Fig. 12.1), oftentimes also termed pericentriolar material (PCM) composed of a lattice of coiled-coil proteins. The centrosomal matrix contains numerous specific centrosomal proteins including the γ-tubulin ring complexes (γ-TuRCs), pericentrin, centrin, and calcium-sensitive fibers [17] (reviewed in [3]). As the centrosome is not membrane bound, cell cycle-specific molecular remodeling is facilitated and includes remodeling of centrosomal matrix proteins. Centrioles, on the other hand, do not significantly change in their molecular composition throughout the cell cycle. Centrioles in mammalian cells are composed of nine outer triplet microtubules forming a barrel-shaped small tube that does not contain central microtubules. Centriole duplication follows a semiconservative duplication pattern by which a younger (daughter) centriole forms perpendicular to the older (mother) centriole. Mother centrioles are structurally and functionally distinguished from daughter centrioles and contain appendages which will be discussed in more detail in Sect. 12.3. In mammalian cells centrioles play important roles in the assembly of specific centrosome proteins and in the duplication of centrosomal material [18]. As mentioned above, numerous centrosomal proteins are associated with the centrosome matrix that undergo cell cycle-specific regulation. As many as 500 centrosomal proteins have been determined in specific cell cycle stages [19] although a large number of these proteins may be classified as centrosome-associated proteins and others may use centrosomes as central hub for cell cycle-specific functions. While the specific composition of centrosomes is still under active investigation, it is clear that about 60 centrosomal proteins are present in a typical somatic cell interphase centrosome (reviewed in [20]) which may be representative of the average centrosome protein quantities that compose interphase centrosomes in typical mammalian somatic cells. Centrosome core proteins are permanently associated with the centrosome structure while others are part of the cell cycle-dependent structural centrosomal changes in most cell systems. Some of these centrosome proteins are overexpressed in cancer cells and play a role in centrosome amplification which will be discussed in Sect. 12.3.

Centrosome proteins that have been identified in purified centrosomes by mass spectrometric analysis include the *structural proteins* (alpha-tubulin, beta-tubulin, gamma-tubulin, gamma-tubulin complex components 1–6, centrin 2 and 3, AKAP450, pericentrin/kendrin, ninein, pericentriolar material 1 (PCM1), ch-TOG protein, C-Nap1, Cep250, Cep2, centriole-associated protein CEP110, Cep1, centriolin, centrosomal P4.1-associated protein (CPAP), CLIP-associating proteins CLASP1 and CLASP 2, ODF2, cenexin, Lis1, Nudel, EB1, centractin, myomegalin); the *regulatory molecules* (cell division protein 2 (Cdc2), Cdk1, cAMP-dependent protein kinase type II-alpha regulatory chain, cAMP-dependent protein

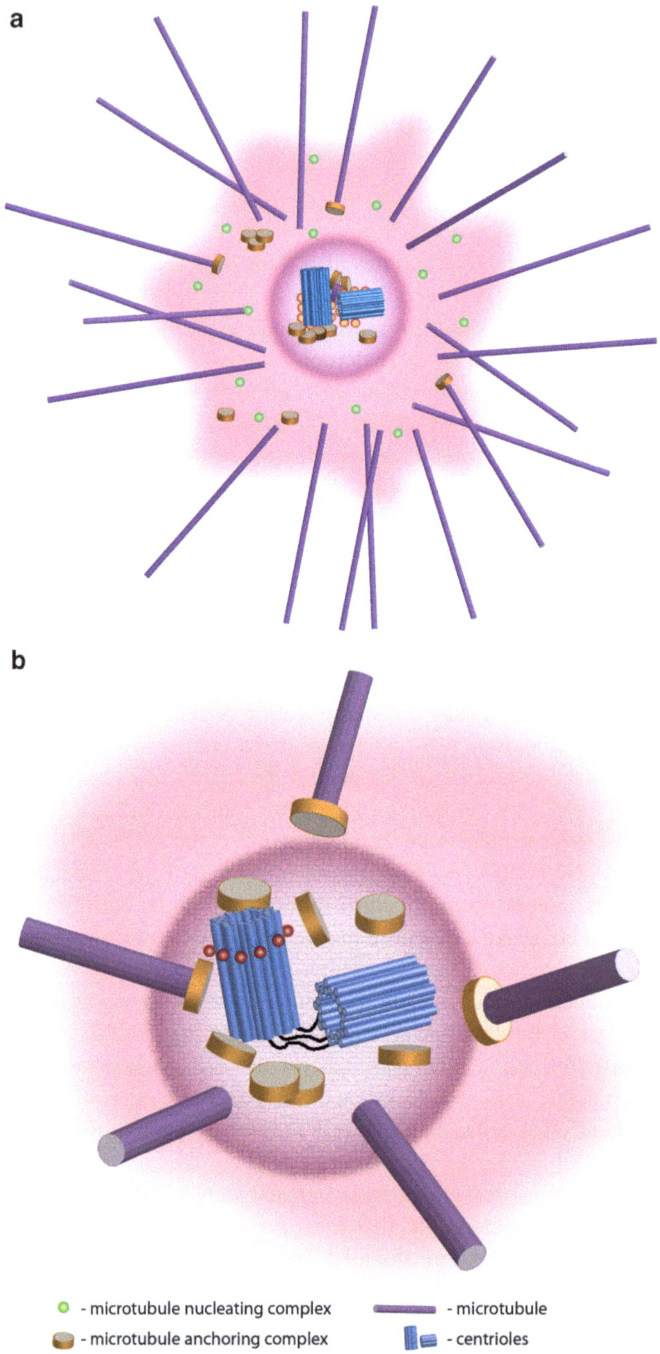

Fig. 12.1 (**a**) A typical centrosome in somatic cells is composed of centrosomal material surrounding two perpendicularly oriented centrioles. The centrosomal material, also referred to as pericentriolar material (PCM), consists of a meshwork of proteins embedded in a matrix of yet undetermined structural composition. Gamma-tubulin and the gamma-tubulin ring complex are embedded in the PCM and nucleate microtubules along with associated proteins. This diagram also shows two complexes within the PCM, the microtubule nucleating complex and the microtubule anchoring complex. The diagram in (**b**) shows in more detail the two centrioles (mother and daughter centriole) surrounded by PCM. Both centrioles are connected by interconnecting fibers. The mother centriole is distinguished from the daughter centriole by distal and subdistal appendages

kinase-alpha catalytic subunit, serine/threonine protein kinase Plk1, serine/threonine protein kinase Nek2, serine/threonine protein kinase Sak, Casein kinase I, delta and epsilon isoforms, protein phosphatase 2A, protein phosphatase 1 alpha isoform, 14-3-3 proteins, epsilon and gamma isoforms); *the motor and motor-related proteins* (dynein heavy chain, dynein intermediate chain, dynein light chain, dynactin 1, p150 Glued, dynactin 2, p50, dynactin 3); and the heat shock proteins, heat shock protein Hsp90, TCP subunits, and heat shock protein Hsp73.

The γ-tubulin ring complex, pericentrin, centrin, and the centrosome-associated protein NuMA (nuclear mitotic apparatus protein) will be discussed in more detail below.

Gamma-tubulin is an essential centrosomal protein that is primarily found in the centrosome core structure, but it can also serve as nucleating sites in areas away from the centrosome. The nucleating complex for microtubules from centrosomes is the ca. 2.2-MDa γ-TuRC that is associated with the centrosome core structure in all cells studied so far [21] and consists of 12 or 14 γ-tubulin molecules. Various other components are needed to anchor the γ-TuRC to the centrosome core structure including the large coiled-coil A-kinase anchoring proteins [22–31] and Cep135 [32]. The microtubule minus-end binding proteins including the γ-TuRC are accumulated at the proximal ends of centrioles, while tubulin polyglutamylation of the centriole walls modulates interaction between tubulin- and microtubule-associated proteins. More detailed information on γ-TuRC is available in a recent paper by Teixidó-Travesa et al. [33]. Microtubule anchoring to the centrosome includes ninein that serves as a microtubule minus-end anchoring protein [34] and dynactin that has a major role in microtubule anchorage at centrosomes as well as at non-centrosomal anchorage sites. It is preferentially localized to the mother centriole and plays a role in microtubule organization [35–37].

In interphase, the γ-TuRC nucleates fewer but longer microtubules, while in mitosis, increased γ-TuRCs become associated with the centrosome, which is part of the centrosome maturation process that takes place from interphase to mitosis. Mitotic microtubules are shorter, larger in number, and highly dynamic. They are regulated by a number of cell cycle-specific proteins that participate in centrosome regulation such as the small GTPase Ran, Aurora A kinase, polo-like kinases, and others that will be discussed in Sect. 12.3.

Pericentrin is a centrosome protein that plays a role in centrosome and spindle organization [24, 25, 38]. It forms a ca. 3-MDa complex with γ-tubulin and depends on dynein for assembly onto centrosomes [38]. Pericentrin is involved in recruiting γ-tubulin to centrosomes [24], and it is part of the pericentrin/AKAP450 centrosomal targeting (PACT) domain [27]. Mutation of the pericentrin gene results in loss of recruitment of several other centrosomal proteins, resulting in diseases of various kinds (reviewed in [5]).

Centrins are small proteins and members of a highly conserved subgroup of the EF-hand superfamily of Ca^{2+}-binding proteins. It is primarily associated with centrioles, but centrin is also an intrinsic component of centrosomes and has an essential role in the duplication of centrosomes [18, 39–41] (reviewed in [42]).

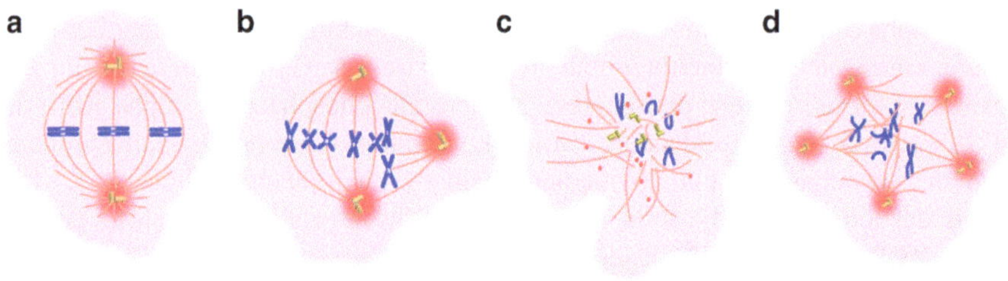

Fig. 12.2 Schematic representation of various structural centriole–centrosome abnormalities in cancer cells. (**a**) Bipolar spindle in noncancer cells. (**b**–**d**) Examples of abnormalities in the spindle and centriole–centrosome complex as a result of centrosomal amplification or centriole abnormalities. (**b**) Tripolar spindle. (**c**) Highly disorganized spindle with scattered centriole–centrosome complexes and scattered chromosomes. (**d**) Multipolar spindle. *Red* = centrosomes; *green* = microtubules; *blue* = chromosomes

NuMA (*nu*clear *m*itotic *a*pparatus protein) is a multifunctional protein (reviewed in [43, 44]) that is essential for the organization of the mitotic apparatus during mitosis. In interphase, NuMA serves as nuclear matrix protein but it is not associated with interphase centrosomes. During nuclear envelope breakdown NuMA becomes dispersed into the cytoplasm; Cdk1/cyclin B-dependent phosphorylation is important for translocation of NuMA from the nucleus into the cytoplasm [45] which allows its association with microtubules using a dynein–dynactin-mediated mechanism for its translocation along microtubules to the centrosome area where it forms an insoluble crescent around centrosomes that tethers microtubules precisely into the bipolar mitotic apparatus [46]. NuMA is important for cross-linking of spindle microtubules and for the organization and stabilization of spindle poles from early mitosis to anaphase. To relocate to the nucleus at the exit from mitosis, NuMA becomes dissociated from the mitotic spindle poles, a process requiring cdc1/cyclin B [47]. Destruction of cyclin B promotes exit from mitosis. Failure of NuMA to relocate to the nucleus will result in cytoplasmic NuMA spots that organize abnormal microtubule asters [47] and may contribute to mitotic abnormalities. NuMA abnormalities are strongly associated with breast cancer. NuMA region on chromosome 11q13 has been associated with breast cancer susceptibility [48] and might serve as potential biomarker for breast cancer. As indicated above, NuMA requires specific signaling for its centrosome-associated functions which includes a signaling function of cyclin B (reviewed in [43]), but it has also been shown that signaling factors can be different in different tissues; for example, NuMA has been shown to respond to hormonal signaling in cervical cancer cells [49]. Specific signaling mechanisms leading to NuMA abnormalities in breast cancer have not yet been determined.

Various structural centrosome abnormalities are observed in cancer tissue and are shown in schematic representation in Fig. 12.2. Morphological differences in cancer cell centrosomes compared to centrosomes in noncancerous healthy cells are clearly apparent and can include increased centrosome number and volume, excess of centrosomal material, supernumerary centrioles, abnormally oriented centrioles, and a host of imbalanced centrosomal proteins surrounding centrioles. The causes for such centrosomal abnormalities are not entirely clear but may be related to imbalances or dysfunctions of structural proteins, regulatory molecules, motor-related molecules, and others perhaps not yet known. In addition, basal bodies derived from dislocated primary cilia [50, 51] (discussed in Sect. 12.4) can form additional nucleation sites for microtubule organization and form microtubule-based asters that participate in chromosome segregation, thereby contributing to division abnormalities [50–53].

12.3 Regulation of the Centriole–Centrosome Complex

The regulation of the centriole–centrosome complex is critically important for its accurate functions, as centrosomes organize microtubules that attach to kinetochores in mitotic cells and are part of a complex molecular machinery involved in the accurate separation of chromosomes to the daughter cells during cell division. Cell cycle abnormalities and their consequences leading to abnormal cell divisions are shown in Fig. 12.3. To assure coordination of centrosome and chromosome dynamics, the duplication cycles of both have to be precisely regulated to yield synchronized centrosome and chromosome duplication through parallel pathways of regulation to form the bipolar mitotic spindle that precisely partitions chromosomes equally to the daughter cells. In cancer cells, this coordination is lost, and centrosomes can form multiple poles (as shown in Figs. 12.2 and 12.3) that do not separate chromosomes equally to the dividing daughter cells but form cells with unequal chromosome numbers and cells that may lack tumor suppressor genes, therefore giving advantage to cancer cell growth and loss of tissue architecture. Cell polarity becomes gradually lost with tumor progression and advanced histological grade.

12.3.1 Centrosome Regulation in Normal Cell Cycles

To understand centrosome misregulation in breast cancer, it is important to understand centrosome regulation in normal cell cycles and determine the origins of centrosome dysfunctions in breast cancer. Any of the misregulated steps may serve as target for the development of potential new breast cancer therapies to correct cellular dysfunctions.

As mentioned above, *centrosome duplication* in a regular cell cycle is typically well synchronized with the DNA cycle, and it is important that centrosomes are duplicated only once during the cell cycle; excellent studies on duplication of centrosomes have been performed and revealed that there is a block to centrosome reduplication which assures that centrosomes are duplicated accurately only once within a normal cell cycle (reviewed in [3, 8–11]).

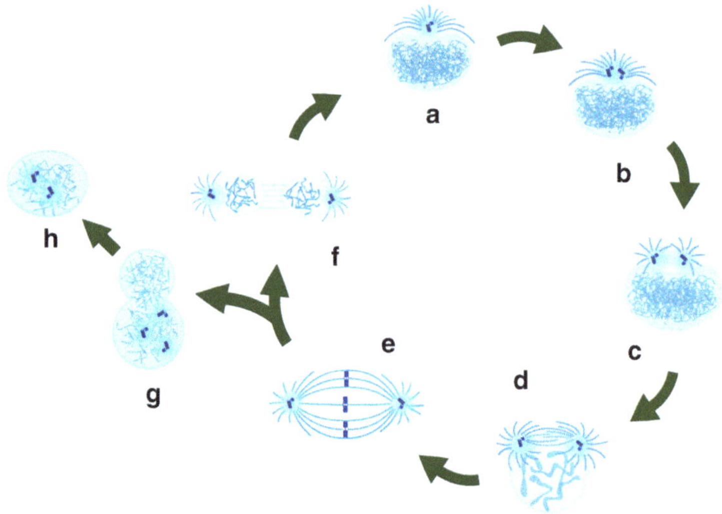

Fig. 12.3 (**a**–**f**) Somatic cell centrosome cycle within the cell cycle. (**a**) The single interphase centrosome containing a pair of centrioles is closely associated with the nucleus and nucleates an array of interphase microtubules. (**b**) Centriole–centrosome duplication occurs during the S phase in synchrony with DNA duplication. (**c**) Separation of the duplicated centriole–centrosome complex toward the opposite spindle poles takes place in the early prophase stage. (**d**) The bipolar mitotic apparatus becomes established when each centriole–centrosome complex has reached the opposite pole, and the nuclear envelope has broken down. During this stage interphase centrosomes mature into mitotic centrosomes acquiring mitosis-associated centrosomal proteins including NuMA that had moved out of the nucleus during nuclear envelope breakdown. (**e**) The metaphase centrosome becomes highly compacted to organize the metaphase spindle with microtubules attached to the kinetochores. (**f**) Telophase is the stage when centrosomal material becomes decompacted again before reorganizing into interphase centrosomes that associate with the nuclei of the separating daughter cells. (**g**–**h**) Centrosomal abnormalities associated with cell cycle dysfunctions. In cancer cells (**g**–**h**), centrosome and centriole numbers can amplify or over-replicate leading to aneuploidy or failure of cytokinesis

Centrosomes duplicate shortly before the G2 cell cycle stage in a precisely orchestrated duplication program. This process begins with disorientation of the pair of centrioles, centriole duplication, and centriole disjunction resulting in sister centriole separation as reviewed in [54] and [55]. While excellent data have been obtained to understand centriole dynamics in the duplication process, we still do not yet fully understand centrosome dynamics in sufficient detail, and we frequently correlate centrosome duplication and separation to the better understood centriole dynamics.

Initiation of centrosome duplication is under cytoplasmic control and driven by cyclin-dependent kinase 2 (Cdk2) complexed with cyclin E or cyclin A that rises during the late G1 stage (reviewed in [56]). It has also been shown that initiation of centrosome duplication requires calcium/calmodulin-dependent kinase II (CaMKII) [57]; CaMKII phosphorylates centrosome proteins in vitro [58] and it is localized to spindle poles [59]. Ubiquitin-mediated proteolysis of centrosomal proteins may be involved in the block to reduplication, as a variety of components of the SCF (Skp1/cullin/F-box; ubiquitin ligase complex) proteolysis pathway as well as the 26S proteasome are localized to centrosomes in human cells [8–10, 14, 60–62].

Centrosome *separation* refers to the spatial separation of centrosome material around the nucleus and during mitosis which in part is driven by plus- and minus-end-directed microtubule motor proteins. Phosphorylation of the centrosomal protein centrin plays a role in centrosome disjunction at the G2/prophase transition, and it has been shown that the Nek2 kinase is involved in centrosome disjunction [63] (reviewed in [64]).

Centrosome duplication and DNA replication both require hyperphosphorylation of the retinoblastoma (RB) protein and activation of Cdk2. Although the centrosome cycle and DNA cycle are normally coupled through cell cycle-dependent checkpoints, it has been possible to dissociate the centrosome cycle from other cell cycle events which may also occur in cells during carcinogenesis (reviewed in [3]).

The mechanisms by which centrosomes become phosphorylated have been well studied for normal cell cycles (reviewed in [8–11]) and include specific centrosome proteins that depend on multiple signaling to allow the transition from G2 to mitosis. While it is clear that G2/M cell cycle transition is critical for centrosome phosphorylation to become division competent and allow cell proliferation (reviewed in [3]), it is not yet well understood why cancer cell centrosomes are abnormally phosphorylated throughout the cell cycle although misguided signal transduction resulting in mistargeting of key cell cycle regulators is likely to play a determining role. As indicated above, cyclin-dependent kinases are critically important for centrosome cycle progression. The G2/M transition requires Cdk1/cyclin B as well as Cdk1/cyclin A (reviewed in [65]). Cdk1 is localized to centrosomes at the onset of mitosis [66, 67], and Cdk1/cyclin B activation is detected in centrosomes during prophase [68].

Significant centrosome remodeling (also referred to as centrosome maturation) takes place during G2/M by acquiring mitotic centrosome proteins including polo-like kinase 1 (Plk1) [69], NuMA [70], and others, while interphase centrosome proteins such as C-Nap1 [71] or Nlp [72] are removed. γ-TuRC recruitment to the centrosome increases, assuring increased nucleation of microtubules for spindle formation.

Several of the important *mitotic cell cycle regulators* are concentrated at the centrosome and include the abovementioned polo and Aurora A kinases [73] and cdc2/cyclin B kinase [68]. Some of these proteins bind to the anaphase promotion complex/cyclosome (APC/C); the activated APC/CCdc20 degrades cyclin B and securin to allow cell cycle exit from mitosis [74–77]. Polo-like kinases are required for multiple stages of mitotic progression and have been implicated in centrosome separation. Specific kinases including Plk1 are involved in centrosome and microtubule organization [78–85]; Plk1 and Plk3 both have been implicated in microtubule and centrosome functions in interphase and in mitosis [86–88]. Loss of Plk3 function has also been associated with loss of cell shape [88], affecting microtubule functions underneath the plasma membrane which may play a role in loss of cellular polarity in cancer cells and tissue.

Microtubule motor proteins are required for the composition of a functional mitotic centrosome, as several centrosomal proteins are shuttled along microtubules to the centrosome core structure including pericentrin, centrin, ninein, and NuMA. Imbalanced or disrupted transport of centrosome proteins can lead to pathologies related to centrosome and microtubule functions as well as to failures in organelle and vesicle distribution. Disruption of transport may either be the cause or effect in a chain of signal transduction events which can result in secondary pathologies.

12.3.2 Centrosome Misregulation Leading to Abnormalities in Breast Cancer

As indicated above, numerous centrosomal abnormalities are implicated in centrosome dysfunctions, and new studies have started to analyze cause and effect of centrosome dysfunctions on molecular levels with the goal to uncover and perhaps also repair centrosomal pathologies. Several causes have been identified so far, some of which are common to a variety of cancers while others are tissue specific. Oncogenic insults may result in a high number of mutation rates in cells predispositioned to tumor, and the mutation rate increases in cells that have reached replicative senescence. Cervical carcinogenesis has strongly been associated with infections by high-risk human papillomavirus (HPVs), and it has been proposed that the HPV E7 oncoprotein may induce primary centrosome duplication errors and act as mitotic mutator (reviewed in [89]). Abnormal multipolar mitoses resulting from supernumerary centrosomes have clearly been associated with HPV-associated lesions, and centrosome abnormalities are already detected in early stages of tumor development. Excellent studies have shown that uncoupling of the cell division cycle from the centrosome cycle subverts centrosome homeostasis [90]. Dissociation of centrosome cycles from DNA cycles may be among the causes for centrosome aberrations and multipolar spindle formations after irradiation. Aberrant hypermethylation has recently been implicated in inactivation of checkpoint genes that may influence cell cycle-dependent centrosome abnormalities as reported for pancreatic cancer [91, 92].

Environmental stress can result in the formation of aggresomes, which are oftentimes localized close to centrosomes and are thought to be the result of misfolded proteins [93–97]. Some of the aggresomes contain γ-tubulin and are associated with disease or disorders including noncancer diseases such as Parkinson's and dementia [98].

In cancer cells, it has clearly been shown that overexpression of specific centrosome proteins results in abnormal centrosome configurations and aneuploidy [53, 99], highlighting the important role of centrosomes in cancer development and progression. Increased centrosome number and volume, supernumerary centrioles, accumulation of increased PCM, and abnormal phosphorylation of centrosomes have all been associated with cancer cell centrosomes followed by loss of cell polarity [5, 7]. Several factors play a role in centrosome misregulation and subsequent abnormal microtubule nucleation, abnormal spindle formation, and chromosomal mis-segregation. As mentioned above, among the factors is the loss of tumor suppressor genes that affect centrosome functions.

Aurora kinases play central roles in the mitotic process and cell division, and Aurora A has been implicated in centrosome amplification in breast cancer. Aurora A kinase (AURKA) is an important centrosome-associated serine/threonine kinase and has been shown to play a role in tumor development when overexpressed in animal models [100, 101]. Aurora A localizes to centrosomes and overexpression of Aurora A causes multipolar mitotic spindles that play a role in early development of mammary tumors. The studies by Wang et al. have further shown that the pro-survival AKT pathway is activated which prevents cell death and allows cell proliferation while generating tetraploid cells with accumulated centrosomes.

Deregulation of centrosome duplication and genes implicated in centrosome amplification are among the processes that lead to cascades of cell cycle-related abnormalities. The loss of the tumor suppressor *p53* is associated with multiple cycles of centrosome duplication in one S phase resulting in multiple centrosome numbers [102]. It has been suggested by numerous investigators that p53 might play a role in the block to reduplication in synchrony with the DNA cycle. Viral oncoproteins that inactivate p53 also result in cells with supernumerary centrosomes as has been shown for the abovementioned papillomavirus (reviewed in [89]). However, different investigators have challenged that loss of p53 directly affects centrosome duplication [103] and attribute centrosome abnormalities to missing checkpoint functions after loss of p53. Regardless, loss of p53 following genotoxic stress or mitogenic stimulation plays a role in the generation of supernumerary centrosomes in breast cancer cells in which the CDK2/cyclin-dependent pathway is implicated [104, 105].

The breast- and ovary-specific tumor suppressor gene *BRCA1* has been implicated in the deregulation of centrosome duplication. It is involved in G2/M checkpoint functions and it plays a role in preventing centrosome overduplication. Loss of BRCA1 may result in loss of the block to centrosome reduplication. Targeted deletion of BRCA1 exon11 leads to centrosome amplification [106]; it causes centrosome overduplication and perhaps centrosome fragmentation in human breast cancer cell lines [107–109] (reviewed in [6]). A model for the regulation of the centrosome by BRCA1 has been presented in an excellent review by Kais and Parvin [14]. In this model BRCA1 ubiquitinates the already duplicated centrosomes to inhibit reduplication. Loss of BRCA1 results in supernumerary centrosomes during the S phase. Overexpression of AURKA mimics the effects of BRCA1 loss. This model also proposes that overexpressed AURKA overrides the spindle checkpoint and thereby contributes to abnormal mitosis. However, the exact mechanism by which BRCA1 affects centrosome duplication remains to be fully clarified.

We know that in normal cell cycles BRCA1 forms a complex with the BRCA1-associated RING domain 1 (BARD1) acting as E3 ubiquitin ligase. The BRCA1–BARD1 complex ubiquitinates γ-tubulin and maintains centrosome homeostasis, preventing abnormal duplication in normal cell cycles and abnormal microtubule nucleation by γ-tubulin. The BRCA1-associated centrosomal ninein-like protein (Nlp) is also involved in centrosome abnormalities. In transgenic mice, spontaneous breast tumorigenesis occurs when Nlp is overexpressed, perhaps mimicking BRCA1 loss [110]. The role of BRCA2 in centrosome functions is less explored than BRCA1 but we do have some data. It is known that the BRCA2-associated protein NPM forms a complex with ROCK2 to maintain numerical centrosome integrity;

aberrant regulation of this protein may result in centrosome overduplication and fragmentation [111]. The centrosomal kinase Nek2 further plays a role in promoting centrosome accumulation which has been shown in breast epithelial cells [112]. Several other centrosomal components are involved in human breast cancer and include Pin1, a prolyl isomerase implicated in the activation of oncogenic pathways and induces centrosome overduplication in S-phase-arrested cell lines [113]. The Y-box binding protein YB-1 is overexpressed in 75 % of human breast cancers [114]. Recently, centrosome overduplication has been reported via nuclear expression of endogenous aryl hydrocarbon receptor (AhR) and cyclin E [115]. A recent large-scale breast cancer risk control study by Olson et al. [116] evaluated centrosome-related genes, genetic variation, and risk of breast cancer, suggesting that single nucleotide polymorphisms (SNPs) play a role in microtubule nucleation from the centrosome and contribute to breast cancer. The study further suggested that the centrosome pathway is highly enriched for SNPs that are associated for breast cancer risk.

As outlined above, while the cause and mechanisms leading to centrosome amplification in breast cancer have been addressed in several recent studies to gain an understanding on centrosome functions and dysfunctions, our understanding is still incomplete. Recent studies have focused on *centrosome clustering* that takes place during mitosis. New studies have shown that centrosome amplification, cell cycle control dysfunctions, and aggregation of centrosomal material for the formation of the spindle poles during mitosis are associated with centrosome clustering abnormalities. This topic has gained increased attention in recent years (reviewed in [117]), as amplified cancer cell centrosomes can cluster into abnormal bipolar spindles that are not easily discernible from regular bipolar spindles with non-amplified centrosomes compared to the obvious tripolar or multipolar mitotic spindles that are most commonly seen in cancer cells. This topic has been reviewed in *Drosophila* cells [118] as well as in cancer cells [13, 117], and it is clear that more research is needed to understand the mechanisms underlying centrosome clustering in normal cells and dysfunctions in breast cancer cells.

Although the mechanisms underlying centrosome clustering into bipolar mitotic centrosomes are not well understood, several components of the cytoskeleton may be involved. Experiments by Kwon et al. [119] showed that the actin cytoskeleton plays a role in centrosome clustering, and our previous experiments on the invertebrate sea urchin model revealed that microtubules and microfilaments are required for centrosome dynamics that may be important for centrosome clustering.

It may be worthwhile considering that other components of the cytoskeleton participate in centrosome clustering taking into account the nature of the centrosomal matrix structure and centrosome clustering mechanisms that have been described for reproductive (germ) cells. While we do not yet have a unifying model for the centrosome matrix structure, we have gained some understanding from invertebrate models. The invertebrate sea urchin egg has allowed us to first understand the formation of tri- and multipolar centrosomes [120, 121] (reviewed in [3]) as a result of multiple sperm incorporation during fertilization that led us to understand abnormal mitosis in cancer cells [1]; we may also take clues from the sea urchin model to understand the mechanisms underlying centrosome clustering. For example, we know that centrosomal foci can be induced through artificial activation in unfertilized sea urchin eggs [122] and in pig oocytes [123, 124], and we further know from these studies that numerous small centrosomal foci can aggregate or cluster into larger and fewer centrosomal foci [122] (reviewed in [51]) which has led to a model in which centrosome clusters are composed of fibrous material and interconnected like string forming multiple yarn clusters. In the *Spisula* model, studies have revealed that the material left after high-salt extraction of centrosomal proteins displayed a fibrous component [125], resembling filament-like fibers that may contain intermediate filament-like components that contribute to the composition of the centrosome matrix. In our earlier studies we used Ah6, a monoclonal antibody to intermediate filament-like proteins to clearly stain centrosomes in sea urchin cells [126]. It may be speculated that a similar filament-like system may exist in somatic cells and function in the process of centrosome clustering. In unfertilized mouse oocytes, numerous centrosomal aggregates (clusters) are found in the MII ooplasm [121, 127, 128] (reviewed in [129]) that aggregate or cluster to form the meiotic spindle in a dynein-dependent process. These studies suggest that microtubule motor protein dynamics play a role in the centrosome clustering process. Such basic studies may provide relevant information on centrosome clustering in breast cancer cells and may allow new approaches for targeting amplified abnormal centrosome clustering to prevent the formation of abnormal bipolar mitosis and induce cell death rather than allow the formation of aneuploid cells. A possible role for microtubule motor proteins in centrosome clustering in somatic cancer and noncancer cells has been discussed by Krämer et al. [117].

Other studies performed in *Drosophila* cells have shown that efficient centrosome clustering requires a large number of proteins associated with a variety of cellular functions [118, 119] which includes genes that play a role in spindle assembly checkpoint (SAC) functions. Spindle tension is necessary for clustering supernumerary centrosomes into a bipolar mitotic apparatus. The studies by Kwon et al. [119] concluded that SAC does not monitor extra centrosomes but that the time during which SAC monitors proper kinetochore–microtubule attachments gives supernumerary centrosomes time for clustering into a bipolar mitotic apparatus. It was also suggested that microtubule-associated proteins (MAPs) are involved in centrosome clustering. It has further been shown [119] that efficient centrosome clustering depends on HSET (kinesin-related protein), as HSET depletion blocks centrosome clustering and promotes multipolar divisions. These studies showed that HSET selectively eliminates cells with supernumerary centrosomes without affecting viability of cells with normal centrosome numbers. As will be discussed in Sect. 12.5, preventing centrosome clustering into an abnormal bipolar mitotic apparatus may provide new targets for breast cancer therapy, as multipolar mitoses undergo fragmented cell divisions with fewer chances for cancer cell viability [130] compared to abnormal bipolar mitotic spindles that can undergo cell division and produce aneuploid cells.

12.4 The Role of Primary Cilia in Breast Cancer

During the past decade the primary cilium has gained increased attention when it was recognized that it plays important roles in signal transduction and that it is directly associated with cell cycle functions (reviewed in [5, 131]).

The primary cilium is a specialized nonmotile single cilium protruding from almost all cells in our body [132–134]. It is formed when dividing cells exit mitosis and the mother (elder) cell centriole becomes located at the cellular surface [135] to form a basal body as important nucleation site for microtubules that grow and build the interior skeleton of the primary cilium (reviewed in [131]). The primary cilium contains 9 outer microtubule doublets with no central microtubule pair ("9 + 0"), and it is covered by a specialized receptor-rich plasma membrane that plays a role in tissue-specific functions. This single cilium (one per cell) has an average diameter of 0.2 μm and a length of ca. 8 μm.

The regulatory relationship between primary cilia functions and the cell cycle has clearly been established (reviewed in [136]). The primary cilium–centriole–centrosome cycle begins during G1 when the distal end of the mother centriole becomes associated with a membrane vesicle (reviewed by [136]). Following this process the axoneme lengthens and the ciliary vesicle enlarges into a sheath to fuse and become continuous with the plasma membrane. Centrioles duplicate and lengthen during the subsequent S phase, achieving mature length in late G2/M. Centriole shortening then occurs at the G2/M transition. The coordination between primary cilia and cellular centrosomes is highly regulated, and dysfunctions in this regulation result in cellular dysfunctions (reviewed in [131]). Closely related proteins are located in all three structures, basal body, primary cilium, and centrosomes, and play a role in cilia-related diseases.

It has clearly been shown that primary cilia functions are coordinated with cell cycle regulation, and new studies have revealed details of signal transduction cascades between primary cilia and the centrosome that are essential for accurate cell cycle progression [3, 133, 134, 137–141]. At least three important pathways require signaling through primary cilia and include the Wnt, hedgehog, and platelet-derived growth factor (PDGF) pathways [142, 143]. Furthermore, MAP kinase signaling through primary cilia has been well documented which is important for centrosome functions. Downstream signaling cascades include phosphorylation and activation of the Akt and Mek1/2–Erk1/2 pathways [144]. For proper primary cilia functions, an intraflagellar transport (IFT) system is important to achieve ciliogenesis and transport of molecules to the cell body.

In cancer cells, the primary cilium becomes dislodged during progressive stages of cancer development [50–52], and it can clearly be visualized by transmission electron microscopy analysis, as the dislodged primary cilium may still contain characteristic cilia components [51]. On genetic and molecular levels, several cilia-associated genes are commonly mutated in breast cancer including Gli1, DNAH9, and RPGR1P1. Gli1 is a component of the hedgehog signaling pathway; DNAH9 is a component of the dynein motor that is important for IFT [145]. RPGR1P1 is located in the basal body and serves as GTPase regulator [146]. The oncogenic Aurora A kinase (Aurora A) is localized to the basal body of primary cilia and either may play a role in primary cilia disassembly or may block primary cilia reassembly along with other interacting proteins [147] which may be perturbed in cancer cells, ascribing a possible role of Aurora A in primary cilia–cell cycle dysfunctions.

The role of primary cilia in *hedgehog signaling* is one of the best studied signaling pathways in which localization of the transmembrane protein, Smoothened (Smo), in the primary cilium plays a role in the activation of the hedgehog pathway; hedgehog-dependent transcription is mediated by the three transcription factors, Gli1, Gli2, and Gli3 [148, 149]. Recent studies have shown a role for hedgehog signaling through primary cilia in cancer progression and propose that the presence or absence of primary cilia may be a critical aspect in the design of therapeutics.

The regulation of the *PDGF signaling* pathway by primary cilia has been well explored by Schneider et al. [144] who have determined that the platelet-derived growth factor receptor-α (PDGFRα) localizes to primary cilia in murine embryonic fibroblast cells; it was further shown that activation of Akt and ERK1/2 pathways by PDGF ligand requires primary cilia. Importantly, expression of PDGFRα is a poor prognostic indicator of breast cancer [150, 151], clearly linking PDGF signaling through primary cilia to breast cancer.

It has been reported that the occurrence of primary cilia is decreased in breast cancer epithelial cells compared with normal breast epithelial cells, which has been explored in cultured cell lines and in breast tissue [152] (reviewed in detail in [152, 153]) for specific breast cancer types and populations. The authors showed that while primary cilia are not frequently seen in breast epithelial cells, they are present in stromal cells in breast cancer. However, this field is still young and further clarifications and detailed studies on the reasons and functional relationships regarding the presence or absence of primary cilia in specific breast and breast cancer cell types and subpopulations are needed to determine the nature of the relationship between loss of primary cilia and their causative or consequential relationship during tumorigenesis.

12.5 Centrosomes as Target for Breast Cancer Therapy and Prevention

Excellent general as well as detailed breast cancer and breast cancer treatment information for patients and health professionals is available from the National Institutes of Health at http://www.cancer.gov/cancertopics/types/breast and addresses many aspects that will not be addressed in this chapter. Significant progress has been made in the diagnosis and treatment of breast cancer which has led to personalized treatment procedures that target different abnormalities based on different types of cancer and different dysfunctions of molecular pathways and molecular abnormalities.

However, the disease is complex and breast cancer is still one of the leading causes of cancer deaths. Still about 50 % of cancer patients die of the disease primarily related to metastasis and development of resistance to chemotherapy. Tissue heterogeneity and genomic instability are among the complex abnormalities associated with cancer that present problems for the design of appropriate effective treatment and may require multiple targeted treatment strategies. Centrosomes are increasingly being discussed as new targets for breast cancer treatment, as centrosomes are central to cell division; abnormal centrosomes are associated with abnormal cell divisions that are hallmark characteristics for cancer cells and tissue.

As centrosome abnormalities play critical roles in cancer cell proliferation and may have multiple causes, several lines of research have started to develop different strategies to target centrosomes for the development of new specific therapies to inhibit cancer cell proliferation.

Targeting cancer cell centrosomes includes targeting abnormal centrosomes directly or targeting signal transduction molecules that play a role in abnormal centrosome formation. Such approaches may target overexpressed centrosome proteins, abnormal centrosome clustering, abnormal primary cilia dynamics, overexpressed phosphorylation such as Aurora A that is implicated in centrosome hyperphosphorylation or other components in the phosphorylation cascade, as well as different molecules that play a role in centrosome function such as the aryl hydrocarbon receptor (AhR) and cyclin E, as reported by Korzeniewski et al. [115]. Such approaches may involve the development of new pharmaceuticals, or it may also be possible to use specific dietary ingredients to control abnormal centrosome dynamics and/or prevent cancer development or reoccurrence after surgical procedures or after radiation treatment during posttreatment recovery.

Pharmaceutical or dietary ingredients can either be used alone or in combination to target several abnormalities that may have been diagnosed for specific subpopulations of breast cancer.

One of the best known plant derivatives that have been developed into a cancer-targeting pharmaceutical is Taxol. Paclitaxel (or Taxol) had originally been isolated from the gymnosperm *Taxus brevifolia*. It is known to target microtubule dynamics by primarily inhibiting depolymerization of microtubules, thereby preventing progression of mitosis and cell division ([154, 155] and others). Interactions of Taxol with microtubules at the centrosome–microtubule nucleation sites have been reported [156, 157], and it had been proposed that centrosomes in Taxol-treated cells may lose their capacity to nucleate microtubules [156]; in addition, Taxol may have other molecular target interactions that are still under investigation. Taxol has been widely used for ovarian and breast cancer treatment, but as with many anticancer drugs, drug resistance can develop [158, 159] which calls for development of new drugs to target the mitotic molecular machinery to prevent mitosis progression or to induce mitotic cell death to inhibit cancer cell division.

Plant-derived drugs are widely used to control a variety of diseases and include the microtubule drug colcemid to control gout. However, unlike colcemid or especially Taxol that has been thoroughly investigated and modified for the use as therapeutic anticancer drug, other plant-derived components have not yet undergone thorough testing and are not yet available for efficient breast cancer treatment. This includes the potential use of *curcumin*, a natural polyphenol found in the rhizomes of *Curcuma longa* (turmeric) that has been shown to prevent or inhibit abnormal signaling cascades in cancer cells [160–164] which may include abnormal signaling to centrosomes. Curcumin has gained increased attention as potent anti-breast cancer drug, and new more potent curcumin analogues are now being developed to increase its antitumor efficiency [165, 166]. Curcumin and curcumin derivatives show potent inhibitory effects on several signaling pathways including the NFκB pathway [166] that plays a role in inflammation and in cancer development and progression.

The potential of using curcumin as anti-breast cancer drug is excellent and highly promising which has generated increased interest in this dietary ingredient for further modifications into an efficient drug for breast cancer treatment. Because of curcumin's poor solubility and stability (in part due to its hydrophobicity) and other factors impacting its bioavailability, new emphasis has been placed on new strategies to increase efficient bioavailability for more potent metabolic impact. These include encapsulation of curcumin into liposomes, polymeric nanoparticles, biodegradable microspheres, curcumin–cyclodextrin complexes, and micellar formulations and hydrogels [162, 163, 167–171]. Recently, still more promising strategies have been proposed including the use of theragnostic curcumin-encapsulated nanoparticles that will increase bioavailability and allow more potent clinical applications [172]. Further specific improvements to increase bioavailability and stability include use of didodecyldimethylammonium bromide (DDAB) and pluronic F127 polymer as a surfactant/stabilizer, in association with the use of a rotating tube microfluidic platform [164]. Such improvements are important and highly promising to modify the dietary ingredient into a potent anti-breast cancer pharmaceutical.

It will further be important to determine the biomolecular mechanisms by which curcumin acts on cancer cells that are still not well understood although several pathways have been implicated [173] and epigenetic effects have also been proposed [174]. Our recent preliminary experiments clearly showed an effect on the microtubule cytoskeleton in several cancer cell lines, and we are currently exploring the effects on centrosomes (Schatten et al. unpublished).

Other studies have focused on the effects of curcumin on the NFκB pathway that may also affect centrosome dynamics although further studies are needed to pursue this line of thought and generate more data. The antimitotic drug griseofulvin that arrests cells at the G2/M transition stage in a concentration-dependent manner may provide indications that the NFκB pathway and centrosome dynamics are connected, as griseofulvin has been shown to interfere with the NFκB pathway [175]. This finding is interesting, as griseofulvin has recently been shown to specifically inhibit supernumerary centrosome clustering in cancer cells which demonstrates its effects on centrosome dynamics (reviewed in [117]).

As mentioned in Sect. 12.3, prevention of amplified centrosome clustering into a bipolar mitotic apparatus is another approach that has been proposed as therapeutic targeting to control breast cancer cell proliferation, as non-clustered centrosomes cannot form an abnormal mitotic apparatus but induce cell fragmentation and cell death rather than allowing formation of aneuploid cells with consequences for genomic instability. This approach is attractive, as centrosome clustering pathways are dispensable in cells with normal centrosome numbers, but centrosome clustering is required for supernumerary centrosomes to be distributed into the bipolar mitotic apparatus.

Griseofulvin has been identified as inhibitor of centrosomal clustering [176]. This drug is intriguing, as it may affect multiple sites in the mitotic process. Griseofulvin is already approved as an effective orally administered antifungal drug that affects microtubule formation and disrupts microtubule dynamics in vivo and in vitro [177–180]. Its mechanisms of action have been investigated in several cell systems ([178, 180–182]; other references therein), and several investigators have shown that griseofulvin induces multipolar mitoses in tumor cells [176, 178, 183, 184]. At present it is not clear how griseofulvin affects centrosome clustering. It may interfere with microtubule minus ends rather than interacting with the centrosome structure directly. Its action as antitumor drug needs further investigation, as determining its molecular mechanisms would allow specific targeting of breast cancer subpopulations and further allow its use in combination with other drugs to allow for more effective actions in combination therapies. However, toxicity studies are still needed to exclude negative effects on vital organs such as the liver.

12.6 Conclusion and Future Directions

While significant progress has been made in the overall diagnosis and treatment of breast cancer including personalized treatment procedures that target specific abnormalities in specific types of breast cancer, there are still no effective drugs for the treatment of breast cancer and new therapeutic approaches are needed. Furthermore, drug resistance to the most commonly used drugs such as paclitaxel can develop and calls for new targets to effectively combat breast cancer. Renewed interest has focused on centrosomes as target for breast cancer therapy, as centrosomes play decisive roles in allowing abnormal cell divisions that are hallmarks for cancer cells and tissue.

Unlike centrosomes in noncancer cells, breast cancer cell centrosomes are misregulated, leading to abnormal centrosome phosphorylation throughout the cell cycle that remain division competent without undergoing phases of cell cycle rest. The abnormal phosphorylation of centrosomes in cancer cells indicates that cancer cell centrosomes have lost phosphorylation control. Misregulation of centrosomes leads to centrosome amplification with consequences for genomic instability and loss of tumor suppressor genes. Several new studies have started to determine mechanisms to control centrosome amplification which includes inhibiting signaling pathways that allow centrosome amplification, some of which are tissue specific. Targeted destruction of centrosome clustering has been another approach to produce fragmented cells that will lose viability and not participate in abnormal cell divisions.

Other approaches include the development of new drug delivery systems to achieve more efficient bioavailability for more potent metabolic impact to destroy cancer cells. Such new developments are highly promising avenues for targeting breast cancer cell centrosomes and prevent or inhibit breast cancer cell growth and potential development into metastasis.

Acknowledgments It is a pleasure to gratefully acknowledge Donald Connor's professional help with the illustrations and support from MU's Research Council for studies on curcumin and fruitful collaborations with the Cancer Research Center.

References

1. Boveri T. Zur Frage der Entstehung maligner Tumoren. Jena: G. Fisher; 1914.
2. Boveri T. Concerning the origin of malignant tumours by Theodor Boveri (Translated and annotated by Henry Harris). J Cell Sci. 2008;121 Suppl 1:1–84.
3. Schatten H. The mammalian centrosome and its functional significance. Histochem Cell Biol. 2008;129:667–86.
4. Schatten H, Sun QY. The functional significance of centrosomes in mammalian meiosis, fertilization, development, nuclear transfer, and stem cell differentiation. Environ Mol Mutagen. 2009;50(8):620–36.
5. Badano JL, Teslovich TM, Katsanis N. The centrosome in human genetic disease. Nat Rev Genet. 2005;6:194–205.
6. Chan JY. A clinical overview of centrosome amplification in human cancers. Int J Biol Sci. 2011;7(8):1122–44.
7. Lingle WL, Lutz WH, Ingle JN, Maihle NJ, Salisbury JL. Centrosome hypertrophy in human breast tumors: implications for genomic stability and cell polarity. Proc Natl Acad Sci USA. 1998;95:2950–5.
8. Fisk HA. Many pathways to destruction: the centrosome and its control by and role in regulated proteolysis. In: Schatten H, editor. The centrosome, Chap 8. New York: Springer Science and Business Media; 2012.
9. Prosser SL, Fry AM. Regulation of the centrosome cycle by protein degradation. In: Schatten H, editor. The centrosome, Chap 9. New York: Springer Science and Business Media; 2012.
10. Fukasawa K. Molecular links between centrosome duplication and other cell cycle associated events. In: Schatten H, editor. The centrosome, Chap 10. New York: Springer Science and Business Media; 2012.
11. Boutros R. Regulation of centrosomes by cyclin-dependent kinases. In: Schatten H, editor. The centrosome, Chap 11. New York: Springer Science and Business Media; 2012.
12. Korzeniewski N, Duensing S. Disruption of centrosome duplication control and induction of mitotic instability by the high-risk human papillomavirus oncoproteins E6 and E7. In: Schatten H, editor. The centrosome, Chap 12. New York: Springer Science and Business Media; 2012.
13. Saladino C, Bourke E, Morrison CG. Centrosomes, DNA damage and aneuploidy. In: Schatten H, editor. The centrosome, Chap 13. New York: Springer Science and Business Media; 2012.
14. Kais Z, Parvin JD. Centrosome regulation and breast cancer. In: Schatten H, editor. The centrosome, Chap 14. New York: Springer Science and Business Media; 2012.
15. Yan B, Chng W-J. The role of centrosomes in multiple myeloma. In: Schatten H, editor. The centrosome, Chap 15. New York: Springer Science and Business Media; 2012.
16. Olivero OA. Centrosomal amplification and related abnormalities induced by nucleoside analogs. In: Schatten H, editor. The centrosome, Chap 16. New York: Springer Science and Business Media; 2012.

17. Salisbury JL. Centrosomes: Sfi1p and centrin unravel a structural riddle. Curr Biol. 2004;14:R27–29.
18. Salisbury JL, Suino KM, Busby R, Springett M. Centrin-2 is required for centriole duplication in mammalian cells. Curr Biol. 2002;12:1287–92.
19. Andersen JS, Wilkinson CJ, Mayor T, Mortensen P, Nigg EA, Mann M. Proteomic characterization of the human centrosome by protein correlation profiling. Nature. 2003;426:570–4.
20. Wilkinson CJ, Andersen JS, Mann M, Nigg EA. A proteomic approach to the inventory of the human centrosome. In: Nigg E, editor. Centrosomes in development and disease. Weinheim: Wiley-VCA; 2004. p. 125–42.
21. Hannak E, Oegema K, Kirkham M, Gonczy P, Habermann B, Hyman AA. The kinetically dominant assembly pathway for centrosomal asters in Caenorhabditis elegans is γ-tubulin dependent. J Cell Biol. 2002;157:591–602.
22. Tassin AM, Celati C, Moudjou M, Bornens M. Characterization of the human homologue of the yeast Spc98p and its association with gamma-tubulin. J Cell Biol. 1998;141:689–701.
23. Murphy S, Urbani L, Stearns T. The mammalian gamma-tubulin complex contains homologues of yeast spindle pole body component sspc97p and spc98p. J Cell Biol. 1998;141:663–74.
24. Dictenberg J, Zimmerman W, Sparks C, Young A, Vidair C, Zheng Y, Carrington W, Fay F, Doxsey SJ. Pericentrin and gamma tubulin form a protein complex and are organized into a novel lattice at the centrosome. J Cell Biol. 1998;141:163–74.
25. Doxsey SJ, Stein P, Evans L, Calarco P, Kirschner M. Pericentrin, a highly conserved protein of centrosomes involved in microtubule organization. Cell. 1994;76:639–50.
26. Flory MR, Davis TN. The centrosomal proteins pericentrin and kendrin are encoded by alternatively spliced products of one gene. Genomics. 2003;82:401–5.
27. Gillingham AK, Munro S. The PACT domain, a conserved centrosomal targeting motif in the coiled-coil proteins AKAP450 and pericentrin. EMBO Rep. 2000;1:524–9.
28. Kawaguchi S, Zheng Y. Characterization of a Drosophila centrosome protein CP309 that shares homology with Kendrin and CG-NAP. Mol Biol Cell. 2004;15:37–45.
29. Keryer G, Di Fiore B, Celati C, Lechtreck KF, Mogensen M, Delouvee A, Lavia P, Bornens M, Tassin AM. Part of Ran is associated with AKAP450 at the centrosome: involvement in microtubule-organizing activity. Mol Biol Cell. 2003;14:4260–71.
30. Steadman BT, Schmidt PH, Shanks RA, Lapierre LA, Goldenring JR. Transforming acidic coiled-coil-containing protein 4 interacts with centrosomal AKAP350 and the mitotic spindle apparatus. J Biol Chem. 2002;277(33):30165–76.
31. Takahashi M, Yamagiwa A, Nishimura T, Mukai H, Ono Y. Centrosomal proteins CG-NAP and kendrin provide microtubule nucleation sites by anchoring gamma-tubulin ring complex. Mol Biol Cell. 2002;13:3235–45.
32. Ohta T, Essner R, Ryu JH, Palazzo RE, Uetake Y, Kuriyama R. Characterization of Cep135, a novel coiled-coil centrosomal protein involved in microtubule organization in mammalian cells. J Cell Biol. 2002;156:87–99.
33. Teixidó-Travesa N, Roig J, Lüders J. The where, when and how of microtubule nucleation – one ring to rule them all. J Cell Sci. 2012;125:4445–56.
34. Mogensen MM, Malik A, Piel M, Bouckson-Castaing V, Bornens M. Microtubule minus-end anchorage at centrosomal and non-centrosomal sites: the role of ninein. J Cell Sci. 2000;113:3013–23.
35. Quintyne NJ, Gill SR, Eckley DM, Crego CL, Compton DA, Schroer TA. Dynactin is required for microtubule anchoring at fibroblast centrosomes. J Cell Biol. 1999;147:321–34.
36. Schroer TA. Microtubules don and doff their caps: dynamic attachments at plus and minus ends. Curr Opin Cell Biol. 2001;13:92–6.
37. Quintyne NJ, Schroer TA. Distinct cell cycle-dependent roles for dynactin and dynein at centrosomes. J Cell Biol. 2002;159:245–54.
38. Young A, Dictenberg JB, Purohit A, Tuft R, Doxsey SJ. Cytoplasmic dynein-mediated assembly of pericentrin and γ tubulin onto centrosomes. Mol Biol Cell. 2000;11:2047–56.
39. Levy YY, Lai EY, Remillard SP, Heintzelman MB, Fulton C. Centrin is a conserved protein that forms diverse associations with centrioles and MTOCs in Naegleria and other organisms. Cell Motil Cytoskeleton. 1996;33:298–323.

40. Salisbury JL. Centrin, centrosomes, and mitotic spindle poles. Curr Opin Cell Biol. 1995;7:39–45.
41. Lutz W, Lingle WL, McCormick D, Greenwood TM, Salisbury JL. Phosphorylation of centrin during the cell cycle and its role in centriole separation preceding centrosome duplication. J Biol Chem. 2001;276:20774–80.
42. Manandhar G, Schatten H, Sutovsky P. Centrosome reduction during gametogenesis and its significance. Biol Reprod. 2005;72:2–13.
43. Sun QY, Schatten H. Multiple roles of NuMA in vertebrate cells: Review of an intriguing multi-functional protein. Front Biosci. 2006;11:1137–46.
44. Sun Q-Y, Schatten H. Centrosome inheritance after fertilization and nuclear transfer in mammals. In: Sutovsky P (ed) Somatic cell nuclear transfer. Landes Bioscience. Adv Exp Med Biol. 2007;591:58–71.
45. Saredi A, Howard L, Compton DA. Phosphorylation regulates the assembly of NuMA in a mammalian mitotic extract. J Cell Sci. 1997;110:1287–97.
46. Merdes A, Cleveland DA. The role of NuMA in the interphase nucleus. J Cell Sci. 1998;111:71–9.
47. Gehmlich KL, Haren L, Merdes A. Cyclin B degradation leads to NuMA release from dynein/dynactin and from spindle poles. EMBO Rep. 2004;5:97–103.
48. Kammerer S, Roth RB, Hoyal CR, Reneland R, Marnellos G, Kiechle M, Schwarz-Boeger U, Griffiths LR, Ebner F, Rehbock J, Cantor CR, Nelson MR, Brown A. Association of the NuMA region on chromosome 11q13 with breast cancer susceptibility. Proc Natl Acad Sci USA. 2005;102(6):2004–9.
49. Yam HF, Wang ZH, Or PC, Wang SW, Li J, Chew EC. Effect of glucocorticoid hormone on nuclear matrix in cervical cancer cells in vitro. Anticancer Res. 1998;18:209–16.
50. Lingle WL, Salisbury JL. Altered centrosome structure is associated with abnormal mitoses in human breast tumors. Am J Pathol. 1999;155:1941–51.
51. Schatten H, Wiedemeier A, Taylor M, Lubahn D, Greenberg NM, Besch-Williford C, Rosenfeld C, Day K, Ripple M. Centrosomes-centriole abnormalities are markers for abnormal cell divisions and cancer in the transgenic adenocarcinoma mouse prostate (TRAMP) model. Biol Cell. 2000;92:331–40.
52. Lingle WL, Salisbury JL. The role of the centrosome in the development of malignant tumors. Curr Top Dev Biol. 2000;49:313–29.
53. Lingle WL, Barrett SL, Negron VC, D'Assoro AB, Boeneman K, Liu W, Whitehead CM, Reynolds C, Salisbury JL. Centrosome amplification drives chromosomal instability in breast tumor development. Proc Natl Acad Sci USA. 2002;99:1978–83.
54. Mack GJ, Ou Y, Rattner JB. Integrating centrosome structure with protein composition and function in animal cells. Microsc Res Tech. 2000;49:409–19.
55. Ou Y, Rattner JB. The centrosome in higher organisms: structure, composition and duplication. Int Rev Cytol. 2004;238:119–82.
56. Sluder G. Centrosome duplication and its regulation in the higher animal cell. In: Nigg E, editor. Centrosomes in development and disease. Weinheim: Wiley-VCA; 2004. p. 167–89.
57. Matsumoto Y, Maller JL. Calcium, calmodulin, and CaMKII requirement for initiation of centrosome duplication in Xenopus egg extracts (comment). Science. 2002;295:499–502.
58. Pietromonaco SF, Seluja GA, Elias L. Identification of enzymatically active Ca^{2+}/calmodulin-dependent protein kinase in centrosomes of hematopoietic cells. Blood Cells Mol Dis. 1995;21:34–41.
59. Ohta Y, Ohba T, Miyamoto E. Ca^{2+}/calmodulin-dependent protein kinase II: localization in the interphase nucleus and the mitotic apparatus of mammalian cells. Proc Natl Acad Sci USA. 1990;87:5341–5.
60. Tugendreich S, Tomkiel J, Earnshaw W, Hieter P. CDC27Hs colocalizes with CDC16Hs to the centrosome and mitotic spindle and is essential for the metaphase to anaphase transition. Cell. 1995;81:261–8.
61. Freed E, Lacey KR, Huie P, Lyapina SA, Deshaies RJ, Stearns T, Jackson PK. Components of an SCF ubiquitin ligase localize to the centrosome and regulate the centrosome duplication cycle. Genes Dev. 1999;13:2242–57.

62. Gstaiger M, Marti A, Krek W. Association of human SCF(SKP2) subunit p19(SKP1) with interphase centrosomes and mitotic spindle poles. Exp Cell Res. 1999;247:554–62.
63. Meraldi Pand Nigg EA. Centrosome cohesion is regulated by a balance of kinase and phosphatase activities. J Cell Sci. 2001;114:3749–57.
64. Fry AM. The Nek2 protein kinase: a novel regulator of centrosome structure. Oncogene. 2002;21:6184–94.
65. Fry AM, Hames RS. The role of the centrosome in cell cycle progression. In: Nigg E, editor. Centrosomes in development and disease. Weinheim: Wiley-VCA; 2004. p. 143–66.
66. Bailly E, Dorée M, Nurse P, Bornens M. P34cdc2 is located in both nucleus and cytoplasm; part is centrosomally associated at G2/M and enters vesicles at anaphase. EMBO J. 1989;8:3985–95.
67. Pockwinse SM, Krockmalnic G, Doxsey SJ, Nickerson J, Lian JB, vanWijnen AJ, Stein JL, Stein GS, Penman S. Cell cycle independent interaction of CDC with the centrosome, which is associated with the nuclear matrix-intermediate filament scaffold. Proc Natl Acad Sci USA. 1997;94:3022–7.
68. Jackman M, Lindon C, Nigg E, Pines J. Active cyclin B1-Cdk1 first appears on centrosomes in prophase. Nat Cell Biol. 2003;5:143–8.
69. Golsteyn RM, Mundt KE, Fry AM, Nigg EA. Cell cycle regulation of the activity and subcellular localization of Plk1, a human protein kinase implicated in mitotic spindle function. J Cell Biol. 1995;129:1617–28.
70. Merdes A, Ramyar K, Vechio JD, Cleveland DW. A complex of NuMA and cytoplasmic dynein is essential for mitotic spindle assembly. Cell. 1996;87:447–58.
71. Fry AM, Mayor T, Meraldi P, Stierhof YD, Tanaka K, Nigg EA. C-Nap1, a novel centrosomal coiled-coil protein and candidate substrate of the cell cycle-regulated protein kinase Nek2. J Cell Biol. 1998;141:1563–74.
72. Casenghi M, Meraldi P, Weinhart U, Duncan PI, Korner R, Nigg EA. Polo-like kinase 1 regulates Nlp, a centrosome protein involved in microtubule nucleation. Dev Cell. 2003;5:113–25.
73. Barr AR, Gergely F. Aurora A: the maker and breaker of spindle poles. J Cell Sci. 2007;120:2987–96.
74. Kramer ER, Scheuringer N, Podtelejnikov AV, Mann M, Peters JM. Mitotic regulation of the APC activator proteins CDC20 and CDH1. Mol Biol Cell. 2000;11:1555–69.
75. Huang J, Raff JW. The disappearance of cyclin B at the end of mitosis is regulated spatially in Drosophila cells. EMBO J. 1999;18:2184–95.
76. Wakefield J, Huang J-Y, Raff JW. Centrosomes have a role in regulating the destruction of cyclin B in early Drosophila embryos. Curr Biol. 2000;10:1367–70.
77. Wei Y, Multi S, Yang CR, Ma J, Zhang QH, Wang ZB, Li M, Wei L, Ge ZJ, Zhang CH, Ouyang YC, Hou Y, Schatten H, Sun QY. Spindle assembly checkpoint regulates mitotic cell cycle progression during preimplantation embryo development. PLoS One. 2011;6(6):e21557.
78. Sun QY, Lai L, Park KW, Kühholzer B, Prather RS, Schatten H. Dynamic events are differently regulated by microfilaments, microtubules, and mitogen-activated protein kinase during porcine oocyte maturation and fertilization in vitro. Biol Reprod. 2001;64:879–89.
79. Sun QY, Lai L, Bonk A, Prather RS, Schatten H. Cytoplasmic changes in relation to nuclear maturation and early embryo developmental potential of porcine oocytes: effects of gonadotropins, cumulus cells, follicular size and protein synthesis inhibition. Mol Reprod Dev. 2001;59:192–8.
80. Sun Q-Y, Lai L, Wu G, Park K-W, Day B, Prather RS, Schatten H. Microtubule assembly after treatment of pig oocytes with taxol: correlation with chromosomes, γ-tubulin and MAP kinase. Mol Reprod Dev. 2001;60:481–90.
81. Sun Q-Y, Wu GM, Lai L, Park KW, Day B, Prather RS, Schatten H. Translocation of active mitochondria during porcine oocyte maturation, fertilization and early embryo development in vitro. Reproduction. 2001;122:155–63.
82. Sun Q-Y, Lai L, Wu G, Bonk A, Cabot R, Park K-W, Day B, Prather RS, Schatten H. Regulation of mitogen-activated protein kinase phosphorylation, microtubule organization, chromatin behavior, and cell cycle progression are regulated by protein phosphatases during pig oocyte maturation and fertilization in vitro. Biol Reprod. 2002;66(3):580–8.

83. Tong C, Fan H-Y, Lian L, Li S-W, Chen D-Y, Schatten H, Sun Q-Y. Polo-like kinase-1 is a pivotal regulator of microtubule assembly during mouse oocyte meiotic maturation, fertilization, and early embryonic mitosis. Biol Reprod. 2002;67:546–54.
84. Tong C, Fan H-Y, Li S-W, Chen D-Y, Song X-F, Schatten H, Sun Q-Y. Effects of MEK inhibitor U0126 on meiotic progression in mouse oocytes: microtuble organization, asymmetric division and metaphase II arrest. Cell Res. 2003;13(5):375–83.
85. Fan H-Y, Tong C, Teng C-B, Lian L, Li S-W, Yang Z-M, Chen D-Y, Schatten H, Sun Q-Y. Characterization of Polo-like kinase-1 in rat oocytes and early embryos implies its functional roles in the regulation of meiotic maturation, fertilization and cleavage. Mol Reprod Dev. 2003;65:318–29.
86. Fenton B, Glover DM. A conserved mitotic kinase active at late anaphase-telophase in syncytial Drosophila embryos. Nature. 1993;363:637–40.
87. Donaldson MM, Tavares AAM, Hagan IM, Nigg EA, Glover DM. The mitotic roles of polo-like kinase. J Cell Sci. 2001;114:2357–8.
88. Wang Q, Xie S, Chen J, Fukasawa K, Naik U, Traganos F, Darzynkiewicz Z, Jhanwar-Uniyal M, Dai W. Cell cycle arrest and apoptosis by human polo-like kinase 3 is mediated through perturbation of microtubule integrity. Mol Cell Biol. 2002;22(10):3450–9.
89. Münger K, Duensing S. Radiation therapy and centrosome anomalities in pancreatic cancer. In: Nigg E, editor. Centrosomes in development and disease. Weinheim: Wiley-VCA; 2004. p. 353–70.
90. Duensing S, Lee LY, Duensing A, Basile J, Piboonniyom S, Gonzalez S, Crum CP, Munger K. The human papillomavirus type 16 E6 and E7 oncoproteins cooperate to induce mitotic defects and genomic instability by uncoupling centrosome duplication from the cell division cycle. Proc Natl Acad Sci USA. 2000;97:10002–7.
91. Ohki R, Nemoto J, Murasawa H, Oda E, Inazawa J, Tanaka N, Taniguchi T. Reprimo, a new candidate mediator of the p53-mediated cell cycle arrest at the G2 phase. J Biol Chem. 2000;275:22627–30.
92. Sato N, Maitra A, Fukushima N, van Heek NT, Matsubayashi H, Iacobuzio-Donahue CA, Rosty C, Goggins M. Frequent hypomethylation of multiple genes overexpressed in pancreatic ductal adenocarcinoma. Cancer Res. 2003;63:4158–66.
93. Ellgaard L, Molinari M, Helenius A. Setting the standards: quality control in the secretory pathway. Science. 1999;286:1882–8.
94. Johnston JA, Ward CL, Kopito RR. Aggresomes: a cellular response to misfolded proteins. J Cell Biol. 1998;143:1883–98.
95. Wojcik C, DeMartino GN. Intracellular localization of proteasomes. Int J Biochem Cell Biol. 2003;35:579–89.
96. Kopito RR. Aggresomes, inclusion bodies and protein aggregation. Trends Cell Biol. 2000;10:524–30.
97. Roth J, Yam GH, Fan J, Hirano K, Gaplovska-Kysela K, Le Fourn V, Guhl B, Santimaria R, Torossi T, Ziak M, Zuber C. Protein quality control: the who's who, the where's and therapeutic escapes. Histochem Cell Biol. 2008;129:163–77.
98. McNaught KS, et al. Impairment of the ubiquitin-proteasome system causes dopaminergic cell death and inclusion body formation in ventral mesencephalic cultures. J Neurochem. 2002;81:301–6.
99. Katayama H, Brinkley WR, Sen S. The Aurora kinases: role in cell transformation and tumorigenesis. Cancer Metastasis Rev. 2003;22:451–64.
100. Goepfert TM, Adigun YE, Zhong L, et al. Centrosome amplification and overexpression of aurora A are early events in rat mammary carcinogenesis. Cancer Res. 2002;62: 4115–22.
101. Wang X, Zhou YX, Qiao W, et al. Overexpression of aurora kinase A in mouse mammary epithelium induces genetic instability preceding mammary tumor formation. Oncogene. 2006;25:7148–58.
102. Pihan GA, Doxsey SJ. The mitotic machinery is a source of genetic instability in cancer. Semin Cancer Biol. 1999;9:289–302.

103. Meraldi P, Honda R, Nigg EA. Aurora-A overexpression reveals tetraploidization as a major route to centrosome amplification in p53–/– cells. EMBO J. 2002;21:483–92.
104. D'Assoro AB, Busby R, Suino K, et al. Genotoxic stress leads to centrosome amplification in breast cancer cell lines that have an inactive G1/S cell cycle checkpoint. Oncogene. 2004;23:4068–75.
105. D'Assoro AB, Busby R, Acu ID, et al. Impaired p53 function leads to centrosome amplification, acquired ERalpha phenotypic heterogeneity and distant metastases in breast cancer MCF-7 xenografts. Oncogene. 2008;27:3901–11.
106. Xu X, Weaver Z, Linke SP, Li C, Gotay J, Wang XW, Harris CC, Ried T, Deng CX. Centrosome amplification and a defective G2-M cell cycle checkpoint induce genetic instability in BRCA1 exon 11 isoform-deficient cells. Mol Cell. 1999;3:389–95.
107. Starita LM, Machida Y, Sankaran S, Elias JE, Griffin K, Schlegel BP, Gygi SP, Parvin JD. BRCA1-dependent ubiquitination of gammatubulin regulates centrosome number. Mol Cell Biol. 2004;24:8457–66.
108. Schlegel BP, Starita LM, Parvin JD. Overexpression of a protein fragment of RNA helicase A causes inhibition of endogenous BRCA1 function and defects in ploidy and cytokinesis in mammary epithelial cells. Oncogene. 2003;22:983–91.
109. Ko MJ, Murata K, Hwang DS, et al. Inhibition of BRCA1 in breast cell lines causes the centrosome duplication cycle to be disconnected from the cell cycle. Oncogene. 2006;25:298–303.
110. Shao S, Liu R, Wang Y, et al. Centrosomal Nlp is an oncogenic protein that is gene-amplified in human tumors and causes spontaneous tumorigenesis in transgenic mice. J Clin Invest. 2010;120:498–507.
111. Wang HF, Takenaka K, Nakanishi A, et al. BRCA2 and nucleophosmin coregulate centrosome amplification and form a complex with the Rho effector kinase ROCK2. Cancer Res. 2011;71:68–77.
112. Hayward DG, Clarke RB, Faragher AJ, et al. The centrosomal kinase Nek2 displays elevated levels of protein expression in human breast cancer. Cancer Res. 2004;64:7370–6.
113. Suizu F, Ryo A, Wulf G, et al. Pin1 regulates centrosome duplication, and its overexpression induces centrosome amplification, chromosome instability, and oncogenesis. Mol Cell Biol. 2006;26:1463–79.
114. Bergmann S, Royer-Pokora B, Fietze E, et al. YB-1 provokes breast cancer through the induction of chromosomal instability that emerges from mitotic failure and centrosome amplification. Cancer Res. 2005;65:4078–87.
115. Korzeniewski N, Wheeler S, Chatterjee P, et al. A novel role of the aryl hydrocarbon receptor (AhR) in centrosome amplification – implications for chemoprevention. Mol Cancer. 2010;9:153.
116. Olson JE, Wang X, Pankratz VS, Fredericksen ZS, Vachon CM, Vierkant RA, Cerhan JR, Couch FJ. Centrosome-related genes, genetic variation, and risk of breast cancer. Breast Cancer Res Treat. 2011;125(1):221–8. doi:10.1007/s10549-010-0950-8.
117. Krämer A, Anderhub S, Maier B. Mechanisms and consequences of centrosome clustering in cancer cells. In: Schatten H, editor. The centrosome, Chap 17. New York: Springer Science and Business Media; 2012.
118. Gergely F, Basto R. Multiple centrosomes: together they stand, divided they fall. Genes Dev. 2008;2008(22):2291–6.
119. Kwon M, Godinho SA, Chandhok NS, Ganem NJ, Azioune A, Thery M, Pellman D. Mechanisms to suppress multipolar divisions in cancer cells with extra centrosomes. Genes Dev. 2008;22:2189–203.
120. Boveri T. Zellen-Studien: Über die Natur der Centrosomen. vol. 28 Jena. Germany: Fisher Z Med Naturw. 1901;28:1–220.
121. Schatten H, Schatten G, Mazia D, Balczon R, Simerly C. Behavior of centrosomes during fertilization and cell division in mouse oocytes and in sea urchin eggs. Proc Natl Acad Sci USA. 1986;83:105–9.
122. Schatten H, Walter M, Biessmann H, Schatten G. Activation of maternal centrosomes in unfertilized sea urchin eggs. Cell Motil Cytoskeleton. 1992;23:61–70.

123. Kim NH, Simerly C, Funahashi H, Schatten G, Day BN. Microtubule organization in porcine oocytes during fertilization and parthenogenesis. Biol Reprod. 1996;54(6):1397–404.
124. Kim NH, Moon SJ, Prather RS, Day BN. Cytoskeletal alteration in aged porcine oocytes and parthenogenesis. Mol Reprod Dev. 1996;43(4):513–8.
125. Schnackenberg BJ, Palazzo RE. Identification and function of the centrosome centromatrix. Biol Cell. 1999;91(6):429–38.
126. Schatten H, Walter M, Mazia D, Biessmann H, Paweletz N, Coffe G, Schatten G. Centrosome detection in sea urchin eggs with a monoclonal antibody against drosophila intermediate filament proteins: characterization of stages of the division cycle of centrosomes. Proc Natl Acad Sci USA. 1987;84:8488–92.
127. Schatten G, Simerly C, Schatten H. Microtubule configurations during fertilization, mitosis and early development in the mouse and the requirement for egg microtubule-mediated motility during mammalian fertilization. Proc Natl Acad Sci USA. 1985;82:4152–6.
128. Maro B, Howlett SK, Webb M. Non-spindle microtubule organizing centers in metaphase II-arrested mouse oocytes. J Cell Biol. 1985;101:1665–72.
129. Schatten H, Sun QY. Centrosome dynamics during mammalian oocyte maturation with a focus on meiotic spindle formation. Mol Reprod Dev. 2011;78(10–11):757–68.
130. Ganem NJ, Godinho SA, Pellman D. A mechanism linking extra centrosomes to chromosomal instability. Nature. 2009;460(7252):278–82.
131. Schatten H, Sun Q-Y. The role of centrosomes in fertilization, cell division and establishment of asymmetry during embryo development. Semin Cell Dev Biol. 2010;21:174–84.
132. Wheatley DN, Wang AM, Strugnell GE. Expression of primary cilia in mammalian cells. Cell Biol Int. 1996;20:73–81.
133. D'Angelo A, Franco B. The dynamic cilium in human diseases. PathoGenetics. 2009;2(3):1–15.
134. Veland IR, Awan A, Pedersen LB, Yoder BK, Christensen ST. Primary cilia and signaling pathways in mammalian development, health and disease. Nephron Physiol. 2009;111:39–53.
135. Sorokin S. Centrioles and the formation of rudimentary cilia by fibroblasts and smooth muscle cells. J Cell Biol. 1962;15:363–77.
136. Pan J, Snell W. The primary cilium: keeper of the key to cell division. Cell. 2007;129: 1255–7.
137. Quarmby LM, Parker JDK. Cilia and the cell cycle? J Cell Biol. 2005;169(5):707–10.
138. Hildebrandt F, Otto E. Cilia and centrosomes: a unifying pathogenic concept for cystic kidney disease? Nat Rev Genet. 2005;6:928–40.
139. Davenport JR, Yoder BK. An incredible decade for the primary cilium: a look at a once-forgotten organelle. Am J Physiol Renal Physiol. 2005;289:F1159–69.
140. Michaud EJ, Yoder BK. The primary cilium in cell signaling and cancer. Cancer Res. 2006;66:6463–7.
141. Satir P, Christensen ST. Structure and function of mammalian cilia. Histochem Cell Biol. 2008;129:687–93.
142. Sharma N, Berbari NF, Yoder BK. Ciliary dysfunction in developmental abnormalities and diseases. Curr Top Dev Biol. 2008;2008(85):371–427.
143. Berbari NF, O'Connor AK, Haycraft CJ, Yoder BK. The primary cilium as a complex signaling center. Curr Biol. 2009;19:R526–535.
144. Schneider L, Clement CA, Teilmann SC, et al. PDGFR alpha signaling is regulated through the primary cilium in fibroblasts. Curr Biol. 2005;15:1861–6.
145. Bartoloni L, Blouin JL, Maiti AK, et al. Axonemal beta heavy chain dynein DNAH9: cDNA sequence, genomic structure, and investigation of its role in primary ciliary dyskinesia. Genomics. 2001;72:21–33.
146. Arts HH, Cremers FP, Knoers NV, Roepman R. Focus on molecules: RPGRIP1. Exp Eye Res. 2009;88:332–3.
147. Inoko A, Matsuyama M, Goto H, Ohmuro-Matsuyama Y, Hayashi Y, Enomoto M, Ibi M, Urano T, Yonemura S, Kiyono T, Izawa I, Inagaki M. Trichoplein and Aurora A block aberrant primary cilia assembly in proliferating cells. J Cell Biol. 2012;197(3):391–405.

148. Kasper M, Regl G, Frischauf AM, Aberger F. GLI transcription factors: mediators of oncogenic Hedgehog signalling. Eur J Cancer. 2006;42:437–45.
149. Haycraft CJ, Banizs B, Aydin-Son Y, et al. Gli2 and gli3 localize to cilia and require the intraflagellar transport protein polaris for processing and function. PLoS Genet. 2005;1:e53.
150. Jechlinger M, Sommer A, Moriggl R, et al. Autocrine PDGFR signaling promotes mammary cancer metastasis. J Clin Invest. 2006;116:1561–70.
151. Carvalho I, Milanezi F, Martins A, Reis RM, Schmitt F. Overexpression of platelet-derived growth factor receptor alpha in breast cancer is associated with tumour progression. Breast Cancer Res. 2005;7:R788–795.
152. Yuan K, Frolova N, Xie Y, Wang D, Cook L, Kwon YJ, Steg AD, Serra R, Frost AR. Primary cilia are decreased in breast cancer: analysis of a collection of human breast cancer cell lines and tissues. J Histochem Cytochem. 2010;58(10):857–70.
153. Yuan K, Serra R, Frost AR. Primary cilia in the breast and breast cancer. Open Breast Cancer J. 2010;2:101–7.
154. Schiff PB, Fant J, Horwitz SB. Promotion of microtubule assembly *in vitro* by taxol. Nature. 1979;277:665–7.
155. Schatten G, Schatten H, Bestor T, Balczon R. Taxol inhibits the nuclear movements during fertilization and induces asters in unfertilized sea urchin eggs. J Cell Biol. 1982;94:455–65.
156. De Brabander M, Geuens G, Nuydens R, Willebrords R, De Mey J. Taxol induces the assembly of free microtubules in living cells and blocks the organizing capacity of the centrosomes and kinetochores. Proc Natl Acad Sci USA. 1981;78:5608–12.
157. Dimitriadis I, Katsaros C, Galatis B. The effect of taxol on centrosome function and microtubule organization in apical cells of Sphacelaria rigidula (Phaeophyceae). Phycol Res. 2001;49:23–34.
158. Barlow JB, Gonzalez-Garay ML, Cabral F. Paclitaxel-dependent mutants have severely reduced microtubule assembly and reduced tubulin synthesis. J Cell Sci. 2002;115:3469–78.
159. Yin S, Bhattacharya R, Cabral F. Human mutations that confer paclitaxel resistance. Mol Cancer Ther. 2010;9:327–35.
160. Safavy A, Raisch PKP, Mantena S, Sanford LL, Sham SW, Krishna NR, Bonner JA. Design and development of water-soluble curcumin conjugates as potential anticancer agents. J Med Chem. 2007;50:6284–8.
161. Yoon H, Liu RH. Effect of selected phytochemicals and apple extracts on NF–kB activation in human breast cancer MCF-7 cells. J Agric Food Chem. 2007;55(8):3167–73.
162. Yallapu MM, Gupta BK, Jaggi M, Chauhan SC. Fabrication of curcumin encapsulated PLGA nanoparticles for improved therapeutic effects in metastatic cancer cells. J Colloid Interface Sci. 2010;351(1):19–29.
163. Yallapu MM, Jaggi M, Chauhan SC. Poly(β-cyclodextrin)/curcumin self-assembly: a novel approach to improve curcumin delivery and its therapeutic efficacy in prostate cancer cells. Macromol Biosci. 2010;10:1141–51.
164. Dev S, Prabhakaran P, Filgueira L, Iyer KS, Raston CL. Microfluidic fabrication of cationic curcumin nanoparticles as an anti-cancer agent. Nanoscale. 2012;4:2575–9.
165. Nagaraju GP, Aliya S, Zafar SF, Basha R, Diazd R, El-Rayes BF. The impact of curcumin on breast cancer. Integr Biol. 2012;4(9):996–1007. doi:10.1039/c2ib20088k.
166. Yamaguchi M, Moore TW, Sun A, Snyder JP, Shoji M. Novel curcumin analogue UBS109 potently stimulates osteoblastogenesis and suppresses osteoclastogenesis through Smad activation and NFκB inhibition. Integr Biol. 2012;4(8):905–13.
167. Anand P, Kunnumakkara AB, Newman RA, Aggarwal BB. Bioavailability of curcumin: problems and promises. Mol Pharm. 2007;4:807–18.
168. Tonnesen HH, Masson M, Loftsson T. Studies of curcumin and curcuminoids. XXVII. Cyclodextrin complexation: solubility, chemical and photochemical stability. Int J Pharm. 2002;244:127–35.
169. Wang D, Veena MS, Stevenson K, Tang C, Ho B, Suh JD, Duarte V, Faull KF, Mehta K, Srivatsan ES, Wang MB. Liposome-encapsulated curcumin suppresses growth of head and neck squamous cell carcinoma in vitro and in xenografts through the inhibition of nuclear factor kappaB by an AKT-independent pathway. Clin. Cancer Res. 2008;14:6228–36.

170. Mukerjee A, Viswanatha JK. Formulation, characterization and evaluation of curcumin-loaded PLGA nanospheres for cancer therapy. Anticancer Res. 2009;29:3867–76.
171. Cartiera MS, Ferreira EC, Caputo C, Egan ME, Caplan MJ, Saltzman WM. Partial correction of cystic fibrosis defects with PLGA nanoparticles encapsulating curcumin. Mol Pharm. 2009;7(1):86–93.
172. Shin SJ, Beech JR, Kelly KA. Targeted nanoparticles in imaging: paving the way for personalized medicine in the battle against cancer. Integr Biol. 2012;5(1):29–42. doi:10.1039/C2IB20047C.
173. Wilken R, Veena MS, Wang MB, Srivatsan ES. Curcumin: a review of anti-cancer properties and therapeutic activity in head and neck squamous cell carcinoma. Mol Cancer. 2011;10(12):1–19.
174. Reuter S, Gupta SC, Park B, Goel A, Aggarwal BB. Epigenetic changes induced by curcumin and other natural compounds. Genes Nutr. 2011;6:93–108.
175. Uen YH, Liu DZ, Weng MS, Ho YS, Lin SY. NF-kappaB pathway is involved in griseofulvin-induced G2/M arrest and apoptosis in HL-60 cells. J Cell Biochem. 2007;101(5):1165–75.
176. Rebacz B, Larsen TO, Clausen MH, Ronnest MH, Loffler H, Ho AD, Krämer A. Identification of griseofulvin as an inhibitor of centrosomal clustering in a phenotype-based screen. Cancer Res. 2007;67:6342–50.
177. Marchetti F, Mailhes JB, Bairnsfather L, Nandy I, London SN. Dose-response study and threshold estimation of griseofulvin induced aneuploidy during female mouse meiosis I and II. Mutagenesis. 1996;11:195–200.
178. Schatten H. Untersuchungen über die Wirkung von Griseofulvin in Seeigeleiern und in Mammalierzellen. Universität Heidelberg; 1977 (Effects of griseofulvin on sea urchin eggs and on mammalian cells. University of Heidelberg; 1977).
179. Wehland J, Herzog W, Weber K. Interaction of griseofulvin with microtubules, microtubule protein and tubulin. J Mol Biol. 1977;111:329–42.
180. Schatten H, Schatten G, Petzelt C, Mazia D. Effects of griseofulvin on fertilization and early development of sea urchins. Independence of DNA synthesis, chromosome condensation, and cytokinesis cycles from microtubule-mediated events. Eur J Cell Biol. 1982;27:74–87.
181. Grisham LM, Wilson L, Bensch KG. Antimitotic action of griseofulvin does not involve disruption of microtubules. Nature. 1973;244:294–6.
182. Miao YL, Zhang X, Zhao JG, Spate L, Zhao MT, Murphy CN, Prather RS, Sun QY, Schatten H. Effects of griseofulvin on in vitro porcine oocyte maturation and embryo development. Environ Mol Mutagen. 2012;53(7):561–6. doi:10.1002/em.21717.
183. Ho YS, Duh JS, Jeng JH, Wang YJ, Liang YC, Lin CH, Tseng CJ, Yu CF, Chen RJ, Lin JK. Griseofulvin potentiates antitumorigenesis effects of nocodazole through induction of apoptosis and G2/M cell cycle arrest in human colorectal cancer cells. Int J Cancer. 2001;91:393–401.
184. Panda D, Rathinasamy K, Santra MK, Wilson L. Kinetic suppression of microtubule dynamic instability by griseofulvin: implications for its possible use in the treatment of cancer. Proc Natl Acad Sci USA. 2005;102:9878–83.

Chapter 13
A New Perspective on Cyclin D1: Beyond Cell Cycle Regulation

Chenguang Wang, Timothy G. Pestell, and Richard G. Pestell

Abstract The canonical function of cyclin D1 in regulating cell cycle progression was well established in the 1990s. The role of cyclin D1 has been expanded in the last decade. The novel functions of cyclin D1 include regulation of hormone signaling through cross talk with nuclear receptors and cofactors, gene transcription through direct interaction with transcription factors, cell migration and invasion, cell death, angiogenesis, mitochondrial metabolism, DNA damage and repair, and chromosomal instability (CIN). High-throughput analysis, including whole-genome expression profiling and deep sequencing, identified cyclin D1 binding sites in the context of local chromatin. Proteomics and protein arrays identified cyclin D1-interacting proteins, which confirmed both known interactive proteins involved in cell cycle regulation and a number of new proteins. These novel experimental approaches revealed previously unrecognized functions of cyclin D1 which may contribute to the role of cyclin D1 in tumorigenesis. Herein, we discuss recent findings on the role of cyclin D1 in regulating metabolism, DNA damage, and CIN and how these processes may guide novel clinical management.

C. Wang (✉) • R.G. Pestell (✉)
Kimmel Cancer Center, Department of Cancer Biology, Thomas Jefferson University, Bluemle Building, Room 1050, 233 South 10th Street, Philadelphia, PA 19107, USA
e-mail: chenguang.wang@jefferson.edu; richard.pestell@jefferson.edu

T.G. Pestell
Kimmel Cancer Center, Department of Cancer Biology, Thomas Jefferson University, 233 South 10th Street, Philadelphia, PA 19107, USA

H. Schatten (ed.), *Cell and Molecular Biology of Breast Cancer*,
DOI 10.1007/978-1-62703-634-4_13, © Springer Science+Business Media New York 2013

13.1 Introduction

Cyclin D1 (BCL1, PRAD1 oncogene) is a 34 kDa regulatory protein identified as a prominent oncogenic driver in several subsets of cancer. Cyclin D1 conveys several essential cellular functions, including cell cycle regulation at the G1/S checkpoint, mitogenesis and metabolism, nuclear receptor hormone signaling, and maintenance of chromosomal stability [1–12]. Alternative splicing generates two isoforms of cyclin D1 (cyclin D1A, cyclin D1B) [13, 14] with distinct carboxyl terminus and cellular localizations. Cyclin D1 was initially identified as the regulatory subunit of the holoenzyme that phosphorylates and inactivates the retinoblastoma protein (pRb) [15, 16] allowing for the release of E2F transcription factors and progression through G_1/S phase of the cell cycle. Cyclin D1 was identified as a candidate oncogene activated in a subset of parathyroid tumors through genetic rearrangement [17]. Cyclin D1 regulates progression through the cell cycle in a stepwise fashion from DNA replication, through cell division and cytokinesis [18]. The identification of cyclin D1-associated proteins including $p21^{CIP1}/p27^{KIP1}$ [19], tumor suppression (BRCA1) [20], transcription factors [9], and co-integrator proteins with histone acetyl transferase (HAT) [7] or histone deacetylase activity (HDAC) [21] further expanded cyclin D1's role in coordinating histone modification and subsequent transcriptional amplification of proliferation and/or differentiation genes [22]. The cyclin D1/Cdk4 complex was also found to phosphorylate non-Rb proteins including nuclear respiratory protein 1 (NRF-1), Smad3, filamin A, BRCA1, and FoxM1 demonstrating transcriptional influence on metabolic biogenesis, carcinogenesis, and cellular migration/invasion [3, 23–26]. Genomic deletion analysis supports the reliance on cyclin D1 for cellular proliferation, angiogenesis [27], and cellular migration [28]. Cyclin D1 is a modulator of transcription co-regulators such as BRCA1 and the co-integrators including SRC1, p300/CBP, and P/CAF [5, 7, 10, 20, 29]. The abundance of cyclin D1 is a key determinant of both human and murine tumorigenesis [30]. As p300/CBP serve as rate-limiting co-integrator of many transcription factors, it is likely that cyclin D1-dependent regulation of p300/CBP function has a broad role in transcriptional regulation. Cyclin D1 and p300 knockout analysis suggested a role for cyclin D1/p300 in DNA-replication fidelity [7]. Understanding the cellular and transcriptional roles of cyclin D1 could elucidate the possible therapeutic targets for cyclin D1 in cancer therapy.

13.2 Structure

The *CCND1* gene is located on human chromosome 11 spanning about 15 kb including five exons and four introns (Fig. 13.1a, b). Northern blot and PCR analysis of cDNA derived from cell lines and tissues identified an alternatively spliced transcript of the *CCND1* gene [31]. The variant transcript shows a failure to splice 3′ of exon 4 and encodes a protein with an altered carboxyl-terminal domain (cyclin D1 transcript B) (Fig. 13.1c).

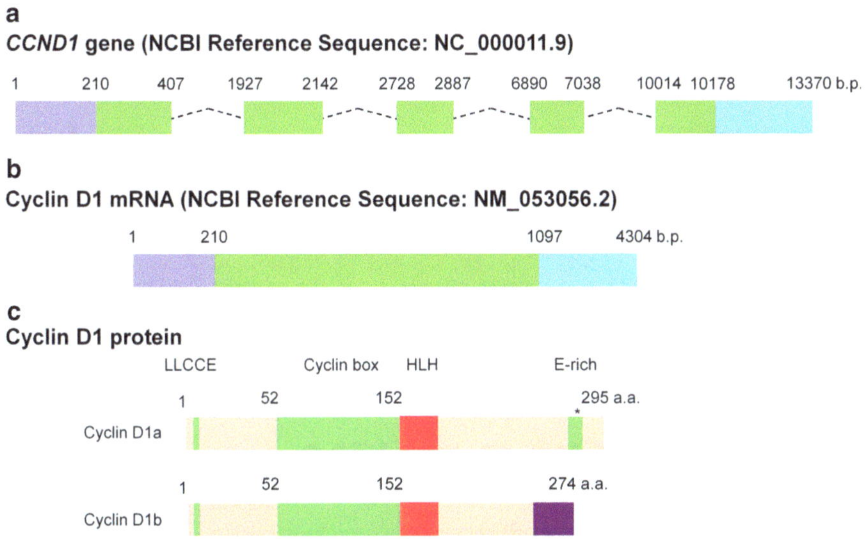

Fig. 13.1 Structure of CCND1 gene and protein product (cyclin D1). (**a**) Schematic representation of genomic structure of cyclin D1 gene. (**b**) Cyclin D1 and alternatively spliced cyclin D1b. (**c**) The functional domains of cyclin D1 protein

The human cyclin D1 protein consisted of 295 amino acids (Fig. 13.1c). In addition to the domains identified for Cdk4 and Rb binding, the HLH (a.a. 142–178, the C terminus of cyclin box) and E-rich (a.a. 272–280) motif have been described [6, 32]. These two motifs were necessary for binding nuclear receptors (AR and PPARγ) and PACSIN 2, respectively. The HLH motif, initially defined as a motif required for cyclin D1 repression of PPARγ transactivation, is important in regulating transcription cofactor binding. Deletion of the HLH region abolished cyclin D1 repression of p300 [7]. The HLH motif also participates in cyclin D1 regulation of BRCA1 function at the ERα [20]. The E-rich motif, conserved between cyclin D1 and D2, is required for PACSIN 2 binding and plays an essential role in regulating cell spreading and migration [32].

The *CCND1* gene promoter and its regulation are well studied using the promoter-driven luciferase reporter system. This system has significantly contributed to the understanding of cyclin D1 gene expression in cancer and other diseases [33]. Many classes of eukaryotic transcriptional regulatory elements were identified within the 1,745-bp upstream of *CCND1* 5′-UTR as illustrated in Fig. 13.2a. Upon the binding of these transcription-responsive DNA elements, the expression of *CCND1* gene is tightly regulated in response to growth signaling [33–42]. The transcriptional regulators of the *CCND1* gene expression have been summarized in our prior publication [43]. A 3,400-bp promoter sequence of *CCND1* gene has been cloned, providing a unique tool to study additional upstream transcriptional regulatory elements of the *CCND1* gene [44].

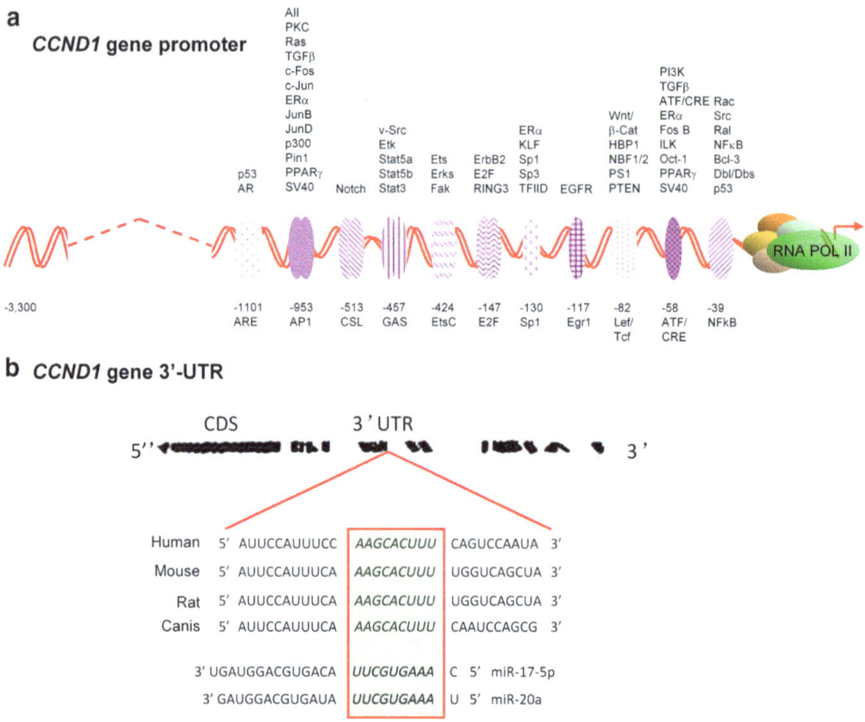

Fig. 13.2 The regulatory networks of cyclin D1 gene expression. (**a**) Schematic representation of *CCND1* gene promoter (−1,745 bp) showing transcription factors and cofactors binding. (**b**) 3′-UTR of *CCND1* gene contains miRNA binding sites

MicroRNAs (miRNAs) are a class of small noncoding RNA molecules (~21–25 nucleotides in length) that regulate gene expression at the posttranscriptional levels. The inverse correlation between the expression of miR-17-5p/miR-20a miRNA cluster and *CCND1* led to the discovery of a miR-17-5p/miR-20a binding site in the 3′-UTR of *CCND1* gene (Fig. 13.2b). miR-17-5p/miR-20a, through binding the *CCND1* 3′-UTR and suppressing cyclin D1 protein translation, inhibited breast cancer cell proliferation and tumor colony formation [45].

13.3 Expression, Cellular Localization, and Regulation

Two forms of *CCND1* mRNA (transcript A and B) are generated in a variety of cells and tissues. Although the function of this polymorphism remains controversial, specific polymorphisms of cyclin D1 correlate with risk or mortality in diseases including non-small cell cancer of the lung [31]; squamous cell carcinoma of the head and neck [46]; epithelial ovarian cancer [47]; hereditary nonpolyposis colorectal cancer

[48]; risk of urinary bladder cancer and prostate cancer [49, 50]; esophageal, gastric, and cardiac carcinoma [51]; and early onset of lung cancer, especially squamous cell carcinoma [52]. Cyclin D1 isoform abundance also predicts response to adjuvant chemotherapy, for example, in colorectal cancer [53]. Cyclin D1 is mainly localized in the nucleus of dividing cells; however, membrane and cytosolic distribution have been reported [32]. Through binding $p21^{CIP1}$ or $p27^{KIP1}$, the cyclin D1/Cdk4/6 is able to translocate to the nucleus and elicit a transcriptional effect. In some studies, the protein encoded by transcript B was constitutively localized in the nucleus, possibly due to the loss of threonine 286, the phosphorylation site of transcript A required for nuclear export [54, 55]. The kinase that phosphorylates threonine 286 is contested as GSK3β seems to have a limited role in some studies [56].

13.4 Biological Functions

13.4.1 Cyclin D1 as a Cell Cycle Regulator

The cyclin-dependent kinase holoenzymes are a family of serine/threonine kinases that control progression through the cell cycle [15, 16, 22]. The cyclins encode regulatory subunits of the kinases, which phosphorylate specific proteins, including the retinoblastoma (Rb) protein, to promote transition through specific cell cycle checkpoints [22, 30, 57, 58]. Cyclin D1 plays a pivotal role in G_1/S-phase cell cycle progression in fibroblasts and is rate limiting in growth factor- or estrogen-induced mammary epithelial cell proliferation [2, 59].

13.4.2 Cyclin D1-Dependent Tumorigenesis

Cyclin D1 is capable of transforming fibroblast and functions in collaborative oncogenesis with E1A, ErbB2, and Ras [60–62]. Cyclin D1 is sufficient for tumorigenesis when targeted to specific organs in transgenic mice [63]. Ectopic expression of a fusion gene between Cdk4 and cyclin D1 immortalized primary REF and collaborated with H-Ras G12V mutant in anchorage-independent growth in vitro and tumor formation in vivo. In contrast, cyclin D1 and H-RasG12V co-expression alone did not lead to transformation, illustrating the importance of Cdk4 in transformation of REFs. A Cdk4 mutant, K35M, that can bind to $p16^{INK4}$ but has lost catalytic activity due to its inability to bind ATP collaborates with H-Ras in transforming primary rat fibroblasts as efficiently as the wild-type Cdk4. The Cdk4/R24C gain of function mutant, which evades binding to $p16^{INK4}$, can't collaborate with cyclin D1 and H-Ras in transformation. Expression of $p16^{INK4}$or $p21^{CIP1}$ inhibits the transformation of REFs by cyclin D1/H-Ras, Cdk4/H-Ras, or c-myc/H-Ras.

Ccnd$^{-/-}$ mice are resistant to oncogene-induced skin and gastrointestinal tumors [64, 65]. Mice with knock-in of a homozygous of cyclin D1 mutant (D1$^{KE/KE}$) that is unable to induce pRb phosphorylation or bind p27^{KIP1} or p21^{CIP1} show normal mammary development in contrast with *Cdk4*$^{-/-}$ mice [66]. The mice with a knock-in of D1$^{KE/KE}$ show resistance to Neu-induced mammary tumorigenesis. The role of cyclin D1-dependent phosphorylation of pRb in senescence, cell cycle progression, and transformation may be cell type specific. S-phase entry due to overexpression of cyclin D1/Cdk4 is independent of pRb phosphorylation [67]. Furthermore, cyclin D1 mutants that evade pRb binding remain oncogenically active in promoting G_1/S cell cycle transition and abrogate pRb function in senescence assays [68]. Abrogation of D1 by siRNA or ectopic expression of a Cdk4 mutant that is kinase dead prevents tumor formation elucidating cyclin D1/Cdk4 reliance in tumorigenesis [69].

13.4.3 Cyclin D1 Regulation of Nuclear Hormone Signaling

Cyclin D1 regulates nuclear hormone signaling resulting in distinct transcription outputs; however, the physiological significance in vivo has at this time been verified primarily for PPARγ. Cyclin D1 binds to the estrogen receptor α (ERα) and enhances ligand-independent reporter gene activity [8, 9]. Cyclin D1 increases the association between ERα and its co-activators [10]. Cyclin D1 also augments ERα signaling via inactivation of BRCA1-mediated repression [20]. In vivo, estradiol (E2) failed to induced progesterone receptor expression effectively in *Ccnd*$^{-/-}$ mice, suggesting cyclin D1 is required for E2 signaling in vivo [70]. P/CAF potentiates ERα activation through direct binding between cyclin D1 and P/CAF, showing the receptor-specific positive regulation [71]. Conversely, androgen receptor (AR) reporter gene activity upon ligand binding is inhibited by cyclin D1 through a competition mechanism by which cyclin D1 recruits corepressors with HDAC activity and disengages AR and co-activator binding [5, 8–10].

Investigation of the in vivo function of cyclin D1 identified PPARγ as a rate-limiting target of cyclin D1 [6]. Using cyclin D1 knockout mice, it was shown that cyclin D1 abundance determines nuclear receptor recruitment to transcription factor binding sites in the context of chromatin in ChIP assays [65]. The mechanisms by which cyclin D1 regulates nuclear receptor function correlate with cyclin D1 occupancy in the context of local chromatin [21]. The endogenous PPARE of *LPL* gene promoter binds cyclin D1 associated with the recruitment of chromatin-modifying proteins (HP1α, Suv29, HDAC1/3, p300). The occupancy of cyclin D1 on gene promoters assessed by ChIP-on-chip technology mapped cyclin D1 to approximately 900 genes [11]. Subsequent genome-wide binding site mapping of cyclin D1 by ChIP-Seq identified cyclin D1 at these sites and additional sites (total 1200 sites) including those bound by nuclear receptors (ERα and PPARγ) [12]. Based on these recent findings, it is time to reexamine the interaction between cyclin D1 and nuclear receptors to determine whether cyclin D1 in the context of local chromatin regulates nuclear receptor function.

13.4.4 Cyclin D1 Promotion of Cell Migration

Given that a subset of human cancer cyclin D1 overexpression correlated with metastasis, we hypothesized that cyclin D1 may promote cell migration. In support of this hypothesis, we found that *Ccnd1*$^{-/-}$ mouse macrophages and embryonic fibroblasts (MEFs) displayed increased cellular adherence, defective motility, and an impaired wound healing response [28]. The mechanism by which cyclin D1 promotes cell migration includes transcriptional repression of thrombospondin-1 (TSP-1) and inhibition of RhoA, and thereby ROCK II kinase, a serine/threonine kinase [72, 73]. Cyclin D1-bound p27^{KIP1} was essential for the interaction with RhoA and involved the residue defined in the cyclin D1 KE mutant. Cyclin D1A, but not the alternatively spliced form cyclin D1B, rescued the migratory defects of *Ccnd1*$^{-/-}$ MEFs, suggesting the C terminus of cyclin D1 contributes to the migration function [74]. Further studies showed that the E-rich motif located in the C terminus of cyclin D1 is required for binding to PACSIN 2, a protein kinase C and casein kinase substrate in neurons [32]. Mass spectrometry identified that cyclin D1/Cdk4 binds to, and phosphorylates, filamin A [24], providing additional evidence in support of cyclin D1's role in cell migration.

13.4.5 Cyclin D1 Regulation of DNA Damage and Cell Death

Earlier studies have shown that cyclin D1 expression is selectively induced in dying neurons [75], indicating a role of cyclin D1 in cell death. γ-irradiation and UV irradiation induce apoptosis, which is potentiated by cyclin D1 deficiency [36, 76]. Cyclin D1 expression inhibited apoptosis in chorionic trophoblast, which was dependent on the presence of p300. Consistent with this, the CH3 region of p300 required for induction of cyclin D1 gene expression was required for the inhibition of apoptosis. p300 inhibited apoptosis in *Ccnd*$^{+/+}$ fibroblasts but increased apoptosis in *Ccnd*$^{-/-}$ [36]. Bcl-2 prevents cells from undergoing apoptosis, while Bcl-2 increased cyclin D1 expression through promoter activation [77]. A recent intriguing finding from the Hinds laboratory was that mammary epithelial cells (MECs) derived from *Ccnd1*$^{KE/KE}$ mice, which are deficient in cyclin D1 kinase activity, exhibit an autophagy-like process, which is resistant to oncoprotein ErbB2-induced senescence both in vivo and in cultured cells [78]. The molecular mechanism by which cyclin D1 regulates autophagy remains to be fully understood.

The role for cyclin D1 in DNA repair was recognized nearly 2 decades ago. Cells in the G_1 phase exposed to UV lost cyclin D1 protein expression, and cyclin D1 overexpression prevented the cells from repairing the damaged DNA [79]. In examining the mechanism by which cyclin D1 expression enhanced the DNA damage response (DDR), Li et al. showed that cyclin D1 enhanced γ-H2AX phosphorylation, facilitated the assembly of DNA repair foci, and recruited specific DNA repair factors to chromatin [80]. Cyclin D1 deletion in fibroblasts or siRNA-mediated reduction of endogenous cyclin D1 reduced the 5-fluorouracil-mediated DDR in colon cancer cells. Induction of the DDR was uncoupled from the role of cyclin D1 in the induction of DNA synthesis and G_1/S transition as the effect of cyclin D1 was observed in G_1-/G_0-arrested cells. Heterozygous loss of the *BRCA1* or *BRCA2* alleles predisposes an individual to developing breast and ovarian cancers. BRCA1 is a multifunctional tumor suppressor protein forming complexes with proteins governing cell cycle, gene transcription, DNA damage response, and chromatin remodeling. Cyclin D1 physically interacts with BRCA1 and antagonizes BRCA1 repression on ERα function in regulating gene expression [20], suggesting a functional network may be involved in DNA damage response and repair. ERα, particularly the membrane-bound form, induces DNA damage response upon ligand activation [81]. Whether cyclin D1 regulation of ERα participates in membrane-bound ERα signaling warrants investigation.

Cyclin D1 enhancement of the DNA damage response involved cyclin D1 induction of RAD51 expression, binding to RAD51, and recruitment of cyclin D1 to the local chromatin at sites of damaged DNA [80] (Fig. 13.3). Subsequent proteomic analyses of cyclin D1 binding proteins identified a network of DNA repair proteins, which also included RAD51 [82]. γ-irradiation was shown to enhance cyclin D1 and RAD51 binding. Cyclin D1 was recruited to DNA damage sites in a BRCA2-dependent manner. BRCA2 recruitment to DNA damage sites did not require cyclin D1; however, cyclin D1 deficiency reduced RAD51 recruitment to damaged DNA [82]. Collectively, these studies illustrate the involvement of cyclin D1 in binding to proteins involved in both homologous and nonhomologous DNA damage repair and the role of cyclin D1 in mediating both the response to DNA damage and the subsequent repair and cell survival.

13.4.6 Cyclin D1 Regulation of Metabolism and Energy Homeostasis

Cell cycle progression and mitochondrial biogenesis are integrated biological processes. The underlying mechanisms were revealed in studies of the *Ccnd1*$^{-/-}$ mice [3]. Cyclin D1 inhibits mitochondria function and size [3, 4]. The kinase function of cyclin D1 was required for repression of mitochondrial activity through direct phosphorylation of nuclear respiratory factor 1 (NRF-1), a transcription factor involved in promoting mitochondrial biogenesis via induction of mitochondria transcription factor A (mtTFA). Cyclin D1 binds to NRF-1 in cells by immunoprecipitation-Western blot (IP-WB) analysis and by mammalian 2-hybrid assays. Cyclin D1 and NRF-1 share substantial overlap in microarray-based gene expression studies. Of the 254 candidate NRF-1-inducible genes and 210 genes repressed by cyclin D1, 73 similar genes and 18 identical genes were shared between the two gene sets [3]. Increased cyclin D1 abundance reduced mitochondrial activity. From these studies, we proposed a new model in which cyclin D1 integrates cell cycle progression with mitochondrial

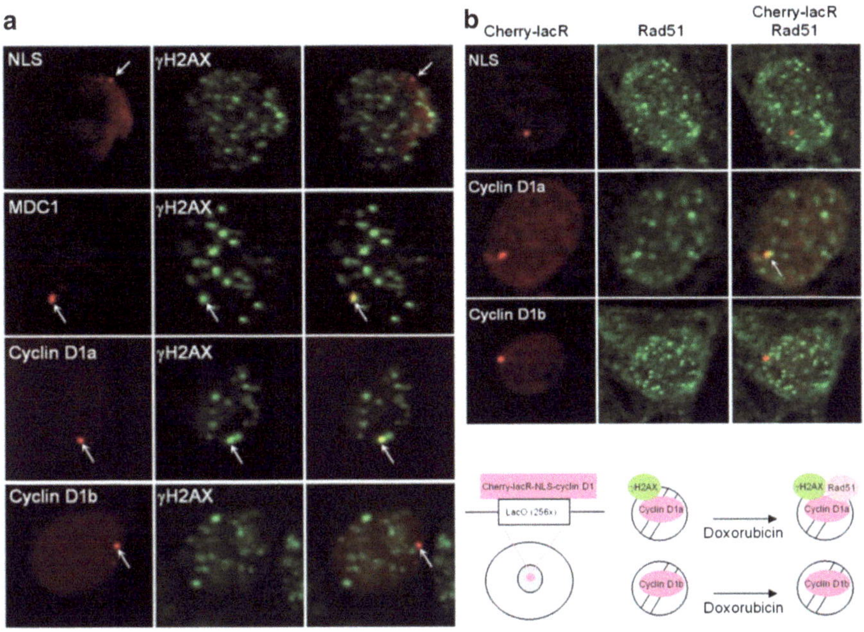

Fig. 13.3 Immobilization of cyclin D1 on chromatin leads to DDR activation. (**a**) Confocal immunofluorescence microscopy of NIH2/4 cells transiently transfected with plasmid DNA encoding cyclin D1 isoforms or repair factor fused to Cherry-lacR-NLS (*red*). Phosphorylation of γH2AX (*green*) indicates DDR activation. (**b**) Cyclin D1a isoform recruits Rad51 to local chromatin in response to DNA damage. NIH2/4 cells were transfected with Cherry-lacR-NLS-cyclin D1a, Cherry-lacR-NLS-cyclin D1b, or vector control Cherry-lacR-NLS by using Nucleofector system. Twenty-four hour later cells were treated with 0.4 μM doxorubicin for 3 h. Then immunofluorescence staining was conducted using specific antibody to Rad51. The *lower panel* shows a model of Cherry-lacR/lacO system and cyclin D1a inducing γH2AX foci in the absence of DNA damage, cyclin D1a recruiting Rad51 to local chromatin in response to DNA damage, and cyclin D1b failing to recruit Rad51 even in the response to DNA damage (This figure was derived from prior publication [80])

biogenesis. Reduced mitochondrial activity from increased cyclin D1 expression shifts glucose metabolism toward cytosolic glycolysis. Such a metabolic shift is known to occur during tumor progression as a component of the Warburg effect.

Tissue-specific inducible cyclin D1 antisense transgenics confirmed that cyclin D1 inhibits mitochondrial biogenesis [4]. Mitochondrial activity was enhanced by antisense inactivation of Ccnd1 expression or small interfering RNA to *Ccnd1*. Global gene expression profiling and functional analysis of mammary gland-targeted cyclin D1 antisense transgenics demonstrated that cyclin D1 inhibits mitochondrial activity and aerobic glycolysis in vivo. Together, these studies suggest that cyclin D1 integrates nuclear DNA synthesis and mitochondrial function by phosphorylating Rb and NRF-1, respectively.

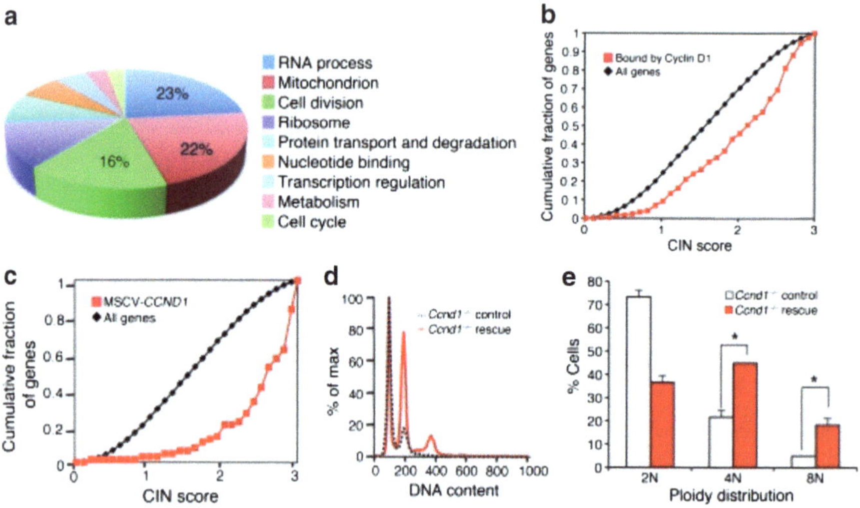

Fig. 13.4 Cyclin D1 rescue of *Ccnd*$^{-/-}$ MEFs induces chromosomal instability (CIN). (**a**) Functional annotation clustering by DAVID of cyclin D1-associated genes, based on percent enrichment score of the top hits. (**b**) Cyclin D1-bound promoter regions (0 to −500 bp) were enriched in genes demonstrating an association with CIN ($p<0.0001$). (**c**) The expression profile for cyclin D1-induced genes identified by microarray [74] was enriched for high CIN score ($p<0.0001$). (**d**) PI staining demonstrated increased polyploidy in cyclin D1-rescued versus control *Ccnd1*$^{-/-}$ MEFs. (**e**) Quantitation of PI staining based on three separate cell lines (mean ± SEM) $p<0.005$ (This figure was derived from our prior publication [12])

13.4.7 Cyclin D1 Promotes Chromosomal Instability

In view of our prior findings that cyclin D1 regulated transcription factor occupancy at target DNA sites by ChIP [65] and that a DNA-bound form of cyclin D1 by ChIP analysis recruited chromatin remodeling enzymes, we conducted a genome-wide interrogation of cyclin D1 binding sites in the context of local chromatin by ChIP-Seq. We identified 3,222 regions (intervals) bound by cyclin D1. Approximately 70 % of these intervals were located in 10 kb of 2,840 genes with a high density located within 500 bp of the transcriptional start site (ATG) (Fig. 13.4a). The transcription factors enriched at the interval region included ERα, Sp1, and Ctcf. Interestingly, among these transcription factors, Ctcf (CCCTC-binding factor), a zinc-finger DNA binding protein that regulates gene transcription enhancer function, is also involved in sister chromatid cohesion. We interrogated the signaling pathways associated with the genes bound by cyclin D1. Cell division emerged to be one of the most enriched terms, which is involved in G_2/M phase and cellular mitosis. Increased abundance of cyclin D1 during G_2/M has previously been described [83]. Given that mis-regulation of genes that govern the mitotic phase often leads to chromosomal instability (CIN), we next determined the functional consequence of cyclin D1 using combined expression profile, fluorescence-activated cell sorting (FACS) analysis, and spectral karyotyping (SKY) approaches. Rescue of cyclin D1 expression in

MEFs using a cyclin D1 retrovirus induced expression of CIN-associated genes (Fig. 13.4b, c). The proportion of polyploid cells increased within three cellular divisions, increasing the relative proportion of 4N and 8N cells by 45 and 15 %, respectively in *Ccnd1*$^{-/-}$ cells (Fig. 13.4e). Abnormal karyotype was induced by cyclin D1 by SKY analysis, with 75 % of *Ccnd1*$^{-/-}$ MEF metaphases having a normal karyotype compared with 30 % of the cyclin D1-rescued *Ccnd1*$^{-/-}$ MEFs.

To further classify the chromosomal abnormalities, we employed spectral karyotyping (SKY), a whole-genome painting assay that can recognize complex genomic rearrangements. In addition to the induction of aneuploidy, cyclin D1 also induced a large number of reciprocal and nonreciprocal translocations. The nonreciprocal translocations can be potently transforming if the DNA fragment involved carries oncogenes or tumor suppressors at the breakpoint. In cyclin D1-rescued *Ccnd1*$^{-/-}$ fibroblasts, over 50 % of the cells exhibited multiple centrosomes that give rise to increased multipolar spindles in prometaphase/metaphase. Together these findings demonstrate that cyclin D1 induces chromosomal abnormalities, which may contribute to oncogenic transformation. Cyclin D1 induction of CIN was confirmed in vivo using tetracycline (Tet)-inducible cyclin D1 transgenic mice. In this mammary gland-targeted Tet-inducible model, a short-term (7 days) induction of cyclin D1 expression promoted CIN [12]. The gene expression profile of tumors derived from MMTV-cyclin D1 mice showed enrichment for the CIN signature. Importantly, cyclin D1 mRNA expression correlates with the highest ranking CIN genes in luminal B subtype of human breast cancer in the analysis of >2,200 breast cancer specimens [12].

13.5 Medical Applications

Cyclin D1 couples signals from cell surface receptors to transcription factors and co-integrator proteins, thereby regulating diverse gene expression networks. Cyclin D1 is expressed at high levels in a variety of tumors, including breast, prostate, and colon cancer. The functional requirement for cyclin D1 in cell cycle progression and cellular proliferation makes it an ideal target for molecular therapeutics. Molecules such as flavopiridol, rapamycin, and 17-allylamino-17-demethoxygeldanamycin have been assessed as inhibitors of cyclin D1 expression. Rapamycin and herbimycin inhibit the translation of cyclin D1 mRNA. Great efforts have been made to develop inhibitors of cyclin D1/Cdk4 kinase activity. This has been challenging due to the nonspecificity of molecules that target the kinase core; however, promising compounds are being tested. The kinase-independent function of cyclin D1 warrants alternative approaches to target cyclin D1 itself. Thus, understanding the regulatory networks that enhance cyclin D1 expression will help in defining the strategies to develop molecular inhibitors. There is interest in the clinic in using drugs that inhibit tumor CIN. The high CIN index associated with cyclin D1 overexpression in luminal B breast cancer [12] provides the basis for such a targeted therapy. Understanding the transcriptional role of cyclin D1 in promoting CIN is of considerable clinical importance given the frequent overexpression of cyclin D1 in a wide variety of human cancers.

Acknowledgments This work was supported in part by Grants R01CA70896, R01CA75503, R01CA137494, and R01CA86072 (to R.G.P.). The Kimmel Cancer Center was supported by the NIH Cancer Center Core Grant P30CA56036 (to R.G.P.). This project is funded in part from the Dr. Ralph and Marian C. Falk Medical Research Trust, the Breast Cancer Research Foundation (to R.G.P.), and a grant from the Pennsylvania Department of Health (to R.G.P. and C.W.). The Department specifically disclaims responsibility for analyses, interpretations or conclusions.

References

1. Matsushime H, Ewen ME, Strom DK, Kato JY, Hanks SK, Roussel MF, et al. Identification and properties of an atypical catalytic subunit (p34PSK-J3/cdk4) for mammalian D type G1 cyclins. Cell. 1992;71:323–34.
2. Baldin V, Lukas J, Marcote MJ, Pagano M, Draetta G. Cyclin D1 is a nuclear protein required for cell cycle progression in G1. Genes Dev. 1993;7:812–21.
3. Wang C, Li Z, Lu Y, Du R, Katiyar S, Yang J, et al. Cyclin D1 repression of nuclear respiratory factor 1 integrates nuclear DNA synthesis and mitochondrial function. Proc Natl Acad Sci USA. 2006;103:11567–72.
4. Sakamaki T, Casimiro MC, Ju X, Quong AA, Katiyar S, Liu M, et al. Cyclin D1 determines mitochondrial function in vivo. Mol Cell Biol. 2006;26:5449–69.
5. Reutens AT, Fu M, Wang C, Albanese C, McPhaul MJ, Sun Z, et al. Cyclin D1 binds the androgen receptor and regulates hormone-dependent signaling in a p300/CBP-associated factor (P/CAF)-dependent manner. Mol Endocrinol. 2001;15:797–811.
6. Wang C, Pattabiraman N, Zhou JN, Fu M, Sakamaki T, Albanese C, et al. Cyclin D1 repression of peroxisome proliferator-activated receptor gamma expression and transactivation. Mol Cell Biol. 2003;23:6159–73.
7. Fu M, Wang C, Rao M, Wu X, Bouras T, Zhang X, et al. Cyclin D1 represses p300 transactivation through a cyclin-dependent kinase-independent mechanism. J Biol Chem. 2005;280:29728–42.
8. Neuman E, Ladha MH, Lin N, Upton TM, Miller SJ, DiRenzo J, et al. Cyclin D1 stimulation of estrogen receptor transcriptional activity independent of cdk4. Mol Cell Biol. 1997;17:5338–47.
9. Zwijsen RM, Wientjens E, Klompmaker R, van der Sman J, Bernards R, Michalides RJ. CDK-independent activation of estrogen receptor by cyclin D1. Cell. 1997;88:405–15.
10. Zwijsen RM, Buckle RS, Hijmans EM, Loomans CJ, Bernards R. Ligand-independent recruitment of steroid receptor coactivators to estrogen receptor by cyclin D1. Genes Dev. 1998;12:3488–98.
11. Bienvenu F, Jirawatnotai S, Elias JE, Meyer CA, Mizeracka K, Marson A, et al. Transcriptional role of cyclin D1 in development revealed by a genetic-proteomic screen. Nature. 2010;463:374–8.
12. Casimiro MC, Crosariol M, Loro E, Ertel A, Yu Z, Dampier W, et al. ChIP sequencing of cyclin D1 reveals a transcriptional role in chromosomal instability in mice. J Clin Invest. 2012;122:833–43.
13. Lu F, Gladden AB, Diehl JA. An alternatively spliced cyclin D1 isoform, cyclin D1b, is a nuclear oncogene. Cancer Res. 2003;63:7056–61.
14. Solomon DA, Wang Y, Fox SR, Lambeck TC, Giesting S, Lan Z, et al. Cyclin D1 splice variants. Differential effects on localization, RB phosphorylation, and cellular transformation. J Biol Chem. 2003;278:30339–47.
15. Ewen ME, Sluss HK, Sherr CJ, Matsushime H, Kato J, Livingston DM. Functional interactions of the retinoblastoma protein with mammalian D-type cyclins. Cell. 1993;73:487–97.
16. Kato J, Matsushime H, Hiebert SW, Ewen ME, Sherr CJ. Direct binding of cyclin D to the retinoblastoma gene product (pRb) and pRb phosphorylation by the cyclin D-dependent kinase CDK4. Genes Dev. 1993;7:331–42.

17. Motokura T, Bloom T, Kim HG, Jüppner H, Ruderman JV, Kronenberg HM, et al. A novel cyclin encoded by a *bcl1*-linked candidate oncogene. Nature. 1991;350:512–5.
18. Fu M, Wang C, Li Z, Sakamaki T, Pestell RG. Minireview: cyclin D1: normal and abnormal functions. Endocrinology. 2004;145:5439–47.
19. Poon RY, Toyoshima H, Hunter T. Redistribution of the CDK inhibitor p27 between different cyclin. CDK complexes in the mouse fibroblast cell cycle and in cells arrested with lovastatin or ultraviolet irradiation. Mol Biol Cell. 1995;6:1197–213.
20. Wang C, Fan S, Li Z, Fu M, Rao M, Ma Y, et al. Cyclin D1 antagonizes BRCA1 repression of estrogen receptor alpha activity. Cancer Res. 2005;65:6557–67.
21. Fu M, Rao M, Bouras T, Wang C, Wu K, Zhang X, et al. Cyclin D1 inhibits peroxisome proliferator-activated receptor gamma-mediated adipogenesis through histone deacetylase recruitment. J Biol Chem. 2005;280:16934–41.
22. Pestell RG, Albanese C, Reutens AT, Segall JE, Lee RJ, Arnold A. The cyclins and cyclin-dependent kinase inhibitors in hormonal regulation of proliferation and differentiation. Endocr Rev. 1999;20:501–34.
23. Zelivianski S, Cooley A, Kall R, Jeruss JS. Cyclin-dependent kinase 4-mediated phosphorylation inhibits Smad3 activity in cyclin D-overexpressing breast cancer cells. Mol Cancer Res. 2010;8:1375–87.
24. Zhong Z, Yeow WS, Zou C, Wassell R, Wang C, Pestell RG, et al. Cyclin D1/cyclin-dependent kinase 4 interacts with filamin A and affects the migration and invasion potential of breast cancer cells. Cancer Res. 2010;70:2105–14.
25. Kehn K, Berro R, Alhaj A, Bottazzi ME, Yeh WI, Klase Z, et al. Functional consequences of cyclin D1/BRCA1 interaction in breast cancer cells. Oncogene. 2007;26:5060–9.
26. Anders L, Ke N, Hydbring P, Choi YJ, Widlund HR, Chick JM, et al. A systematic screen for CDK4/6 substrates links FOXM1 phosphorylation to senescence suppression in cancer cells. Cancer Cell. 2011;20:620–34.
27. Hanai J, Dhanabal M, Karumanchi SA, Albanese C, Waterman M, Chan B, et al. Endostatin causes G1 arrest of endothelial cells through inhibition of cyclin D1. J Biol Chem. 2002;277:16464–9.
28. Neumeister P, Pixley FJ, Xiong Y, Xie H, Wu K, Ashton A, et al. Cyclin D1 governs adhesion and motility of macrophages. Mol Biol Cell. 2003;14:2005–15.
29. Inoue I, Shino K, Noji S, Awata T, Katayama S. Expression of peroxisome proliferator-activated receptor alpha (PPAR alpha) in primary cultures of human vascular endothelial cells. Biochem Biophys Res Commun. 1998;246:370–4.
30. Sherr CJ. Cancer cell cycles. Science. 1996;274:1672–7.
31. Betticher DC, Thatcher N, Altermatt HJ, Hoban P, Ryder WD, Heighway J. Alternate splicing produces a novel cyclin D1 transcript. Oncogene. 1995;11:1005–11.
32. Meng H, Tian L, Zhou J, Li Z, Jiao X, Li WW, et al. PACSIN 2 represses cellular migration through direct association with cyclin D1 but not its alternate splice form cyclin D1b. Cell Cycle. 2011;10:73–81.
33. Albanese C, Johnson J, Watanabe G, Eklund N, Vu D, Arnold A, et al. Transforming $p21^{ras}$ mutants and c-Ets-2 activate the cyclin D1 promoter through distinguishable regions. J Biol Chem. 1995;270:23589–97.
34. Brown JR, Nigh E, Lee RJ, Ye H, Thompson MA, Saudou F, et al. Fos family members induce cell cycle entry by activating cyclin D1. Mol Cell Biol. 1998;18:5609–19.
35. Westwick JK, Lee RJ, Lambert QT, Symons M, Pestell RG, Der CJ, et al. Transforming potential of Dbl family proteins correlates with transcription from the cyclin D1 promoter but not with activation of Jun NH_2-terminal kinase, p38/Mpk2, serum response factor, or c-Jun. J Biol Chem. 1998;273:16739–47.
36. Albanese C, D'Amico M, Reutens AT, Fu M, Watanabe G, Lee RJ, et al. Activation of the cyclin D1 gene by the E1A-associated protein p300 through AP-1 inhibits cellular apoptosis. J Biol Chem. 1999;274:34186–95.
37. Guttridge DC, Albanese C, Reuther JY, Pestell RG, Baldwin Jr AS. NF-kappaB controls cell growth and differentiation through transcriptional regulation of cyclin D1. Mol Cell Biol. 1999;19:5785–99.

38. Lee RJ, Albanese C, Stenger RJ, Watanabe G, Inghirami G, Haines 3rd GK, et al. pp 60(v-src) induction of cyclin D1 requires collaborative interactions between the extracellular signal-regulated kinase, p38, and Jun kinase pathways. A role for cAMP response element-binding protein and activating transcription factor-2 in pp60(v-src) signaling in breast cancer cells. J Biol Chem. 1999;274:7341–50.
39. Matsumura I, Kitamura T, Wakao H, Tanaka H, Hashimoto K, Albanese C, et al. Transcriptional regulation of the cyclin D1 promoter by STAT5: its involvement in cytokine-dependent growth of hematopoietic cells. EMBO J. 1999;18:1367–77.
40. Shtutman M, Zhurinsky J, Simcha I, Albanese C, D'Amico M, Pestell R, et al. The cyclin D1 gene is a target of the beta-catenin/LEF-1 pathway. Proc Natl Acad Sci USA. 1999;96:5522–7.
41. Wang C, Fu M, D'Amico M, Albanese C, Zhou JN, Brownlee M, et al. Inhibition of cellular proliferation through IkappaB kinase-independent and peroxisome proliferator-activated receptor gamma-dependent repression of cyclin D1. Mol Cell Biol. 2001;21:3057–70.
42. Zhao J, Pestell R, Guan JL. Transcriptional activation of cyclin D1 promoter by FAK contributes to cell cycle progression. Mol Biol Cell. 2001;12:4066–77.
43. Wang C, Li Z, Fu M, Bouras T, Pestell RG. Signal transduction mediated by cyclin D1: from mitogens to cell proliferation: a molecular target with therapeutic potential. Cancer Treat Res. 2004;119:217–37.
44. Rao M, Casimiro MC, Lisanti MP, D'Amico M, Wang C, Shirley LA, et al. Inhibition of cyclin D1 gene transcription by Brg-1. Cell Cycle. 2008;7:647–55.
45. Yu Z, Wang C, Wang M, Li Z, Casimiro MC, Liu M, et al. A cyclin D1/microRNA 17/20 regulatory feedback loop in control of breast cancer cell proliferation. J Cell Biol. 2008;182:509–17.
46. Matthias C, Branigan K, Jahnke V, Leder K, Haas J, Heighway J, et al. Polymorphism within the cyclin D1 gene is associated with prognosis in patients with squamous cell carcinoma of the head and neck. Clin Cancer Res. 1998;4:2411–8.
47. Dhar KK, Branigan K, Howells RE, Musgrove C, Jones PW, Strange RC, et al. Prognostic significance of cyclin D1 gene (CCND1) polymorphism in epithelial ovarian cancer. Int J Gynecol Cancer. 1999;9:342–7.
48. Kong S, Wei Q, Amos CI, Lynch PM, Levin B, Zong J, et al. Cyclin D1 polymorphism and increased risk of colorectal cancer at young age. J Natl Cancer Inst. 2001;93:1106–8.
49. Wang L, Habuchi T, Mitsumori K, Li Z, Kamoto T, Kinoshita H, et al. Increased risk of prostate cancer associated with AA genotype of cyclin D1 gene A870G polymorphism. Int J Cancer. 2003;103:116–20.
50. Wang L, Habuchi T, Takahashi T, Mitsumori K, Kamoto T, Kakehi Y, et al. Cyclin D1 gene polymorphism is associated with an increased risk of urinary bladder cancer. Carcinogenesis. 2002;23:257–64.
51. Zhang J, Li Y, Wang R, Wen D, Sarbia M, Kuang G, et al. Association of cyclin D1 (G870A) polymorphism with susceptibility to esophageal and gastric cardiac carcinoma in a northern Chinese population. Int J Cancer. 2003;105:281–4.
52. Qiuling S, Yuxin Z, Suhua Z, Cheng X, Shuguang L, Fengsheng H. Cyclin D1 gene polymorphism and susceptibility to lung cancer in a Chinese population. Carcinogenesis. 2003;24:1499–503.
53. Myklebust MP, Li Z, Tran TH, Rui H, Knudsen ES, Elsaleh H, et al. Expression of cyclin D1a and D1b as predictive factors for treatment response in colorectal cancer. Br J Cancer. 2012;107:1684–91.
54. Solomon DA, Wang Y, Fox SR, Lambeck TC, Giesting S, Lan Z, et al. Cyclin D1 splice variants. Differential effects on localization, RB phosphorylation, and cellular transformation. J Biol Chem. 2003;278:30339–47.
55. Diehl JA, Cheng M, Roussel MF, Sherr CJ. Glycogen synthase kinase-3beta regulates cyclin D1 proteolysis and subcellular localization. Genes Dev. 1998;12:3499–511.
56. Yang K, Guo Y, Stacey WC, Harwalkar J, Fretthold J, Hitomi M, et al. Glycogen synthase kinase 3 has a limited role in cell cycle regulation of cyclin D1 levels. BMC Cell Biol. 2006;7:33.

57. Hunter T, Pines J. Cyclins and cancer II: cyclin D and CDK inhibitors come of age. Cell. 1994;79:573–82.
58. Weinberg RA. The retinoblastoma protein and cell cycle control. Cell. 1995;81:323–30.
59. Zwijsen RML, Klompmaker R, Wientjens EBHGM, Kristel PMP, van der Burg B, Michalides RJAM. Cyclin D1 triggers autonomous growth of breast cancer cells by governing cell cycle exit. Mol Cell Biol. 1996;16:2554–60.
60. Hinds PW, Dowdy SF, Eaton EN, Arnold A, Weinberg RA. Function of a human cyclin gene as an oncogene. Proc Natl Acad Sci USA. 1994;91:709–13.
61. Lee RJ, Albanese C, Fu M, D'Amico M, Lin B, Watanabe G, et al. Cyclin D1 is required for transformation by activated Neu and is induced through an E2F-dependent signaling pathway. Mol Cell Biol. 2000;20:672–83.
62. Yu Q, Geng Y, Sicinski P. Specific protection against breast cancers by cyclin D1 ablation. Nature. 2001;411:1017–21.
63. Wang TC, Cardiff RD, Zukerberg L, Lees E, Arnold A, Schmidt EV. Mammary hyperplasia and carcinoma in MMTV-cyclin D1 transgenic mice. Nature. 1994;369:669–71.
64. Robles AI, Rodriguez-Puebla ML, Glick AB, Trempus C, Hansen L, Sicinski P, et al. Reduced skin tumor development in cyclin D1-deficient mice highlights the oncogenic *ras* pathway in vivo. Genes Dev. 1998;12:2469–74.
65. Hulit J, Wang C, Li Z, Albanese C, Rao M, Di Vizio D, et al. Cyclin D1 genetic heterozygosity regulates colonic epithelial cell differentiation and tumor number in ApcMin mice. Mol Cell Biol. 2004;24:7598–611.
66. Landis MW, Pawlyk BS, Li T, Sicinski P, Hinds PW. Cyclin D1-dependent kinase activity in murine development and mammary tumorigenesis. Cancer Cell. 2006;9:13–22.
67. Leng X, Connell-Crowley L, Goodrich D, Harper JW. S-phase entry upon ectopic expression of G1 cyclin-dependent kinases in the absence of retinoblastoma protein phosphorylation. Curr Biol. 1997;7:709–12.
68. Baker GL, Landis MW, Hinds PW. Multiple functions of D-type cyclins can antagonize pRb-mediated suppression of proliferation. Cell Cycle. 2005;4:330–8.
69. Yu Q, Sicinska E, Geng Y, Ahnstrom M, Zagozdzon A, Kong Y, et al. Requirement for CDK4 kinase function in breast cancer. Cancer Cell. 2006;9:23–32.
70. Casimiro MC, Wang C, Li Z, DiSanti G, Willmart NE, Addya S, et al. Cyclin d1 determines estrogen signaling in the mammary gland in vivo. Mol Endocrinol. 2013;27(9):1415–1428.
71. McMahon C, Suthiphongchai T, DiRenzo J, Ewen ME. P/CAF associates with cyclin D1 and potentiates its activation of the estrogen receptor. Proc Natl Acad Sci USA. 1999;96:5382–7.
72. Burbelo P, Wellstein A, Pestell RG. Altered Rho GTPase signaling pathways in breast cancer cells. Breast Cancer Res Treat. 2004;84:43–8.
73. Li Z, Wang C, Jiao X, Lu Y, Fu M, Quong AA, et al. Cyclin D1 regulates cellular migration through the inhibition of thrombospondin 1 and ROCK signaling. Mol Cell Biol. 2006;26:4240–56.
74. Li Z, Wang C, Jiao X, Katiyar S, Casimiro MC, Prendergast GC, et al. Alternate cyclin D1 mRNA splicing modulates p27KIP1 binding and cell migration. J Biol Chem. 2008;283:7007–15.
75. Freeman RS, Estus S, Johnson Jr EM. Analysis of cell cycle-related gene expression in post-mitotic neurons: selective induction of Cyclin D1 during programmed cell death. Neuron. 1994;12:343–55.
76. Agami R, Bernards R. Distinct initiation and maintenance mechanisms cooperate to induce G1 cell cycle arrest in response to DNA damage. Cell. 2000;102:55–66.
77. Lin HM, Lee YJ, Li G, Pestell RG, Kim HR. Bcl-2 induces cyclin D1 promoter activity in human breast epithelial cells independent of cell anchorage. Cell Death Differ. 2001;8:44–50.
78. Brown NE, Jeselsohn R, Bihani T, Hu MG, Foltopoulou P, Kuperwasser C, et al. Cyclin D1 activity regulates autophagy and senescence in the mammary epithelium. Cancer Res. 2012;72:6477–89.
79. Pagano M, Theodoras AM, Tam SW, Draetta GF. Cyclin D1-mediated inhibition of repair and replicative DNA synthesis in human fibroblasts. Genes Dev. 1994;8:1627–39.

80. Li Z, Jiao X, Wang C, Shirley LA, Elsaleh H, Dahl O, et al. Alternative cyclin D1 splice forms differentially regulate the DNA damage response. Cancer Res. 2010;70:8802–11.
81. Pedram A, Razandi M, Evinger AJ, Lee E, Levin ER. Estrogen inhibits ATR signaling to cell cycle checkpoints and DNA repair. Mol Biol Cell. 2009;20:3374–89.
82. Jirawatnotai S, Hu Y, Michowski W, Elias JE, Becks L, Bienvenu F, et al. A function for cyclin D1 in DNA repair uncovered by protein interactome analyses in human cancers. Nature. 2011;474:230–4.
83. Stacey DW. Cyclin D1 serves as a cell cycle regulatory switch in actively proliferating cells. Curr Opin Cell Biol. 2003;15:158–63.

Chapter 14
Gene Signatures of Inflammatory Breast Cancer: Epithelial Plasticity and a Cancer Stem Cell Phenotype

Fredika M. Robertson, Khoi Chu, Sandra V. Fernandez, Zaiming Ye, Sanford H. Barsky, and Massimo Cristofanilli

Abstract Inflammatory breast cancer (IBC) is the most lethal variant of locally advanced breast cancer and carries with it a very low survival rate of 40 % at 5 years. IBC does not present as a lump but rather mimics characteristics of an inflammation that first appears as swelling of the breast, with edema, redness, and common lymph node involvement. The physical changes in the breast are associated with the presence of nests of aggregated tumor cells, defined as tumor emboli that are encircled by lymphatic vessels, effectively blocking lymphatic drainage. Little is understood about IBC, in part due to the lack of preclinical models that recapitulate its distinct characteristics. This chapter provides an overview of our studies that have profiled all available preclinical models of IBC, including two new models recently developed, to elucidate the molecular underpinnings of this lethal variant of breast cancer. Our studies demonstrate that IBC is enriched for cells that express

F.M. Robertson (✉) • K. Chu • Z. Ye
Department of Experimental Therapeutics, The University of Texas MD Anderson Cancer Center, Unit 1950, 1901 East Road, South Campus Research Building 4; Office 3.1009, P.O. Box 301429, Houston, TX 77230-1429, USA
e-mail: frobertson@mdanderson.org

S.V. Fernandez • M. Cristofanilli
Department of Medical Oncology, Thomas Jefferson University, Philadelphia, PA 19107, USA

S.H. Barsky
Department of Pathology, The University of Nevada School of Medicine, Reno, NV 89557, USA

H. Schatten (ed.), *Cell and Molecular Biology of Breast Cancer*,
DOI 10.1007/978-1-62703-634-4_14,

$CD44^{+}$ and $CD133^{+}$ and have aldehyde dehydrogenase-1 (ALDH1) activity, supporting a cancer stem cell/tumor initiating phenotype, associated with a very high metastatic potential to multiple distant organ sites. IBC has a distinct gene signature including E-cadherin expression with associated loss of expression of *ZEB*1, a transcriptional repressor of E-cadherin. IBC is also characterized by loss of expression of genes within the transforming growth factor-beta (TGFβ) signaling pathway, which is permissive for cohesive invasion by IBC tumor emboli. Taken together, these studies suggest that IBC is a very distinct variant of breast cancer characterized by epithelial plasticity, enrichment of a stem cell phenotype, and cohesive invasion as an adaptive survival mechanism, consistent with the definition of IBC as the most metastatic variant of breast cancer.

14.1 Introduction

Inflammatory breast cancer (IBC) is the most lethal form of locally advanced breast cancer (reviewed in [1]). Although primary IBC is less commonly diagnosed than other types of breast cancer, accounting for an estimated 2–5 % of all breast cancers in the United States and an estimated 13 % worldwide, IBC is responsible for a disproportionate number of breast cancer-related deaths (7 %) that occur each year worldwide [2, 3].

A clinical diagnosis of IBC is based on the combination of the physical appearance of the affected breast, a careful medical history, physical examination, and pathological findings from a skin biopsy and/or needle or core biopsy to confirm the diagnosis of carcinoma [4]. IBC does not occur as a lump but rather is characterized by a very rapid onset of changes in the skin overlying the involved breast that include edema, redness, and swelling in over one half to two thirds of the breast and that may also include a wrinkled, orange peel appearance of the skin, defined as "peau d'orange" [1, 5, 6]. The changes in the skin of the involved breast of IBC patients are the first clinical signs of IBC and are due to the presence of tumor emboli lodged within dermal lymphatic vessels, which is one of the classical histopathological findings in IBC [7–9]. While their presence is not a requirement for a diagnosis of IBC, approximately 75 % of IBC patients have tumor emboli, and they serve as one of most distinctive characteristics of IBC.

Research to elucidate the molecular mechanisms that underlie the rapid metastasis exhibited by IBC has been hampered by the relatively rarity of the disease and an associated lack of preclinical models that recapitulate the human disease. This chapter describes the characteristics of all of the available preclinical models of IBC, including two new models that we have developed using cells derived from IBC patients that we have designated as FC-IBC01 and FC-IBC02. We describe the use of these preclinical IBC models to validate the cancer stem cell phenotype that is a characteristic of IBC. This chapter also describes gene profiling studies that have elucidated genomic signatures of IBC which provide insight into the molecular basis for the aggressive metastasis that characterize this lethal variant of breast cancer.

14.2 Characteristics of Preclinical Models of IBC

As with non-IBC breast cancers, IBC tumors can be categorized into intrinsic subtypes including luminal A, luminal B, ErbB2/Her-2^{+}, and triple-negative breast cancers which lack ER/PR and ERBB2/Her-2. Although IBC tumors can be of any subtype, the predominant molecular subtypes are either triple-negative or ERBB2^{+} tumors [10–13], which is also reflected in the preclinical models of IBC.

There are currently seven preclinical models of IBC, which include SUM149, SUM190, KPL-4, MDA-IBC-3, Mary-X, and two newly developed models, FC-IBC01 and FC-IBC02. Of these preclinical models, four of these are classified as triple negative based on their lack of expression of ER, PR, and the ErBB2/Her-2 oncogene. The triple-negative IBC models include the most well-characterized IBC cell line, SUM149, as well as Mary-X, FC-IBC01, and FC-IBC02. The remaining IBC models, SUM190, MDA-IBC-3, and KPL-4 cell lines, all express the Her-2 oncogene. The prevalence of triple negative and Her-2^{+} in preclinical IBC models mirrors the prevalence of these subtypes in the tumors of IBC patients. The heat map of the levels of expression of genes used to subtype breast cancers, including estrogen receptor (ER), progesterone receptor (PR), and the Her-2/ERBB2 oncogene, in preclinical models of IBC is shown in Fig. 14.1.

Using an expanded database of IBC patient tumors, recent studies reported that IBC patients with luminal A subtype (ER^{+}/PR^{+}/Her-2^{-}) have shorter, distant-metastases-free survival intervals compared with their non-luminal IBC counterparts [10]. This is in direct contrast to observations in non-IBC breast cancers, in which patients with luminal A tumors have improved survival compared to all other subtypes of breast cancer [14]. The preclinical models of IBC are either triple

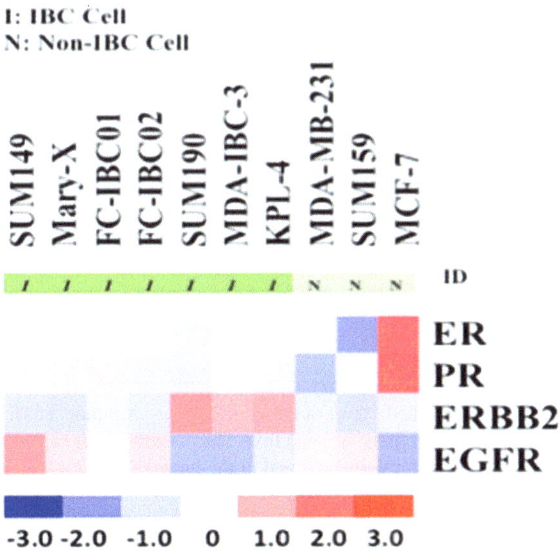

Fig. 14.1 Heat map of IBC breast cancer cell lines showing the expression of ER, PR, Her-2, and EGFR (This figure was originally published in J Clin Exp Pathol. 2012;2:119. doi: 10.4172/2161-0681.1000119)

negative or Her-2 amplified, which is reflective of the most prevalent subtypes of this variant of breast cancer. The observation that the worst prognosis is associated with those with a luminal A subtype of IBC underscores the need for further development of preclinical models of IBC, which will be critical to define the molecular basis for the observation of this differential lower overall survival of patients with luminal A subtype IBC.

14.2.1 Characteristics of Triple-Negative Preclinical Models of IBC

The Mary-X IBC model is a stable transplantable xenograft model developed by Dr. Sanford H. Barsky from the primary tumor of an IBC patient with triple-negative basal-like breast cancer [15–17]. Prior to 2011, Mary-X was the only preclinical model of IBC that recapitulated the formation of tumor emboli and encircling lymphatic vessels that commonly occurs in IBC patients. Mary-X grows as xenografts that appear red and highly vascularized (Fig. 14.2a). Examination of tumor tissues isolated from mice bearing Mary-X reveals the presence of multiple tumor emboli within the dermis (Fig. 14.2b, inset, and c), and metastasis readily occurs, primarily in lung (Fig. 14.2d). Mary-X tumors and tumor emboli that invade into the dermal layer of the skin (Fig. 14.2e, f) and metastatic lesions in lung (Fig. 14.2g–i) have very robust expression of the surface glycoprotein, E-cadherin. Mary-X tumor cells can only be optimally propagated as 3-dimensional tumor spheroids (Fig. 14.2j, k) and have not been successfully maintained as 2-dimensional adherent cultures on plastic. Mary-X tumor spheroids retain expression of E-cadherin (Fig. 14.2l). Our gene profiling studies identified only a ten gene difference between Mary-X tumor emboli isolated by laser capture microdissection, and Mary-X tumor spheroids [18], suggesting that the Mary-X tumor spheroids provide a convenient in vitro model with which to study characteristics of tumor emboli as well as a method to determine the ability of therapeutic agents to target tumor spheroids as surrogates of tumor emboli.

In addition to the Mary-X preclinical model of IBC, we have developed two new models of IBC, designated as FC-IBC01 and FC-IBC02, derived from tumor cells isolated following thoracentesis of IBC patients who had developed metastatic pleural effusions [19, 20]. As was observed with Mary-X cells, FC-IBC01 and FC-IBC02 cells spontaneously form 3-dimensional tumor spheroids in vitro and are optimally propagated for short periods of time in low-adherence culture. When FC-IBC01 and

Fig. 14.2 (continued) layer of the skin. (**d**) Mary-X tumor cells form lung metastasis visible as tumor emboli. (**e**) Tumor tissue isolated from Mary-X has robust expression of E-cadherin protein present in tumor emboli in the dermis (**f**) and in the lungs (**g–i**). (**j**) and (**k**) Mary-X tumor cells isolated from tumor xenografts spontaneously form 3-dimensional tumor spheroids when placed into low-adherence culture conditions. (**l**) Tumor spheroids in culture retain robust expression of E-cadherin (This figure was originally published in J Clin Exp Pathol. 2012;2:119. doi:10.4172/2161-0681.1000119)

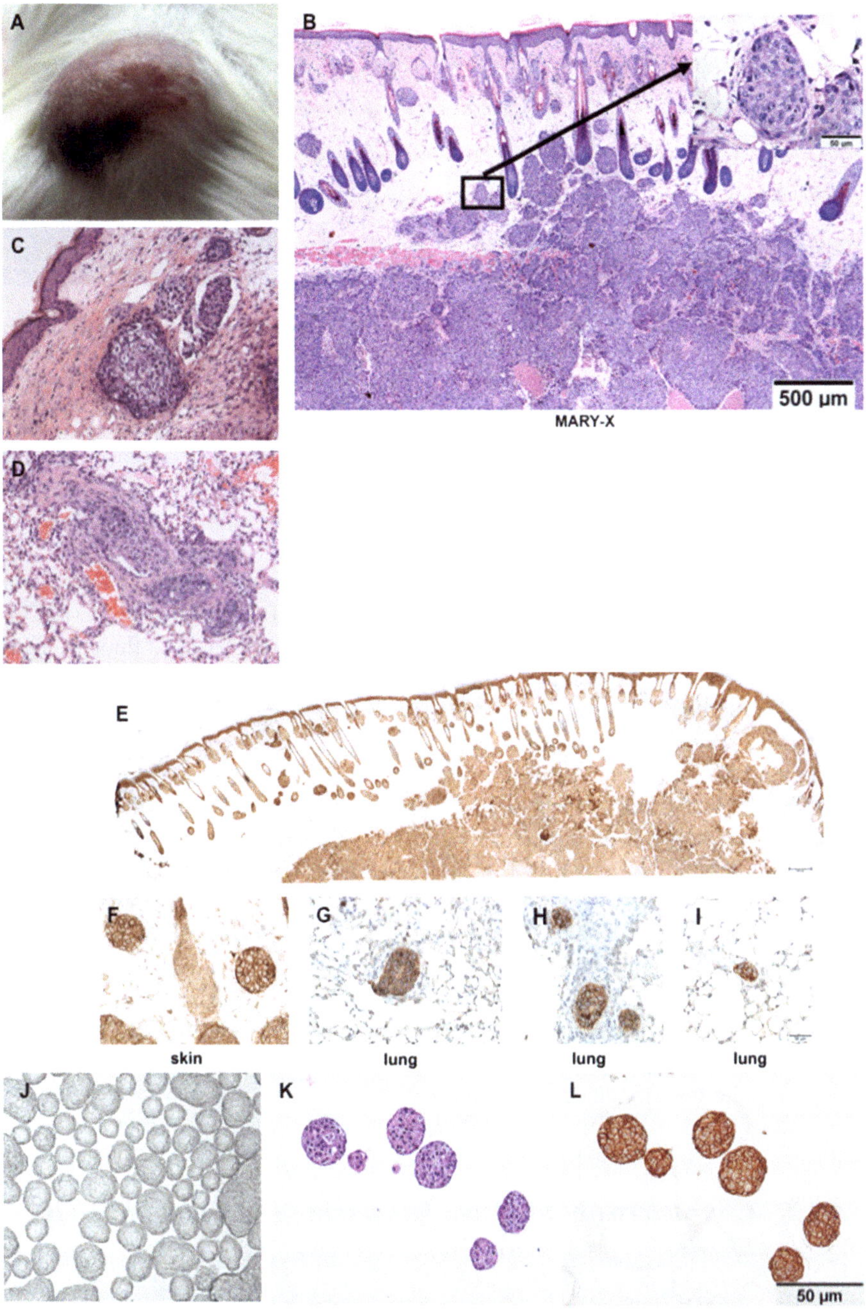

Fig. 14.2 Histological characteristics of Mary-X preclinical model of IBC. (**a**) Mary-X tumor xenograft grows as large tumor with visible vascularization. (**b**) Light micrograph of H&E-stained tumor tissue isolated from Mary-X. *Insets* show the presence of tumor emboli in the dermis. (**c**) Higher magnification of Mary-X tumor tissue which contains numerous tumor emboli within the dermal

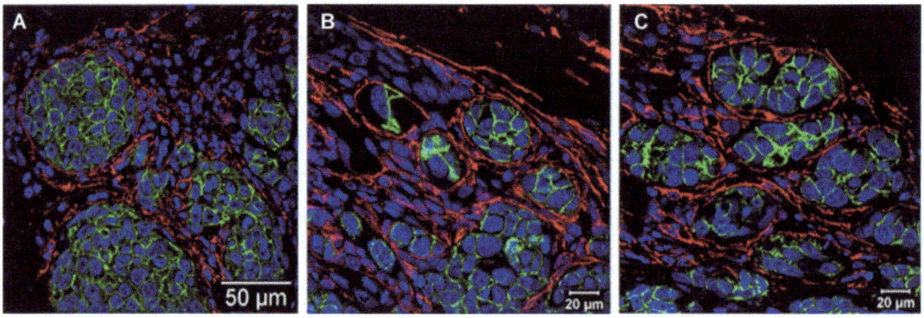

Fig. 14.3 (**a**)–(**c**) Triple-color immunofluorescence and fluorescence microscopy identified tumor emboli in tissues isolated from xenografts of Mary-X (**a**), FC-IBC01 (**b**), and FC-IBC02 (**c**) that express E-cadherin protein (*green fluorescence*) that are encircled by lymphatic vessels that selectively express podoplanin (*red fluorescence*). Topro-3 was used to identify nuclear DNA, shown as *blue fluorescence* (This figure was originally published in J Clin Exp Pathol. 2012;2:119. doi:10.4172/2161-0681.1000119)

FC-IBC02 tumor cells are injected into immunocompromised mice, tumors develop within 2–3 weeks, with the formation of IBC tumor emboli that invade into the skin and rapid metastasis to multiple sites [19, 20]. Mary-X, FC-IBC01, and FC-IBC02 tumor cells form tumor emboli that express E-cadherin protein (green fluorescence) that are encircled by lymphatic vessels that express podoplanin, used as a selective marker of lymphatic endothelium (red fluorescence) (Fig. 14.3a–c, respectively). Topro-3 was used as a marker of nuclear DNA (blue fluorescence).

The SUM149 cell line was developed in the laboratory of Dr. Stephe Ethier and colleagues [21, 22], and has been used for the vast majority of IBC studies. SUM149 cells were developed from a patient with invasive ductal carcinoma prior to receiving chemotherapy. As shown in Fig. 14.1, SUM149 cells are classified as triple-negative subtype based on the lack of ER, PR, and the Her-2 oncogene. Note that all of the triple-negative IBC cell lines, including SUM149, express EGFR. SUM149 cells do not form tumor emboli when grown as xenografts in immunocompromised mice in vivo; however, these cells do rapidly form primary tumors as well as commonly form metastatic lesions at multiple sites, including bone (Fig. 14.4a, b), lung (Fig. 14.4c), lymph nodes, liver, and soft tissues when injected via the intracardiac injection route into immunocompromised mice.

14.2.2 *Characteristics of Her-2 Expressing Preclinical Models of IBC*

The SUM190 IBC cell line was also developed from a patient with invasive ductal carcinoma prior to receiving chemotherapy [21], has amplified Her-2 [21, 22], and proliferates at a much slower rate compared with SUM149 cells, with a doubling time of ~42 h. However, when SUM190 cells are cultured under low-adherence

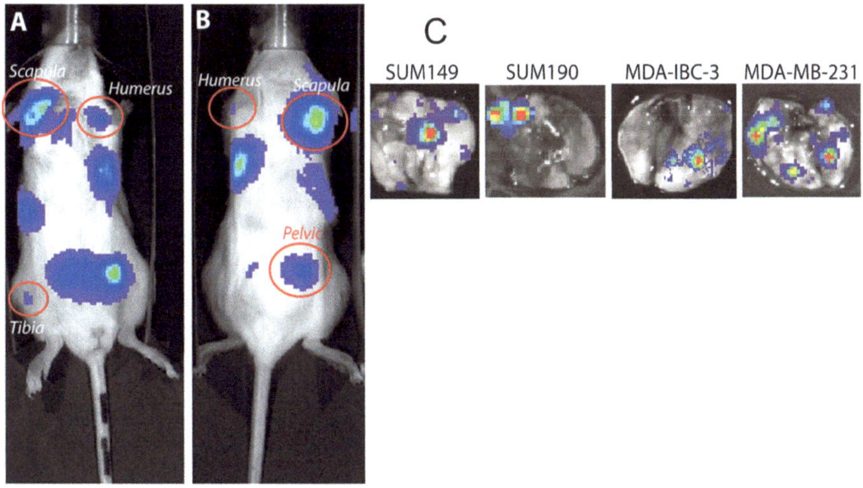

Fig. 14.4 Identification of sites of metastasis in SUM149, SUM190, and MDA-IBC-3 preclinical xenograft models of IBC. (**a**) and (**b**) Luciferase-based images of the presence of metastatic lesions detected at 21–60 days following injection of SUM149 cells (5×10^6) tagged with a luciferase bioluminescent tag into mammary fat pads of NOD. Cg-Prkdc^{scid} Il2rg^{tm1Wjl}/SzJ mice, identifying multiple sites of metastasis of SUM149 to scapula, humerus, tibia, and pelvis. (**c**) Images of the presence of lung metastasis following injection of SUM149, SUM190, MDA-IBC-3, and MDA-MB-231 cells that were tagged with a luciferase bioluminescent tag, detected by bioluminescent signal in cells within isolated lungs (This figure was originally published in J Clin Exp Pathol. 2012;2:119. doi: 10.4172/2161-0681.1000119)

conditions which supports formation of tumor spheroids and then injected into immunocompromised mice, SUM190 produces primary tumors as well as metastatic lesions, primarily to the lung but this occurs less frequently (Fig. 14.4c).

KPL-4 is an additional IBC cell line derived from tumor cells isolated from pleural effusion of a patient with inflammatory skin metastasis [22]. KPL-4 cells have 15-fold amplification of Her-2. When grown as xenografts, KPL-4 cells induced cachexia, which is associated with production of interleukin-6 (IL-6) [23–25], leading to the use of this model to examine the effects of therapeutic agents that block IL-6 production [23–25]. More recently, the KPL-4 model has been used to demonstrate the benefit of combining trastuzumab (Herceptin®; Genentech, South San Francisco, CA) with fluoropyrimidines or a taxane in circumstances where trastuzumab resistance has developed [26, 27]. Due to the very high passage number of KPL-4, this cell line is less commonly used in studies to identify the molecular basis of IBC.

MDA-IBC-3 cells were developed from an IBC patient with pleural effusion by serial transplantation [28]. MDA-IBC-3 cells express Her-2, and although they have a very slow doubling time (~76 h) when cultured as adherent cultures on plastic substrates, they readily form tumor spheroids when placed in low-adherence conditions and grow as xenografts when injected into immunocompromised mice. In our hands, MDA-IBC-3 cells form primary tumors but very rarely form metastatic lung tumors (Fig. 14.4c); however, they do not form tumor emboli when grown as xenografts.

14.3 IBC Is Enriched for Cells with a Cancer Stem Cell Phenotype

Studies using the Mary-X preclinical model of IBC made the initial seminal observation that 3-dimensional tumor spheroids derived from Mary-X tumor tissue have characteristics in common with embryonal blastocysts including expression of transcription factors OCT-4, NANOG, and SOX-2, which are all associated with stem cell self-renewal and embryonic pluripotency, as well as addiction to NOTCH3 [29]. Mary-X spheroids were also shown to express markers of cancer stem cells characterized by the presence of $CD44^{+}/CD24^{-/low}$ and the presence of cells positive for aldehyde dehydrogenase-1 activity (ALDH1) and that express CD133. Additional evidence offered in support of IBC being enriched for cells with a cancer stem cell phenotype came from the demonstration that Mary-X tumor cells expressed genes within stem cell signaling pathways such as NOTCH3 and as few as 100 Mary-X tumor cells could recapitulate the IBC phenotype of tumor emboli formation within the skin. This study also reported that the stem cell phenotype exhibited by Mary-X was also exhibited by the lymphovascular emboli of human IBC cases regardless of the molecular subtype of the tumor, which was the first indication that IBC may be a tumor type enriched for cells with a stem cell phenotype [29]. Following these studies in the Mary-X preclinical model of IBC, Charafe-Jauffret et al. demonstrated that IBC is enriched for cells with a cancer stem cell phenotype based on detection of ALDH1 positive cells [30], which supports the clinical observation that IBC is a disease characterized by resistance to chemotherapy, early disease recurrence, metastasis, and poor clinical outcomes [31]. Our studies were the first to describe the bipotency of IBC tumor cells and also illustrated the self-renewal potential of these IBC tumor cells [32]. The results of our analysis of preclinical models of IBC revealed the presence of cancer stem cell marker CD44, which is expressed primarily by the triple-negative IBC models (Fig. 14.5a, b) and expression of CD133 by Mary-X and FC-IBC02 models of IBC (Fig. 14.5c) [32, 33]. These studies demonstrate that the triple-negative models of IBC are highly enriched for cells with a cancer stem cell phenotype compared to the $Her\text{-}2^{+}$ models of IBC. We are currently using these preclinical models of IBC to define the effectiveness of agents that may target cancer stem cells, with the goal of moving the best candidates into clinical trials.

The expression of cancer stem cell markers coincides with the metastatic capabilities of the preclinical models of IBC. Each of the seven preclinical IBC cell systems grows as orthotopic tumors when implanted into the mammary fat pad and grows as xenografts when injected into the hind flank, with robust tumor emboli formation in the dermal lymphatics by Mary-X, FC-IBC01, and FC-IBC02 (Fig. 14.3) [33], which we did not observe in the other preclinical IBC cell systems. Additionally, we have found that all of the preclinical models of IBC form metastatic lesions at multiple sites (Fig. 14.4a–c).

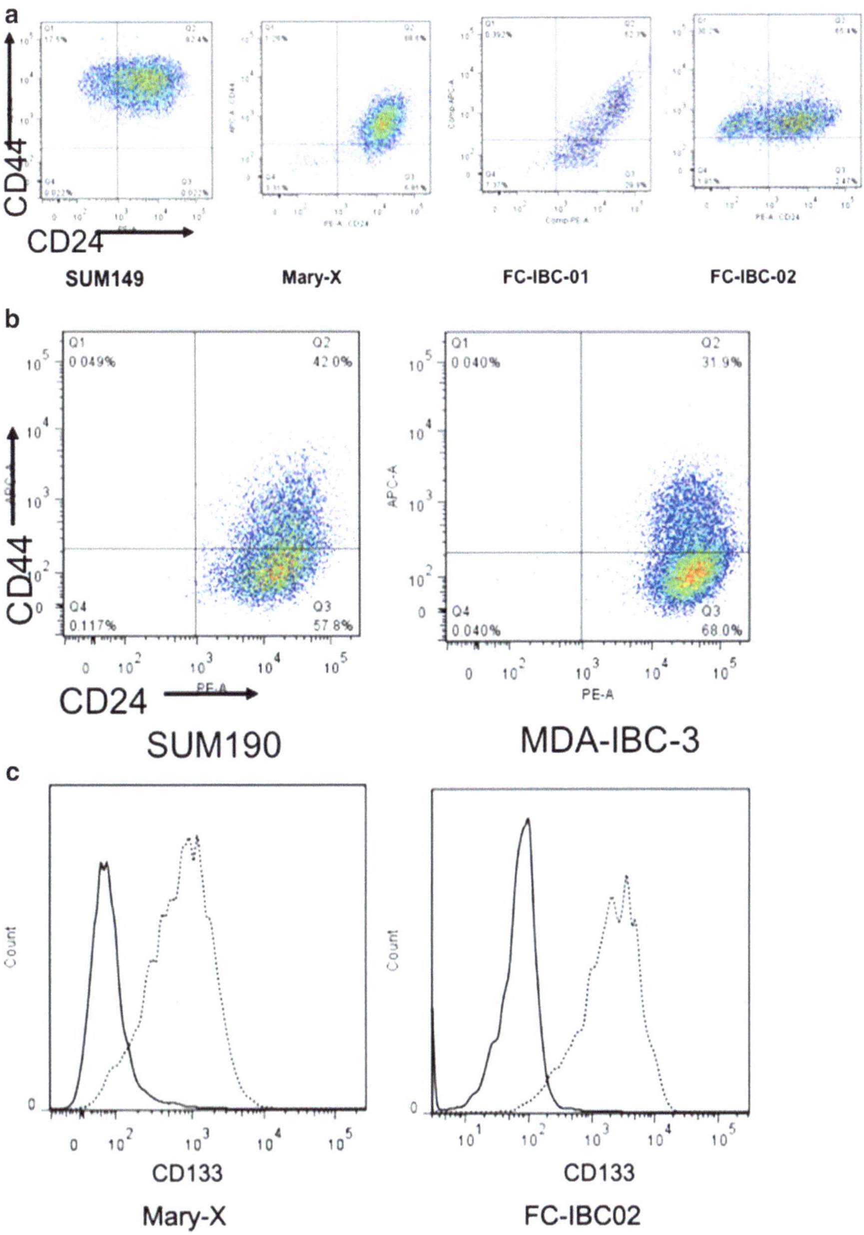

Fig. 14.5 Flow cytometric detection of cancer stem cell surface markers. (**a**, **b**) Flow cytometric detection of CD44/CD24 expressed by (**a**) SUM149, Mary-X, FC-IBC01, and FC-IBC02 cells and (**b**) SUM190 and MDA-IBC-3 cells. (**c**) Flow cytometric detection of CD133 expression by Mary-X and FC-IBC02 cells (This figure was originally published in J Clin Exp Pathol. 2012;2:119. doi: 10.4172/2161-0681.1000119)

14.4 Gene Signatures of IBC

As we have described above, the primary pathological hallmark of IBC is the presence of tumor emboli, which are multicellular aggregates of cells that invade into the dermal lymphatics and retain an epithelial phenotype, characterized by expression of E-cadherin (Figs. 14.2 and 14.3) [7–9, 15, 18, 33, 34]. With the goal of identifying other genes that may be associated with E-cadherin, we evaluated expression of genes involved in cell–cell aggregation and the adherens junctions. Based on whole unbiased transcriptional analysis of preclinical models of IBC, we validated the expression of *CDH1*, which encodes for E-cadherin with the exception of KPL-4 cells. We found that E-cadherin expression coincided with the expression by IBC tumor models of a cassette of genes including gamma (γ) catenin/*JUP*, p120/delta (δ) catenin, and *DSC2* (Fig. 14.6a). Collectively, these genes regulate the tight homotypic aggregation of tumor cells, forming adherens junctions [35, 36], such as occurs in the cell aggregates that comprise IBC tumor emboli [33, 34]. Figure 14.6b shows the Western blot results of studies evaluating the presence of E-cadherin, DSC2, γ-catenin/JUP, and p120/δ catenin proteins in all of the preclinical models of IBC.

The observation that E-cadherin is retained in IBC cells is, at first glance, paradoxical to the current hypothesis that invasion and metastasis occurs with the loss of E-cadherin, as part of the process of the epithelial-mesenchymal transition (EMT) [37, 38]. The loss of E-cadherin during EMT favors a mesenchymal motile phenotype that is associated with acquisition of characteristics of cancer stem cells regulated by transcription factors such as *TWIST1* [39]. In contrast, previous studies in models of IBC using dominant negative molecular approaches or neutralizing antibodies demonstrated that inhibiting the function of E-cadherin effectively blocked the invasion and survival of IBC tumor cells in vitro [15] and diminished the integrity of tumor emboli in vivo [15, 40]. Although E-cadherin has been demonstrated to be necessary for survival of IBC tumor cells, little is known about the transcriptional program that supports the retention of E-cadherin while exhibiting a program of accelerated metastasis.

We performed gene profiling to evaluate the genes associated with the process of EMT that were expressed by preclinical models of IBC (Fig. 14.7a). We found that the expression of CDH1, which encodes for E-cadherin, was accompanied by a loss of the zinc finger E-box binding homeobox gene, *ZEB1*, a transcriptional repressor of *CDH1* (Fig. 14.6a), which we validated by Western blot analysis (Fig. 14.7b). In contrast, the preclinical models of IBC expressed transcription factors *SNAI1* and *SNAI2* and *TWIST1* (Fig. 14.7a) that are known to be involved in maintenance of a stem cell phenotype [33]. To validate the loss of ZEB1 in IBC, we used laser capture microdissection techniques to specifically isolate IBC patient tumor emboli from six samples of IBC core biopsies and four samples taken from skin punch biopsies and also performed laser capture microdissection of tumor emboli taken from the Mary-X preclinical model of IBC. These studies demonstrated the loss of *ZEB1* in tumor emboli and in primary tumor biopsy tissues, which was mirrored by the loss of *ZEB1* in IBC cell lines (Fig. 14.6c, Table 14.1).

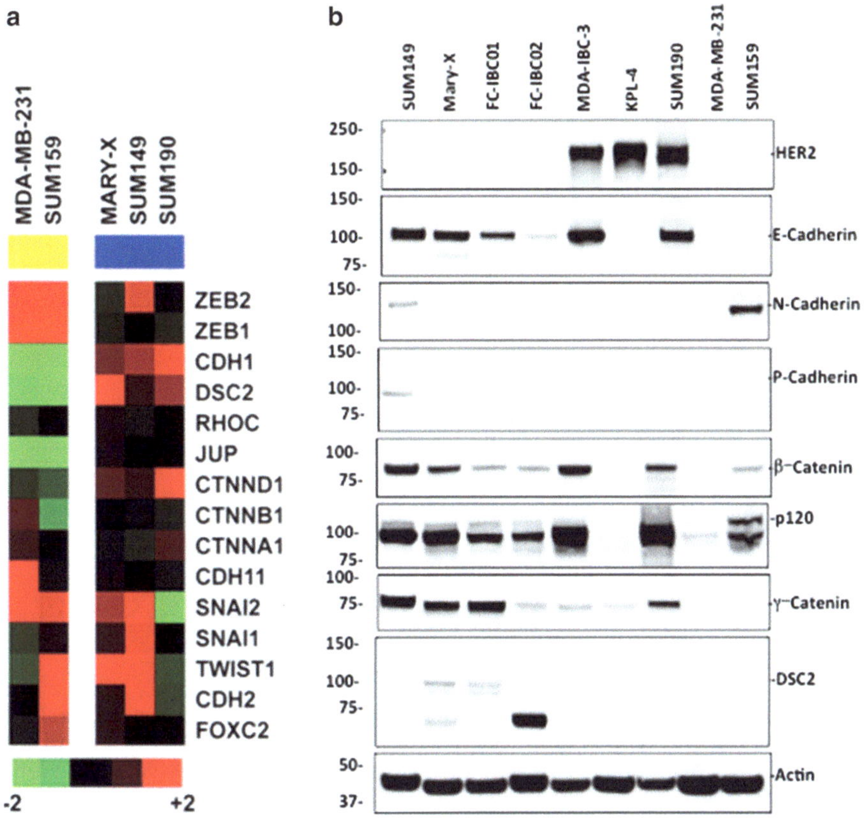

Fig. 14.6 Analysis of genes and proteins associated with cell–cell aggregation in preclinical models of IBC. (**a**) Heat map showing expression of genes involved in cell–cell aggregation in preclinical models of IBC. (**b**) Western blot analysis of proteins involved in cell–cell aggregation in preclinical models of IBC (This figure was originally published in J Clin Exp Pathol. 2012;2:119. doi: 10.4172/2161-0681.1000119)

Taken together, the gene signature and phenotypic characteristics that we have defined suggest that IBC exhibits characteristics of epithelial plasticity, where the tumor cells retain an epithelial phenotype through E-cadherin expression, while simultaneously expressing markers consistent with cells with a cancer stem cell phenotype and the associated expression of the transcription factors *TWIST1*, *SNAI1*, and *SNAI2*. The triple-negative IBC cells also express *VIM*, which encodes for vimentin, usually associated with the process of EMT.

While the process of EMT has become an accepted dogma, there is currently a controversy about how necessary this process is to invasion and metastasis, with very few examples of EMT identified in human tumor specimens. Recent studies suggest that EMT is not required for metastatic spread, but more often the cells express cancer stem cell markers regulated by transcription factors while maintaining an epithelial phenotype [41]. These observations are consistent with our result

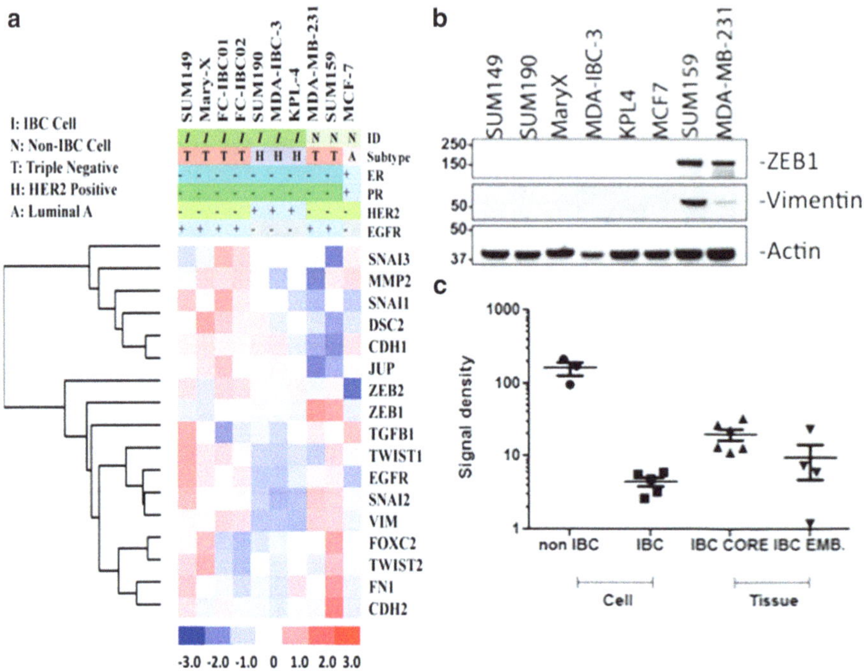

Fig. 14.7 Analysis of EMT-associated genes and proteins in preclinical models of IBC. (**a**) Heat map showing expression of genes involved in EMT in preclinical models of IBC. (**b**) Western blot analysis of ZEB1 and vimentin proteins in preclinical models of IBC compared with non-IBC breast tumor cell lines. (**c**) Gene expression of *ZEB1* in IBC tumor cells, non-IBC tumor cells, and tissue from primary tumor biopsy obtained from six IBC patients isolated by laser capture microdissection and in tissues isolated from tumor emboli isolated from skin punch biopsies of four IBC patients isolated by laser capture microdissection (This figure was originally published in J Clin Exp Pathol. 2012;2:119. doi: 10.4172/2161-0681.1000119)

Table 14.1 Statistical analysis of the comparative differences in *ZEB1* gene expression from cells and IBC patient tissues from studies shown in Fig. 14.7

Non- IBC Cells vs IBC Cells	IBC Cells vs IBC CORE	IBC Cells vs IBC EMBOLI	IBC CORE vs IBC EMBOLI
P<0.0007	P<0.004	N.S.	N.S.

Table was originally published in J Clin Exp Pathol. 2012; 2:119. doi: 10.4172/2161-0681.1000119

demonstrating simultaneous robust expression of E-cadherin and other genes associated with tight homotypic aggregation while maintaining a stem cell phenotype associated with the expression of transcription factors that is involved in maintenance of the stem cell characteristics [33, 34].

14.5 IBC Characterized by Loss of TGFβ Signaling in IBC

One of the growth factors known to induce the process of EMT is transforming growth factor-beta (TGFβ) [42]. Recent studies demonstrated that TGFβ is a key factor in the reversible regulation of motility by single cells, which in its absence, allows cells to revert to a process that has been defined as "cohesive invasion" [43]. Our gene profiling studies demonstrated that, in addition to retention of E-cadherin, preclinical models of IBC have a loss of expression of genes involved in TGFβ signaling (Fig. 14.7, Table 14.2), which is consistent with our collaborative studies reporting similar changes in IBC patient tumor tissues [10]. Since IBC tumor cells are characterized by formation of highly motile aggregates of tumor cells that migrate and invade into the dermis as a collective unit, it is perhaps not surprising that IBC is characterized by a loss of genes within the TGFβ signaling pathway and gain of genes such as SMAD6 and RUNX3 (Table 14.2), which suppress the process of EMT induced by TGFβ. The simultaneous retention of an epithelial phenotype with a lack of TGFβ signaling activation in IBC may be the molecular basis for the ability of aggregates of tumor cells to migrate into lymphatic vessels, as we have demonstrated occurs in the Mary-X, FC-IBC01, and FC-IBC02 preclinical models of IBC [19, 20, 33] and as has been demonstrated in IBC patient skin punch biopsy tissues [1, 7–9].

Table 14.2 List of genes showing the relative difference in expression of genes within the TGFβ signaling pathway in preclinical models of IBC

Symbol	Gene name	Fold changes
INHBA	Inhibin, beta A	1.2581344902386116
BMPR1B	Bone morphogenetic protein receptor, type IB	1.2714285714285716
VDR	Vitamin D (1,25- dihydroxyvitamin D3) receptor	1.292483660130719
TGIF	TGFB-induced factor homeobox 1	1.2964509394572026
MAPK13	Mitogen-activated protein kinase 13	1.3747680890538034
TLX2	T-cell leukemia homeobox 2	1.3917525773195878
MAP2K6	Mitogen-activated protein kinase kinase 6	1.4218749999999998
Smad6	SMAD family member 6	1.5411764705882354
MAPK11	Mitogen-activated protein kinase 11	1.6130952380952381
Runx3	Runt-related transcription factor 3	2.760330578512397
PAI-1	Serpin peptidase inhibitor, clade E	−3.0973451327433628
GSC	Goosecoid homeobox	−3.021276595744681
Bcl-2	B-cell CLL/lymphoma 2	−2.1538461538461537
MAPK12	Mitogen-activated protein kinase 12	−1.7128712871287128
Runx2	Runt-related transcription factor 2	−1.5742753623188408
TGFB1	transforming growth factor, beta 1	−1.5596868884540114
AMHR2	Anti-Mullerian hormone receptor, type II	−1.5166666666666666
Nkx2.5	NK2 homeobox 5	−1.5108225108225108
TCF	Hepatocyte nuclear factor 4, alpha	−1.5104895104895106
BMP2K	BMP2 inducible kinase	−1.4548872180451127

Table was originally published in J Clin Exp Pathol. 2012;2:119. doi: 10.4172/2161-0681.1000119

Interestingly, the same study reporting the loss of TGFβ signaling as permissive for cohesive invasion demonstrated that tumor cells exhibiting collective invasion due to a lack of TGFβ signaling invaded into lymphatic vessels but were incapable of hematogenous metastasis [43]. This is the pattern of metastatic spread exhibited in IBC, which favors lymphatic dissemination leading to locoregional recurrence prior to the occurrence of systemic metastasis. Collectively, the results of our gene profiling studies are consistent with observations that IBC tumor emboli primarily invade into dermal lymphatic vessels, providing a route of metastasis to local lymph nodes, which is the pattern of disease progression that is commonly observed in IBC patients. Our studies suggest that IBC tumor cells display plasticity in their gene signature that allows them to retain the epithelial phenotype, while modulating specific signaling pathways that program them to a specific lymphatic route of metastasis. Based on the high degree of plasticity exhibited in IBC tumor cells, it may be that they are capable of expressing genes within different signaling pathways, depending upon whether they display a propensity for a lymphatic or blood-borne route of metastasis.

While loss of E-cadherin has been associated with the phenotypic program of EMT characterized by increased motility and invasion similar to that of fibroblasts [37, 38], the ability to revert from a mesenchymal phenotype back to an epithelial phenotype is now believed to be necessary for colonization to form distant metastasis. Using the triple-negative breast cancer cell line MDA-MB-231, one study demonstrated that the microenvironment of the lung induced reexpression of E-cadherin associated with what was defined as mesenchymal-to-epithelial reverting transition (MErT) [44]. This phenotypic change of the MDA-MB-231 cells, which usually have a mesenchymal cell phenotype, was associated with altered cell behavior and was critical to their survival at the sites of metastasis. A recent review pointed out that there are a number of tumor types including IBC, ovarian carcinoma, and glioblastoma, which all exhibit an accelerated program of metastasis and are characterized by retention of E-cadherin and exhibiting cohesive invasion. These studies suggested that the role of E-cadherin in metastasis may be currently unappreciated [45].

Thus, the apparent dichotomy of the gene signatures of IBC that includes expression of transcription factors involved in maintenance of a stem cell phenotype while retaining an epithelial phenotype may be explained by the extreme plasticity exhibited by IBC tumor cells as one of their adaptive mechanisms for survival and accelerated rates of metastasis to multiple sites. We have identified the lack of *ZEB1* expression and the loss of genes involved in the TGFβ signaling pathway that induce EMT, with expression of genes that suppress TGFβ–mediated EMT to be characteristic of all of the preclinical models of IBC. Additionally, the lack of *ZEB*1 expression, a known transcriptional repressor of E-cadherin, may be the molecular basis for the retention of robust E-cadherin expression in IBC, which we have previously demonstrated to be associated with the presence of microRNA 200c, which regulates E-cadherin expression [34]. Additionally, our gene profiling studies also demonstrated that preclinical models of IBC express a specific repertoire of transcription factors, including Snail and *TWIST1*, that allows maintenance of a cancer stem cell phenotype, which may confer a survival advantage in the face of chemotherapy and radiation as well as conferring a tumor initiating capability [33, 34].

14.6 Conclusions

Taken together, the simultaneous expression of genes in IBC that support an epithelial phenotype with suppression of expression of genes associated with a mesenchymal phenotype such as *ZEB1* and TGFβ allows IBC tumor emboli to migrate as aggregates of cells into lymphatic vessels, providing a conduit for IBC tumor emboli to rapidly colonize regional lymph nodes. This program of simultaneous gain and loss of specific gene programs may be the basis for the metastatic phenotype exhibited in IBC patient, which is recapitulated in preclinical models of IBC, especially those that are triple negative and can be accurately characterized by a high degree of epithelial plasticity.

Collectively, our studies are among the first to identify that retention of E-cadherin expression in preclinical models of IBC was associated with the suppression of genes within the TGFβ signaling pathway and lack of or low expression of the *ZEB1* transcription factor that are both known to be involved in the process of EMT (Fig. 14.8). These results are among the first to shed light on molecular mechanisms underlying the retention of E-cadherin observed in IBC patient tumors and on a signaling pathway that supports the retention of an epithelial phenotype, in the face of enrichment of cells with a cancer stem cell phenotype and a program of

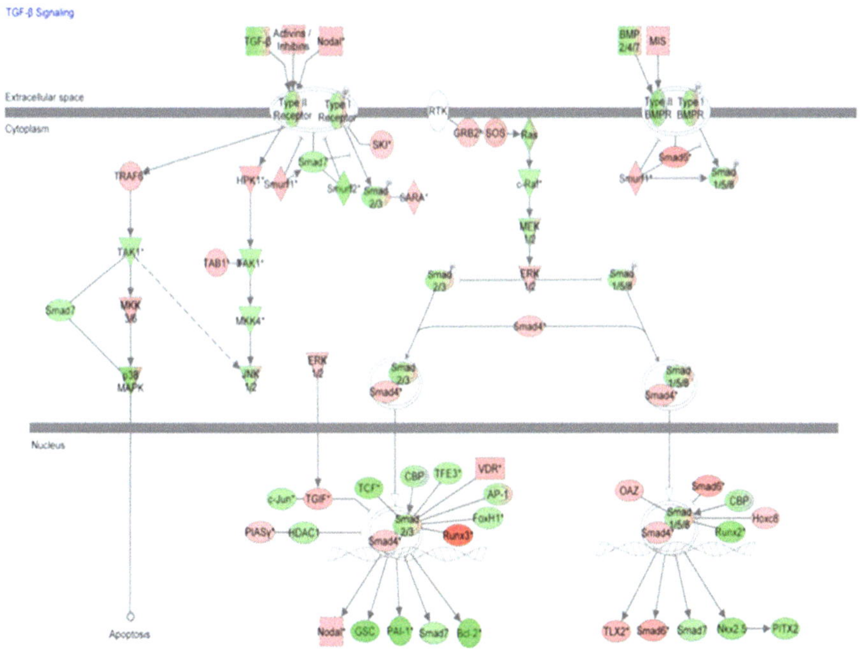

Fig. 14.8 Pathway map of genes within the TGFβ signaling pathway in preclinical models of IBC, with upregulated genes shown in *red* and downregulated genes shown in *green* (This figure was originally published in J Clin Exp Pathol 2:119. doi: 10.4172/2161-0681.1000119)

accelerated metastasis that is consistent with the disease in a patient with an IBC diagnosis. The observations for the suppression of expression of genes within the TGFβ signaling pathway are consistent with results of gene profiling of IBC patient tumors [11]. Collectively, these data provide first-time evidence that IBC is characterized by a signature of epithelial plasticity with enrichment for cancer stem cells, similar to the observations made in collaborative studies in IBC patient tumors.

Acknowledgements This project was supported by the KG081287 American Airlines-Susan G. Komen for the Cure Organization (FMR, MC).

References

1. Robertson FM, Bondy M, Yang W, Yamauchi H, Wiggins S, Kamrudin S, Krishnamurthy S, Le-Petross H, Bidaut L, Player AN, Barsky SH, Woodward WA, Buchholz T, Lucci A, Ueno N, Cristofanilli M. Inflammatory breast cancer: the disease, the biology, the treatment. CA Cancer J Clin. 2010;60(6):351–75.
2. Hance KW, Anderson WF, Devesa SS, Young HA, Levine PH. Trends in inflammatory breast carcinoma incidence and survival: the surveillance, epidemiology, and end results program at the National Cancer Institute. J Natl Cancer Inst. 2005;97:966–75.
3. Levine PH, Veneroso C. The epidemiology of inflammatory breast cancer. Semin Oncol. 2008;35:11–6.
4. Dawood S, Merajver SD, Viens P, Vermeulen PB, Swain SM, Buchholz TA, Dirix LY, Levine PH, Lucci A, Krishnamurthy S, Robertson FM, Woodward WA, Yang WT, Ueno NT, Cristofanilli M. International expert panel on inflammatory breast cancer: consensus statement for standardized diagnosis and treatment. Ann Oncol. 2011;22(3):515–23.
5. Cristofanilli M, Valero V, Buzdar AU, Kau SW, et al. Inflammatory breast cancer (IBC) and patterns of recurrence: understanding the biology of a unique disease. Cancer. 2007;110(7):1436–44.
6. Singletary SE, Cristofanilli M. Defining the clinical diagnosis of inflammatory breast cancer. Semin Oncol. 2008;35(1):7–10.
7. Kleer CG, van Golen KL, Braun T, Merajver SD. Persistent E-cadherin expression in inflammatory breast cancer. Mod Pathol. 2001;14:458–64. 561–574.
8. Vermeulen PB, van Golen KL, Dirix LY. Angiogenesis, lymphangiogenesis, growth pattern, and tumor emboli in inflammatory breast cancer: a review of the current knowledge. Cancer. 2010;116(11 Suppl):2748–54.
9. Kleer CG, van Golen KL, Merajver SD. Molecular biology of breast cancer metastasis. Inflammatory breast cancer: clinical syndrome and molecular determinants. Breast Cancer Res. 2000;2:423–9.
10. Van Laere SJ, Ueno NT, Finetti P, Vermeulen PB, Lucci A, et al. An integrated analysis of three distinct IBC/nIBC Affymetrix gene expression data sets further unveils the molecular biology of IBC. Cancer Res. 2011;71 Suppl 24:33s.
11. Iwamoto T, Bianchini G, Qi Y, Cristofanilli M, Lucci A, et al. Different gene expressions are associated with the different molecular subtypes of inflammatory breast cancer. Breast Cancer Res Treat. 2011;125(3):785–95.
12. Van Laere SJ, Van der Auwera I, Van den Eynden GG, Trinh X, Van Hummelen P, et al. Confirmation of the distinct molecular phenotype of inflammatory breast cancer compared to non-inflammatory breast cancer using Affymetrix based genome-wide gene expression analysis. J Clin Oncol. 2007;25(18S):21055.
13. Van Laere SJ, Ueno NT, Finetti P, Vermeulen PB, Lucci A, et al. An integrated analysis of three distinct IBC/non-IBC affymetrix gene expression data sets to study the transcriptional heterogeneity both between IBC and non-IBC and within IBC. J Clin Oncol. 2011;29:10571.

14. Voduc KD, Cheang MC, Tyldesley S, Gelmon K, Nielsen TO, Kennecke H. Breast cancer subtypes and the risk of local and regional relapse. J Clin Oncol. 2012;28(10):1684–91.
15. Tomlinson JS, Alpaugh ML, Barsky SH. An intact overexpressed E-cadherin/alpha, beta-catenin axis characterizes the lymphovascular emboli of inflammatory breast carcinoma. Cancer Res. 2001;61(13):5231–41.
16. Alpaugh ML, Tomlinson JS, Shao ZM, Barsky SH. A novel human xenograft model of inflammatory breast cancer. Cancer Res. 1999;59(20):5079–84.
17. Mahooti S, Porter K, Alpaugh ML, Ye Y, Xiao Y, et al. Breast carcinomatous tumoral emboli can result from encircling lymphovasculogenesis rather than lymphovascular invasion. Oncotarget. 2010;1(2):131–47.
18. Ye Y, Tian H, Lange AR, Yearsley K, Robertson FM, Barsky SH. The genesis and unique properties of the lymphovascular tumor embolus are because of calpain-regulated proteolysis of E-cadherin. Oncogene. 2012;32:1702–13. doi:10.1038/onc.2012.180.
19. Chu K, Mu Z, Alpaugh RK, Fernandez S, Freiter EM, et al. Development and comparative characterization of metastasis in newly developed pre-clinical models of inflammatory breast cancer. Cancer Res. 2011;71(24 Suppl):439s.
20. Fernandez SV, Robertson FM, Pei J, Aburto-Chumpitaz L, Mu Z, Chu K, Alpaugh RK, Huang Y, Cao Y, Ye Z, Cai KQ, Boley KM, Klein-Szanto AJ, Devarajan K, Addya S, Cristofanilli M. Inflammatory breast cancer (IBC): Clues for targeted therapies. Breast Cancer Res Treat. 2013;140(1):23–33. doi:10.1007/s10549-013-2600-4.
21. Forozan F, Veldman R, Ammerman CA, Parsa NZ, Kallioniemi A, et al. Molecular cytogenetic analysis of 11 new breast cancer cell lines. Br J Cancer. 1999;81(8):1328–34.
22. Ignatoski KM, Ethier SP. Constitutive activation of pp125fak in newly isolated human breast cancer cell lines. Breast Cancer Res Treat. 1999;54(2):173–82.
23. Kurebayashi J, Otsuki T, Tang CK, Kurosumi M, Yamamoto S, et al. Isolation and characterization of a new human breast cancer cell line, KPL-4, expressing the Erb B family receptors and interleukin-6. Br J Cancer. 1999;79(5–6):707–17.
24. Kurebayashi J, Yamamoto S, Otsuki T, Sonoo H. Medroxyprogesterone acetate inhibits interleukin 6 secretion from KPL-4 human breast cancer cells both in vitro and in vivo: a possible mechanism of the anticachectic effect. Br J Cancer. 1999;79(3–4):631–6.
25. Kurebayashi J. Regulation of interleukin-6 secretion from breast cancer cells and its clinical implications. Breast Cancer. 2000;7(2):124–9.
26. Fujimoto-Ouchi K, Sekiguchi F, Tanaka Y. Antitumor activity of combinations of anti-HER-2 antibody trastuzumab and oral fluoropyrimidines capecitabine/5′-dFUrd in human breast cancer models. Cancer Chemother Pharmacol. 2002;49(3):211–6.
27. Fujimoto-Ouchi K, Sekiguchi F, Yamamoto K, Shirane M, Yamashita Y, Mori K. Preclinical study of prolonged administration of trastuzumab as combination therapy after disease progression during trastuzumab monotherapy. Cancer Chemother Pharmacol. 2010;66(2):269–76.
28. Klopp AH, Lacerda L, Gupta A, Debeb BG, Solley T, et al. Mesenchymal stem cells promote mammosphere formation and decrease E-cadherin in normal and malignant breast cells. PLoS One. 2010;5(8):e12180.
29. Xiao Y, Ye Y, Yearsley K, Jones S, Barsky SH. The lymphovascular embolus of inflammatory breast cancer expresses a stem cell-like phenotype. Am J Pathol. 2008;173(2):561–74.
30. Charafe-Jauffret E, Ginestier C, Iovino F, Tarpin C, Diebel M, Esterni B, Houvenaeghel G, Extra JM, Bertucci F, Jacquemier J, Xerri L, Dontu G, Stassi G, Xiao Y, Barsky SH, Birnbaum D, Viens P, Wicha MS. Aldehyde dehydrogenase 1-positive cancer stem cells mediate metastasis and poor clinical outcome in inflammatory breast cancer. Clin Cancer Res. 2010;16(1):45–55.
31. Van Laere S, Limame R, Van Marck EA, Vermeulen PB, Dirix LY. Is there a role for mammary stem cells in inflammatory breast carcinoma? a review of evidence from cell line, animal model, and human tissue sample experiments. Cancer. 2010;116(11 Suppl):2794–805.
32. Robertson FM, Ogasawara MA, Ye Z, Chu K, Pickei R, Debeb BG, Woodward WA, Hittelman WN, Cristofanilli M, Barsky SH. Imaging and analysis of 3D tumor spheroids enriched for a cancer stem cell phenotype. J Biomol Screen. 2010;15(7):820–9.

33. Robertson, FM, Chu, K, Fernandez, S, Mu, Z, Zhang, X, Liu, H, Boley, KM, Alpaugh, RK, Ye, Z, Wright, MC, Luo, A, Moraes, R, Wu, H, Zook, M, Barsky, SH and Cristofanilli, M. Genomic profiling of pre-clinical models of inflammatory breast cancer identifies a signature of mesenchymal-epithelial plasticity and suppression of TGF beta signaling. J Clin Exp Pathol 2(5), 2012. http://dx.doi.org/10.4172/2161-0681.1000119.
34. Robertson FM, Chu K, Circo R, Wulfkuhle J, Krishanmurthy S. Genomic and proteomic pathway reveals signatures of mesenchymal epithelial plasticity in inflammatory breast cancer. In: Done SJ, editor. Breast cancer: recent advances in biology, imaging and therapeutics. New York: Intech; 2011. ISBN 978-953-307-0730-7.
35. David JM, Rajasekaran AK. Dishonorable discharge: the oncogenic role of cleaved E-cadherin fragments. Cancer Res. 2012;72(12):2917–23. ©2112 AACR.
36. Harris TJ, Tepass U. Adherens junctions: from molecules to morphogenesis. Nat Rev Mol Cell Biol. 2010;11(7):502–14. Review.
37. Thiery JP, Acloque H, Huang RY, Nieto MA. Epithelial-mesenchymal transitions in development and disease. Cell. 2009;139(5):871–90.
38. Kalluri R, Weinberg RA. The basics of epithelial-mesenchymal transition. J Clin Invest. 2009;119:1420–8.
39. Mani SA, Guo W, Liao MJ, Eaton EN, Ayyanan A, et al. The epithelial-mesenchymal transition generates cells with properties of stem cells. Cell. 2008;133(4):704–15.
40. Dong HM, Liu G, Hou YF, Wu J, Lu JS, et al. Dominant negative E-cadherin inhibits the invasiveness of inflammatory breast cancer cells in vitro. J Cancer Res Clin Oncol. 2007;133(2):83–92.
41. Chui MH. Insights into cancer metastasis from a clinicopathologic perspective: epithelial mesenchymal transition is not a necessary step. Int J Cancer. 2013;132(7):1487–95. doi:10.1002/ijc.27745.
42. Fuxe J, Karlsson MC. TGF-β-induced epithelial-mesenchymal transition: a link between cancer and inflammation. Semin Cancer Biol. 2012;22:455–61.
43. Giampieri S, Manning C, Hooper S, Jones L, Hill CS, Sahai E. Localized and reversible TGFbeta signalling switches breast cancer cells from cohesive to single cell motility. Nat Cell Biol. 2009;11(11):1287–96.
44. Chao YL, Shepard CR, Wells A. Breast carcinoma cells re-express E-cadherin during mesenchymal to epithelial reverting transition. Mol Cancer. 2010;9:179.
45. Rodriguez FJ, Lewis-Tuffin LJ, Anastasiadis PZ. E-cadherin's dark side: possible role in tumor progression. Biochim Biophys Acta. 2012;1826(1):23–31.

Chapter 15
An Integrated Human Mammary Epithelial Cell Culture System for Studying Carcinogenesis and Aging

Martha R. Stampfer, Mark A. LaBarge, and James C. Garbe

Abstract Experimental examination of the agents and processes that may propel or prevent human breast carcinogenesis can be facilitated by in vitro model systems of transformation, starting with normal cells, that accurately reflect the in vivo biology. Model systems that can replicate the types of alterations seen during in vivo progression offer the potential to understand the mechanisms underlying progression and to examine possible means of individualized prevention and treatment. To this end, we have developed an experimentally tractable human mammary epithelial cell (HMEC) culture system that has been used to examine the normal processes governing HMEC growth, differentiation, aging, and senescence and how these normal processes are altered during immortal and malignant transformation. Isogenic cells at different stages of multistep carcinogenesis were generated by exposing normal finite lifespan HMEC to a variety of oncogenic agents that may play an etiologic role in breast cancer. Examination of the molecular alterations present at each stage has indicated that this model is consistent with observed multistep carcinogenesis in vivo. We have seen that varying target cell type, and oncogenic agents used, can lead to multiple distinct molecular pathways of transformation, although the full diversity of human breast cancer cell types has not yet been generated in culture models. Using this integrated system, we have formulated a comprehensive model of the proliferative barriers normal HMEC must overcome to gain immortality and malignancy. Our data provide insights on acquisition of cancer-associated properties and suggest that the most crucial step in breast cancer progression involves the transition from a finite to an indefinite lifespan. For example, we see that genomic instability originates in finite lifespan HMEC when telomeres become critically short and engage in telomeric associations and is then maintained in resultant immortalized and malignant lines. Direct genomic targeting of the tumor-suppressive

M.R. Stampfer (✉) • M.A. LaBarge • J.C. Garbe
Life Sciences Division, Lawrence Berkeley National Laboratory, 1 Cyclotron Road, Bldg 977, Berkeley, CA 94720, USA
e-mail: mrstampfer@lbl.gov

H. Schatten (ed.), *Cell and Molecular Biology of Breast Cancer*,
DOI 10.1007/978-1-62703-634-4_15,

senescence barriers can produce lines lacking gross genomic errors, supporting the hypothesis that genomic instability is a mechanism to generate cancer-causing errors, but is not necessary per se. Immortalization through telomerase reactivation was also associated with acquisition of resistance to TGFβ growth inhibition and to oncogene-induced senescence (OIS) and with large-scale changes in gene expression and epigenetic marks. Being able to examine the progressive changes that fuel malignancy, starting with normal cells, provides an integrated perspective that can reveal novel information on the origins and consequences of individual cancer-associated aberrations.

15.1 Introduction

Human breast carcinomas exhibit great diversity in phenotypic expression, correlated with differences in clinical parameters [1–4]. The factors that contribute to transforming normal breast cells to cancer, and give rise to the observed diversity of breast cancer subtypes, are currently not well defined. The interplay between initial target cell, genomic alterations that overcome tumor-suppressive barriers and confer malignancy, and cell–stromal interaction is thought to be the main variable that influence the transition of normal cells to the different malignant phenotypes.

A wealth of recent information derived from direct examination of human breast tissues is providing new insights about the pathways and alterations associated with breast carcinogenesis and the extent of inter- and intra-tumor heterogeneity [5–9]. However, determining cause and effect relationships about factors and aberrations that may propel or halt human breast carcinogenesis is constrained by the limitations of in vivo human analyses. The use of animal model systems as in vivo models of human breast cancer also has limitations. Many significant differences in processes implicated in aging and carcinogenesis exist between human cells and cells derived from commonly used rodent model systems, for example, in regulation of telomerase activity and immortalization [10] and in the roles of the cyclin-dependent kinase inhibitors (CKI) $p16^{INK4a}$ and $p14^{ARF}$ [11]. Similarly, many significant differences in biological processes exist between epithelial and mesenchymal cells, for example, responses to chemical carcinogens and TGFβ, mechanisms of senescence, and expression of miRNAs [12–17]. Since the large majority of human cancers derive from epithelial cells, we believed that a full understanding of human carcinogenesis would require the ability to examine human epithelial cells in culture. Further, in order to understand deranged human cellular processes, we believed it would be necessary to have normal cells available for comparison. We therefore developed an experimentally tractable human mammary epithelial cell (HMEC) culture system that supports vigorous growth of normal HMEC of multiple lineages and has generated isogenic cultures that range from normal, to aberrant but still finite, to nonmalignant immortal, and to malignant immortal. Our extensive integrated system allows examination of the progressive changes that fuel malignancy, starting with normal cells, thereby providing a comprehensive perspective that can offer insight on the origins, consequences, and interactions of individual cancer-associated aberrations. This in vitro system can also complement in vivo findings by supporting experimental evaluation of factors that may promote or inhibit malignancy at different stages in progression.

This review will describe the HMEC culture system we have developed and how it has been employed to gain an integrated overview of the central processes associated with human breast carcinoma development. To place the various cell cultures generated within the context of multistep carcinogenesis, we first review our model of the tumor-suppressive senescence barriers that need to be bypassed or overcome for malignant progression in cultured HMEC to proceed.

15.2 Senescence Barriers Encountered by Cultured HMEC

Based on our studies of normal HMEC grown under different culture conditions and exposed to various oncogenic agents (see below), we have generated a new model of the tumor-suppressive senescence barriers that prevent normal cells from becoming immortally and malignantly transformed [15, 18]. Figure 15.1 outlines the generation of our various cultures with respect to growth medium, oncogenic agents employed, and the senescence barriers, and Table 15.1 compares the phenotypes of HMEC and isogenic human mammary fibroblasts (HMF) arrested at distinct senescence barriers. We observe that cultured HMEC encounter at least two mechanistically distinct barriers to indefinite proliferation, stasis (stress-associated senescence) and telomere dysfunction due to telomere attrition. Finite lifespan HMEC are also vulnerable to oncogene-induced senescence (OIS). Some HMEC may cease growth as a consequence of terminal differentiation. Importantly, the model presented here is consistent with observations of in vivo breast cancer progression. We also note that the phenotype of senescent isogenic HMF resembles that of HMEC at stasis rather than at telomere dysfunction (Table 15.1).

Stasis is a stress-associated barrier mediated by the retinoblastoma (RB) pathway and is independent of telomere length and extent of replication [15]. The onset of stasis in cultured HMEC correlates with increased expression of p16, which prevents inactivation of RB [15, 18–20]. Cells at stasis express senescence-associated β-galactosidase (SA-βGal) activity and have a senescent morphology. The number of population doublings (PD) achieved prior to stasis varies with culture conditions; we have observed a range of ~10–60 PD [15, 18, 21, 22]. Molecular correlates that can identify stasis, in addition to p16 expression, include arrest in G1, low labeling index (LI), noncritically short telomeres, and normal karyotypes (Table 15.1) [15, 18, 20]. These parameters are consistent with an RB-mediated arrest and the absence of a DNA damage response (DDR). Stasis can be bypassed or overcome in cultured HMEC by multiple types of individual alterations (genetic and/or epigenetic) in pathways governing RB and does not require loss of p53 function [19, 22–25]. Overcoming stasis may correlate with hyperplasia/atypical hyperplasia in vivo, which can display clonal growth. Errors in the RB pathway (e.g., loss of p16

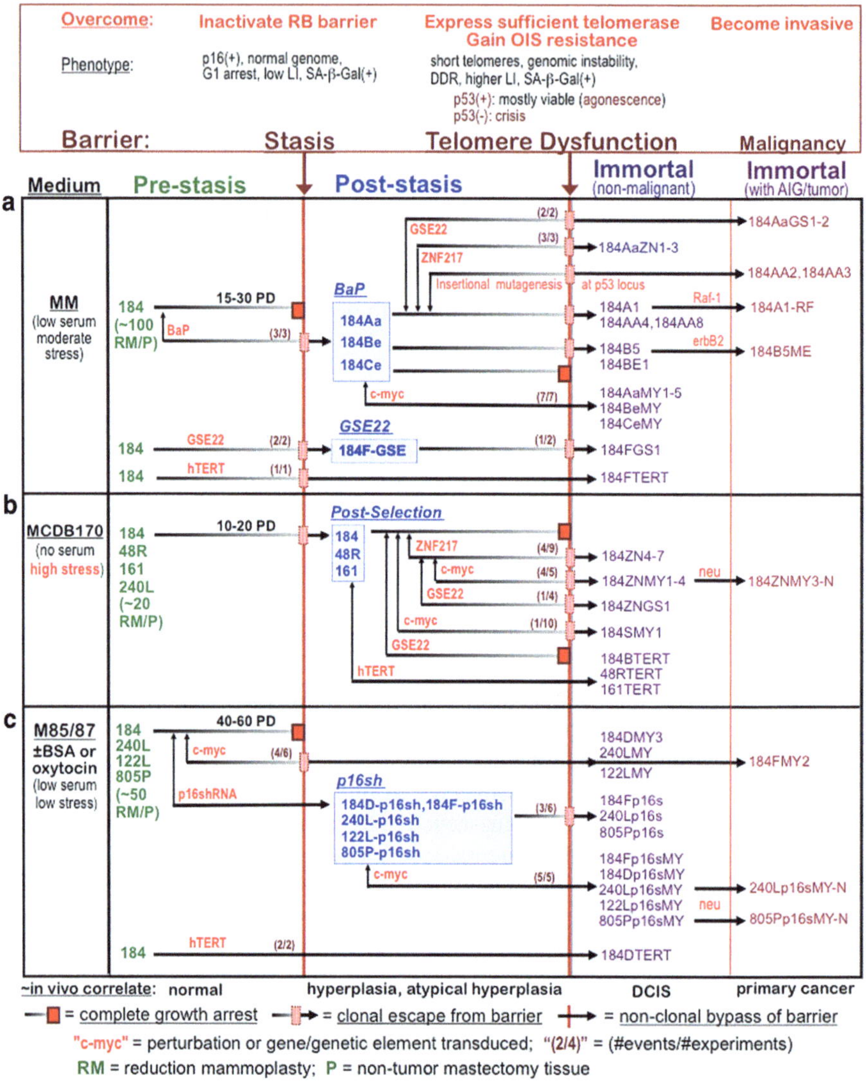

Fig. 15.1 Model of senescence barriers encountered by cultured HMEC and derivation of transformed HMEC cultures. Primary cultures obtained from reduction mammoplasties (RM) or non-tumor mastectomy tissues (P) were initiated in three different types of medium (panels **a, b, c**). All unperturbed cells grown in any serum-containing medium ceased proliferation at the stasis barrier (panels **a, c**). Exposure of pre-stasis cultures to various oncogenic insults (*red*) induced cells to overcome or bypass stasis and become post-stasis by different means. Further alterations were required to overcome the telomere dysfunction barrier, gain telomerase expression, and become immortal. Cells from post-selection post-stasis cultures all ceased proliferation at the telomere dysfunction barrier in the absence of additional oncogenic exposures (panel **b**). Nonmalignant immortal lines were no longer sensitive to OIS, and transduction of a number of different oncogenes conferred AIG (anchorage-independent growth) and/or tumorigenicity

Table 15.1 Molecular correlates of HMEC senescence barriers

Senescence Barrier	Morphology	SA-βGal	p16	p53	Arrest	LI (%)	Karyotype	Mean TRF	53BP1 foci	ser15P p53
Stasis (stress associated)	Senescent[a]	+	+	IND	Viable in G1	<2	Normal	~6–9 kb	+	∓
Telomere dysfunction: *agonescence*	Senescent	+	–	+	Mostly viable, all phases, some death	~15	Abnormal	~4–5 kb	+++	++
Telomere dysfunction: *crisis*	Growing, dying, debris	+	–	–	Eventual massive death	~40	Abnormal	<5 kb	ND	ND
OIS	Sick, dying, debris	–/?[b]	IND	IND	All phases	~15	IND	IND	ND	ND
Isogenic *fibroblast* senescence	Senescent	+	+	IND	Viable in G1	<2	Normal	~6 kb	+	–

IND senescence barrier independent of that property, *ND* not determined

[a]Pre-stasis HMEC arrested at stasis in serum-free MCDB 170 have a morphology showing abundant stress fibers

[b]Negative in pre-stasis HMEC, could not be determined in post-selection HMEC

The expression of 53BP1 foci and serine 15 phosphorylated p53 are markers of a DDR

expression by mutation or epigenetic silencing, mutated RB, overexpressed cyclin D1, mutated cdk4) are common in human carcinomas [19, 26–29]. Gross genomic aberrations are not common at this stage in vivo [30] and are not associated with overcoming stasis in vitro [15, 20].

Although neither cultured HMEC nor HMF express p21 at stasis, we postulate that stasis can also be enforced by p53-dependent p21 in response to DNA-damaging stresses such as oxidative damage or radiation. Other cell types, such as keratinocytes and foreskin fibroblasts, may be more vulnerable to DNA damage-inducing stresses in culture, express p21, and show greater evidence of a DDR at stasis. The presence of a DDR and telomeric damage foci in these cell types is not by itself evidence of senescence due to telomere erosion but can result from the decreased ability of telomeric ends to repair DNA damage [31]. HMEC in vivo may also experience p53-inducing stresses. This p53-dependent type of stasis arrest does not require critically short telomeres or genomic instability, and inactivation of p53 or p21 function may facilitate overcoming this barrier [32–34]. Reactivation of telomerase is neither necessary nor sufficient to overcome stasis; however, ectopic overexpression of hTERT prior to the onset of stasis in HMEC can bypass stasis and produce immortalization (Garbe and Stampfer, unpublished).

In post-stasis HMEC (cells that have bypassed or overcome stasis), ongoing replication in the absence of sufficient telomerase activity produces progressively shortened telomeres. Telomere dysfunction due to telomere attrition (i.e., replicative senescence) occurs when telomeres become critically short (mean TRF $\leq$ 5 kb), and uncapped telomeres elicit genomic instability and a DDR. Where wild-type p53 is present, most cells can mount a viable p53-dependent arrest; this barrier has been termed agonescence [18, 20, 35]. Karyotypic analysis of HMEC at agonescence has shown that virtually all metaphases exhibit gross chromosomal abnormalities, predominantly telomere associations [20, 36]. This result indicates that the p53-dependent senescence arrest due to telomere attrition does not occur as soon as one uncapped telomere is present [37, 38]. When p53 is nonfunctional the cells cannot mount a viable arrest, and crisis-associated massive cell death occurs [18]. Agonescence can be distinguished from stasis in HMEC by the presence of critically short telomeres and genomic instability, higher LI (~15 %), arrest at all phases of the cell cycle, and presence of a DDR (Table 15.1). HMEC at agonescence as well as at stasis display a senescent morphology and SA-βGal, so these properties do not readily distinguish between these two molecularly distinct senescence barriers. Crisis can be distinguished from agonescence by a higher LI (~40 %) and the absence of a viable arrest. Since most human epithelial and fibroblast cells induced to transform in culture had p53 function inactivated to overcome stasis (e.g., using viral oncogenes or inhibitors of p53 function), only crisis was observed in such cultures at the telomere dysfunction barrier.

The telomere dysfunction barrier can be overcome by the expression of sufficient telomerase to maintain stable telomere lengths. Overcoming telomere dysfunction may correlate with DCIS in vivo, which commonly displays short telomeres, genomic instability, and telomerase reactivation [30, 39–43].

Cultured finite lifespan HMEC are additionally vulnerable to OIS, which produces rapid growth inhibition and death [14]. The mechanism underlying OIS in HMEC is not yet fully delineated but, unlike many rodent and fibroblast cells, does not require functional p16 or p53 [14, 16]. The molecular correlates of OIS in HMEC differ from those seen in cells at stasis or telomere dysfunction and are consistent with a DDR (Table 15.1). HMEC that have attained immortality via reactivation of endogenous telomerase are no longer vulnerable to OIS and show gain of malignancy-associated properties when exposed to oncogenes such as Raf-1, Ras, or ErbB2 [14, 16, 44, 45]. HMEC immortalized by exogenous hTERT transduction retain some vulnerability to oncogene exposure [14, 46], but recent studies indicate that unlike finite HMEC they may maintain proliferative capacity [47]. Gaining resistance to OIS may be critical to acquiring malignant properties in vivo.

15.3 Normal HMEC in Culture

Normal and aberrant human mammary cells can be readily obtained from surgical discard tissues (reduction mammoplasties, mastectomies, benign tumors) and milk fluids. Most reduction mammoplasty material is from women in their late teens to early 30s, but tissues from older women are also available. These tissues show the range of age-appropriate pathologies, with increasing presence of mild to atypical hyperplasia with age. From mastectomies, non-tumor tissues are available that can provide material from older women. Peripheral mastectomy tissue is not assumed to be normal, as there may be microtumors within the tissue or field effects from the tumor or environmental exposures; similarly, tissues obtained from contralateral or prophylactic mastectomies are not considered normal. Milk fluids are valuable as a source of functionally differentiated cells. Our early studies developed methods to separate the epithelial cells from the bulk surgical tissues [48]. Epithelial organoids, free of surrounding mesenchymal tissue, were purified by enzymatic digestion, collected on filters, and stored frozen [48, 49]. The digestion process also yielded a single cell population in the filtrate from which isogenic HMF could be obtained for culture and comparison to the HMEC. Our HMEC Bank contains frozen organoids from ~300 individuals ranging in age from 11 to 91.

15.3.1 Pre-stasis Finite Lifespan HMEC

We have grown HMEC derived from reduction mammoplasties, milk, benign tumors, and non-tumor mastectomy tissues in three different types of media: serum-containing (MM and M85/M87A) or serum-free [MCDB 170 (commercial MEGM)] [15, 21, 22, 48, 50]. Depending upon the medium and culture conditions, active proliferation ceased at stasis after ~10–60 PD (Fig. 15.2). Our original medium,

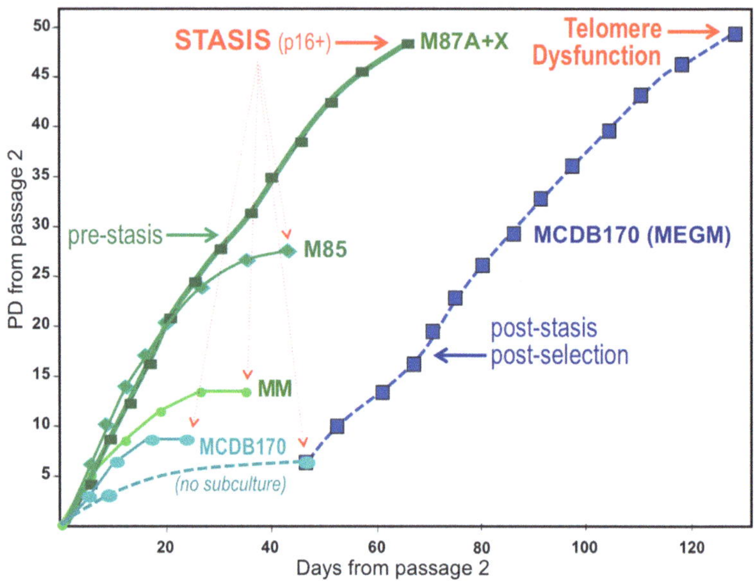

Fig. 15.2 Population doubling potential of pre-stasis HMEC in different media. Primary cultures from reduction mammoplasty specimen 184 were initiated from organoids in different media and subjected to partial trypsinizations. The number of PD in primary culture cannot be accurately determined; growth is shown starting from passage 2. Depending upon the medium, all proliferation stopped in HMEC grown in serum-containing media [MM, M85, and M87A with oxytocin (X)] after 10–50 PD beyond passage 2. The extensive proliferative potential in M87A+X supports generation of large batches of early passage pre-stasis HMEC from individual donors. HMEC initiated in serum-free MCDB170 (commercial MEGM) show rapid induction of p16 and cessation of growth. When cultures are allowed to sit without subculture for 2–3 weeks, post-selection post-stasis HMEC emerge and maintain growth to agonescence. If cultures are repeatedly subcultured, fewer to no post-selection cells may emerge

MM, supported ~15–30 PD, with HMEC with predominantly myoepithelial lineage markers present by second passage (p) [51]. Our more recent M85/M87A media will support long-term pre-stasis growth of ~60 PD. Early passage populations from reduction mammoplasty and non-tumor mastectomy tissues contain a mixture of cells with markers of myoepithelial, luminal, and progenitor lineages, but the luminal cells do not maintain active growth with long-term passage [15, 52]. HMEC grown in the serum-free MCDB 170 medium achieved only ~10–20 PD before stasis [22]. In media that support fewer PD, levels of p16 expression increase earlier. Although the mechanisms responsible for p16 induction have not been defined, it appears to reflect a cellular response to certain types of environmental stress. Virtually all cells express p16 at stasis in all media used [15, 19]. Figure 15.3a illustrates the gradually increasing p16 and SA-βGal expression and decreasing LI of HMEC growing in M85, and Fig. 15.3b, c illustrates expression of luminal lineage markers in M85-grown HMEC.

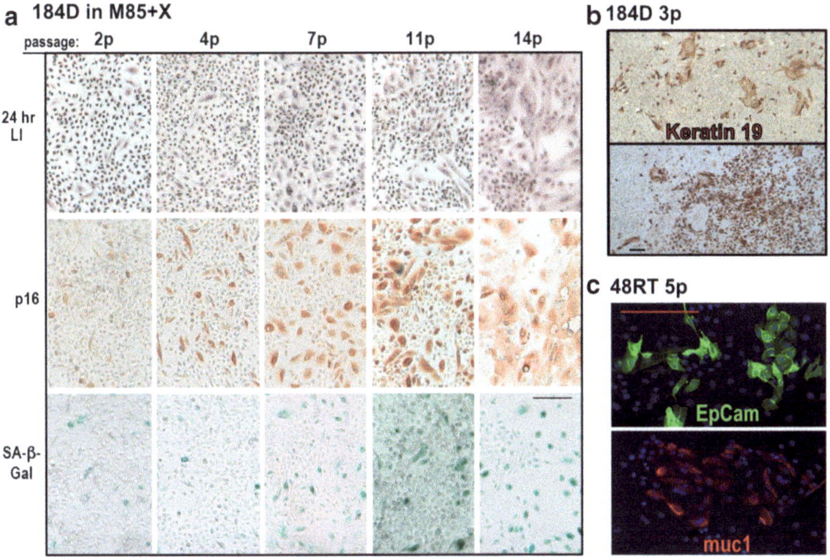

Fig. 15.3 Characterization of pre-stasis HMEC grown in M85 with oxytocin. (**a**) Expression of markers associated with proliferation (LI) and senescence (p16, SA-βGal) in pre-stasis 184 HMEC with increasing passage. Note the reciprocal relationship between the small cells with a positive LI and the larger, often vacuolated cells (senescent morphology) that are positive for p16 and SA-βGal and negative for LI. (**b**) Immunohistochemistry expression of luminal marker K19 in pre-stasis 184 HMEC. (**c**) Immunofluorescence expression of luminal markers EpCam and Muc1 in pre-stasis 48R HMEC. Size marker = 200 μm (Modified from Garbe et al. [15])

With the development of the M85/M87A media, it is now possible to generate and store frozen large quantities of normal HMEC. Using our protocol of repeated partial trypsinization of the primary organoid cultures [48–50], we can make large standardized HMEC batches from individual specimen donors at passages 2–5. These uniform batches permit reproducible large-scale and high-throughput experimentation with normal HMEC from multiple individuals of different ages. Currently, we have initiated cultures from ~50 individuals (reduction mammoplasty, non-tumor mastectomy, milk) of women ranging in age from 14 to 91.

The ability to grow heterogeneous HMEC populations in the M85/M87A medium has enabled us to experimentally examine potential differences in HMEC lineage composition and differentiation as a function of age. In recent studies [52], pre-stasis HMEC strains from 18 young (<30) and 18 older (>55) women were analyzed by FACS and immunofluorescence (IF) for lineage composition (Fig. 15.4). In cultured pre-stasis strains at 4p and in cells from uncultured dissociated organoids, increasing age was associated with a decline of myoepithelial cells and an increase of luminal cells that exhibited molecular features usually ascribed to myoepithelial cells (increased expression of integrin alpha 6 and keratin (K)14) (Fig. 15.4b, c). The tyrosine kinase receptor c-Kit has been postulated to be a marker

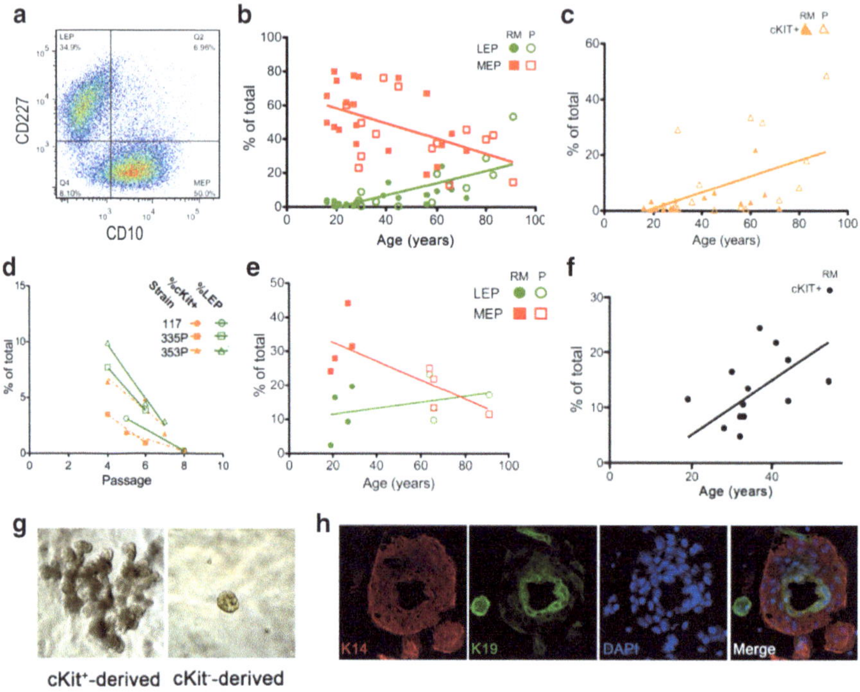

Fig. 15.4 Age-associated alterations in lineage markers in pre-stasis HMEC grown in M87A with oxytocin. (**a**) Representative FACS analyses of CD227 and CD10 expression in 240L HMEC at 4p. Myoepithelial cells (MEP) are CD10(+) CD227(−), while luminal cells (LEP) are CD10(−) CD227(+). (**b**) Linear regression showing changes in proportions of LEP and MEP in HMEC strains at 4p as a function of age ($n=36$ individuals). LEP and MEP from RM-derived strains are shown with *filled circles* or *boxes* and from P-derived strains with *open circles* or *boxes*, respectively. (**c**) Linear regression of proportions of LEP and MEP in dissociated uncultured organoids as a function of age ($n=8$ individuals). (**d**) Changes in proportions of LEPs and c-Kit(+) HMEC in three representative strains as a function of passage. (**e**) Linear regression of proportions of c-Kit(+) HMEC in strains at 4p as a function of age ($n=36$ individuals). (**f**) Linear regression of proportions of c-Kit(+) cells in dissociated uncultured organoids as a function of age ($n=11$). (**g**) Phase images of representative structures derived from c-Kit(+) (*left*) and c-Kit(−) (*right*) cells cultured in laminin-rich basement membrane for 14 days. (**h**) Immunofluorescence of a transverse frozen section that shows K14 (*red*) and K19 (*green*) protein expression in a duct of a c-Kit(+)-derived TDLU-like structure from 3D culture. Nuclei were stained with DAPI (*blue*); the three-color merged image is shown at *right* (Modified from Garbe et al. [52])

of luminal progenitors in humans [53]. With increasing age, the proportion of c-Kit-expressing HMEC increased in pre-stasis strains at 4p and in dissociated reduction mammoplasty samples (Fig. 15.4e, f). HMEC FACS enriched for c-Kit+ cells at 4p and cultured for three additional passages showed self-maintenance and multipotent differentiation. Primary and 4p c-Kit+ cells embedded in 3D laminin-rich ECM cultures gave rise to terminal ductal lobular (TDLU)-like structures that were composed of an inner core of K19-expressing luminal cells surrounded by K14-expressing myoepithelial cells, supporting the hypothesis that c-Kit+ cells are progenitors bearing multipotent activity (Fig. 15.4g, h). However, similar to cells with luminal markers, the absolute percentage of c-Kit+ cells decreased with increasing passage (Fig. 15.4d). These data suggest the exciting possibility that the observed age-associated increase in luminal breast cancer may be connected to changes that occur normally with aging in the human breast. The significant age-dependent changes to the mammary epithelium that we observed could make older women more vulnerable to malignant progression and underlay the increased luminal breast cancer incidence in women >55 years. Myoepithelial cells are thought to be tumor-suppressive and progenitors are putative etiologic roots of some breast cancers. Thus, during the aging process, the potential target cell population may increase, while there is a simultaneous decrease in the cells thought to suppress tumorigenic activity.

Although normal HMEC can be obtained from reduction mammoplasties, growth in 2D on plastic does not recapitulate the complex normal in vivo situation, where cell structure and polarity and cell–cell and cell–matrix interactions play important biological roles [54–56]. The reduction of progenitor and luminal cell types with increasing passage likely reflects the limitations of the 2D culture conditions. Of note, despite growth for several passages on plastic dishes, placement of a heterogeneous HMEC population into constrained 3D conditions, such as micropatterned microwells [57] or laminin-rich ECM [49, 52], leads to 3D structures with correct lineage organization, with internally localized luminal cells surrounded by myoepithelial cells.

Pre-stasis HMEC remain genomically stable even when they reach stasis, consistent with the noncritically short telomeres at stasis and the absence of significant evidence of a DDR [15]. The proliferative arrest of cultured HMEC at stasis can be attributed to the rise in p16 expression, as transduction of shRNA to p16 (p16sh) into pre-stasis HMEC can allow them to bypass the stasis arrest [25, 58]. An outstanding question is how the HMEC perceive stress-inducing conditions and signal that information to promote induction of p16. HMEC arrested at stasis share a similar molecular profile regardless of their PD potential or growth media (Table 15.1), with one noticeable difference. HMEC grown in the serum-containing media have a typical senescent morphology of large flat vacuolated cells (Fig. 15.3a), whereas HMEC grown in serum-free MCDB 170 exhibit a more elongated morphology showing abundant stress fibers [15, 22, 50]. We believe this difference is due to the serum-free medium being more stressful for cultured HMEC, consistent with the early rise of p16 and the low PD potential of HMEC initiated in MCDB 170 [19, 22]. This difference in morphology at stasis may have led other investigators to consider this stasis arrest distinct and refer to it as "M0" [59, 60]. We do not yet know the molecular basis by which differentiated luminal cells may cease proliferation prior to stasis. Studies with HMEC from older women (below) suggest that p16sh transduction is not sufficient to maintain active growth to agonescence in portions of their normal population. Stasis arrest in vivo may also result from p53-inducing stresses such as radiation or DNA-damaging agents.

To enhance the usefulness of our HMEC resources, and as part of our studies on malignant progression, we have been characterizing the normal pre-stasis HMEC for a variety of molecular properties, including gene transcript profiling, global promoter methylation, DDR, and lineage markers as a function of passage [15, 25, 52, 58]. These properties in the normal pre-stasis HMEC have been compared to HMEC that have overcome the stasis and/or telomere dysfunction senescence barriers and are described in more detail below. We have noted in these, and some additional assays [61], that as expected from human populations, interindividual differences can be seen. We also note, as referred to above and further illustrated in Fig. 15.7 below, that HMEC with luminal versus myoepithelial lineage markers may express significant biological differences. For example, EGFR is more highly expressed in late passage pre-stasis HMEC compared to early passage or milk-derived cells that have greater luminal cell representation, consistent with the greater dependence of basal versus luminal breast cancer subtypes on the EGF/MEK versus the IGF/PI3K signaling pathways [62–64]. These considerations become relevant when using normal HMEC as controls for cancer cells. Thus, it is important to examine cells from multiple individuals, and of different ages, to gain a more complete picture of normal HMEC physiology. Further, it may be of value to use lineage-enriched normal population for more accurate comparison to cancer cells with distinct lineage profiles [65].

15.4 Post-stasis Finite Lifespan HMEC

15.4.1 Generation of Post-stasis Finite Lifespan HMEC

Normal HMEC are capable of sensing stress-inducing environments and responding with an RB-mediated growth arrest. In vivo, stasis may serve to eliminate from the proliferative pool cells that have been damaged by stress exposures. Many types of errors in the RB pathway can produce post-stasis HMEC; consequently, post-stasis populations may exhibit significantly different biological properties. For example, a cell that overcame stasis by mutation of RB, a molecular hub, would have more profound alterations than a cell that only lost p16 expression. Bypassing or overcoming stasis is therefore one of the early stages at which the molecular alterations leading to malignancy can diverge. One of our long-term objectives has been to model the different types of pathways a normal HMEC may follow during transformation. We postulate that the combination of stasis-overcoming error and type of target cell affected establishes the initial pathway of cancer progression.

We have used several methods to obtain post-stasis cultured HMEC, focusing on perturbations that could play an etiologic role in human breast cancer. Although we have most commonly observed loss of p16 expression as the means used to overcome stasis, p16 loss can result from diverse mechanisms, and our different p16(−) post-stasis populations show distinct biological properties. Figure 15.1 charts the emergence of varying post-stasis populations from HMEC grown in the different media and subjected to differing oncogenic exposures.

Pre-stasis HMEC from over 150 individuals cultured in our serum-containing media have not shown any instance of a cell spontaneously overcoming the stasis barrier. However, early experiments that exposed primary cultures of specimen 184 HMEC grown in MM to the chemical carcinogen benzo(a)pyrene (BaP) resulted in the emergence of HMEC colonies that maintained growth after the bulk of the cultures ceased proliferation at stasis [23, 66]. Examination of three independently derived BaP-exposed post-stasis populations (initially called Extended Life) showed loss of p16 expression, associated with either mutations or promoter silencing at the p16 locus [19, 29]. BaP post-stasis cultures ceased growth after an additional 10–40 PD, with the exception of very rare cells that became immortal cell lines (see below). Later experiments with MM-grown 184 HMEC transduced with GSE22, a peptide that inhibits p53 function [67], also yielded clonal post-stasis populations. In two experiments, almost all cells ceased growth at stasis, but a few colonies maintained growth [18]. This result indicates that the loss of p53 per se was not sufficient to overcome stasis in these cultured HMEC; the GSE22 post-stasis population presumably sustained errors secondary to the p53 loss that enabled overcoming stasis. The GSE22 post-stasis HMEC exhibit a low level of p16 expression by immunohistochemistry (IHC).

Stasis can be readily bypassed by exposure to p16sh [25, 58]. Cells grown in M85 or M87A+X, from specimens 184, 240L, 805P, and 122L, have been transduced with p16sh at early passages, giving rise to p16sh post-stasis populations that maintained active growth until agonescence [58]. In several instances (see below) clonal immortal lines appeared around the period of agonescence. Recently, we have also observed clonal escape from stasis when c-Myc was transduced into M85/M87A+X-grown 184, 240L, and 122L HMEC [58].

When the HMEC are cultured in the stressful serum-free MCDB 170 medium, rare cells are able to overcome stasis in the absence of additional oncogenic exposures (Fig. 15.2) [22]. We originally called the emergence of these post-stasis cells "selection" and this class of post-stasis HMEC "post-selection." We now recognize that selection (what other labs later termed "M0") is a stasis arrest. Post-selection post-stasis cells express wild-type p53 that is present in a stable form [18, 68, 69] but show DNA methylation of the p16 promoter and absence of p16 expression, as well as nearly 200 other changes in promoter DNA methylation [19, 25]. Most of the differentially methylated regions (DMR) present in post-selection HMEC are also seen in breast cancer-derived HMEC [25], indicating that these changes are associated with malignant progression. A recent publication suggests that post-selection HMEC may be on a transformation pathway leading to metaplastic cancer [70]. Although pre-stasis populations may be heterogeneous with respect to a cell's ease in silencing p16 to become post-selection [71], our data indicate that post-selection cells are induced by growth in the stressful (oncogenic) serum-free MCDB 170 medium and are not present in the starting normal cultures. As mentioned above, we have never seen a post-stasis cell emerge from unperturbed normal pre-stasis HMEC grown in any of our serum-containing media. Additionally, the emergence of post-selection HMEC from pre-stasis HMEC grown in MCDB 170 can be reduced or eliminated by small changes in media composition or methodology. HMEC grown in MCDB 170 cease growth by passages 3–4, but after 2–3 weeks,

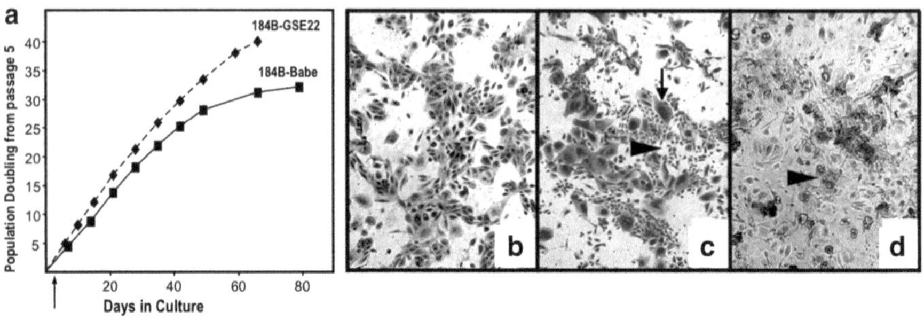

Fig. 15.5 Growth and morphology of post-stasis post-selection 184 HMEC with and without functional p53. 184 HMEC, batch B, were transduced with GSE22-containing or control (Babe) vectors at passage 5. (**a**) Growth curves of 184B-Babe and 184B-GSE22. Note the additional PD in the cultures lacking functional p53. We believe growth rates are similar ± p53, but the absence of p53-mediated growth inhibition allows more cells to continue to proliferate to crisis, leading to apparent faster growth of the population as cells near telomere dysfunction. (**b**) 184B-Babe at agonescence, 2 months after plating at passage 15, contains mostly large, flat cells with some vacuolization; the cell population can retain this morphology and viability for over a year. (**c**) 184-GSE22, 2 weeks after plating at passage 15, shows areas of small proliferating cells and many very large flat cells (*arrows*). (**d**) 184B-GSE22, 4 months after plating at passage 15, shows mostly large multinucleated, vacuolated cells and abundant cell debris. All photographs are at the same magnification (Modified from Garbe et al. [18])

small colonies of post-selection cells suddenly appear. Subculturing prior to the robust appearance of the post-selection cells greatly reduces the number of subsequent emergent colonies. We presume that the induction of the p16(−) cells is occurring during this time when the population is nonproliferative. The nature of the molecular mechanisms responsible for selection remains unknown.

As described further below, post-selection post-stasis HMEC display numerous aberrancies compared to normal pre-stasis cells. This is important to note since post-selection HMEC are being sold commercially as "normal primaries" although they are neither normal nor primaries. Since they have already acquired changes along the pathway to malignancy, they are not an accurate normal control for comparison to cancer cells. Given that it is now possible to grow large quantities of normal pre-stasis HMEC, we strongly recommend that studies aimed at understanding normal HMEC behavior use normal pre-stasis HMEC.

Post-selection p16(−) HMEC grow actively for an additional ~30–70 PD, depending on the individual. As they near agonescence, they exhibit a senescent morphology, SA-βGal activity, a DDR, and an increasing genomic instability [18, 20]. If p53 function is inactivated (e.g., using GSE22), cells continue to proliferate for an additional ~2–4 passages, with increasing evidence of cell death and debris characteristic of crisis (Fig. 15.5) [18]. The telomere dysfunction barrier is very stringent. We have never seen any unperturbed post-selection cell at agonescence spontaneously immortalize. We have also never seen any immortalization at crisis in post-selection HMEC with p53 function inactivated by GSE22, but rare immortalization at crisis using dominant-negative p53 constructs has been reported by others [72, 73]. This stringency is likely due to the molecular nature of this barrier;

cells that fail to maintain a G1 or G2 arrest with critically short telomeres will eventually die or become nonproliferative as a consequence of the genomic instability and mitotic catastrophes. Unlike an arrest based upon blocking cell cycle progression (e.g., elevated levels of CKIs at stasis), the widespread chromosomal derangements present at telomere dysfunction are not reversible. Overcoming this barrier also differs from overcoming stasis in that escaped cells will have acquired genomic abnormalities and may retain some degree of genomic instability [30].

Rare post-stasis cells likely preexist in some breast tissues; methylated p16 promoters in HMEC have been seen in apparently normal breast tissues in vivo [71]. These rare cells have been called vHMEC; the nature of the error(s) leading to the silencing of p16 in vHMEC in vivo is not known. The term vHMEC has also been used by others to refer to p16(−) post-stasis cells in culture that are specifically post-selection [70, 74].

15.4.2 Phenotypes of Post-stasis Versus Pre-stasis HMEC

We have begun molecular analyses to characterize and compare the different types of post-stasis HMEC cultures and post-stasis versus pre-stasis HMEC. These ongoing studies have thus far indicated that post-stasis HMEC have significant differences both from normal pre-stasis cells and among post-stasis types and that the post-selection type of post-stasis appears to be the most deviant from normal. Although not normal, for some experimental purposes post-selection or other post-stasis HMEC may be preferable, e.g., examining the requirements for and mechanisms of overcoming the telomere dysfunction barrier or assaying cells at different stages in progression.

By definition, unlike normal cells, post-stasis HMEC have lost their ability to respond to p16-inducing stresses with growth arrest and have overcome the first tumor-suppressive senescence barrier. Distinct from normal pre-stasis population, which contain cells with markers of multiple lineages [15], most post-stasis types examined have shown predominantly myoepithelial or basal lineage markers (e.g., K5/14, CD10, vimentin), although low levels of some luminal-associated markers can be seen and may increase with passage in post-selection populations (e.g., K8/18, Muc1) [15, 51]. Recent studies from the Kuperwasser lab have shown that post-selection HMEC can differentiate along an epidermal pathway [70]. However, our lineage studies were initially performed with post-stasis cells derived from younger women. Preliminary studies using HMEC from older women, and alternative methods of bypassing stasis, indicate that it is possible to obtain post-stasis HMEC in culture with markers of progenitor or luminal lineage. An important distinction between all post-stasis types and normal pre-stasis HMEC is the gradual increase in genomic instability as post-stasis cells approach the telomere dysfunction barrier, inserting potential unknown changes into these populations, whereas pre-stasis HMEC maintain a normal karyotype, even at stasis [15, 20].

As noted above, compared to pre-stasis cells, post-selection HMEC display a large number of DMR in addition to the p16 promoter locus. In contrast, the BaP

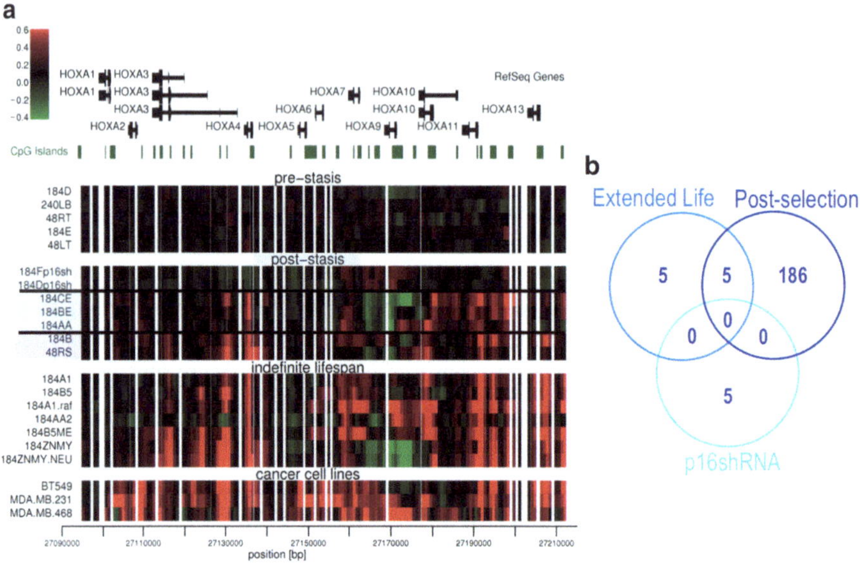

Fig. 15.6 DNA methylation changes in HMEC during malignant progression. (**a**) Progression of DNA methylation in the HOXA gene family cluster from finite lifespan HMEC to malignantly transformed breast cancer cells. *Top*, map of the RefSeq genes of the HOXA cluster followed by the location of CpG islands. *Bottom*, heat map of DNA methylation state of the HOXA gene family cluster based on the microarray data from a custom array with 11,328 probes. *Green*, hypomethylated sites; *red*, hypermethylated sites. Nucleotide position along chromosome 7 is shown below the heat map. Note the differences among the distinct post-stasis types. (**b**) Venn diagram illustrating common and different DMR among the distinct post-stasis types examined (Modified from Novak et al. [25])

and GSE22 post-stasis populations showed only ~10 DMR, including ones in the HOXA cluster also seen in post-selection and tumor-derived HMEC, while the p16sh post-stasis populations showed only ~5 DMR, which did not overlap with the other post-stasis cultures or tumor-derived cells (Fig. 15.6) [25]. Since most of the post-selection DMR are associated with breast cancer cells, post-selection HMEC may be further advanced along the pathway of malignant progression than post-stasis BaP, GSE22, or p16sh types.

Gene expression profiling has compared post-stasis post-selection BaP and p16sh types to normal pre-stasis HMEC and HMF [15, 58, 75] and immortal lines [75]. Gene transcripts from growing and senescent pre-stasis and post-selection HMEC show similarities as well as many differences; there are also a few interindividual differences (Fig. 15.7). Some genes predominantly expressed in pre-stasis HMEC (blue boxes) such as K19, EpCam (TACSTD1), and Prom1 (CD133) are luminal markers, with expression greatly reduced at the higher passages lacking luminal cells. Others, like Muc1, do not have obvious correlation with passage level. Some genes (pink box) appear to be more prevalent in myoepithelial cells, as they are low in the growing milk-derived 250MK luminal cells and reduced in most early passage pre-stasis cultures. The EGF receptor falls into this category, consistent with lower expression of the EGFR in luminal versus basal or metaplastic types

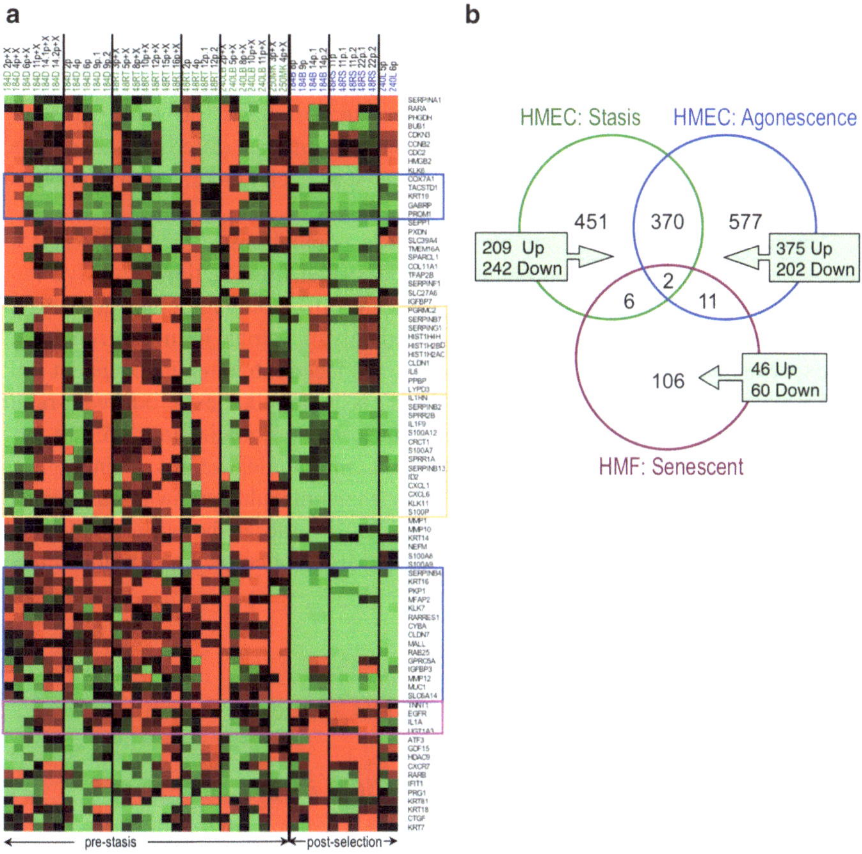

Fig. 15.7 Gene transcript profiles in pre-stasis and post-selection HMEC from different individuals. (**a**) Hierarchical clustering (by *rows*) of gene transcript profiles in growing and senescent pre-stasis and post-selection HMEC. Pre-stasis 184D, 48RT, and 240LB HMEC are shown in *columns* with increasing passage (p) up to stasis; (X) indicates growth in oxytocin. Post-selection 184B and 48RS HMEC are growing and agonescent populations. Genes shown are 77 selected for the greatest variance across all samples, plus a few selected lineage- or differentiation-associated genes. (**b**) Venn diagram of genes modulated at HMEC stasis using growing pre-stasis as baseline, at HMEC agonescence using growing post-selection as baseline, or at HMF senescence with growing fibroblast as baseline. Diagram depicts the number of genes unique to each group and the number that overlaps between and among the groups (Modified from Garbe et al. [15])

of breast cancers [62–64]. Many genes showed increased expression at higher passage levels (yellow boxes); these could be further separated into transcripts expressed at both stasis and agonescence (top yellow), mainly at stasis (lower yellow), and mainly at agonescence. Other genes were preferentially expressed in growing populations (turquoise box). Venn diagrams of transcripts differentially expressed in HMEC at stasis or agonescence and HMF at senescence, compared to their growing populations, illustrate these distinctions more clearly (Fig. 15.7b). The majority of genes modulated at stasis and agonescence were distinct, although there was also significant overlap. These data also highlight the nearly complete lack of overlap between HMEC and HMF senescence-associated transcripts. Such results illustrate strong cell type specificity in biological processes associated with senescence and suggest caution in extrapolating properties of fibroblast senescence to epithelial cells. Preliminary studies comparing post-stasis BaP and p16sh cultures to pre-stasis and post-selection HMEC show significant differences among the post-stasis types, with the p16sh and BaP post-stasis cells more similar to normal pre-stasis HMEC than the post-selection cells.

Another major difference among post-stasis populations is their responses to overexpressed c-Myc [58]. Earlier work showed that transduction of c-Myc into post-selection HMEC did not have a significant effect on telomerase activity, as measured using the TRAP assay, and produced only one clonal immortalized line in ten experiments [58, 76]. In contrast, c-Myc transduced into three independent BaP post-stasis cultures, and all tested p16sh post-stasis populations, produced a rapid increase in TRAP activity and apparently uniform immortalization. We are currently investigating the molecular properties that might underlie this distinction.

Collectively, these data demonstrate how the molecular pathways associated with different types of transformed HMEC can diverge at the earliest stages of malignant progression, in still finite lifespan HMEC, when they bypass/overcome the stasis barrier.

15.5 Immortally Transformed HMEC Lines

15.5.1 Generation of Immortal HMEC Lines

The telomere dysfunction barrier can be overcome or bypassed by the expression of sufficient telomerase to maintain stable telomere lengths. Based on our experience and the reports of others, reactivation of sufficient telomerase in normal finite lifespan HMEC is difficult to achieve using agents thought to play a role in breast cancer etiology (i.e., not ectopic hTERT transduction or viral oncogenes) and may require multiple errors. This situation may reflect the fact that long-lived animals such as humans have evolved mechanisms for stringent repression of telomerase in normal adult non-stem cells, presumably for tumor suppression. In contrast, normal cells from short-lived mammals such as mice do not show such stringent telomerase repression and readily immortalize [10, 77]. We have postulated that telomerase reactivation and immortalization may be a rate-limiting step in human epithelial carcinogenesis and so believe that great caution should be exercised in extrapolating mechanisms of murine malignant progression to humans, since this critical barrier to malignancy is not present in the commonly used rodent cells. Overcoming telomere dysfunction may correlate with DCIS in vivo, which commonly display short telomeres and genomic instability, and may show telomerase reactivation [30, 39–43]. We have hypothesized that the genomic instability associated with agonescence and crisis can give rise to errors permissive for telomerase reactivation [18, 76] and that the generation of breakage-fusion-bridge (BFB) cycles prior to immortalization may underlie some of the genomic instability seen in many carcinomas [30]. Additionally, the extensive genomic instability during telomere dysfunction may introduce unknown errors that can contribute to the ultimate cancer cell phenotype, including level of aggressiveness. Our hypotheses are consistent with recent publications indicating that many properties of invasive tumors are already present in their preinvasive DCIS lesions, such as tumor markers, gene expression profiles, gene methylation, PIK3CA mutations, and genomic errors [78–82].

Overcoming the telomere dysfunction barrier, with its associated genomic instability, is an additional point where molecular pathways to transformation may diverge. As described in more detail below, our studies have led us to speculate that at least two distinct sets of alterations may be involved in HMEC immortalization during malignant progression. One set of changes is needed for increased hTERT expression, and a subsequent set may be needed for maintenance of short stable telomeres.

We have generated a variety of immortally transformed lines, initially from specimen 184 and more recently from additional specimens, using various oncogenic agents (Fig. 15.1) [14, 18, 23–25, 58, 66, 83–87]. Most of these lines were derived from post-stasis cultures, although in a few instances (involving hTERT or c-Myc transduction) lines emerged following perturbations of pre-stasis populations. Our first immortal lines were obtained from the BaP post-stasis cultures, 184Aa and 184Be [23, 24, 66, 83]; extremely rare lines appeared at agonescence (184A1, 184AA4, 184AA8, 184B5, 184BE1) (Fig. 15.1, panel A). These cells had been exposed to BaP and likely harbor errors in addition to the loss of p16 expression. We hypothesize that rare errors produced by the genomic instability at agonescence can complement preexisting errors to allow telomerase reactivation. More frequent but still rare clonal lines appeared at agonescence following transduction of the breast cancer-associated oncogene ZNF217 into the BaP post-stasis 184Aa population (184AaZN1-3) [85], while inactivation of p53 in 184Aa using GSE22 produced frequent clonal immortalization at crisis (184AaGS1-2) [18]. Uniform immortalization was obtained following transduction of c-Myc into three different BaP cultures (184AaMY1, 184BeMY1, 184CeMY1) [58, 84].

No post-selection HMEC has been observed to spontaneously immortalize. Rare clonal lines appeared, usually around agonescence, following overexpression of ZNF217 (184ZN4-7) (Fig. 15.1, panel B) [84, 85]. One clonal line appeared in ten experiments where c-Myc was overexpressed (184SMY1) [58]. We hypothesize that rare errors generated by the genomic instability at agonescence may complement ZNF217 or c-Myc to allow telomerase reactivation. Overexpression of both c-Myc and ZNF217 in post-selection HMEC was able to produce immortal lines in repeat experiments (184ZNMY1-4, unpublished). Some of these lines immortalized early, prior to agonescence, and showed no chromosomal copy-number changes by comparative genomic hybridization (CGH) (Chin, Stampfer, Garbe, unpublished). However, Southern analysis of the retroviral insertion site indicates that these lines are also clonal.

More recently, we have targeted early passage pre-stasis cells grown in M85/M87A for transformation (Fig. 15.1, panel C) [58]. Our preliminary studies indicate that rare clonal lines may emerge following overexpression of c-Myc. If cultures are

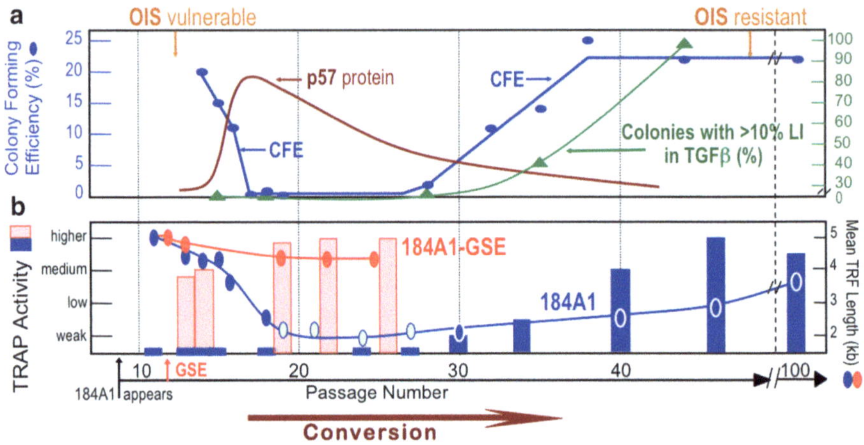

Fig. 15.8 Conversion of newly immortal p53(+) HMEC lines is associated with changes in many key properties. (**a**) The p53(+) 184A1 line undergoing conversion exhibits changes in growth capacity (CFE) and expression of p57 and becomes resistant to TGFβ growth inhibition and OIS. (**b**) 184A1 undergoing conversion exhibits changes in expression of telomerase activity and mean TRF length; *light blue ovals* indicate faint TRF signals. When pre-conversion 184A1 is transduced with GSE22, there is a rapid increase in telomerase activity associated with stabilization of TRF length (Adapted from Stampfer et al. [24, 83] and Nijjar et al. [88])

first transduced with shRNA to p16 and then c-Myc, apparently uniform immortalization occurs. In some instances, clonal lines have emerged around the time of agonescence from the cultures that received the p16sh alone. These new M85/M87A-derived lines are not yet well characterized. The ability of c-Myc to efficiently confer widespread, non-clonal immortalization to BaP and p16sh post-stasis populations provides a reproducible method of immortalization that can facilitate determining the mechanisms involved in HMEC immortalization and assaying agents that may prevent immortalization.

15.5.2 The Conversion Process During Immortalization

Observations of our immortally transformed lines possessing functional p53 led us to describe a process we called conversion [14, 24, 83–88]. Conversion has been most extensively studied in the first immortal line we derived, 184A1 (Fig. 15.8), but remains poorly understood. Tantalizing, but limited data, suggests that this little-known process could be involved in maintaining the short stable telomeres found in most carcinoma cells and in human epithelial cell lines immortalized by reactivation of endogenous telomerase activity [24, 83, 89].

Our basic observations have been that newly immortal clonal p53(+) HMEC lines, which have overcome agonescence and gained the potential to express telomerase, initially displayed little TRAP activity and had ongoing telomere erosion with proliferation. When telomeres became extremely short (≤3 kb), the conversion process ensued. Expression of the CKI p57^{Kip2} initially abruptly increased and then slowly declined, associated with initial slow-heterogeneous growth and then gradual reattaining of uniform good growth. Telomerase activity gradually increased, and the faint very short telomeres seen during conversion gradually became stabilized with a mean TRF of ~3–7 kb. Curiously, HMEC lines during conversion exhibit little evidence of telomere dysfunction and telomeric associations despite their extremely short mean TRF lengths and can emerge post-conversion with no or few additional gross genomic alterations [30, 58, 90]. We have speculated that the initial rapid rise in p57 levels, and associated growth inhibition, may provide protection from potentially catastrophic mitoses until sufficient telomerase becomes available to maintain telomeric ends. Associated with the increased telomerase activity, the immortal lines gradually developed the ability to maintain growth in the presence of TGFβ [86]. Inactivation of p53 function (using GSE22) in pre-conversion populations led to a rapid increase in endogenous TRAP activity, rapid reduction of existing p57 levels, gaining the ability to maintain growth in TGFβ, and lines with short, stable telomeres [24]. GSE22 transduction into the finite lifespan precursors of the immortal lines did not induce significant TRAP activity, indicating that abrogation of p53 function alone is not sufficient for telomerase reactivation in post-stasis HMEC. Therefore, functional p53 appears capable of repressing telomerase expression in newly immortal HMEC lines until conversion-associated change commence when the mean TRF declines to ≤3 kb.

Immortal HMEC lines that lack functional p53 (e.g., 184AA2, 184AA3) showed some initial TRAP activity that rapidly increased, showed no p57 expression, and quickly attained good uniform growth ± TGFβ. Their mean TRF length stabilized at ~4–5 kb and never declined to the very low levels seen in the p53(+) lines [24]. The rapid conversion and telomerase expression by p53(−) lines may explain why the process of conversion has not been commonly reported or investigated. Most in vitro immortalized human epithelial lines have been obtained using agents that inactivate p53. However, the majority of breast cancers express wild-type p53; a slower p53(+) conversion process may be relevant to early stage breast carcinogenesis in vivo. We have speculated that the low levels of telomerase expression coupled with extremely short telomeres could make newly immortal p53(+) breast cancers particularly vulnerable to therapeutic interventions targeting telomere dynamics.

In contrast to transduction of GSE22, ectopic overexpression of hTERT in p53(+) pre-conversion cells produces rapid full immortalization with TGFβ resistance but precludes conversion-induced reactivation of sufficient endogenous telomerase; the resultant populations did not exhibit short stable telomeres [86]. Similarly, when hTERT was transduced into post-stasis post-selection HMEC, rapid full immortalization with TGFβ resistance occurred, but the population also exhibited longer telomeres than 184A1 ± GSE22 [86]. Therefore, high telomerase expression by itself is sufficient to render these HMEC immortal and resistant to the growth inhibitory effects of TGFβ, although they remain sensitive to TGFβ induction of ECM-related molecules [91]. However, hTERT-induced immortalization did not lead to short stable telomeres.

A significant change associated with conversion is gaining resistance to OIS. Our post-conversion lines ± functional p53 are no longer vulnerable to OIS, but pre-conversion lines remained sensitive [14]. The TERT-immortalized lines from post-stasis and pre-conversion HMEC initially appear sensitive to OIS; SA-βGal is expressed and little proliferation is observed; however, growth can be maintained [47]. These results suggest that the process of endogenous telomerase reactivation during conversion is connected to changes in the pathways that govern OIS, a barrier relevant to in vivo human carcinogenesis [92–94].

Our current hypothesis is that conversion may reflect a need to alter chromosome conformation at the telomeres when cells transition from a finite state (no stable telomere length maintenance) to one where sufficient telomerase maintains short stable telomeres. Functional p53 may present a partial barrier to this process until very short telomeres provoke a structural change at the telomeric ends. As well studied in yeast, immortal cells can have "counting" mechanisms to maintain telomeres within a limited size range [95]. Since most human carcinoma cells, as well as our immortal HMEC lines, maintain telomeres within a short range (mean TRF~3–7 kb) [83, 89], some type of "counting" mechanism likely is involved. Short stable telomeres are not seen in normal telomerase-expressing human cells such as stem cells and lymphocytes [96], suggesting that active processes maybe required for conversion to the distinct telomeric state seen in the immortalized and cancer-derived cells. The longer mean TRF lengths of TERT-immortalized HMEC, and TERT-transduced pre-conversion 184A1 (lines that do not undergo conversion) [86] as well as their distinct OIS responses, indicate that lines immortalized by hTERT may not accurately reflect important biological properties and behaviors of carcinoma lines. Preliminary studies using the 184A1 line have shown that some of the epigenetic alterations seen in later passage 184A1 occur during the process of conversion and are not present in pre-conversion 184A1 populations or in later passage TERT-transduced pre-conversion 184A1 (Vrba, Novak, Stampfer, Futscher unpublished). Conversion may potentially represent a promising therapeutic target. If epigenetic alterations are required to allow stable telomere maintenance, interference with this process at the premalignant stage might prevent further progression.

15.5.3 Generation of Malignant HMEC Lines

Once the HMEC are immortally transformed and no longer vulnerable to OIS, the introduction of one or two oncogenes can further transform these cells towards malignancy (e.g., anchorage-independent growth, disorganized growth in Matrigel, growth factor independence, and/or tumorigenicity in immunosuppressed mice) (Fig. 15.1) [14, 44, 45, 97]. This property makes immortally transformed lines such as 184A1 and MCF10A useful for examining agents that can propel cells from the stage of nonmalignant immortal to malignancy and the mechanisms responsible for this transition [98–102]. The same oncogenes overexpressed in finite lifespan HMEC (both normal pre-stasis and abnormal post-stasis) do not confer malignancy and commonly induce senescence. Thus, in marked contrast to normal or finite cells, nonmalignant immortally transformed lines have acquired the errors that allowed them to escape multiple tumor-suppressive senescence barriers and be only one oncogene away from malignancy. The acquisition of OIS resistance upon immortalization likely contributes to the observation that immortality is the most common alteration from normal associated with human carcinomas. Nonmalignant immortal lines, having undergone many significant transformations from the normal state, do not constitute accurate "normal" controls nor do they represent the starting point of early stage carcinogenesis. They can control for changes associated with immortality when comparing immortal malignant tumor lines with nonmalignant immortal cells. We view nonmalignant immortal lines as at a stage similar to abnormal telomerase (+) cells in DCIS, which recent studies have shown already possess many of the errors found in breast cancers [78–82]. Since the aggressive phenotype of breast cancers may be predetermined early, at the premalignant stage, a better understanding of early stage progression, i.e., the steps from normal pre-stasis HMEC to nonmalignant immortal cells, may offer new insight into both the mechanisms of carcinogenesis and possibilities for therapeutic intervention in this progression.

15.5.4 Phenotypes of Immortal Versus Finite Lifespans HMEC

As previously noted, immortal HMEC lines differ from finite lifespan HMEC in their ability to maintain growth in the presence of TGFβ and their resistance to OIS. Most immortal lines, having undergone the period of genomic instability during telomere dysfunction, also exhibit gross genomic errors and ongoing genomic instability [24, 30, 58, 90]. We have further compared immortal, post-stasis, and pre-stasis HMEC for DMR and gene transcript profiles (Figs. 15.6 and 15.9) [25, 58, 75]. As mentioned above, post-stasis HMEC vary widely in number of DMR. When representative immortal lines were examined, a total of ~500 DMR were observed, most of which are also found in breast tumor-derived cells [25]. DMR found in post-selection post-stasis HMEC were also seen in immortal lines that derived from BaP post-stasis cultures lacking these DMR. An unsupervised clustering of DMR (Fig. 15.9a) shows most nonmalignant immortal lines clustering with tumor-derived lines and post-selection post-stasis cultures. Notably, the non-clonal lines derived from Myc-immortalized BaP post-stasis cultures had fewer DMR than the finite post-selection cells, although still many more DMR than seen in their post-stasis precursors. Venn diagrams (Fig. 15.9b) illustrate the overlaps and distinctions in DMR among the normal to tumor cells. A good example of stepwise DNA methylation changes during HMEC transformation can be seen in the HOXA gene cluster, known to undergo aberrant methylation during breast carcinogenesis (Fig. 15.6a) [103, 104]. Methylation increases towards the 3′ end of the cluster as normal HMEC transition to malignancy, with HMEC at different stages in our transformation model showing appropriate intermediate levels of methylation. Altogether, our epigenetic data indicate that cancer-associated DMR can occur at the earliest stages of transformation,

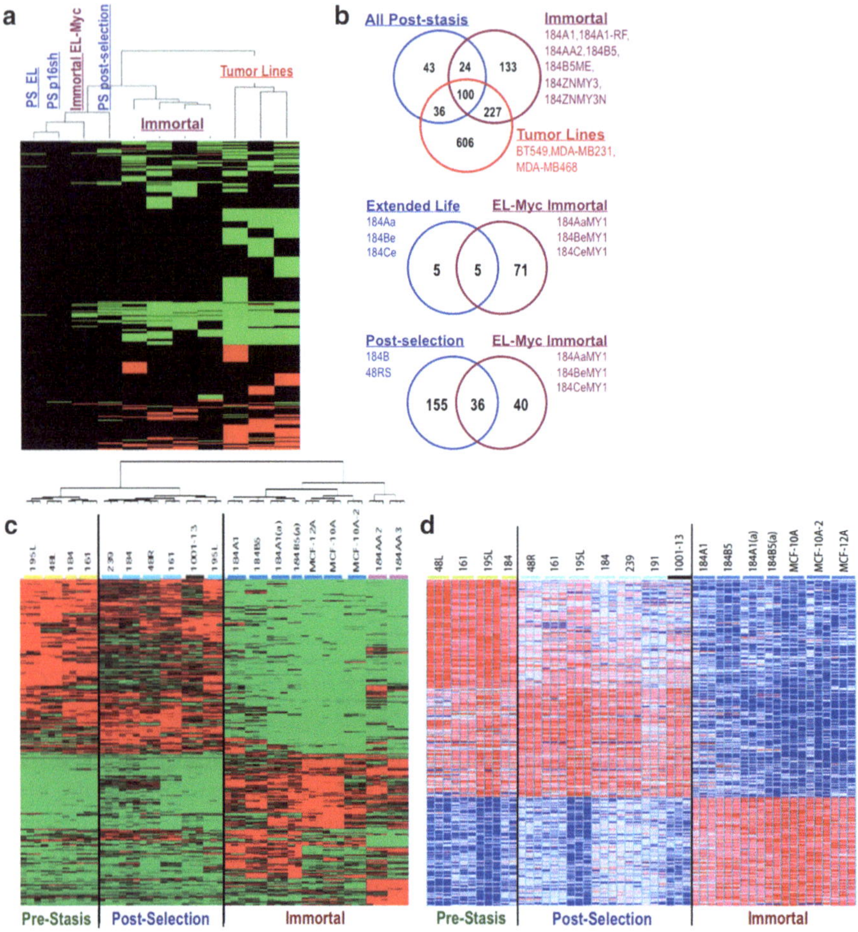

Fig. 15.9 Comparison of DMR and gene transcript profiles among finite (pre- and post-stasis) and immortal (nonmalignant and malignant) HMEC types. (**a**) Heat map showing clustering of DMR in post-stasis to malignant HMEC types, based on the microarray data from a custom array with 11,328 probes. *Red*, hypomethylated sites; *green*, hypermethylated sites. (**b**) Venn diagrams illustrating common and different DMR among post-stasis and immortal (nonmalignant and malignant) HMEC types. (**c**) Unsupervised clustering of gene transcript profiles of pre-stasis, post-selection, and immortalized HMEC. All genes (of 2,319) that changed expression in one or more samples were used to cluster the cell types and lines by overall similarity. Sample 1001-13 was HMEC, advertised as normal, obtained from Clonetics (Lonza). Samples of 184A1 and 184B5 designated by (**a**) were obtained from ATCC. (**d**) Supervised clustering of pre-stasis, post-selection, and immortalized HMEC. Gene expression values were normalized and characterized for the significance of overexpression in one group relative to other groups in the comparison. The top 200 genes (of 1,342) that are significantly overexpressed in one group are shown (**c** and **d** modified from Li et al. [75])

groups of DNA methylation changes can arise concurrently, and malignant progression is associated with progressive DMR changes. We are currently examining a larger range of our immortal HMEC to determine if specific epigenetic changes may correlate with specific parameters of the immortalization pathway.

Using both unsupervised and supervised clustering of gene transcript data, our immortal lines show similar expression as MCF10 and MCF12A and are clearly distinguished from the finite HMEC, although differences exist between pre-stasis and post-selection finite cells (Fig. 15.9c, d). HMEC advertised as normal that were obtained from Lonza/Clonetics (1001-13) sort cleanly with our aberrant post-selection post-stasis HMEC. Preliminary studies examining a greater range of our immortalized lines, post-stasis types (post-selection, BaP, and p16sh), and recent pre-stasis cultures (shown in Fig. 15.7a) are consistent with these earlier data.

Collectively, these data demonstrate that the transition to immortality is accompanied by major molecular alterations in gene expression, epigenetic marks, and other parameters associated with malignancy and show how nonmalignant immortal HMEC lines more closely resemble tumor cells than normal HMEC.

15.5.5 Lineage Characterization of Immortalized HMEC Lines

Human breast carcinomas exhibit a wide range of molecular properties, correlated with distinct clinical behaviors [1–4]. In general, we have seen that our different methods of inducing transformation can yield cell lines with significantly different properties; however, most in vitro immortalized lines thus far generated have lineage markers similar to the basal subtype of human breast cancers [1]. This subtype has a poor prognosis but represents only a minority of breast cancer. We now hypothesize that prior difficulty in developing a greater variety of transformed lines was due to (1) poor culture systems for growing normal human HMEC with luminal or progenitor properties, (2) use of cells from young reduction mammoplasty tissues, and (3) use of a limited number of oncogenic agents. We have begun characterization of lineage markers in some of our newly developed lines (Fig. 15.1, panel C). Lines from the younger women (184, 240L) again show a predominantly basal phenotype using FACS, IF, and IHC analyses, although variation is observed in CD24/CD44 ratios and EpCam expression [58]. The lines vary in expression of EMT- and stem cell-associated properties, gene transcript profiles, genomic errors, and other phenotypes [58] (Garbe, Vrba, Futscher, LaBarge, Stampfer, unpublished). Excitingly, our first experiments using HMEC from older women (805P, 122L) have yielded lines that express luminal and progenitor markers (Garbe, Stampfer, LaBarge, unpublished). Now that we can grow and FACS enrich progenitor and luminal cell populations, future studies can assess the relative contributions of target cell type and oncogenic agents employed in affecting the phenotype of resultant transformed lines.

The generation of transformed lines more representative of in vivo breast cancer types may enhance our understanding of the etiology and properties of a wider range of breast cancers. Thus far, a very limited number of nonmalignant immortal cell lines (mainly MCF10A, 184A1, MCF12A, 184B5, HMT-3522 S1) have been used in a large number of studies examining the transition from nonmalignant to malignant immortal. However, these lines all exhibit a basal or claudin-low phenotype and may consequently not accurately reflect properties of the majority luminal breast cancer types. For example, basal versus luminal HMEC may differ in their relative usage of the EGF/MEK versus IGF/PI3K signaling pathways and in EMT-associated properties.

15.6 Integrated Analysis of HMEC Model System

The work reviewed thus far illustrates how we have developed a wide-ranging cell culture system for investigating HMEC transformation. Starting with pre-stasis HMEC, exposure to a variety of oncogenic agents has generated cells at different stages of multistep carcinogenesis, providing isogenic cultures for examining the molecular alterations associated with progression. Such an integrated system avoids many of the variability problems inherent in comparing normal and transformed cells not only from different individuals but also from different organ sites, tissue types, and/or species and facilitates focus on changes due to the process of transformation. Being able to analyze the range of alterations from normal finite lifespan pre-stasis cells to malignantly transformed cells provides a comprehensive overview that assists understanding how the many alterations associated with carcinogenesis collaborate molecularly and temporally to produce cancerous cells. Below, we provide some examples of how this integrated HMEC model of transformation offers insight into processes associated with human breast carcinogenesis.

15.6.1 Genomic Instability, Telomeres, and Telomerase Expression

Normal human epithelial cells retain a stable genotype in vitro and in vivo; however, carcinomas usually express genomic instability and aneuploidy. Our model system has allowed us to examine when the transition to genomic instability occurs during the process of transformation in culture [15, 20, 24, 30, 90]. Proliferative finite lifespan HMEC undergoing telomere erosion due to insufficient telomerase activity maintain genomic stability until telomeres become critically short, which then leads to uncapped telomeres and telomeric associations. Cells arrested at stasis have a normal karyotype and noncritically short mean TRF (>5 kb), but virtually all HMEC arrested at agonescence display abnormal metaphases, with a preponderance of telomeric associations, and a mean TRF <5 kb [20, 105]. These results are consistent with in vivo data that show normal karyotypes in atypical ductal hyperplasia but genomic instability, abnormal karyotypes, and short telomeres at the DCIS stage [30, 43]. We have consequently hypothesized that the genomic instability naturally encountered in finite cells with eroded telomeres may contribute to the errors that allow these cells to reactivate sufficient telomerase activity and become immortal. However, in the vast majority of instances, in the absence of preexisting immortality-disposing errors, this genomic instability leads to cell death or proliferative arrest, providing an effective tumor-suppressive senescence barrier. The presence of short telomeres and genomic instability in a large percentage of DCIS argues against the proposition that the initial target cell for transformation in these cases already possessed sufficient telomerase activity.

We have further suggested that this inherent genomic instability resulting from eroded telomeres may be a significant contributor to the observed instability and resultant aneuploidy in breast cancer-derived cells. All our immortally transformed lines examined that encountered telomere dysfunction display CGH alterations and/or karyotypic abnormalities. Genomic alterations can be observed before and during the process of immortalization and conversion; however, once sufficient telomerase activity is present, the level of instability can decrease in p53(+) lines [24, 30, 105]. Presumably, telomerase allows telomere capping, preventing the formation of new telomeric associations, but the already present chromosomal derangements lead to ongoing cycles of BFB. In the absence of functional p53, lines that immortalized after undergoing genomic instability may show increasing instability [24]. Most of the genomic alterations generated by telomere dysfunction will be unrelated to the requirements of immortalization or carcinogenesis, but some could affect the clinical properties of a resultant malignant cell. It is therefore possible that the genomic instability in premalignant cells may be the source of many of the "passenger" mutations present in carcinomas, as well as of "driver" mutations that influence prognosis. This hypothesis is consistent with recent publications that suggest that the invasive phenotype of breast cancer is already genetically programmed at the preinvasive stages of disease progression [82].

Another possible corollary of our data is that the timing of telomerase reactivation during the period of telomere dysfunction may affect subsequent instability. Our immortal lines derived from cultures in the midst of telomere dysfunction contain many more genomic errors than lines derived before widespread instability ensued [24]. For example, the 184A1 line, which immortalized at ~7–8 p with a >5 kb mean TRF, has few errors, no BFB, and, unlike most in vitro transformed HMEC lines, can remain genomically stable upon passage. In contrast, the 184AA4 line, also derived from post-stasis 184Aa, immortalized at ~12–13 p when the population was experiencing telomere dysfunction, and exhibits numerous genomic errors and ongoing instability. Possibly, breast cancers with diploid karyotypes reflect cells that underwent immortalization prior to extensive telomere dysfunction-induced genomic instability.

We have recently begun examination of the CGH profiles and karyology of the non-clonal immortal lines generated by direct targeting of the stasis and telomere dysfunctional barriers using transduction of p16sh and c-Myc. The three lines thus far tested contained cells with normal karyotypes at early passage [58]. This result supports our model of the tumor-suppressive senescence barriers and the hypothesis that genomic instability functions to generate errors critical for transformation, but is not essential per se. If the stasis and telomere dysfunction senescence barriers are bypassed by direct targeting, generation of genomic errors may be unnecessary. We therefore believe that no specific mutator genes are required to account for the genomic instability seen in breast carcinomas, although mutations that do increase instability could also be present and selected for during malignant progression, and contribute to this phenotype. Rather, development of genomic instability is inherent in the process of malignant progression, particularly at the stage of telomere dysfunction. The further development of aneuploidy has been proposed to result from dysfunctional telomeres interfering with the completion of cytokinesis [105].

15.6.2 Immortalization and Responses to TGFβ

Normal cultured HMEC are growth inhibited by TGFβ and show induction of ECM- and proteolysis-related molecules (e.g., fibronectin, collagen IV, laminin, type IV collagenase, uPA, and PAI-1) [13, 24, 86, 91]. Our studies comparing the responses to TGFβ of HMEC ranging from normal pre-stasis to transformed [13, 24, 86, 91, 106–108] have indicated that the expression of telomerase activity (from either endogenous reactivation or transduction of hTERT) is sufficient by itself to allow HMEC to maintain growth in TGFβ while also remaining responsive to TGFβ-mediated protein induction. These results were among the first to demonstrate that multiple TGFβ actions can operate via divergent pathways, since the effects on cell growth could be dissociated from stimulation of ECM components. Immortal and malignant lines can maintain growth in TGFβ, although some may exhibit a slightly decreased growth rate, likely reflecting the metabolic price exacted by the increased synthesis and secretion; we do not consider this reflective of a direct growth inhibition. The mechanism by which telomerase activity prevents TGFβ from inhibiting growth is still unknown; however, our results indicate that immortally transformed HMEC expressing hTERT do not require additional errors to become TGFβ growth resistant.

In contrast to normal HMEC, many human carcinomas, including breast, can maintain growth in the presence of TGFβ while retaining other metabolic responses, similar to our immortalized lines, although some carcinomas have lost all responsiveness [109]. However, only rare mutations in the TGFβ pathway have been found in breast cancers [109, 110]. Based on our in vitro data, we suggest that during in vivo carcinogenesis no additional errors beyond acquisition of immortality may be needed to confer resistance to TGFβ growth inhibition. During malignant progression in vivo, it could be beneficial to cancer cells to retain the capacity for TGFβ-inducible ECM- or EMT-related functions while avoiding the growth inhibition. However, it might be advantageous in some circumstances to avoid the growth inhibition prior to full immortalization or to forgo the additional metabolic expenditures, accounting for situations where mutations are observed and all responses to TGFβ are lost.

15.6.3 Immortalization and OIS

Another significant alteration associated with the process of telomerase expression and immortalization is the acquisition of resistance to OIS. For malignant progression to proceed, the normal ability of cells to respond to inappropriate oncogenic expression by cessation of growth needs to be abrogated. We and colleagues have seen that ectopic overexpression of oncogenes such as Raf-1, Ras, and ErbB2 in finite lifespan HMEC can produce rapid growth inhibition, whereas similarly exposed nonmalignant immortal lines maintain growth and acquire malignancy-associated properties such as anchorage-independent growth and reduced growth factor requirements [14, 16, 44, 45]. The transition to OIS resistance is a critical alteration and emphasizes the importance of immortalization and particularly the conversion step in tumorigenesis. The mechanism by which HMEC and other human epithelial cells gain OIS resistance is currently unclear but differs from what has been reported for most fibroblast or rodent cells in lacking a requirement for functional p53, p16, ATM, or CHK2 [14, 16]. A curious and potentially important observation, although thus far not further explored, suggests the possibility that the differences in OIS responses of finite versus immortal HMEC may be based on different levels of expression of the oncogenes. We noted that basal and 4-HT-induced expression levels of the Raf-1:ER transgene, as well as phosphorylated MEK, were consistently reduced in post-conversion immortal 184A1 compared to finite HMEC or pre-conversion 184A1 [14]. When post-conversion 184A1-Raf-1:ER was sorted to obtain cells showing the highest levels of Raf-1, immediate assay showed decreased survival upon induction with 4-HT. However, when these sorted cells were amplified and reexamined, Raf-1 levels were again reduced. Conversely, when post-selection HMEC transduced with Raf-1:ER were exposed to low levels of 4-HT so that expression levels were similar to those seen in 184A1, not only were the cells not growth inhibited, they showed increased growth capacity in the absence of EGF. Possibly, abnormally high levels of oncogene expression can trigger OIS, while lower levels can confer malignant properties, and immortal HMEC may have a mechanism to prevent ongoing high-level expression.

Altogether, these examples emphasize the critical role the immortalization step plays in HMEC carcinogenesis.

15.7 Conclusions

The development of an extensive, integrated culture system for examining normal and aberrant HMEC behavior has allowed us to take a comprehensive overview of how the processes functioning in normal HMEC become subverted during transformation and the relationship of individual alterations incurred with resultant transformed phenotype. Several generic conclusions can be drawn from these studies.

First, a major caveat needs to be considered when evaluating these data. Most of this work was performed with cells growing in two dimensions on plastic substrates, whereas normal and aberrant epithelial cell processes in vivo involve complex interactions of polarized cells within three-dimensional organ systems. As others have elegantly shown [54–56], many important cellular behaviors will differ when cells are placed in culture environments that support cell polarity and provide ECM material and stromal interaction. In developing our HMEC culture system, we tried to balance the goal of being amenable to widespread use with the goal of optimizing the system to reflect in vivo biology. We consequently focused on standard tissue culture technology in order to generate sufficient HMEC to support large-scale, reproducible investigation. HMEC cultures have subsequently been examined using 3D culture systems such as Matrigel or micropatterned wells [49, 52, 57], but such studies have thus far been limited. Importantly, normal HMEC placed in 3D environments have demonstrated appropriate self-organization, indicating that the needed lineage-specific properties have been retained. However, data obtained in 2D culture may not accurately reflect the in vivo biology. A situation where this issue may be most relevant involves the transition from a nonmalignant immortal cell (DCIS) to a malignant primary cancer cell. In vivo, this transition likely involves epithelial-stromal interactions, a hypoxic environment, and selection for errors that promote malignancy-associated properties, such as invasiveness and angiogenesis. Our in vitro selection is only for immortality and is unlikely to recapitulate the changes associated with this transition in vivo. A more accurate approximation of in vivo 3D biology can be expected to offer new and better insights into the processes underlying carcinogenesis and aging.

A common thread in our studies is the extent of diversity and heterogeneity among normal and abnormal cell types and how this manifests in significantly different molecular processes. In terms of understanding human carcinoma progression, HMEC and epithelial cells in general have many significant biological differences compared to HMF and fibroblasts in general. An outstanding example involves mechanisms of senescence. Among many distinctions, there was almost no overlap between genes modulated at HMEC senescence (stasis and telomere dysfunction) and genes modulated in senescent HMF; there were differences in molecules modulated by HMEC and HMF during OIS; normal HMEC are growth inhibited by TGFβ exposure, while isogenic HMF respond with a slight growth stimulation [13–16]. These distinctions are important in light of the common use of fibroblasts to study mechanisms of senescence, often with an implied assumption that the results obtained are generic to "cells." Our development of robust culture conditions for normal HMEC should encourage increased usage of human epithelial cells to understand what is distinct about their senescence mechanisms, which in turn play prominent roles in suppressing carcinogenesis.

Significant differences are found comparing human and rodent epithelial cells in mechanism relevant to carcinogenesis. Most important, rodents lack stringent repression of telomerase in adult cells and thus the crucial telomere dysfunction senescence barrier. They also differ in the relative roles of the CKIs $p19/14^{ARF}$ and p16 in stasis and immortalization. While rodent models offer the ability to perform in vivo experimentation, the critical errors required by human epithelial cells for immortalization will not be amenable for discovery using rodent models. The immortalization step presents a potentially valuable therapeutic target, since almost all breast cancers, regardless of subtype, exhibit telomerase reactivation and are dependent upon immortalization for malignant progression. Further, unlike signaling pathways where extensive redundancy contributes to development of therapeutic resistance, the use of alternate (ALT) pathways for telomere maintenance is extremely rare in human epithelial cells and in breast cancers [111]. Efforts to clinically exploit the requirement for immortalization-promoting errors will be enhanced by the availability of human epithelial cell culture systems that support experimental examination of genomic, epigenomic, and gene expression alterations associated with immortalization.

Heterogeneity exists among normal HMEC in vivo and in vitro. Cells with luminal versus myoepithelial lineage markers may have differences, such as signaling pathway usage, that carry over to observed differences between luminal and basal tumor cell lines. Multiple types of progenitor populations are also present, with distinctions presumably based on epigenetic marks and other properties. Identification and characterization of the different normal HMEC types may be relevant for identification of the initial target cell types of the different breast cancer subtypes and how the properties of the target cell influence cancer progression and treatment. The ability to grow and FACS enrich these diverse lineages in our cultured pre-stasis HMEC can facilitate studies that assess the effects of various oncogenic exposures on differing initial target populations. Additionally, FACS-enriched normal populations may serve as more accurate normal controls for type-specific breast cancer cells than an unsorted heterogeneous population. Proliferative normal HMEC in culture can in some instances provide more relevant controls for proliferative cancer cells than comparisons of normal and tumor tissue in vivo. Normal HMEC in vivo have low proliferation rates, and properties associated with a proliferative state may be erroneously assessed to be tumor-specific based on examination of in vivo tissues.

During carcinogenesis, heterogeneity is amplified by driver and passenger alterations acting on the initial target cells, resulting in the diversity of breast cancer subtypes with corresponding diverse clinical parameters. Since a goal of personalized medicine is matching therapeutic modalities with the specific errors present in individual tumors, the accuracy with which experimental models in vitro match in vivo molecular parameters will influence the usefulness of such models for evaluating potential therapeutics. Such considerations underscored our use of oncogenic agents thought to play a role in breast cancer etiology in our HMEC transformation models. For example, most breast cancers express wild-type p53 and retain functional RB [112]. Loss of these key molecular hubs will have much greater consequences on a cancer cell's behavior than impairment of one sub-pathway, such as loss of p16. Immortalization of HMEC achieved by use of viral oncogenes SV40T or HPVE6 and E7 not only inactivates p53 and RB function but also produces many other undefined changes; cells transformed by such methods are unlikely to provide accurate models for exploring potential breast cancer type-specific therapeutics. Lines immortalized by ectopic overexpression of hTERT will lack the critical alterations associated with reactivation of endogenous telomerase, including, as discussed above, the process of conversion and related changes in telomere dynamics and OIS responses. Additionally, transformation systems employing hTERT and viral oncogenes are not amenable for understanding the mechanisms of HMEC immortalization during in vivo carcinogenesis or therefore examination of agents that might prevent this step in progression. By using pathologically relevant oncogenic agents, we have obtained transformed cells that share many of the properties seen during in vivo breast carcinogenesis, such as retaining wild-type p53 and RB. By requiring cells to reactivate endogenous telomerase activity, we can examine the crucial immortalization step during cancer progression. Nonetheless, until recently, we have not been able to model most of the phenotypes observed in actual breast cancers.

Most published in vitro transformed HMEC lines have exhibited a basal, triple-negative, or claudin-low phenotype, while the majority of breast cancers belong to luminal subtypes. A large number of studies are being performed on a very limited set of immortalized lines that are not representative of most breast cancers and may not be reflective of most breast cancers' behavior. Consequently, our more recent efforts have been directed towards generating transformed lines more reflective of the range of breast cancers in vivo. These studies are currently in progress but indicate that using HMEC from older women as target cells, and employing additional agents to bypass stasis, can lead to transformed cells lines with luminal lineage markers.

The other main conclusion from our integrated model system is the crucial role of telomere dysfunction and the immortalization step in human breast cancer progression. The changes associated with overcoming telomere dysfunction support multiple aspects of tumor progression. In addition to the advantages provided by unlimited proliferative potential, the immortalization step also promotes genomic instability, changes OIS into oncogenic promotion of malignancy, and abrogates TGFβ-induced growth inhibition while leaving cells responsive to TGFβ-induced tumor promotion and EMT. Our comparisons of isogenic finite and immortal HMEC indicate that the transition from finite to immortal is associated with the greatest extent of changes in epigenomic marks and gene expression. The requirement of immortalization for malignancy and the lack of easy redundant alternatives to telomerase reactivation suggest that immortalization may be a valuable target for clinical intervention. While there has been significant effort to develop pharmacologic agents that could interfere with telomerase action, other errors necessary to attain or maintain immortalization could also be valuable targets. Our limited understanding of the mechanisms underlying human epithelial cell immortalization, and the absence of accurate rodent models of this step, has held up exploration of this possibility. Our development of reproducible methods for non-clonal immortalization using pathologically relevant agents may open up new way to explore potential novel therapeutics targeting this step.

The processes implicated in human epithelial cell senescence and carcinogenesis in vivo are complex, involving alterations both within a cell's genome and physiology, and in relationship to its immediate and whole-body environment. Many recent exciting publications that directly examine human breast tissues are providing large quantities of information about the pathways and derangements associated with breast carcinogenesis and illuminating the extent of inter- and intra-tumor heterogeneity during various stages of tumor development [5–9]. However, determining cause and effect relationships, identifying driver abnormalities among the hundreds of other changes, and testing potential therapeutics are constrained using only in vivo approaches. An in vitro HMEC model system, although also limited, offers an experimentally tractable approach to investigate the effects of individual perturbations in HMEC at different stages in transformation along distinct transformation pathways. It can be expected that the closer such model systems reflect the processes occurring in vivo, the more accurate they will be for assessing potential clinical interventions. We have presented an overview of our integrated HMEC model system for such experimentation and highlighted some of the ways in which our comprehensive culture system has provided novel insight into these complex processes. Importantly, our model system starts with normal finite lifespan pre-stasis HMEC, allowing examination of the critical early stage changes that occur as normal cells transition to immortality. Ongoing improvements in HMEC model systems, including better modeling of 3D and microenvironmental conditions and of the range of pathways to and phenotypes of transformed cells, can greatly assist efforts to delineate the different pathways a normal HMEC can take to become malignant and enable investigation into potential therapeutic approaches to prevent malignant progression.

Acknowledgements This work has been supported by NIH grants CA24844, CA112970, AG033176, and AG040081, Department of Defense BCRP grants BC030946, BC060444, and W81XWH-04-1-0580, and the Office of Energy Research, Office of Health and Biological Research, U.S. Department of Energy under Contract No. DE-AC02-05CH11231.

References

1. Perou CM, Sorlie T, Eisen MB, van de Rijn M, Jeffrey SS, Rees CA, Pollack JR, Ross DT, Johnsen H, Akslen LA, Fluge O, Pergamenschikov A, et al. Molecular portraits of human breast tumours. Nature. 2000;406:747–52.
2. Sorlie T, Perou CM, Tibshirani R, Aas T, Geisler S, Johnsen H, Hastie T, Eisen MB, van de Rijn M, Jeffrey SS, Thorsen T, Quist H, et al. Gene expression patterns of breast carcinomas distinguish tumor subclasses with clinical implications. Proc Natl Acad Sci USA. 2001;98:10869–74.
3. Jacquemier J, Ginestier C, Rougemont J, Bardou VJ, Charafe-Jauffret E, Geneix J, Adelaide J, Koki A, Houvenaeghel G, Hassoun J, Maraninchi D, Viens P, et al. Protein expression profiling identifies subclasses of breast cancer and predicts prognosis. Cancer Res. 2005;65:767–79.
4. Troester MA, Hoadley KA, Sorlie T, Herbert BS, Borresen-Dale AL, Lonning PE, Shay JW, Kaufmann WK, Perou CM. Cell-type-specific responses to chemotherapeutics in breast cancer. Cancer Res. 2004;64:4218–26.
5. Shah SP, Roth A, Goya R, Oloumi A, Ha G, Zhao Y, Turashvili G, Ding J, Tse K, Haffari G, Bashashati A, Prentice LM, et al. The clonal and mutational evolution spectrum of primary triple-negative breast cancers. Nature. 2012;486:395–9.
6. Curtis C, Shah SP, Chin SF, Turashvili G, Rueda OM, Dunning MJ, Speed D, Lynch AG, Samarajiwa S, Yuan Y, Graf S, Ha G, et al. The genomic and transcriptomic architecture of 2,000 breast tumours reveals novel subgroups. Nature. 2012;486:346–52.
7. Nik-Zainal S, Alexandrov LB, Wedge DC, Van Loo P, Greenman CD, Raine K, Jones D, Hinton J, Marshall J, Stebbings LA, Menzies A, Martin S, et al. Mutational processes molding the genomes of 21 breast cancers. Cell. 2012;149:979–93.
8. Nik-Zainal S, Van Loo P, Wedge DC, Alexandrov LB, Greenman CD, Lau KW, Raine K, Jones D, Marshall J, Ramakrishna M, Shlien A, Cooke SL, et al. The life history of 21 breast cancers. Cell. 2012;149:994–1007.
9. Stephens PJ, Tarpey PS, Davis H, et al. The landscape of cancer genes and mutational processes in breast cancer. Nature. 2012;486:400–4.

10. Prowse KR, Greider CW. Developmental and tissue-specific regulation of mouse telomerase and telomere length. Proc Natl Acad Sci USA. 1995;92:4818–22.
11. Gil J, Peters G. Regulation of the INK4b-ARF-INK4a tumour suppressor locus: all for one or one for all. Nat Rev Mol Cell Biol. 2006;7:667–77.
12. Bartley J, Bartholomew JC, Stampfer MR. Metabolism of benzo(a)pyrene by human epithelial and fibroblastic cells: metabolite patterns and DNA adduct formation. J Cell Biochem. 1982;18:135–48.
13. Hosobuchi M, Stampfer MR. Effects of transforming growth factor-β on growth of human mammary epithelial cells in culture. In Vitro. 1989;25:705–12.
14. Olsen CL, Gardie B, Yaswen P, Stampfer MR. Raf-1-induced growth arrest in human mammary epithelial cells is p16-independent and is overcome in immortal cells during conversion. Oncogene. 2002;21:6328–39.
15. Garbe JC, Bhattacharya S, Merchant B, Bassett E, Swisshelm K, Feiler HS, Wyrobek AJ, Stampfer MR. Molecular distinctions between stasis and telomere attrition senescence barriers shown by long-term culture of normal human mammary epithelial cells. Cancer Res. 2009;69:7557–68.
16. Cipriano R, Kan CE, Graham J, Danielpour D, Stampfer M, Jackson MW. TGF-beta signaling engages an ATM-CHK2-p53-independent RAS-induced senescence and prevents malignant transformation in human mammary epithelial cells. Proc Natl Acad Sci USA. 2011;108:8668–73.
17. Vrba L, Garbe JC, Stampfer MR, Futscher BW. Epigenetic regulation of normal human mammary cell type-specific miRNAs. Genome Res. 2011;21:2026–37.
18. Garbe JC, Holst CR, Bassett E, Tlsty T, Stampfer MR. Inactivation of p53 function in cultured human mammary epithelial cells turns the telomere-length dependent senescence barrier from agonescence into crisis. Cell Cycle. 2007;6:1927–36.
19. Brenner AJ, Stampfer MR, Aldaz CM. Increased p16INK4a expression with onset of senescence of human mammary epithelial cells and extended growth capacity with inactivation. Oncogene. 1998;17:199–205.
20. Romanov S, Kozakiewicz K, Holst C, Stampfer MR, Haupt LM, Tlsty T. Normal human mammary epithelial cells spontaneously escape senescence and acquire genomic changes. Nature. 2001;409:633–7.
21. Stampfer MR. Cholera toxin stimulation of human mammary epithelial cells in culture. In Vitro. 1982;18:531–7.
22. Hammond SL, Ham RG, Stampfer MR. Serum-free growth of human mammary epithelial cells: rapid clonal growth in defined medium and extended serial passage with pituitary extract. Proc Natl Acad Sci USA. 1984;81:5435–9.
23. Stampfer MR, Bartley JC. Induction of transformation and continuous cell lines from normal human mammary epithelial cells after exposure to benzo[a]pyrene. Proc Natl Acad Sci USA. 1985;82:2394–8.
24. Stampfer MR, Garbe J, Nijjar T, Wigington D, Swisshelm K, Yaswen P. Loss of p53 function accelerates acquisition of telomerase activity in indefinite lifespan human mammary epithelial cell lines. Oncogene. 2003;22:5238–51.
25. Novak P, Jensen TJ, Garbe JC, Stampfer MR, Futscher BW. Step-wise DNA methylation changes are linked to escape from defined proliferation barriers and mammary epithelial cell immortalization. Cancer Res. 2009;67:5251–8.
26. Geradts J, Wilson PA. High frequency of aberrant $p16^{INK4A}$ expression in human breast cancer. Am J Pathol. 1996;149:15–20.
27. Loughran O, Malliri A, Owens D, Gallimore PH, Stanley MA, Ozanne B, Frame MC, Parkinson EK. Association of CDKN2A/p16INK4A with human head and neck keratinocyte replicative senescence: relationship of dysfunction to immortality and neoplasia. Oncogene. 1996;13:561–8.
28. Baylin SB, Herman JG, Graff JR, Vertino PM, Issa JP. Alterations in DNA methylation: a fundamental aspect of neoplasia. Adv Cancer Res. 1998;72:141–96.
29. Brenner AJ, Paladugu A, Wang H, Olopade OI, Dreyling MG, Aldaz CM. p16 at senescence and extended life. Clin Cancer Res. 1996;2:1993–8.

30. Chin K, Ortiz de Solorzano C, Knowles D, Jones A, Chou W, Rodriguez E, Kuo W-L, Ljung B-M, Chew K, Krig S, Garbe J, Stampfer M. In situ analysis of genome instability in breast cancer. Nat Genet. 2004;36:984–8.
31. Fumagalli M, Rossiello F, Clerici M, Barozzi S, Cittaro D, Kaplunov JM, Bucci G, Dobreva M, Matti V, Beausejour CM, Herbig U, Longhese MP, et al. Telomeric DNA damage is irreparable and causes persistent DNA-damage-response activation. Nat Cell Biol. 2012;14:555.
32. Brown JP, Wei W, Sedivy JM. Bypass of senescence after disruption of p21CIP1/WAF1 gene in normal diploid human fibroblasts. Science. 1997;277:831–4.
33. Rheinwald JG, Hahn WC, Ramsey MR, Wu JY, Guo Z, Tsao H, De Luca M, Catricala C, O'Toole KM. A two-stage, p16^{INK4a}-and p53-dependent keratinocyte senescence mechanism that limits replicative potential independent of telomere status. Mol Cell Biol. 2002;22:5157–72.
34. Evans RJ, Wyllie FS, Wynford-Thomas D, Kipling D, Jones CJ. A P53-dependent, telomere-independent proliferative life span barrier in human astrocytes consistent with the molecular genetics of glioma development. Cancer Res. 2003;63:4854–61.
35. Tlsty TD, Romanov SR, Kozakiewicz BK, Holst CR, Haupt LM, Crawford YG. Loss of chromosomal integrity in human mammary epithelial cells subsequent to escape from senescence. J Mammary Gland Biol Neoplasia. 2001;6:235–43.
36. Soler D, Genesca A, Arnedo G, Egozcue J, Tusell L. Telomere dysfunction drives chromosomal instability in human mammary epithelial cells. Genes Chromosomes Cancer. 2005;44:339–50.
37. Zou Y, Sfeir A, Gryaznov SM, Shay JW, Wright WE. Does a sentinel or a subset of short telomeres determine replicative senescence? Mol Biol Cell. 2004;15:3709–18.
38. Ouelette MM, Liao M, Herbert B, Johnson M, Holt SE, Liss HS, Shay JW, Wright WE. Subsenescent telomere lengths in fibroblasts immortalized by limiting amounts of telomerase. J Biol Chem. 2000;275:10072–6.
39. Bednarek A, Sahin A, Brenner AJ, Johnston DA, Aldaz CM. Analysis of telomerase activity in breast cancer: positive detection at the in situ breast carcinoma stage. Clin Cancer Res. 1997;3:11–6.
40. Poremba C, Shroyer KR, Frost M, Diallo R, Fogt F, Schafer K-L, Burger H, Shroyer AL, Dockhorn-Dworniczak B, Boecker W. Telomerase is a highly sensitive and specific molecular marker in fine-needle aspirates of breast lesions. J Clin Oncol. 1999;17:2020–6.
41. Shpitz B, Zimlichman S, Zemer R, Bomstein Y, Zehavi T, Liverant S, Bernehim J, Kaufman Z, Klein E, Shapira Y, Klein A. Telomerase activity in ductal carcinoma *in situ* of the breast. Breast Cancer Res Treat. 1999;58:65–9.
42. Sugino T, Yoshida K, Bolodeoku J, Tahara H, Buley I, Manek S, Wells C, Goodison S, Ide T, Suzuki T, Tahara E, Tarin D. Telomerase activity in human breast cancer and benign breast lesions: diagnostic applications in clinical specimens, including fine needle aspirates. Int J Cancer. 1996;69:301–6.
43. Meeker AK, Beckman-Hicks JL, Argani P, De Marzo AM. Dramatic telomere shortening occurs in breast luminal epithelial cells in subsets of non-malignant, normal-appearing lobules and small ducts: a potential early molecular mechanism underlying breast tumorigenesis. Proc Am Assoc Cancer Res. 2003;44:182.
44. Clark R, Stampfer M, Milley B, O'Rourke E, Walen K, Kriegler M, Kopplin J. Transformation of human mammary epithelial cells by oncogenic retroviruses. Cancer Res. 1988;48:4689–94.
45. Pierce JH, Arnstein P, DiMarco E, Artrip J, Kraus MH, Lonardo F, DiFiore PP, Aaronson SA. Oncogenic potential of *erb*B-2 in human mammary epithelial cells. Oncogene. 1991;6:1189–94.
46. Borgdorff V, Lleonart ME, Bishop CL, Fessart D, Bergin AH, Overhoff MG, Beach DH. Multiple microRNAs rescue from Ras-induced senescence by inhibiting p21(Waf1/Cip1). Oncogene. 2010;29:2262–71.
47. Sherman MY, Meng L, Stampfer MR, Gabai VL, Yaglom JA. Oncogenes induce senescence with incomplete growth arrest and suppress the DNA damage response. Aging Cell. 2011;10(6):949–61.
48. Stampfer MR, Hallowes R, Hackett AJ. Growth of normal human mammary epithelial cells in culture. In Vitro. 1980;16:415–25.

49. LaBarge MA, Garbe JC, Stampfer MR. Processing of human reduction mammoplasty and mastectomy tissues for cell culture. J Vis Exp. 2013;71:e50011.
50. Stampfer MR. Isolation and growth of human mammary epithelial cells. J Tissue Cult Methods. 1985;9:107–16.
51. Taylor-Papadimitriou J, Stampfer MR, Bartek J, Lane EB, Lewis A. Keratin expression in human mammary epithelial cells cultured from normal and malignant tissue: relation to in vivo phenotypes and influence of medium. J Cell Sci. 1989;94:403–13.
52. Garbe JC, Pepin F, Pelissier F, Sputova K, Fridriksdottir AJ, Guo DE, Villadsen R, Park M, Petersen OW, Barowsky A, Stampfer MR, Labarge MA. Accumulation of multipotent progenitors with a basal differentiation bias during aging of human mammary epithelia. Cancer Res. 2012;72:3687–701.
53. Lim E, Vaillant F, Wu D, Forrest NC, Pal B, Hart AH, Asselin-Labat ML, Gyorki DE, Ward T, Partanen A, Feleppa F, Huschtscha LI, et al. Aberrant luminal progenitors as the candidate target population for basal tumor development in BRCA1 mutation carriers. Nat Med. 2009;15:907–13.
54. Han J, Chang H, Giricz O, Lee GY, Baehner FL, Gray JW, Bissell MJ, Kenny PA, Parvin B. Molecular predictors of 3D morphogenesis by breast cancer cell lines in 3D culture. PLoS Comput Biol. 2010;6:e1000684.
55. Kenny PA, Lee GY, Myers CA, Neve RM, Semeiks JR, Spellman PT, Lorenz K, Lee EH, Barcellos-Hoff MH, Petersen OW, Gray JW, Bissell MJ. The morphologies of breast cancer cell lines in three-dimensional assays correlate with their profiles of gene expression. Mol Oncol. 2007;1:84–96.
56. Weaver VM, Lelievre S, Lakins JN, Chrenek MA, Jones JC, Giancotti F, Werb Z, Bissell MJ. beta4 Integrin-dependent formation of polarized three-dimensional architecture confers resistance to apoptosis in normal and malignant mammary epithelium. Cancer Cell. 2002;2:205–16.
57. Chanson L, Brownfield G, Garbe JC, Kuhn I, Stampfer MR, Bissell MJ, Labarge MA. Self-organization is a dynamic and lineage-intrinsic property of mammary epithelial cells. Proc Natl Acad Sci USA. 2011;108:3264–9.
58. Garbe JC, Vrba L, Sputova K, Fuchs L, Novak P, Jackson MW, Chin K, LaBarge MA, Watts GS, Futscher BW, Stampfer MR. Efficient immortalization of normal human mammary epithelial cells using two pathologically relevant agents does not require gross genomic alterations. In prep.
59. Foster SA, Galloway DA. Human papillomavirus type 16 E7 alleviates a proliferative block in early passage human mammary epithelial cells. Oncogene. 1996;12:1773–9.
60. Ramirez RD, Morales CP, Herbert BS, Rohde JM, Passons C, Shay JW, Wright WE. Putative telomere-independent mechanisms of replicative aging reflect inadequate growth conditions. Genes Dev. 2001;15:398–403.
61. Bartley JC, Stampfer MR. Factors influencing benzo(a)pyrene metabolism in human mammary epithelial cells in culture. Carcinogenesis. 1985;6:1017–22.
62. Reis-Filho JS, Milanezi F, Carvalho S, Simpson PT, Steele D, Savage K, Lambros MB, Pereira EM, Nesland JM, Lakhani SR, Schmitt FC. Metaplastic breast carcinomas exhibit EGFR, but not HER2, gene amplification and overexpression: immunohistochemical and chromogenic in situ hybridization analysis. Breast Cancer Res. 2005;7:R1028–35.
63. Hoadley KA, Weigman VJ, Fan C, Sawyer LR, He X, Troester MA, Sartor CI, Rieger-House T, Bernard PS, Carey LA, Perou CM. EGFR associated expression profiles vary with breast tumor subtype. BMC Genomics. 2007;8:258.
64. Siziopikou KP, Cobleigh M. The basal subtype of breast carcinomas may represent the group of breast tumors that could benefit from EGFR-targeted therapies. Breast. 2007;16:104–7.
65. Lapuk A, Marr H, Jakkula L, Pedro H, Bhattacharya S, Purdom E, Hu Z, Simpson K, Pachter L, Durinck S, Wang N, Parvin B, et al. Exon-level microarray analyses identify alternative splicing programs in breast cancer. Mol Cancer Res. 2010;8:961–74.
66. Stampfer MR, Bartley JC. Human mammary epithelial cells in culture: differentiation and transformation. Cancer Treat Res. 1988;40:1–24.

67. Ossovskaya VS, Mazo IA, Chernov MV, Chernova OB, Strezoska Z, Kondratov R, Stark GR, Chumakov PM, Gudkov AV. Use of genetic suppressor elements to dissect distinct biological effects of separate p53 domains. Proc Natl Acad Sci USA. 1996;93:10309–14.
68. Lehman T, Modali R, Boukamp P, Stanek J, Bennett W, Welsh J, Metcalf R, Stampfer M, Fusenig N, Rogan E, Reddel R, Harris C. p53 mutations in human immortalized epithelial cell lines. Carcinogenesis. 1993;14:833–9.
69. Delmolino L, Band H, Band V. Expression and stability of p53 protein in normal human mammary epithelial cells. Carcinogenesis. 1993;14:827–32.
70. Keller PJ, Arendt LM, Skibinski A, Logvinenko T, Klebba I, Dong S, Smith AE, Prat A, Perou CM, Gilmore H, Schnitt S, Naber SP, et al. Defining the cellular precursors to human breast cancer. Proc Natl Acad Sci USA. 2012;109:2772–7.
71. Holst CR, Nuovo GJ, Esteller M, Chew K, Baylin SB, Herman JG, Tlsty TD. Methylation of p16(INK4a) promoters occurs in vivo in histologically normal human mammary epithelia. Cancer Res. 2003;63:1596–601.
72. Gao Q, Hauser SH, Liu X-L, Wazer DE, Madoc-Jones H, Band V. Mutant p53-induced immortalization of primary human mammary epithelial cells. Cancer Res. 1996;56: 3129–33.
73. Gollahon LS, Shay JW. Immortalization of human mammary epithelial cells transfected with mutant p53 (273^{his}). Oncogene. 1996;12:715–25.
74. Zhang J, Pickering CR, Holst CR, Gauthier ML, Tlsty TD. p16INK4a modulates p53 in primary human mammary epithelial cells. Cancer Res. 2006;66:10325–31.
75. Li Y, Pan J, Li JL, Lee JH, Tunkey C, Saraf K, Garbe JC, Whitley MZ, Jelinsky SA, Stampfer MR, Haney SA. Transcriptional changes associated with breast cancer occur as normal human mammary epithelial cells overcome senescence barriers and become immortalized. Mol Cancer. 2007;6:7.
76. Stampfer MR, Yaswen P. Human epithelial cell immortalization as a step in carcinogenesis. Cancer Lett. 2003;194:199–208.
77. Gomes NM, Ryder OA, Houck ML, Charter SJ, Walker W, Forsyth NR, Austad SN, Venditti C, Pagel M, Shay JW, Wright WE. Comparative biology of mammalian telomeres: hypotheses on ancestral states and the roles of telomeres in longevity determination. Aging Cell. 2011;10:761–8.
78. Warnberg F, Nordgren H, Bergkvist L, Holmberg L. Tumour markers in breast carcinoma correlate with grade rather than with invasiveness. Br J Cancer. 2001;85:869–74.
79. Muggerud AA, Ronneberg JA, Warnberg F, Botling J, Busato F, Jovanovic J, Solvang H, Bukholm I, Borresen-Dale AL, Kristensen VN, Sorlie T, Tost J. Frequent aberrant DNA methylation of ABCB1, FOXC1, PPP2R2B and PTEN in ductal carcinoma in situ and early invasive breast cancer. Breast Cancer Res. 2010;12:R3.
80. Miron A, Varadi M, Carrasco D, Li H, Luongo L, Kim HJ, Park SY, Cho EY, Lewis G, Kehoe S, Iglehart JD, Dillon D, et al. PIK3CA mutations in in situ and invasive breast carcinomas. Cancer Res. 2010;70:5674–8.
81. Ma XJ, Salunga R, Tuggle JT, Gaudet J, Enright E, McQuary P, Payette T, Pistone M, Stecker K, Zhang BM, Zhou YX, Varnholt H, et al. Gene expression profiles of human breast cancer progression. Proc Natl Acad Sci USA. 2003;100:5974–9.
82. Espina V, Liotta LA. What is the malignant nature of human ductal carcinoma in situ? Nat Rev Cancer. 2011;11:68–75.
83. Stampfer MR, Bodnar A, Garbe J, Wong M, Pan A, Villeponteau B, Yaswen P. Gradual phenotypic conversion associated with immortalization of cultured human mammary epithelial cells. Mol Biol Cell. 1997;8:2391–405.
84. Stampfer MR, Yaswen P. Immortal transformation and telomerase reactivation of human mammary epithelial cells in culture. In: Mattson M, Pandita T, editors. Telomerase, aging and disease, vol. 8. Amsterdam: Elsevier; 2001. p. 103–30.
85. Nonet G, Stampfer MR, Chin K, Gray JW, Collins CC, Yaswen P. The *ZNF217* gene amplified in breast cancers promotes immortalization of human mammary epithelial cells. Cancer Res. 2001;61:1250–4.

86. Stampfer MR, Garbe J, Levine G, Lichtsteiner S, Vasserot AP, Yaswen P. Expression of the telomerase catalytic subunit, hTERT, induces resistance to transforming growth factor beta growth inhibition in p16INK4A(-) human mammary epithelial cells. Proc Natl Acad Sci USA. 2001;98:4498–503.
87. Garbe J, Wong M, Wigington D, Yaswen P, Stampfer MR. Viral oncogenes accelerate conversion to immortality of cultured human mammary epithelial cells. Oncogene. 1999;18:2169–80.
88. Nijjar T, Wigington D, Garbe JC, Waha A, Stampfer MR, Yaswen P. p57/KIP2 loss of heterozygosity and expression during immortal conversion of human mammary epithelial cells. Cancer Res. 1999;59:5112–8.
89. van Steensel B, de Lange T. Control of telomere length by the human telomeric protein TRF1. Nature. 1997;385:740–3.
90. Walen K, Stampfer MR. Chromosome analyses of human mammary epithelial cells at stages of chemically-induced transformation progression to immortality. Cancer Genet Cytogenet. 1989;37:249–61.
91. Stampfer MR, Yaswen P, Alhadeff M, Hosoda J. TGFβ induction of extracellular matrix associated proteins in normal and transformed human mammary epithelial cells in culture is independent of growth effects. J Cell Physiol. 1993;155:210–21.
92. Bartkova J, Horejsi Z, Koed K, Kramer A, Tort F, Zieger K, Guldberg P, Sehested M, Nesland JM, Lukas C, Orntoft T, Lukas J, et al. DNA damage response as a candidate anti-cancer barrier in early human tumorigenesis. Nature. 2005;434:864–70.
93. Gorgoulis VG, Vassiliou LV, Karakaidos P, Zacharatos P, Kotsinas A, Liloglou T, Venere M, Ditullio Jr RA, Kastrinakis NG, Levy B, Kletsas D, Yoneta A, et al. Activation of the DNA damage checkpoint and genomic instability in human precancerous lesions. Nature. 2005;434:907–13.
94. Halazonetis TD, Gorgoulis VG, Bartek J. An oncogene-induced DNA damage model for cancer development. Science. 2008;319:1352–5.
95. Shore D, Bianchi A. Telomere length regulation: coupling DNA end processing to feedback regulation of telomerase. EMBO J. 2009;28:2309–22.
96. Aubert G, Baerlocher GM, Vulto I, Poon SS, Lansdorp PM. Collapse of telomere homeostasis in hematopoietic cells caused by heterozygous mutations in telomerase genes. PLoS Genet. 2012;8:e1002696.
97. Frittitta L, Vigneri R, Stampfer MR, Goldfine ID. Insulin receptor overexpression in 184B5 human mammary epithelial cells induces a ligand-dependent transformed phenotype. J Cell Biochem. 1995;57:666–9.
98. Avdulov S, Li S, Michalek V, Burrichter D, Peterson M, Perlman DM, Manivel JC, Sonenberg N, Yee D, Bitterman PB, Polunovsky VA. Activation of translation complex eIF4F is essential for the genesis and maintenance of the malignant phenotype in human mammary epithelial cells. Cancer Cell. 2004;5:553–63.
99. Isakoff SJ, Engelman JA, Irie HY, Luo J, Brachmann SM, Pearline RV, Cantley LC, Brugge JS. Breast cancer-associated PIK3CA mutations are oncogenic in mammary epithelial cells. Cancer Res. 2005;65:10992–1000.
100. Atwood AA, Sealy L. Regulation of C/EBPbeta1 by Ras in mammary epithelial cells and the role of C/EBPbeta1 in oncogene-induced senescence. Oncogene. 2010;29:6004–15.
101. Iliopoulos D, Hirsch HA, Wang G, Struhl K. Inducible formation of breast cancer stem cells and their dynamic equilibrium with non-stem cancer cells via IL6 secretion. Proc Natl Acad Sci USA. 2011;108:1397–402.
102. Leung CT, Brugge JS. Outgrowth of single oncogene-expressing cells from suppressive epithelial environments. Nature. 2012;482:410–3.
103. Novak P, Jensen T, Oshiro MM, Wozniak RJ, Nouzova M, Watts GS, Klimecki WT, Kim C, Futscher BW. Epigenetic inactivation of the HOXA gene cluster in breast cancer. Cancer Res. 2006;66:10664–70.
104. Novak P, Jensen T, Oshiro MM, Watts GS, Kim CJ, Futscher BW. Agglomerative epigenetic aberrations are a common event in human breast cancer. Cancer Res. 2008;68:8616–25.

105. Pampalona J, Frias C, Genesca A, Tusell L. Progressive telomere dysfunction causes cytokinesis failure and leads to the accumulation of polyploid cells. PLoS Genet. 2012;8:e1002679.
106. Valverius EM, Walker-Jones D, Bates SE, Stampfer MR, Clark R, McCormick F, Dickson RB, Lippman ME. Production of and responsiveness to transforming growth factor β in normal and oncogene transformed human mammary epithelial cells. Cancer Res. 1989;49:6407–11.
107. Slingerland JM, Hengst L, Pan C-H, Alexander D, Stampfer MR, Reed SI. A novel inhibitor of cyclin-cdk activity detected in transforming growth factor β-arrested epithelial cells. Mol Cell Biol. 1994;14:3683–94.
108. Sandhu C, Garbe J, Bhattacharya N, Daksis JI, Pan C-H, Yaswen P, Koh J, Slingerland JM, Stampfer MR. TGF-β stabilizes p15INK4B protein, increases p15INK4B/cdk4 complexes and inhibits cyclin D1-cdk4 association in human mammary epithelial cells. Mol Cell Biol. 1997;17:2458–67.
109. Dumont N, Arteaga CL. Targeting the TGF beta signaling network in human neoplasia. Cancer Cell. 2003;3:531–6.
110. Seoane J. Escaping from the TGFbeta anti-proliferative control. Carcinogenesis. 2006;27: 2148–56.
111. Subhawong AP, Heaphy CM, Argani P, Konishi Y, Kouprina N, Nassar H, Vang R, Meeker AK. The alternative lengthening of telomeres phenotype in breast carcinoma is associated with HER-2 overexpression. Mod Pathol. 2009;22:1423–31.
112. Shackney SE, Silverman JF. Molecular evolutionary patterns in breast cancer. Adv Anat Pathol. 2003;10:278–90.

Chapter 16
New Breast Cancer Treatment Considerations: A Brief Review of the Use of Genetically Modified (Attenuated) Bacteria as Therapy for Advanced and Metastatic Breast Cancer

Robert A. Kazmierczak, Alison Dino, Abraham Eisenstark, and Heide Schatten

Abstract Breast cancer is still among the most common life-threatening cancers that affects one out of eight women, and it further affects a small percent of the male population. While early detection has been helpful to reduce the mortality rate, we currently still do not have cures for advanced and metastatic breast cancer. In recent years, new strategies have been proposed to treat breast cancers with poor prognosis by utilizing genetically modified bacteria, including *Salmonella typhimurium*, that preferentially replicate within solid tumors (1,000:1 and up to 10,000:1 compared to noncancerous tissue) destroying cancer cells without causing septic shock that is typically associated with wild-type *S. typhimurium* infections. Furthermore, these bacteria have the potential to be utilized as drug delivery systems to more effectively target different subpopulations of breast tumor cells. This chapter reviews progress in using genetically modified *S. typhimurium* for destruction of breast cancer cells in culture and in solid breast cancer tissue. We discuss the potential and future prospects for applications in clinical trials as novel breast cancer therapy for advanced stages of the disease. We further discuss potential combination therapies for optimal destruction of breast cancer cells.

16.1 Introduction

Metastatic breast cancer is still among the most difficult cancers to treat and traditional treatments such as surgery, radiation, chemotherapy, or hormone treatments have been employed with only limited success. In addition, resistance to anticancer

R.A. Kazmierczak • A. Dino • A. Eisenstark
Cancer Research Center, Columbia, MO 65201, USA

H. Schatten (✉)
Department of Veterinary Pathobiology, University of Missouri-Columbia,
1600 E Rollins Street, Columbia, MO 65211, USA
e-mail: SchattenH@missouri.edu

H. Schatten (ed.), *Cell and Molecular Biology of Breast Cancer*,
DOI 10.1007/978-1-62703-634-4_16, © Springer Science+Business Media New York 2013

drugs is among the primary reasons for breast cancer mortality. There is a compelling need for new approaches to treat breast cancer especially in women with advanced tumor progression or breast cancer metastasis. Several investigations have focused on the potential use of genetically modified *Salmonella* to treat breast cancer in advanced and metastatic stages. This approach is based on findings that genetically modified bacteria such as *Salmonella typhimurium* preferentially replicate within solid tumors (1,000:1 or greater, compared to noncancerous tissue) destroying cancer cells without causing septic shock that is typically associated with wild-type *S. typhimurium* infections. Furthermore, these bacteria may be utilized as drug delivery systems to effectively target different subpopulations of tumor tissue. Direct targeting of tumor tissue has several advantages over intravenous or oral drug administration, as it is more effective and it would also eliminate site effects that are oftentimes caused when non-tumor cells are destroyed by the anticancer drugs that are not directly reaching the tumor tissue.

In addition, advances in genetic engineering of bacteria allow specific tumor-targeting capabilities and increased effectiveness to deliver anticancer agents into solid tumors. Studies so far have shown that it will be particularly important to determine the most effective strains of bacteria, as different strains of genetically modified bacteria may target different cellular components which can be utilized for optimal and effective tumor therapy. Several thousands of *S. typhimurium* strains are currently already available that differ from each other genetically and can be utilized for further modifications. This allows engineering of strains that are optimal in appropriate lipopolysaccharide surface for attachment and penetration into tumor cells, antibiotic sensitivity, metabolic poverty of entire pathways, as well as individual nutrients. Such genetically modified bacteria might also be able to deliver nano-encapsulated drugs directly into the tumor where drugs are released to destroy cancer cells.

16.2 From Anecdotal Information to Our Current State of Knowledge and Progress Toward Human Clinical Trials

The ability of bacteria to cause tumor regression had been recognized over 100 years ago [1], and more studies followed when physicians noticed that tumor patients exposed to *Salmonella* contaminations showed regression of their tumors [2]. The casual observation was possible because requirements for appropriate hygiene had not yet advanced to our current standards and knowledge about the need for hygiene was still in infant stages. Patients with tumors were placed in beds that had previously been occupied by patients with *Salmonella* infections without considering contamination by the infectious agent. However, the casual observations resulted in the development of Coley's toxin, a bacterial extract that stimulated a general immune response [2–4]. The idea of stimulating the patient's immune response to fight cancer has prevailed to our current time, and different lines of research have focused on fighting cancers by boosting our own immune responses [5, 6].

Subsequent investigators started to explore the use of bacteria as anticancer therapeutic agents [7, 8] taking advantage of the discoveries that tumors contain anoxic regions [9] which inspired research on using the obligate anaerobe *Clostridium* to evoke tumor regression in mice. In a subsequent small clinical trial, oncolysis was observed in three out of five patients that had been injected with *Clostridium butyricum* [10]. Further efforts to use bacteria as anticancer agents included *Salmonella* [11], *Escherichia* [12], *Clostridium* [7, 8], *Bifidobacterium* [13], *Caulobacter* [14], *Listeria* [15, 16], *Proteus* [17], and *Streptococcus* [18], yielding varying results.

For anaerobic bacteria, cancer tissue represents an ideal environment in which nutrient enrichment provides an optimal source for bacterial growth and cell division. In addition, tumor tissue vasculature contains large intercapillary distances allowing bacteria to easily reach the tumor tissue while they are more restricted from noncancer tissue with "non-leaky" vasculature [19].

Current research has focused on *Salmonella* for its anticancer effectiveness, as *Salmonella* was shown to cause effective tumor regression [11, 20] in multiple individual research approaches. Targeted bioengineering of *Salmonella* has resulted in significant progress in destroying tumor cells in culture [21–24] and in tumor tissue of animal models including melanoma, colon, lung, and breast cancers [20, 25–32]. It was also shown that *Salmonella* evokes immune sensitization and antitumor immune responses, inducing neutrophil infiltration into tumors [33]. Numerous studies followed to generate optimal strains of *Salmonella,* and the attenuated *Salmonella* strain termed VNP 20009 with a modified lipid A component was first used in human clinical trials in patients with metastatic melanoma; VNP 20009 is nontoxic and had shown tumor colonization [34] in prior animal studies. In dogs, administration of the *Salmonella* strain VNP 20009 resulted in complete cures of 4 out of 35 animals [35]. However, clinical trials in humans showed limited success, and it was recognized that further research is needed to optimally design bacterial strains to target tumor in humans.

Since then multiple laboratories have focused their efforts on designing new strains of genetically modified *Salmonella* with the goal to satisfy the following criteria for optimal strain development aimed at (1) entrapment of bacteria in the vasculature of tumor [19], (2) infiltration of tumors following inflammation [36], (3) chemotaxis toward compounds produced by tumors [37, 38], (4) preferential growth in tumor-specific microenvironments [37, 39], and (5) protection from recognition by the immune system [40]. These approaches utilize specific knowledge of bacteria-tumor interactions and include, for example, chemotaxis of *Salmonella* toward small molecular gradients such as serine, aspartase, and ribose or others that are enriched in tumor tissue. Tumor-specific auxotroph *Salmonella* have been created that require leucine and arginine for bacterial survival; these nutrients are available for bacterial metabolism from dying tumor tissue [31, 39]. The leaky tumor vasculature [41, 42] and bacterial motility allow for diagnosis and treatment of cancers in distal and hard-to-image regions of the tumors which is especially important for the treatment of metastatic tumors. Bacteria can move from the vasculature to distant tumor tissue by using flagellum-driven motility [43].

Other research has focused on generating strains for transfection with genes for therapeutic molecules including toxins [44–46], cytokines [5, 47], tumor antigens [48], apoptosis-inducing factors [49–53], and several others that are reviewed in more detail by [54].

Most of the studies so far have been conducted in the mouse which allowed fast advances in the field, and it further allowed advances using genetically modified *Salmonella* for tumor imaging and therapy. New research has also focused on using bacteria for diagnosis and better determination of tumor for imaging tumor tissue and metastases. Labeled bacteria can be detected by light microscopy [12, 39, 55] including multiphoton tomography, magnetic resonance imaging (MRI) [56], and positron emission tomography (PET) [57–59]. The use of quantum dot- and nanoparticle-labeled bacteria will further add to improved diagnosis as well as therapy. Quantum dots combined with nano-encapsulation of drugs is an attractive novel approach for tumor imaging and drug delivery.

Other approaches to use bacteria for tumor imaging and identification employ magnetotactic bacteria that could be detected by MRI and include the use of *Magnetospirillum magneticum*, a bacterium that produces magnetic particles. *M. magneticum* has been shown to accumulate in tumors [56]. Numerous variations of the abovementioned approaches are possible utilizing our current knowledge of bacterial gene transfer and bacterial gene manipulations to create nontoxic bacterial strains for simultaneous tumor imaging, diagnosis, and treatment. The use of quantum dots in combination with new strain development is particularly attractive in this regard. By using our current technologies, it is easy to generate bioluminescent bacteria by transformation with plasmid containing the lucCDABE operon from *Photobacterium leiognathi* [12, 44, 60–62] or generating fluorescent bacteria by transformation with plasmids containing the gene for fluorescent protein [12, 39, 55] which has been used for imaging and identifying tumors in mice using whole-mouse imaging [12, 39, 44, 55, 60–62].

Such imaging approaches would also benefit patients with multiple tumor foci in breast cancer as well as in other cancers such as pancreatic and ovarian cancers that are hard to detect through early diagnosis. The use of such genetically modified *Salmonella* would also allow reaching pockets of cancer that are not treatable by conventional methods and include targeting of tumor stem cells [63] which comprise a small population of cells that can cause tumor growth reinitiation resulting in tumor reoccurrence.

The abovementioned advances along with new studies to target tumor stem cell niches with *Salmonella* to prevent the reoccurrence of cancer are all aimed at producing strains of genetically modified *Salmonella* for specific tumor targeting. An additional advantage of custom-designing *Salmonella* strains for specific tumor targeting is the fact that multiple strains with different genetic makeup can be utilized to target different subpopulations of the heterogeneous cell populations within a specific tumor. It would allow personalized designs of strains that would find use in newly pursued personalized medicine approaches.

16.3 New Strategies and New Hopes for More Efficient Breast Cancer Therapy Including Nanotechnology Combination Approaches

Effective delivery of drugs into the tumor tissue is still among the most challenging aspects in tumor therapy. New developments in nanotechnology have advanced this goal significantly and revealed three aspects that are important for effective drug delivery into target tissue which are (a) efficient encapsulation of the drugs, (b) successful delivery of the drugs to the targeted region in the body, and (c) successful release of the drug. However, while nano-encapsulation of drugs has resulted in progress toward effective targeting of tumor tissue, the methods for effective drug delivery are still in the exploratory stages. In several cases, phagocytosis and encapsulation into liposomes have been used successfully for drug transport and delivery into tumor tissue, and animal studies are underway in a number of laboratories to take this approach for successful tumor-targeting drug administration ([64]; reviewed by [65]).

Successful nano-encapsulation of breast cancer drugs includes new developments of nanoparticles that are loaded with doxorubicin or taxol (abraxane) (reviewed in [65, 66]). Engineering of bacteria to deliver such drugs has been proposed but not yet moved into practical applications. However, the potential of prodrug or drug delivery by bacteria is significant, and the prospects of using bacteria as nanoparticle carriers that specifically target tumor tissue are highly promising [54].

Bacterially derived 400 nm *Salmonella* minicells labeled with tumor-specific antibody receptors have successfully delivered doxorubicin [67] and siRNA therapies [68] to implanted breast tumors in mouse models.

The therapeutic effectiveness of *Salmonella* producing inhibiting RNA, protein, vaccines, and prodrug enzymes at tumor sites has recently been reported. It was further shown that *Salmonella* successfully targeted a MTDH/AEG-1-based DNA vaccine to orthotopically implanted 4T1 breast cancer tumors, resulting in enhanced doxorubicin activity which inhibited tumor growth and metastatic spread [69].

Taken together, these new advances of using nontoxic genetically modified bacteria for breast cancer therapy offer new realistic hopes to successfully treat advanced and metastatic tumors.

16.4 Conclusions and Future Perspectives

Genetically modified bacteria offer new hope to effectively combat breast cancer, especially in advanced and metastatic stages for which optimal treatment strategies are not yet available. Several lines of research are aimed at determining the best suited genetically modified bacteria, as many avenues are possible that have not yet been explored at this time. Determining the most effective strain of genetically modified bacteria that will effectively invade tumor tissue and destroy cancer cells without causing side effects is within reach and is actively pursued in several laboratories worldwide. In addition to endogenously produced cancer therapeutics, encapsulating drugs through nanotechnology is also pursued in several laboratories worldwide and will allow more efficient tumor targeting than the currently available administration methods. The methods for drug delivery are in exploratory stages and have already been tested in animal models including phagocytosis, minicell, and liposome utilization. The use of nontoxic genetically modified bacteria for delivery of encapsulated drugs directly into tumor tissue is an attractive avenue to pursue. These approaches would allow increase in drug efficiency by only reaching and affecting the tumor-infested organ, thereby preventing site effects that are mainly caused when noncancer cells in healthy tissue are damaged by the antitumor drugs before reaching the tumor tissue.

Acknowledgment This work was supported by the Cancer Research Center.

References

1. Coley WB. Late results of the treatment of inoperable sarcoma by the mixed toxins of Erysipelas and Bacillus prodigosus. Am J Med Sci. 1906;131:375–430.
2. Nauts HC, Swift WE, Coley BL. The treatment of malignant tumors by bacterial toxins as developed by the late William B. Coley, MD, reviewed in the light of modern research. Cancer Res. 1946;6:205–16.
3. Hall SS. A commotion in the blood: life, death, and the immune system. New York: Henry Holt; 1997.
4. Coley WB. Contribution to the knowledge of sarcoma. Ann Surg. 1981;14:199–220.
5. Loeffler M, Le'Negrate G, Krajewska M, Reed JC. Attenuated *Salmonella* engineered to produce human cytokine LIGHT inhibit tumor growth. Proc Natl Acad Sci USA. 2007;104:12879–83.
6. Fensterle J, Bergmann B, Yone CL, Hotz C, Meyer SR, Spreng S, Goebel W, Rapp UR, Gentschev I. Cancer immunotherapy based on recombinant *Salmonella enterica* serovar Typhimurium *aroA* strains secreting prostate-specific antigen and cholera toxin subunit B. Cancer Gene Ther. 2008;15:85–93.
7. Parker RC, Plummer HC, Siebenmann CO, Chapman MG. Effect of histolyticus infection and toxin on transplantable mouse tumors. Proc Soc Exp Biol Med. 1947;66:461–7.
8. Malmgren RA, Flanigan CC. Localization of the vegetative form of *Clostridium tetani* in mouse tumor following intravenous spore administration. Cancer Res. 1955;15:473–8.
9. Mottram JC. Factors of importance in radiosensitivity of tumors. Br J Radiol. 1936;9:606–14.
10. Carey RW, Holland JF, Whang HY, Neter E, Bryant B. Clostridial oncolysis in man. Eur J Cancer. 1967;3:37–46.
11. Pawelek JM, Low KB, Bermudes D. Tumor targeted *Salmonella* as a novel anticancer vector. Cancer Res. 1997;57:4537–44.
12. Yu YA, Shabahang S, Timiryasova TM, Zhang Q, Beltz R, Gentschev I, Goebel W, Szalay AA. Visualization of tumors and metastases in live animals with bacteria and vaccinia virus encoding light-emitting proteins. Nat Biotechnol. 2004;22:313–20.
13. Kohwi Y, Imai K, Tamura Z, Hashimoto Y. Antitumor effect of *Bifidobacterium* infantis in mice. Gann. 1978;69:613–8.
14. Bhatnagar PK, Awasthi A, Nomellini JF, Smit J, Suresh MR. Anti-tumor effects of the bacterium caulobacter crescentus in murine tumor models. Cancer Biol Ther. 2006;5:485–91.

15. Pan ZK, Weiskirch LM, Paterson Y. Regression of established B16F10 melanoma with a recombinant *Listeria monocytogenes* vaccine. Cancer Res. 1999;59:5264–9.
16. Kim SH, Castro F, Paterson Y, Gravekamp C. High efficacy of a *Listeria*-based vaccine against metastatic breast cancer reveals a dual mode of action. Cancer Res. 2009;69:5860–6.
17. Arakawa M, Sugiura K, Reilly HC, Stock CC. Oncolytic effect of *Proteus mirabilis* upon tumor bearing animals. II. Effect on transplantable mouse and rat tumors. Gann. 1968;59:117–22.
18. Maletzki C, Linnebacher M, Kreikemeyer B, Emmrich J. Pancreatic cancer regression by intratumoural injection of live *Streptococcus pyogenes* in a syngeneic mouse model. Gut. 2008;57:483–91.
19. Forbes NS, Munn LL, Fukumura D, Jain RK. Sparse initial entrapment of systemically injected *Salmonella typhimurium* leads to heterogeneous accumulation within tumors. Cancer Res. 2003;63:5188–93.
20. Low KB, Ittensohn M, Le T, Platt J, Sodi S, Amoss M, Ash O, Carmichael E, Chakraborty A, Fischer J, Lin SL, Luo X, Miller SI, Zheng L, King I, Pawelek JM, Bermudes D. Lipid A mutant *Salmonella* with suppressed virulence and TNFα induction retain tumor targeting in vivo. Nat Biotechnol. 1999;17:37–41.
21. Schatten H, Eisenstark A. Destruction of human breast cancer cells by tumor-targeted Salmonella. World and Ehrlich Conference on Magic Bullets. Germany: Nürnberg; 2004.
22. Schatten H, Fea A, French W, Eisenstark A. Breast tumor targeting with genetically altered Salmonella. Microsc Microanal. 2005; 11(Suppl S02):932–3.
23. Eisenstark A, Kazmierczak RA, Fea A, Khreis R, Newman D, Schatten H. Development of *Salmonella* strains as cancer therapy agents and testing in tumor cell lines. In: Schatten H, Eisenstark A, editors. Methods in molecular biology, vol. 253: Salmonella protocols. Totowa, NJ: Humana; 2007. p. 321–53.
24. Zhong Z, Kazmierczak RA, Fea A, Khreis R, Eisenstark A, Schatten H. *Salmonella*-host cell interactions, changes in host cell architecture, and destruction of prostate tumor cells with genetically altered *Salmonella*. Microsc Microanal. 2007;13(5):372–83.
25. Bermudes D, Low KB, Pawelek J, Feng M, Belcourt M, Zheng LM, King I. Tumour-selective *Salmonella* based cancer therapy. Biotechnol Genet Eng Rev. 2001;18:219–33.
26. Bermudes D, Zheng LM, King IC. Live bacteria as anticancer agents and tumor-selective protein delivery vectors. Curr Opin Drug Discov Devel. 2002;5:194–9.
27. Pawelek JM, Sodi S, Chakraborty AK, Platt JT, Miller S, Holden DW, Hensel M, Low KB. *Salmonella* pathogenicity island-2 and anticancer activity in mice. Cancer Gene Ther. 2002;9:813–8.
28. Pawelek JM, Low KB, Bermudes D. Bacteria as tumour-targeting vectors. Lancet Oncol. 2003;4:548–56.
29. Chakrabarty AM. Microorganisms and cancer: quest for a therapy. J Bacteriol. 2003;185: 2683–6.
30. Saltzman DA. Cancer immunotherapy based on the killing of *Salmonella typhimurium-infected* tumour cells. Expert Opin Biol Ther. 2005;5:443–9.
31. Zhao M, Yang M, Ma HY, Li XM, Tan XY, Li SK, Yang ZJ, Hoffman RM. Targeted therapy with a Salmonella typhimurium leucine-arginine auxotroph cures orthotopic human breast tumors in nude mice. Cancer Res. 2006;66:7647–52.
32. Yu B, Yang M, Shi L, Yao Y, Jiang Q, Li X, Tang LH, Zheng BJ, Yuen KY, Smith DK, Song E, Huang JD. Explicit hypoxia targeting with tumor suppression by creating an "obligate" anaerobic Salmonella Typhimurium strain. Sci Rep. 2012;2:436.
33. Lee CH, Wu CL, Shiau AL. *Salmonella choleraesuis* as an anticancer agent in a syngeneic model of orthotopic hepatocellular carcinoma. Int J Cancer. 2008;122:930–5.
34. Toso JF, Gill VJ, Hwu P, Marincola FM, Restifo NP, Schwartzentruber DJ, Sherry RM, Topalian SL, Yang JC, Stock F, Freezer LJ, Morton KE, Seipp C, Haworth L, Mavroukakis S, White D, MacDonald S, Mao J, Sznol M, Rosenberg SA. Phase I study of the intravenous administration of attenuated *Salmonella typhimurium* to patients with metastatic melanoma. J Clin Oncol. 2002;20:142–52.

35. Thamm DH, Kurzman ID, King I, Li ZJ, Sznol M, Dubielzig RR, Vail DM, MacEwen EG. Systemic administration of an attenuated, tumor-targeting Salmonella typhimurium to dogs with spontaneous neoplasia: phase I evaluation. Clin Cancer Res. 2005;11:4827–34.
36. Leschner S, Westphal K, Dietrich N, Viegas N, Jablonska J, Lyszkiewicz M, Lienenklaus S, Falk W, Gekara N, Loessner H, Weiss S. Tumor invasion of *Salmonella enterica* serovar Typhimurium is accompanied by strong hemorrhage promoted by TNF-α. PLoS One. 2009;4:e6692.
37. Kasinskas RW, Forbes NS. *Salmonella typhimurium* specifically chemotax and proliferate in heterogeneous tumor tissue in vitro. Biotechnol Bioeng. 2006;94:710–21.
38. Kasinskas RW, Forbes NS. *Salmonella typhimurium* lacking ribose chemoreceptors localize in tumor quiescence and induce apoptosis. Cancer Res. 2007;67:3201–9.
39. Zhao M, Yang M, Li XM, Jiang P, Baranov E, Li SK, Xu MX, Penman S, Hoffman RM. Tumor targeting bacterial therapy with amino acid auxotrophs of GFP-expressing *Salmonella typhimurium*. Proc Natl Acad Sci USA. 2005;102:755–60.
40. Sznol M, Lin SL, Bermudes D, Zheng LM, King I. Use of preferentially replicating bacteria for the treatment of cancer. J Clin Invest. 2000;105:1027–30.
41. Jain RK. The next frontier of molecular medicine: delivery of therapeutics. Nat Med. 1998;4:655–7.
42. Vaupel P, Kallinowski F, Okunieff P. Blood flow, oxygen and nutrient supply, and metabolic microenvironment of human tumors: a review. Cancer Res. 1989;49:6449–65.
43. Toley BJ, Forbes NS. Motility is critical for effective distribution and accumulation of bacteria in tumor tissue. Integr Biol (Camb). 2012;4(2):165–76.
44. Nguyen VH, Kim HS, Ha JM, Hong Y, Choy HE, Min J-J. Genetically engineered Salmonella typhimurium as an imageable therapeutic probe for cancer. Cancer Res. 2010;70:18–23.
45. Jiang SN, Phan TX, Nam TK, Nguyen VH, Kim HS, Bom HS, Choy HE, Hong Y, Min JJ. Inhibition of tumor growth and metastasis by a combination of *Escherichia coli*mediated cytolytic therapy and radiotherapy. Mol Ther. 2010;18(3):635–42.
46. Ryan RM, Green J, Williams PJ, Tazzyman S, Hunt S, Harmey JH, Kehoe SC, Lewis CE. Bacterial delivery of a novel cytolysin to hypoxic areas of solid tumors. Gene Ther. 2009;16:329–39.
47. Loeffler M, Le'Negrate G, Krajewska M, Reed JC. *Salmonella typhimurium* engineered to produce CCL21 inhibit tumor growth. Cancer Immunol Immunother. 2009;58:769–75.
48. Gentschev I, Fensterle J, Schmidt A, Potapenko T, Troppmair J, Goebel W, Rapp UR. Use of a recombinant Salmonella enterica serovar Typhimurium strain expressing C-Raf for protection against CRaf induced lung adenoma in mice. BMC Cancer. 2005;5:15.
49. Ganai S, Arenas RB, Forbes NS. Tumour targeted delivery of TRAIL using *Salmonella typhimurium* enhances breast cancer survival in mice. Br J Cancer. 2009;101:1683–91.
50. Loeffler M, Le'Negrate G, Krajewska M, Reed JC. Inhibition of tumor growth using *Salmonella* expressing Fas ligand. J Natl Cancer Inst. 2008;100:1113–6.
51. Theys J, Nuyts S, Landuyt W, Van Mellaert L, Dillen C, Bohringer M, Durre P, Lambin P, Anne J. Stable *Escherichia coli–Clostridium acetobutylicum* shuttle vector for secretion of murine tumor necrosis factor α. Appl Environ Microbiol. 1999;65:4295–300.
52. Nuyts S, Theys J, Landuyt W, van Mellaert L, Lambin P, Anne J. Increasing specificity of antitumor therapy: cytotoxic protein delivery by non-pathogenic clostridia under regulation of radio-induced promoters. Anticancer Res. 2001;21:857–61.
53. Nuyts S, Van Mellaert L, Theys J, Landuyt W, Bosmans E, Anne J, Lambin P. Radio-responsive recA promoter significantly increases TNFα production in recombinant clostridia after 2 Gy irradiation. Gene Ther. 2001;8:1197–201.
54. Forbes NS. Engineering the perfect (bacterial) cancer therapy. Nat Rev Cancer. 2010;10(11):785–94.
55. Hoffman RM, Zhao M. Whole-body imaging of bacterial infection and antibiotic response. Nat Protoc. 2006;1:2988–94.
56. Benoit MR, Mayer D, Barak Y, Chen IY, Hu W, Cheng Z, Wang SX, Spielman DM, Gambhir SS, Matin A. Visualizing implanted tumors in mice with magnetic resonance imaging using magnetotactic bacteria. Clin Cancer Res. 2009;15:5170–7.

57. Tjuvajev J, Blasberg R, Luo X, Zheng LM, King I, Bermudes D. *Salmonella*-based tumor-targeted cancer therapy: tumor amplified protein expression therapy (TAPETTM) for diagnostic imaging. J Control Release. 2001;74:313–5.
58. Soghomonyan SA, Doubrovin M, Pike J, Luo X, Ittensohn M, Runyan JD, Balatoni J, Finn R, Tjuvajev JG, Blasberg R, Bermudes D. Positron emission tomography (PET) imaging of tumor-localized *Salmonella* expressing HSV1-TK. Cancer Gene Ther. 2005;12:101–8.
59. Brader P, Stritzker J, Riedl CC, Zanzonico P, Cai S, Burnazi EM, Ghani ER, Hricak H, Szalay AA, Fong Y, Blasberg R. *Escherichia coli* Nissle 1917 facilitates tumor detection by positron emission tomography and optical imaging. Clin Cancer Res. 2008;14:2295–302.
60. Min JJ, Kim HJ, Park JH, Moon S, Jeong JH, Hong YJ, Cho KO, Nam JH, Kim N, Park YK, Bom HS, Rhee JH, Choy HE. Noninvasive real-time imaging of tumors and metastases using tumor-targeting light-emitting *Escherichia coli*. Mol Imaging Biol. 2008;10:54–61.
61. Mi JJ, Nguyen VH, Kim HJ, Hong YJ, Choy HE. Quantitative bioluminescence imaging of tumor-targeting bacteria in living animals. Nat Protoc. 2008;3:629–36.
62. Cheng CM, Lu YL, Chuang KH, Hung WC, Shiea J, Su YC, Kao CH, Chen BM, Roffler S, Cheng TL. Tumor-targeting prodrug-activating bacteria for cancer therapy. Cancer Gene Ther. 2008;15:393–401.
63. Chang WW, Kuan YD, Chen MC, Lin ST, Lee CH. Tracking of mouse breast cancer stem-like cells with *Salmonella*. Exp Biol Med. 2012;237:1189–96.
64. Smith AM, Duan H, Mohs AM, Nie S. Bioconjugated quantum dots for in vivo molecular and cellular imaging. Adv Drug Deliv Rev. 2008;60:1226–40.
65. Wagner V, Dullaart A, Bock AK, Zweck A. The emerging nanomedicine landscape. Nat Biotechnol. 2006;24(10):1211–7.
66. Porsch C, Zhang Y, Östlund Å, Damberg P, Ducani C, Malmström E, Nyström AM. In vitro evaluation of non-protein adsorbing breast cancer theranostics based on ^{19}F-polymer containing nanoparticles. Part Part Syst Charact. 2013;2013:1–10.
67. MacDiarmid JA, Mugridge NB, Weiss JC, Phillips L, Burn AL, Paulin RP, Haasdyk JE, Dickson KA, Brahmbhatt VN, Pattison ST, James AC, Al Bakri G, Straw RC, Stillman B, Graham RM, Brahmbhatt H. Bacterially derived 400 nm particles for encapsulation and cancer cell targeting of chemotherapeutics. Cancer Cell. 2007;11:431–45.
68. MacDiarmid JA, Amaro-Mugridge NB, Madrid-Weiss J, Sedliarou I, Wetzel S, Kochar K, Brahmbhatt VN, Phillips L, Pattison ST, Petti C, Stillman B, Graham RM, Brahmbhatt H. Sequential treatment of drug-resistant tumors with targeted minicells containing siRNA or a cytotoxic drug. Nat Biotechnol. 2009;27:643–51.
69. Qian BJ, Yan F, Li N, Liu QL, Lin YH, Liu CM, Luo YP, Guo F, Li HZ. MTDH/AEG-1-based DNA vaccine suppresses lung metastasis and enhances chemosensitivity to doxorubicin in breast cancer. Cancer Immunol Immunother. 2011;60(6):883–93.

Index

H. Schatten (ed.), *Cell and Molecular Biology of Breast Cancer*,
DOI 10.1007/978-1-62703-634-4, © Springer Science+Business Media New York 2013

C

F

MIX
Papier aus verantwortungsvollen Quellen
Paper from responsible sources
FSC® C105338

If you have any concerns about our products,
you can contact us on
ProductSafety@springernature.com

In case Publisher is established outside the EU,
the EU authorized representative is:
Springer Nature Customer Service Center GmbH
Europaplatz 3, 69115 Heidelberg, Germany

Printed by Libri Plureos GmbH
in Hamburg, Germany